# CARBONIUM IONS

**VOLUME IV
MAJOR TYPES (CONTINUED)**

# REACTIVE INTERMEDIATES IN ORGANIC CHEMISTRY

**Edited by GEORGE A. OLAH**
**Case Western Reserve University**

*A series of collective volumes and monographs on the chemistry of all the important reactive intermediates of organic reactions:*

**CARBONIUM IONS**
Edited by George A. Olah of Case Western Reserve University and Paul v. R. Schleyer of Princeton University: Vol. I (1968), Vol. II (1970), Vol. III (1972), Vol. IV (1973), Vol. V (in preparation)

**RADICAL IONS**
Edited by E. T. Kaiser of the University of Chicago and L. Kevan of the University of Kansas (1968)

**NITRENES**
Edited by W. Lwowski of New Mexico State University (1970)

**CARBENES**
Edited by Maitland Jones, Jr., of Princeton University and Robert A. Moss of Rutgers University, the State University of New Jersey: Vol. I (1973), Vol. II (in preparation)

**FREE RADICALS**
Edited by J. K. Kochi of Indiana University: Vol. I (1973), Vol.II (1973)

Planned for the Series

**CARBANIONS**
By M. Szwarc of the State University of New York, College of Forestry, Syracuse

**ARYNES**
Edited by M. Stiles of the University of Michigan

# CARBONIUM IONS

Edited by
**GEORGE A. OLAH**
Department of Chemistry
Case Western Reserve University
Cleveland, Ohio

**PAUL von R. SCHLEYER**
Department of Chemistry
Princeton University
Princeton, New Jersey

**VOLUME IV**

*Major Types (continued)*

WILEY—INTERSCIENCE
a division of John Wiley & Sons, Inc., New York · London · Sydney · Toronto

Copyright © 1973, by John Wiley & Sons, Inc.

All rights reserved. Published simultaneously in Canada.

No part of this book may be reproduced by any means, nor transmitted, nor translated into a machine language without the written permission of the publisher.

Library of Congress Catalog Card Number: 67-13956

ISBN 0-471-65337-3

Printed in the United States of America.

10 9 8 7 6 5 4 3 2 1

## Introduction to the Series

Reactive intermediates have always occupied a place of importance in the spectrum of organic chemistry. They were, however, long considered only as transient species of short life-time. With the increase in chemical sophistication many reactive intermediates have been directly observed, characterized, and even isolated. While the importance of reactive intermediates has never been disputed, they are usually considered from other points of view, primarily relative to possible reaction mechanism pathways based on kinetic, stereochemical, and synthetic chemical evidence. It was felt that it would be of value to initiate a series that would be primarily concerned with the reactive intermediates themselves and their impact and importance in organic chemistry. In each volume, critical, but not necessarily exhaustive coverage is anticipated. The reactive intermediates will be discussed from the points of view of: formation, isolation, physical characterization, and reactions.

The aim, therefore, is to create a forum wherein all the resources at the disposal of experts in the field could be brought together to enable the reader to become acquainted with the reactive intermediates in organic chemistry and their importance.

As the need arises, it is anticipated that supplementary volumes will be published to present new data in this rapidly developing field.

GEORGE A. OLAH

# Editors' Preface to Volume I

Three years ago we undertook the task of trying to organize a survey of the field of carbonium ion chemistry in the form of an advanced monograph. Although we both have been active in research on carbonium ions, we were not fully aware, at the time the project was begun, of the vast dimensions to which the field had grown, particularly in the past two decades. To cover the entire area a very substantial effort was needed and the cooperation of a large number of colleagues and friends was necessary. We have been very gratified by the favorable response of research workers in the field to our invitations to contribute chapters in their own specialities. As the project proceeded, it became quite obvious that it would be impossible to cover all the material, as originally projected, in fewer than four volumes, if these were to be kept to manageable size.

Thought has been given to the best arrangement of chapters in these volumes. Some authors were able to produce their manuscripts more rapidly than others, and it might have been possible to assemble the volumes from unrelated chapters, in order of receipt. We have felt it better to have each individual volume maintain a coherence and identity by bringing together chapters related in subject matter to one another. Thus, a historical introduction, general aspects, and methods of investigation of carbonium ions are included in this first volume. Major types of such ions will be featured in the second volume, and the third will deal with the classical–nonclassical ion problem. The last volume will contain chapters devoted to diverse types of carbonium ions, and it will have comprehensive subject and author indexes. In addition, each volume will have short indexes.

We have not tried to alter in any way the individual author's contributions, but, hopefully, we have provided some coordination of topics to minimize overlap and to make a multiauthored book, as much as possible, an organized entity. Consequently, the different chapters reflect their authors' standpoints and philosophy. We felt particularly that exposure of the reader to views from opposing sides of topics in which there is controversy not only would be stimulating, but would further an understanding of the topic. Often, one's opinions and attitudes are influenced by a particularly persuasive argument. It is to the good to have the alternative interpretations available conveniently together, to facilitate comparison. The editors have attempted to survey the fascinating field of carbonium ions impartially by affording the many active workers an opportunity to summarize their viewpoint and evidence. To what extent we have been successful will be decided by the readers.

We should like to express our sincere appreciation not only to the authors whose contributions made this book possible, but also to the numerous colleagues who commented on individual chapters, provided us with additional information (frequently not yet published), and called attention to many aspects of carbonium ions, whose importance we had not fully realized.

As the preparation of a monograph of this size takes a long time, we must apologize to our readers and especially to our co-authors that the unavoidable manuscript collecting and publishing delays have made it impossible to cover the literature to the last minute. In this first volume reference to the literature through 1966 is general, and in frequent instances citations to books published as late as summer, 1967, were added in the galley proof stage.

As the first volume of our monograph is going into production, the second and third volumes are practically complete and in the printer's hands. It is hoped that the whole project will be finished by the end of next year. The cooperation and excellent work of the publisher's staff made the preparation of these volumes a much more pleasant task than we could possibly have anticipated originally.

GEORGE A. OLAH<br>PAUL V. R. SCHLEYER

*January 1968*

# Editors' Note to Volume IV

As we no longer consider the classical–nonclassical ion problem to be of special nature, the original outline of the series has been changed somewhat. Volume IV continues the presentation of major types of ions. Because space limitations precluded publication of all chapters originally planned, Volume V will complete the series. Volume V will include an author index and a comprehensive subject index.

Progress in this very active field of chemistry continues to be rapid. It has become clear that the original title of our series, *Carbonium Ions*, in its proper usage should cover only one class of the positive ions of carbon compounds, that is, the highest valency penta- or tetracoordinated "onium" ions exemplified by $CH_5^+$. The generic name "carbocations" has been suggested for all positive ions of carbon compounds, just as the related negative ions are called carbanions. Carb*onium* and carb*enium* ions (trivalent) would then be the major subcategories. We have considered changing the title of our series to "Carbocations," but decided that this would not be appropriate at this time. It is hoped that the Nomenclature Committees of the American Chemical Society and the International Union of Pure and Applied Chemistry will eventually settle existing difficulties in the naming of organic cations.

Whereas delays continue to plague projects of this size, most chapters were revised to bring them up to date. We thank our authors for their understanding and willingness to cope with this problem.

GEORGE A. OLAH
PAUL V. R. SCHLEYER

*November 1972*

# Authors of Volume IV

J. C. BARBORAK, *Department of Chemistry, Princeton University, Princeton, New Jersey*

RICHARD N. BUTLER, *Department of Chemistry, University College, Cork, Ireland*

RAYMOND C. FORT, JR., *Department of Chemistry, Kent State University, Kent, Ohio*

H. H. FREEDMAN, *The Dow Chemical Company, Eastern Research Laboratory, Wayland, Massachusetts*

KENNETH M. HARMON, *Oakland University, Rochester, Michigan*

RONALD E. LEONE, *Research Laboratories, Eastman Kodak Company, Rochester, New York*

DANIEL H. O'BRIEN, *Department of Chemistry, Texas A & M University, College Station, Texas*

GEORGE A. OLAH, *Department of Chemistry, Case Western Reserve University, Cleveland, Ohio*

PAUL V. R. SCHLEYER, *Department of Chemistry, Princeton University, Princeton, New Jersey*

FRANCIS L. SCOTT, *Department of Chemistry, University College, Cork, Ireland*

A. M. WHITE, *Department of Chemistry, Research Department, Imperial Oil Ltd., Sarnia, Ontario, Canada*

# Contents

## Volume I. General Aspects and Methods of Investigation

1. HISTORICAL OUTLOOK. *By Costin D. Nenitzescu* . . . . 1
2. THERMODYNAMIC ASPECTS. *By J. L. Franklin* . . . . . 77
3. CYROSCOPIC AND CONDUCTIMETRIC MEASUREMENTS IN SULFURIC ACID. *By R. J. Gillespie and E. A. Robinson* . . 111
4. ELECTROLYTIC CONDUCTIVITY. *By Norman N. Lichtin* . . . 135
5. ELECTRONIC SPECTRA. *By George A. Olah, Charles U. Pittman, Jr., and Martyn C. R. Symons* . . . . . . 153
6. VIBRATIONAL SPECTRA. *By J. C. Evans* . . . . . . . 223
7. NUCLEAR MAGNETIC RESONANCE SPECTRA. *By Gideon Fraenkel and Donald G. Farnum* . . . . . . . . 237
8. MASS SPECTROMETRIC INVESTIGATIONS OF GASEOUS CATIONS. *By Maurice M. Bursey and Fred W. McLafferty* . . . 257
9. ISOTROPIC TRACERS IN THE INVESTIGATION OF CARBONIUM IONS. *By Clair J. Collins* . . . . . . . . . . 307
10. CARBONIUM IONS ON THE SURFACE OF ACID CATALYSTS. *By Harry P. Leftin* . . . . . . . . . . . . . 353
11. ION CYCLOTRON RESONANCE SPECTROSCOPY. *By J. D. Baldeschwieler* . . . . . . . . . . . . . 413
12. HEATS OF FORMATION BY SOLUTION CALORIMETRY. *By E. M. Arnett and J. W. Larsen* . . . . . . . . . 441
    SUBJECT INDEX, VOLUME I . . . . . . . . 457

## Volume II. Methods of Formation and Major Types

13. INTERMOLECULAR HYDRIDE SHIFTS. *By C. D. Nenitzescu* . . 463
14. INTRAMOLECULAR HYDRIDE SHIFTS. *By G. J. Karabatsos and J. L. Fry* . . . . . . . . . . . . . 521
15. FREE CARBONIUM IONS. *By J. T. Keating and P. S. Skell* . . 573
16. CARBONIUM ION FORMATION FROM DIAZONIUM IONS. *By L. Friedman* . . . . . . . . . . . . . . 655
17. ALKYLCARBONIUM IONS. *By G. A. Olah and J. A. Olah* . . 715
18. ENYLIC CATIONS. *By N. C. Deno* . . . . . . . . 783
19. DIENYLIC AND POLYENYLIC CATIONS. *By T. S. Sorensen* . . 807
20. CYCLOHEXADIENYL CATIONS (ARENONIUM IONS). *By D. M. Brouwer, E. L. Mackor, and C. MacLean* . . . . . 837
21. DISUBSTITUTED CARBONIUM IONS. *By H. G. Richey and J. M. Richey* . . . . . . . . . . . . . . 899
    SUBJECT INDEX, VOLUME II . . . . . . . . 959

## Volume III. Major Types (Continued)

22. NONCLASSICAL IONS AND HOMOAROMATICITY. *By S. Winstein.*    965
23. HOMOALLYLIC AND HOMOAROMATIC CATIONS. *By Paul R. Story and Benjamin C. Clark, Jr.* . . . . . . . . 1007
24. THE 2-NORBORNYL CATION. *By G. Dann Sargent* . . . . 1099
25. CYCLOPROPYLCARBONIUM IONS. *By Herman G. Richey, Jr.* . 1201
26. CYCLOPROPYLCARBINYL AND CYCLOBUTYL CATIONS. *By Kenneth B. Wiberg, B. Andes Hess, Jr., and Arthur J. Ashe, III* . . . . . . . . . . . . . . 1295
27. PHENONIUM IONS: THE SOLVOLYSIS OF $\beta$-ARYLALKYL SYSTEMS. *By Charles J. Lancelot, Donald J. Cram, and Paul v. R. Schleyer* . . . . . . . . . . . 1347
    SUBJECT INDEX, VOLUME III    . . . . . . . 1485

## Volume IV. Major Types (Continued)

28. ARYLCARBONIUM IONS. *By H. H. Freedman.* . . . . . 1501
29. CYCLOHEPTATRIENYLIUM (TROPENYLIUM) IONS. *By Kenneth M. Harmon* . . . . . . . . . . . . . . 1579
30. AZACARBONIUM IONS. *By Francis L. Scott and Richard N. Butler.* . . . . . . . . . . . . . . 1643
31. PROTONATED HETEROALIPHATIC COMPOUNDS. *By George A. Olah, A. M. White, and Daniel H. O'Brien* . . . . . 1697
32. BRIDGEHEAD CARBONIUM IONS. *By Raymond C. Fort, Jr.* . . 1783
33. DEGENERATE CARBONIUM IONS. *By Ronald E. Leone, J. C. Barborak, and Paul v. R. Schleyer* . . . . . . 1837
    SUBJECT INDEX, VOLUME IV . . . . . . . . . 1941

## Volume V. Miscellaneous Ions. (In Preparation)

# CARBONIUM IONS

**VOLUME IV
MAJOR TYPES (CONTINUED)**

# CHAPTER 28

# Arylcarbonium Ions

### H. H. FREEDMAN
*The Dow Chemical Company, Eastern Research Laboratory, Wayland, Massachusetts*

I. Introduction . . . . . . . . . . 1501
   A. Historical . . . . . . . . . . 1501
   B. Carbonium Ion Stabilization by the Aryl Group . . . . 1502
II. The Ionization of Arylmethyl Derivatives . . . . . . 1504
   A. Investigations in the Solid State . . . . . . 1504
      1. X-Ray Investigations . . . . . . . 1505
      2. Infrared Studies . . . . . . . . 1506
      3. Miscellaneous . . . . . . . . 1508
   B. Formation of Arylcarbonium Ions in Solution . . . . 1508
      1. Influence of the Solvent . . . . . . . 1509
      2. Influence of the Leaving Group . . . . . 1515
      3. Influence of Lewis Acids . . . . . . 1517
III. The Effect of Structure on Arylcarbonium Ion Stability . . . 1525
   A. General Background . . . . . . . . 1525
   B. Quantitative Determination of Carbonium Ion Stabilization Energies
      ($\Delta F_R^\circ{}^+$) . . . . . . . . . . 1526
      1. Conductivity Measurements . . . . . . 1526
      2. Electronic Spectrophotometry . . . . . . 1528
      3. Ionization of Alcohols ($pK_R{}^+$) . . . . . . 1530
      4. Emf Measurements. . . . . . . . 1537
      5. The Nmr Method . . . . . . . . 1538
      6. Miscellaneous Methods . . . . . . . 1542
   C. The Influence of Steric Factors on $\Delta F_R^\circ{}^+$ . . . . 1543
   D. The Influence of Electronic Factors on $\Delta F_R^\circ{}^+$ . . . 1548
IV. The Conformation of Trityl and Related Cations . . . . 1554
   A. Possible Structures and their Relative Merits . . . . 1554
   B. Chirality in Arylcarbonium Ions . . . . . . 1561
V. Acknowledgments . . . . . . . . . 1573
References . . . . . . . . . . . 1574

## I. INTRODUCTION

### A. Historical

It is of historical interest that the triphenylmethyl (trityl) cation was isolated as a stable crystalline species in 1901 (1), many years prior to the common acceptance of carbonium ion intermediates in organic reactions.

The elucidation of the exact nature of the highly colored salts obtained from triarylmethyl halides and Lewis acids took approximately 40 years and paralleled the development of many of the theoretical aspects of carbonium ion theory and the effect of structure on reactivity. It is fair to say that these stable salts, undoubtedly containing a trivalent positive carbon atom, did much to overcome the resistance of chemists to the acceptance of carbonium ions as transient reaction intermediates whose presence was detectable only through subtle kinetic schemes. The knowledge that tri- and diaryl carbonium ions could be reversibly equilibrated with their corresponding halides assuredly was instrumental in the development of the underlying principles of the $S_N1$ mechanism, one of the cornerstones of theoretical organic chemistry.

In a very real sense then, the stable arylcarbonium salts, or their precursors, are the *sine qua non* of carbonium ion chemistry. Therefore, it is not surprising that many areas concerning their formation, structure, and properties have been previously dealt with in reviews covering the general topic of carbonium ion chemistry (2). There are also available summaries of specific topics in which aryl-stabilized carbonium ion intermediates play a key role; considerable pertinent information is available from reviews on solvolysis kinetics (3), conductivity measurements (4), electronic spectra (5), Friedel-Crafts reactions (6), etc. No attempt is made to assimilate this enormous mass of data (although it is hoped to at least summarize the pertinent facts and arguments and to point out potentially productive areas for further development). This is especially true for the kinetic investigations of solvolytic processes in which arylmethyl cations have been postulated to be intermediates. Much of this area, especially solvolysis in nonpolar media, is highly complex and controversial, and it is hardly possible to do justice to a subject which Ingold has described as "largely a collection of paradoxes" (7) and still adequately cover the topic as a whole. Similarly, it is impossible to encompass all the intricacies of the role of the various kinetically distinguishable ion pairs in solvolytic processes, and the reader is referred to a recent review on the subject (8). When dealing with isolable species, rather than transient intermediates, the emphasis must necessarily be on the triarylcarbonium ions. The relative synthetic accessibility of these and related compounds is attested to by the availability, as early as 1914, of sufficient data to justify publication of a complete monograph by Schmidlin (9), augmented in 1924 by Pfeiffer (10).

## B. Carbonium Ion Stabilization by the Aryl Group

Electron-deficient carbon (i.e., a carbonium ion) is inherently an unstable species from both a thermodynamic and kinetic point of view. To increase the stability of such ions it is necessary to delocalize this unit positive charge by the substitution of suitable stabilizing groups at the positive site. This charge delocalization effectively spreads out the charge concentration in the ion so that no single carbon atom is excessively cationic. Simultaneously, the peripheral charge distribution acts to increase the size of the ion, an important stabilizing factor especially in solution where solvation forces become important, and in this respect carbonium ions do not differ from inorganic ionic species.

The size of the phenyl ring coupled with its ability to delocalize charge by a resonance mechanism, primarily to its *ortho* and *para* positions, makes it an extremely efficient group for the stabilization of trigonally hybridized ($sp^2$) carbon. Thus replacement of hydrogen by phenyl in the extremely energetic and nonisolable methyl cation **1** leads to the benzyl cation **2**, a kinetically established (11) but not yet isolable intermediate, in which the size of the ion is approximately six-fold greater and the charge is distributed over four carbons. Further stability is imparted by substitution of a second phenyl ring; the diphenylmethyl cation **3** is a well-recognized species in solution and can even be isolated as stable, although highly reactive, crystalline salts (12). Finally, replacement of all three hydrogens of **1** by phenyl gives the trityl cation, in which, as structure **4** suggests, charge is distributed over ten carbon atoms and the resulting, unusually large ion is now stable enough to be readily isolable as a partner to a large number of nonnucleophilic anions. Using only the ease of isolation of crystalline salts of cations **2–4** as a criterion of stability, it is obvious that the triphenyl carbonium ion is orders of magnitude more stable than the diphenyl, which in turn is similarly more stable than the monophenyl. Detailed quantitative data to support this conclusion are presented here.

Substituents in the aromatic rings of ions **2–4** all act to increase the ion size, but may either increase or decrease its stability, depending on the ability of the substituent to delocalize the positive charge. A gross misunderstanding of the principles involved led to a lively controversy concerning the structure of the trityl ion and its substituted analogs. One school, championed primarily by Gomberg and his students (13), maintained that these salts were primarily quinonoid in nature, as in **5**,

$$Ar_2C=\underset{(5)}{\underset{}{\bigcirc}}\oplus$$

whereas Dilthey et al. (14) preferred the currently accepted benzenoid carbonium ion structure. It cannot be said that any single piece of evidence discredited one theory in favor of the other; rather, the quinonoid structure expired slowly as theoretical progress, particularly in the valence-bond resonance theory, was made. We now recognize **5** as a resonance form of **4** and, as such, a convenient shorthand form for portraying charge delocalization. However, with strongly electron-donating *para* substituents such as methoxy and dimethylamino, contributions from the quinoidal resonance form (**6**) may become so important that, in fact, the

(6)

(**6a**) X = OR;   (**6b**) X = N(CH$_3$)$_2$

compound is better described as a quinoidal oxonium ion (**6a**) or an iminium ion (**6b**), rather than a benzenoidal carbonium ion. For the time being we must leave this very fundamental question unanswered, except to point out that resonance theory assures us that neither the benzenoid nor the quininoid structure can exclude the other; both structures contributing to the overall stability of the ion.

## II. THE IONIZATION OF ARYLMETHYL DERIVATIVES

### A. Investigations in the Solid State

In contrast to solution studies, the methods for studying the extent of ionization of carbonium ions precursors in the solid state is quite limited. When applicable, however, they often offer incontrovertible structural information free from ambiguities due to solvation equilibria, ion-pairing phenomena, rapid exchange processes, and the multiplicity of species that are so often present in the solution studies. This is particularly true of the

X-ray method which, when carried out to a suitable degree of refinement, affords not only exact structural information but accurate bond angles, distances, and conformational data, all having important theoretical implications.

### 1. X-Ray Investigations

A single-crystal three-dimensional X-ray analysis of trityl perchlorate at 85°, refined to a reliability index ($R$ value) of 8.4% has been carried out (15). The carbonium ion character of the central carbon atom is clearly established by its three equivalent coplanar, central bonds and the central carbon–phenyl bond distance of 1.454 ($\pm 0.018$) Å. These results may be compared with those obtained by an electron-diffraction study of the normal covalent compound triphenylmethane (131) in which the aliphatic-to-aromatic carbon bond distance is 1.53 Å and the tetrahedral angle is 112°, both fairly typical of normal $sp^3$ hybridization. Trityl fluoroborate at 110° is isomorphous with the perchlorate and it may be assumed to have an identical structure (15). A similar attempt to accurately determine the molecular parameters of tri-*p*-methoxytrityl hydrogen dichloride tetrahydrate was not successful because of the statistical distribution of the hydrated anion, ($HCl_2^- \cdot 4H_2O$), although it was ascertained that the central bonds were coplanar, or nearly so (16).

Two additional X-ray structures of aryl-stabilized carbonium ions have been reported. The "Hückel aromatic" *syn*-triphenylcyclopropenium perchlorate (**7**) first synthesized by Breslow (17) is not directly analogous to the aryl carbonium ions; however, it also has a phenyl ring directly attached to an $sp^2$-hybridized carbon and this is reflected in the average exocyclic carbon–carbon bond distance of 1.436 Å (18). Because of steric compression of the *ortho* hydrogens, the aryl rings cannot be coplanar with the plane of the three-membered ring as was previously theorized (19) and the molecular assumes a propellerlike arrangement with the phenyl groups twisted an average of 13.6° (18). An analogous twisting of the rings in the trityl cations is discussed in detail in Section IV. Bryan (20) has reported the preliminary structural details on 3-chloro-1,2,3,4-tetraphenyl-

(7)   (8)

cyclobutenium pentachlorostannate (8). In this molecule the positive charge is allylically distributed to C-2 and C-4 and their associated phenyls, resulting in a coplanar arrangement of these phenyl groups. On the other hand, the relative absence of charge at C-3 is suggested by the phenyl at this position, which is rotated 57° out of the cyclobutene plane, and the complete absence of charge delocalization by the phenyl at the $sp^3$ carbon is demonstrated by its inclination of 83° to the planes of the phenyls at C-2 and C-4. Corroboration for the carbonium character of C-2 and C-4 is found in the considerably shorter bond distances for the phenyl–carbon bond at these positions (1.41 and 1.39 Å) as compared with the corresponding bond at C-1 (1.47 Å), although further refinement is required and is presumably in progress (21).

## 2. Infrared Studies

Solid state infrared spectroscopy is a powerful and convenient method for obtaining structural information, and it is particularly valuable in assessing whether the substituent on the potential carbonium site is covalently or ionically bound. Investigations in this area were pioneered by Sharp and Sheppard (22), who characterized a number of covalent and ionic trityl compounds in the crystalline state. Aside from conclusions concerning the conformation of the trityl ion, which agreed well with later X-ray results (see below), the ionic nature of the species could be confirmed or rejected on the basis of the presence or absence of characteristic absorption bands. This was relatively simple to accomplish when the infrared absorption of the potentially anionic substituent itself was significantly different for the covalent and ionic species. Thus $BF_4^-$ anion absorbs at 1032 and 1058 cm$^{-1}$ in its ionic alkali metal salts and appears as a broad band at 1050 and 1097 cm$^{-1}$ in trityl tetrafluoroborate (22) and at 1060 cm$^{-1}$ in tri-$p$-methoxytrityl tetrafluoroborate (23). Similarly, ionic perchlorates (24) absorb at 1081–93 cm$^{-1}$, and this characteristic band is found at 1093 and 1090 cm$^{-1}$ in trityl (25) and tri-$p$-methoxy trityl perchlorates (23), respectively. On the other hand, trityl trifluoroacetate has a normal covalent ester carbonyl absorption at 1785 cm$^{-1}$ and is therefore not a carbonium salt in the solid state (22).

Even when the anion (or potential anion) does not absorb in the accessible infrared region, the dramatic differences shown by the spectra of covalent and ionic trityl compounds may be used to unequivocally distinguish one from the other. In contrast to the covalent compounds, trityl cations that differ only in the anionic portion show identical infrared spectra (except for bands due to the anion) as has been documented for trityl (22,26) tri-$p$-methoxytrityl (23) and tri-$p$-methyl trityl (27) salts. A detailed vibrational frequency assignment for all the major absorption bands of trityl alcohol, chloride, tetrafluoroborate, and perchlorate and their $^{13}$C and deuteriated analogs has been recently reported (28), and this comprehensively extends the original interpretations by Sharp and Sheppard (22).

An elaborate discussion of the infrared spectral differences between the crystalline covalent and ionic compounds would be out of place here; however, it is appropriate to summarize our own observations in this area. In general, the infrared spectra of the covalent compounds are considerably more complex than the ionic materials, particularly in the 8–15 μ "fingerprint" region, and the intensity of the out-of-plane carbon–hydrogen deformation bands is considerably decreased in the latter compounds. In the aryl carbonium salts there invariably are present two bands at 1585 ± 10 and 1355 ± 10 cm$^{-1}$, which very often are the strongest bands in the cationic portion of the spectrum. The 1585-cm$^{-1}$ band is undoubtedly one of the characteristic benzenoid C=C skeletal stretching modes; invariably it is very weak in nonconjugated aryl compounds and becomes medium to strong when conjugation with a double bond (C=O, C=C, NO$_2$, etc.) is introduced (29). It is thus reasonable that this band, barely distinguishable in the $sp^3$-hybridized covalent compounds, becomes intense in the $sp^2$-hybridized conjugated carbonium ion.

The origin of the 1355-cm$^{-1}$ band is less clear, and neither Bellamy (29) nor the authors of other infrared texts list this region as containing a characteristic benzenoid absorption. It should be noted that the symmetrical CH deformation frequency of the carbon–methyl group is found in this vicinity, but it is considerably weaker and, except for unjustifiably dilute Nujol mulls, is readily distinguished from the band in question. Consideration of the unusually short carbon–phenyl bond in the trityl cation and its resulting effect on the carbon–carbon stretching force constants has led to the assignment of the strong band at 1359 cm$^{-1}$ in trityl tetrafluoroborate to a carbon–phenyl stretching vibration (28), and this has been supported by $^{13}$C and deuterium labeling. This assignment seems reasonable.

To summarize, it appears that intense characteristic bands at 1585 and 1355 ± 10 cm$^{-1}$ in the carbonium ion salts, absent in their covalent analogs, arise as a direct consequence of the overlap of the vacant π orbital of the carbonium center with the π orbitals of the aromatic rings. Charge delocalization into the ring is the essence of the stabilization of trivalent carbon by aromatic rings and, as will be amply demonstrated, is reflected in all the ground state properties of the cation. The characteristic "carbonium ion bands" at 1585 and 1355 cm$^{-1}$ should therefore be common to all aryl-conjugated carbonium ions, and this is supported by the observation (from this laboratory) that, without exception, these bands are unmistakably present in the solid state infrared spectra of various substituted trityl, diphenylmethyl, xanthenyl, and tetraphenylcyclobutenyl cations. Further work in the infrared area, which has been understandably overshadowed by the rapid developments in nmr spectroscopy (30), could yield valuable data. Of particular potential interest would be correlations involving the absolute intensity of the 1585-cm$^{-1}$ absorption band and a comparison of solid state and solution spectra.

### 3. Miscellaneous

A distinction between covalent and ionic species in the solid state has been made on the basis of single-crystal optical and electrical properties by Evans and Yoffe (31). By comparing the absorption spectra, photoconductivity, and low-frequency dielectric constants of trityl perchlorate and trityl azide, they concluded that the former is ionic and the latter essentially covalent.

### B. Formation of Aryl Carbonium Ions in Solution

In discussing the ionizability of a functional group attached to arylsubstituted carbon, we use the terminology of Fuoss (32) and define an ionophore as a substance that is ionic in its crystalline state but may be associated to some degree in solution; an ionogen is one that is covalent in the crystal but ionizes to a detectable degree in solution. Although the following discussion is generally applicable to all ionogenic compounds, most of the examples are taken from the trityl cation literature, for which the greatest amount of data is available. Much of the physical and spectroscopic evidence for the formation and detection of carbonium ions has been discussed in detail elsewhere; therefore nmr, visible-ultraviolet spectroscopy, cryoscopy, conductivity, etc., are mentioned only with reference to a particular structural feature or analytical method. Since we are primarily concerned with factors that influence the equilibrium concentration of ionic species, kinetic data are utilized only when the appropriate thermodynamic data are not available. In general, the extent of ionization depends on a number of discrete factors, which are briefly summarized and then discussed in greater detail.

$$Ar_3C-X \underset{k_r}{\overset{k_f}{\rightleftharpoons}} Ar_3C^+X^- \qquad (1)$$

Ionization of a triarylmethyl compound in the absence of extraneous nucleophiles is an equilibrium process (eq. 1) in which the magnitude of the equilibrium constant ($K_{eq} = k_f/k_r$) will depend on the relative efficacies of the forward process leading to ionization ($k_f$), and the reverse process ($k_r$) in which the ionic species recombine to form the covalent starting compound. Considering first the factors that influence $k_f$, we must include (a) the "ionizing power" of the solvent, (b) the "ionizability" of the leaving group X, and (c) the "stability" of the resulting carbonium ion. For the reverse process $k_r$, we must consider (d) the "nucleophilicity" of the X group and (e) the Lewis acidity of the carbonium ion. In practice it is not possible to rigorously assign individual quantitative parameters to any one of the factors placed in quotation marks and we must be content with a measure of their overall effect on the equilibrium constant for the ionization process.

To obtain an insight into the relative influence of factors (a)–(e) we must keep in mind that they are mutually interrelated and that, in assessing the importance of any one component, it is vital to maintain all the others constant. However, even when this constancy is maintained, the situation may be further complicated by effects due to specific solvation, ion pairing, and steric factors, to enumerate just a few of the possible pitfalls. Nevertheless, it is instructive to discuss these factors individually.

## 1. Influence of the Solvent

The vague term "ionizing power" includes the various properties of a solvent, summarizing its ability to solvate ions. Since creation of charge from ionogenic compounds is favored by solvents of high "ionizing power," equation 1 will shift to the right in such solvents, $K_{eq}$ increasing with increasing "ionizing power," and vice versa. Since we are forming a relatively reactive species, we must take special pains to avoid the presence of nucleophiles that would irreversibly displace equilibrium 1 to the right. It is at once obvious that some solvents with great ionizing power (water, alcohols, etc.) are precisely those which react irreversibly with the carbonium ion and thus are unsuitable for the study of equilibrium processes. However, they may be eminently suitable for kinetic studies where the (rate-determining) ionization step is not too fast.

In general terms, the solvent ($S$:) which interacts with and solvates the ions produced will increase $K_{eq}$ of Equation 1 by both increasing $k_f$ and decreasing $k_r$. Regardless of whether the resulting solvated species is a solvent-separated ion pair (as shown) or symmetrically solvated free ions, the net result is to decrease the accessibility of the positive site to its

$$Ar_3C-Y \rightleftharpoons Ar_3C^+Y^- \overset{S:}{\rightleftharpoons} [Ar_3C:S]^+Y^-$$
covalent species — ion pair — solvent-separated ion pair

anionic partner ($Y^-$), both by decreasing the charge concentration and by steric shielding of the electrophilic center. This dispersal of charge and shielding of the reaction site results in a slowing down of ion-pair collapse to covalent species with a corresponding decrease in $k_r$ and net increase in $K_{eq}$.

Since solvation of the developing ions is a primary function of the solvent and since the carbonium ion is an electrophilic species (Lewis acid) it readily interacts with nucleophilic solvents (Lewis bases). For a given carbonium ion, the extent of interaction will, to a first approximation, be determined by the basicity (pK) of the solvent, a measurable quantity.

Table I lists the $pK_a$ of various solvents in order of decreasing basicity. In every case the availability of the lone-pair electrons of the solvent heteroatom (oxygen or nitrogen) determines the basicity of the solvent, and this varies by approximately $10^{20}$ as we go from a basic amine to a weakly complexing nitro compound. For a given carbonium ion, the strength of the bond between the electron-deficient carbonium site and the electron-rich solvent site will depend on the $pK_a$ of the solvent-base and the gradations realized will vary from a completely covalent bond with the strongly basic solvents (pK > 5), relatively weak oxonium-type complex with the intermediate oxygen bases, to a weak donor-acceptor complex with the weakly basic solvents (pK < −10). These three possibilities are illustrated for the ionization of a triaryl halide in a strongly basic amine, weakly basic ether, and very weakly complexing nitrile solvents.

## TABLE I
### Average Basicity of Representative Organic Solvents[a]

| Solvent | $pK_a$ |
|---|---|
| Trimethylamine | 9.8 |
| Aniline | 4.6 |
| Dimethylformamide | ∼ −1.5 |
| Dimethylsulfoxide | −2.5 |
| Methanol | −2 |
| Tetrahydrofuran | −2.1 |
| Diethyl ether | −3.6 |
| Acetic acid | −6.1 |
| Ethyl acetate | ∼ −6.5 |
| Phenol | −6.7 |
| Acetone | −7.2 |
| Acetonitrile | −10.1 |
| Nitrobenzene | −11.3 |
| Nitromethane | −11.9 |

[a] Data from Ref. 40.

$$[\text{Ar}_3\text{C}]^+\text{Y}^- \xrightleftharpoons[\text{RC}\equiv\text{N:}]{\substack{\text{R}_3\text{N:} \\ \text{R}_2\text{O:}}} \begin{array}{l} [\text{Ar}_3\text{C}-\overset{\oplus}{\text{N}}\text{R}_3]\text{Y}^- \\ [\text{Ar}_3\text{C}-\overset{\oplus}{\text{O}}\text{R}_2]\text{Y}^- \\ [\text{Ar}_3\text{C} \leftarrow \text{N}\equiv\text{CR}]^{\oplus}\text{Y}^- \end{array}$$

We may conclude with the help of Table I that if we wish to observe the properties characteristic of only the carbonium ion and not of its subsequent reaction product with the solvent, we must choose a polar solvent with a p$K$ more negative than $-10$. The three solvents in Table I which meet this requirement are nitromethane, nitrobenzene, and acetonitrile—all the others are more or less unsuitable in that they must lead to other than purely carbonium products. In point of fact, as we document shortly, even solvents of p$K$ down to $-6$ can be extremely useful, especially when they act to solvate the leaving group rather than the carbonium ion.

As with all acid-base reactions, complex formation between the carbonium ion and the solvent is a reversible process that in practice is rendered irreversible because of the enormous molar excess of the latter. The reversibility of complex formation is readily demonstrated when comparable concentrations of carbonium ion precursor and complexing solvent are investigated in an inert diluent. When a basic solvent is added to the deep yellow solution characteristic of an ionized trityl salt in an inert solvent, diminution of the color occurs as charge delocalization into the aryl rings of the carbonium ion is replaced by charge localization on the heteroatoms of the basic solvent. The relatively strong base pyridine will form with trityl bromide the isolable colorless, crystalline salt triphenylpyridinium bromide (33). In 1,1,2,2-tetrachloroethane solution, with varying amounts of reagents, a combination of spectroscopic and conductimetric measurements has shown (33) that the association constant $K_1$ for this equilibrium is $3 \times 10^5$ at 20° and that, therefore, the reaction

$$\phi_3\text{C}-\text{Br} + \text{Py} \rightleftharpoons \phi_3\text{C}^+ \cdots \text{Br}^- + \text{Py} \xrightleftharpoons{K_{eq}} [\phi_3\text{C}-\text{Py}]^+\text{Br}^-$$
$$K_{eq} = 3 \times 10^5$$

proceeds substantially to completion even with only small amounts of pyridine present. Similarly, addition of diethyl ether to a yellow solution of partly ionized trityl chloride in nitromethane causes a diminution of the 400-mµ carbonium ion band and an analysis of the variation of this band

$$\phi_3\text{C}^+\text{Cl}^- + \text{Et}_2\text{O:} \rightleftharpoons \phi_3\overset{+}{\text{C}}\text{OEt}_2\text{Cl}^-$$
$$K_{eq} = 0.8 \times 10^{-4}$$

with varying reagent concentration reveals (34) an association constant for the 1:1 trityl ion–ether complex of $0.8 \times 10^{-4}$. This is dramatically smaller than that for the corresponding pyridine reaction, as would be predicted from the relative basicities of pyridine and ether. Furthermore, in contrast to the isolable pyridinium salt, we would expect the colorless oxonium complex to be readily dissociated to trityl cation and ether, and Sharp and Sheppard (22) note that trityl fluoroborate gives a colorless solution in ether which deposits starting material on evaporation of the solvent.

With solvents of intermediate basicity ($pK \sim -7$) we are dealing with borderline cases in which it is difficult to predict to what extent coordination with solvent will affect carbonium ion properties. Alcohols ($pK = \sim -2$) readily form oxonium complexes with trityl halides that lead to covalent trityl ethers on loss of a proton.

$$Ar_3Cl \rightleftharpoons Ar_3C^+Cl^- \overset{ROH}{\rightleftharpoons} [Ar_3\overset{+}{C}ORH]Cl^- \rightleftharpoons Ar_3COR + HCl$$

$$R = alkyl$$

With $m$-cresol ($pK = \sim -7$), however, no covalent ether is formed (35) and we may therefore assume that phenols are too weakly basic to form an appreciable equilibrium amount of the oxonium ion complex. On the other hand, phenols have a high ionizing power for the trityl halides despite their low dielectric constant, and Evans (35) has clearly demonstrated that trityl chloride is substantially ionized ($\sim 35\%$) in this solvent. It was further shown that tri-$p$-tolylmethylchloride has the same absorption spectrum in $m$-cresol as does tri-$p$-tolymethanol in 98% sulfuric acid (35), and inasmuch as the latter system is the commonly accepted standard for complete ionization, the former must also be completely ionized. This remarkable ionizing power of $m$-cresol has been attributed (35) to its ability to form a hydrogen bond between the phenolic OH group and the halide ion, i.e.,

$$Ar_3C^+ \cdots Cl^- \cdots H{-}OAr$$

Thus it is "solvation" of the halide ion by the acidic hydroxyl group of the phenol, rather than solvation of the cation by the basic oxygen, which explains the ability of this solvent to promote ionization.

Acetic and formic acids also appear to assist ionization of trityl halides via the formation of strong hydrogen bonds (36). This is especially true for the latter which, although it has a much higher dielectric constant than all the solvents previously mentioned, ionizes trityl halides far beyond its expected capacity based on dielectric constant alone. It is interesting to note that solvation of the electron-rich halide by the acidic hydroxyl group of the solvent (i.e., the hydrogen bond) is precisely the opposite of the previous solvation mechanism in which the electron-rich heteroatom of the solvent interacted with the acidic carbonium ion carbon. However, it appears reasonable to expect both mechanisms to operate when possible, and the high ionization power of both formic acid and $m$-cresol is probably best explained as simultaneous solvation of both the positive and the negative sites of the ion by the basic and acidic sites of the solvent, respectively.

$$Ar_3C-Cl \rightleftharpoons Ar_3C^+\cdots Cl^-\cdots H-OR$$
$$\overset{\vdots}{ROH}$$

$$R = H\overset{O}{\overset{\|}{C}}- \text{ or } Ar$$

When the leaving group of the trityl compounds is considerably more basic than halide, such as OH or OR, then an acidic solvent can donate a proton to it, rather than merely forming a hydrogen bond, and the resulting protonated oxonium species readily loses water or alcohol respectively, to form the desired carbonium ion.

$$Ar_3C-\overset{..}{O}R + HCOOH \rightleftharpoons [Ar_3C\overset{\overset{H}{|}}{O}R]^+HCOO^- \rightleftharpoons Ar_3C^+\bar{O}OCH + ROH$$
$$R = H \text{ or alkyl}$$

This has been shown to occur with trityl alcohol and trityl ether, which are both 89% ionized in formic acid (37). The ionization of trityl alcohols is discussed further in conjunction with the measurement of carbonium ion stabilities.

A number of solvents of low dielectric constant and extremely low basicity have proven to be very useful for studying processes involving arylcarbonium ions and their precursors. These include chlorinated aliphatics—chloroform (38), methylene chloride, 1,2-di-chloroethane, *sym*-tetrachloroethene (39)—and, surprisingly, aromatic hydrocarbons such as benzene and toluene. Although the basicity of these solvents is too low to be measured by the same $H_0$ techniques employed to measure those of Table I, they can be placed in a relative order by infrared spectroscopic techniques and the energy of formation of donor-acceptor complexes (40). In this laboratory we have found that methylene chloride is an especially useful solvent for carbonium salts and their precursors and one which is readily purified. The latter criterion is an extremely important one, especially when dealing with low concentrations of ions of the order of stability of trityl, which react avidly with traces of nucleophiles. The literature is replete with examples of erroneous conclusions directly attributable to solvent impurity; nitrated solvents are especially guilty in this respect, and it is unfortunate that a large portion of elaborate pioneering spectroscopic studies of triaryl carbonium ions in nitroalkanes by Evans and his co-workers (41) has been questioned by other equally competent investigators (42,43) because of known or suspected complications arising from purity of solvent.

No discussion of solvent effects would be complete without mentioning liquid sulfur dioxide, a remarkable solvent, not for its dissolving power, but for its unique ability to complex halide ion and thus promote ionization of trityl halides by the formation of ion pairs as well as by free ion dissociation.

$$\phi_3CCl \underset{}{\overset{SO_2}{\rightleftharpoons}} \phi_3C^+\cdots Cl^-(SO_2) \rightleftharpoons \phi_3C^+ + Cl^-(SO_2)$$

This makes sulfur dioxide perhaps the very best nonnucleophilic solvent for the study of the extent of ionization-dissociation of trityl halides and similar compounds by a combination of spectroscopic and conductiometric techniques. No further details on this unique solvent are given here inasmuch as all aspects of this topic have been reviewed by one of the foremost investigators in this area (4).

Table II summarizes much of the available data for the comparative efficiency of diverse solvents in promoting the ionization of trityl chloride.

### TABLE II
**Ionization Equilibrium Constants $K_{eq}$ of Trityl Chloride in Various Solvents at 17–25°**

| Solvent | $D^a$ | $K_{eq} \times 10^4$ | $\Delta F^{\circ\,b}$ kcal/mole | Ref. |
|---|---|---|---|---|
| Acetonitrile | 37.5 | ~0 | — | 55 |
| Nitrobenzene | 34.5 | Too low to measure | | 98 |
| Methylene chloride | 9.1 | 0.07 | 7.1 | c |
| sym-Tetrachloroethane | 11.9 | 0.48 | 5.8 | 39 |
| 1,2-Dichloroethane | 10.4 | 0.56 | 5.7 | 39 |
| Nitromethane | 37.4 | 4.4 | 4.5 | 63 |
| | | 2.7 | — | 43 |
| Sulfur dioxide[d] | 15.4 | 146 | 2.3 | 4,99 |
| Formic acid | 58.5 | 3100 | 0.7 | 36 |
| m-Cresol | 12.4 | 5600 | 0.3 | 35 |

[a] Dielectric constant at ~20°.
[b] Free energy at appropriate temperature.
[c] Private communication from R. Waack and M. A. Doran.
[d] At 0°.

From these data, in which $K_{eq}$ varies by at least a factor of $10^5$, it is abundantly clear that any comparison of rates or equilibria involving the ionization of carbonium ion precursors (ionogenic compounds) must refer to identical solvent conditions and solvent purity, and this must be taken into account when comparing data from different laboratories.

Although the preceding discussion on solvents may appear to be unnecessarily detailed, a strong case can be established for a direct correlation between progress in physical-organic research and the discovery of new solvent systems. This is especially obvious in the field of anion chemistry, where solvents such as tetrahydrofuran and dimethylsulfoxide each opened up new vistas, and it is no less true in carbonium ion chemistry where, e.g., the highly acidic solvent systems developed by Olah (44) makes possible the direct spectroscopic examination of exceedingly unstable cationic species.

## 2. Influence of the Leaving Group

The extent of the reversible ionization of a given ionogen in a given solvent is determined primarily by the difference in the efficacy of the ionizing group as a leaving group on the one hand and as a nucleophile on the other. For an ionogenic trityl compound, $Ar_3C-Y$, we may obtain an insight into the ease of breaking of the C–Y bond from thermodynamic data. In the absence of solvation and other extraneous effects, the sum of the dissociation energy $D$ of the C–Y bond to form the radical $Y\cdot$ and the electron affinity $E$ of $Y\cdot$ to form $Y^-$ should afford a rough relative measure of the ease of ionization.

$$\begin{array}{ll} R_3C-Y \longrightarrow (R_3C\cdot) + Y\cdot & \text{dissociation energy } d \\ Y\cdot + e^- \longrightarrow Y^- & \text{electron affinity } e \\ \hline R_3C-Y + e^- \longrightarrow (R_3C\cdot) + Y^- & \end{array}$$

Table III incorporates these data, and the last column affords a reasonable order for the relative ease of ionization of a covalent carbonium ion precursor when the effects due to the carbonium ion and solvent either remain constant or are eliminated. The values range from $-20$ kcal for (presumably) the most readily ionizable group, iodine, to $+73$ kcal for the least ionizable group, hydrogen. It is amusing and completely fortuitous that the negative values, the usual thermodynamic criterion for exothermicity, should occur only with those halogens, which have been shown to ionize to a measurable extent in suitable solvents (see below).

No ionization equilibrium constants for groups other than halide in Table III have been reported, and we can assume that they are too small to measure. However, we can obtain some useful information by noting the effect of the leaving group on the rates of $S_N1$ solvolysis of trityl compounds

## TABLE III
### Sum of Dissociation Energy and Electron Affinity for C–Y Bonds

| Bond | Dissociation energy $D^a$, kcal/mole | Electron affinity $E^b$, kcal/mole | $D + E$, kcal/mole |
|---|---|---|---|
| C—I   | 54  | −74      | −20 |
| C—Br  | 68  | −81      | −13 |
| C—Cl  | 81  | −85      | −4  |
| C—F   | 107 | −88      | 19  |
| C—CN  | 103 | −74$^c$  | 29  |
| C—OH  | 90  | −58      | 32  |
| C—NH$_2$ | 81 | −28$^c$ | 53 |
| C—OR  | 77  | ∼ −24    | ∼53 |
| C—CH$_3$ | 84 | −25     | 59  |
| C—H   | 105 | −28      | 73  |

$^a$ C. T. Mortimer, *Reaction Heats and Bond Strengths*, Pergamon Press, New York, 1962, Chapter 7.

$^b$ R. W. Kiser, "Tables of Ionization Potentials," U.S. Atomic Energy Commission, 1960.

$^c$ R. S. Neale, *J. Phys. Chem.*, **68**, 143 (1964).

if we make the usual assumption that rates and equilibrium parallel each other. Apparently trityl iodide and bromide solvolyze too fast for convenient study, and no rate data are available. The $D + E$ values of chloride and fluoride suggest that the former should ionize at a very much faster rate, and this is confirmed by Swain, who has reported that trityl fluoride in aqueous acetone solvolyzes one million times slower than the chloride (45). Furthermore, although acetate is usually considered a good leaving group, the rate of solvolysis of trityl acetate is about the same as that of the fluoride (46). To observe ionization with oxygen-containing leaving groups such as OH or OR, acid conditions are necessary. Protonation of the oxygen results in a much more readily ionizable group, mainly because of the increased electron affinity of the resulting cation.

$$Ar_3C-OR \underset{}{\overset{H^+}{\rightleftharpoons}} Ar_3C-\overset{H}{\underset{\oplus}{O}}R \longrightarrow Ar_3C\cdot + [R\dot{O}H]^{\oplus} \xrightarrow{e^-} R\ddot{O}H$$

R = H or alkyl

Assuming that the leaving group does indeed readily form an anion (and it should be noted that borderline groups such as fluoride and acetate may ionize to a significant extent with potential carbonium ions more stable

than unsubstituted trityl), we must still be concerned with the nucleophilicity of this anion which, in principle, can either vastly increase or decrease $K_{eq}$ by its effect on $k_r$. Inasmuch as there is no simple correlation between ionizability and nucleophilicity and the latter is influenced by a number of complex factors (47,48), it is no trivial matter to predict which effect will predominate under a given set of circumstances, and no attempts to do so are made here. On the basis of their nucleophilic activities, as measured by their "nucleophilic constants," (48) we would expect $K_{eq}$ for the trityl halides to increase in the order $F > Cl > Br > I$, whereas the actual order is the reverse of this. The known large nucleophilic activity of the other common groups such as $OH^-$, $CN^-$, and $OCH_3^-$ reinforces their small ionizability, and both effects operate to decrease $K_{eq}$ in nonacidic solution. Azide ion ($N_3^-$) is a particularly good reagent for trapping trityl ion and converting it to undissociated trityl azide; Swain (49) has effectively used this excellent nucleophile to explore the kinetic behavior of the trityl ion.

With leaving groups of large steric bulk, an additional factor is introduced—relief of strain in the carbonium ion. This is discussed in detail in Section III-C, but it should be noted that this factor will act to increase $k_f$ and $K_{eq}$ in the order $Br > Cl > F$ and thus compensate for the reverse order of nucleophilicities.

Finally, a number of covalent trityl derivatives have been described whose extreme hydrolytic sensitivity suggest that $K_{eq}$ in solution would be reasonably large, although again, quantitative data are not available. These covalent derivatives include trityl nitrate (50) and trityl formate (51), which are both quantitatively hydrolyzed to trityl alcohol on exposure to moist air, even in the solid state. It would also be of interest to examine $K_{eq}$ for trityl tosylate; to date this compound has not been synthesized, but on the basis of the foregoing discussion, it should be readily ionizable.

### 3. Influence of Lewis Acids

The importance of arylcarbonium ion formation by complexation with a Lewis acid has been mentioned previously and a number of examples have been summarized by Olah (6) and by Pfeiffer (10). As before, the overall reaction is an equilibration process and can be treated in an entirely analogous fashion to the spontaneous ionization of an ionogenic trityl species. The reaction of a trityl halide with a Lewis acid (MX = metal halide), is usually written as equation 2

$$\phi_3C-X + MX \longrightarrow \phi_3C^+(MX_{n+1})^- \qquad (2)$$

and implies that a direct attack on covalent halide by the Lewis acid provides the entire driving force for the resulting formation of ionic species. However, this view is probably an oversimplification and it appears possible, although experimental data are not yet available, that, as with other trityl reactions, the rate-determining step is an $S_N1$ ionization of the halide (eq. 3), followed by capture of the halide ion by the Lewis acid (eq. 4)

$$\phi_3C-X \underset{k_r'}{\overset{k_f'}{\rightleftarrows}} \phi_3C^+X^- \qquad (3)$$

$$\phi_3C^+X^- + MX_n \underset{k_r'}{\overset{k_f'}{\rightleftarrows}} \phi_3C^+(MX_{n+1})^- \qquad (4)$$

The formation of an ionized salt by this $S_N1$ mechanism should have a rate constant ($k_f$) independent of the Lewis acid and dependent only on the ionogenic species and the solvent. Therefore, this process would be kinetically identical to the well-known $S_N1$ ionization of trityl chloride in the presence of a good nucleophile such as azide (46). Presumably both processes have the identical rate-determining step, but the presence of $N_3^-$ leads to a covalent trityl azide by attack at the cationic center, whereas the Lewis acid antimony pentachloride, with its high affinity for chloride, leads to the ionic hexachloroantimonate by attack at the anionic center.

Returning to our thermodynamic approach to the influence of Lewis acids on the formation of arylcarbonium ions, we note that although the mechanistic validity for a bimolecular process is open to question, the equilibrium constant is equal to the product of both unimolecular processes; therefore, regardless of the kinetic order, the thermodynamic results are equivalent. As before, we may regard the overall process (eq. 2) as an acid-base type of equilibration where $K_{eq}$ will be determined by the relative acidities of the Lewis acid and the carbonium ion. Therefore, $K_{eq}$ will increase when we either decrease the electrophilic character (acidity) of the carbonium ion (by increasing its stability) or increase the acidity of the Lewis acid.

With few exceptions (some are mentioned below), the Lewis acid is primarily a halide acceptor; and by forming a complex with the halide, it effectively reduces the rate of collapse of the ionic species $\phi_3C^+X^-$ to covalent halide and correspondingly increases $K_{eq}$ for overall ionization. The function of the Lewis acid, then, is to complex with and reduce the nucleophilicity of the halide leaving group; in this context it acts in a similar fashion to sulfur dioxide, which also forms complexes with halide ion.

For a given carbonium ion and solvent, the extent of ionization of its covalent halide precursor by the Lewis acid will be directly related to the equilibrium amount of halide liberated in solution by dissociation of the complex anion

$$[MX_{n+1}]^- \rightleftharpoons MX_n + X^-$$

and this, of course, is a direct measure of the strength (halide affinity) of the Lewis acid. The intricate question of what determines the strengths of inorganic Lewis acids has yet to be answered. There have been only very limited quantitative studies on this general subject and these have been reviewed by Gillespie (52). Similarly, very few systematic, quantitative studies involving the equilibrium ionization of arylcarbonium ions with varying Lewis acids have been reported; this appears to be a promising area for future developments.

It is common practice to synthesize and isolate arylcarbonium ions from their covalent halides by using either (*a*) a large excess of Lewis acid, (*b*) a solvent from which the carbonium salt precipitates, or (*c*) the silver salt of the Lewis acid. In every case this displaces equilibrium 1 to the right, especially in (*c*), where the combination of "silver-assisted solvolysis" and the insolubility of the resulting silver halide is especially effective. Many techniques have been applied to probe the ionic nature of the resulting crystalline carbonium ion salts (cf. Section II-A), but mainly electronic spectroscopy has been used in investigating the extent of association of the ionophoric salt when redissolved in solution. In essence, this technique involves the following assumptions: (*1*) that ionization of the ionogenic alcohol of a stable carbonium ion is complete in concentrated sulfuric acid, (*2*) that the characteristic electronic absorption band of the resulting carbonium ion is approximately independent of both anion and solvent, and that the extinction coefficient follows Beer's law, and (*3*) that the free ions and ion pairs produced have identical spectral properties. These assumptions have been justified for the association of $Ph_3C^+HgCl_3^-$ in nitromethane (53) and presumably hold for the similar studies mentioned later.

Fairbrother (54) was the first to quantitatively determine the extent of association of a trityl salt stabilized with an anion formed from a relatively weak Lewis acid, $SnBr_4$.

$$\phi_3C\text{---}Br + SnBr_4 \xrightleftharpoons{K_1} \phi_3C^+SnBr_5^- \xrightleftharpoons{K_2} \phi_3C^+ + SnBr_5^-$$

Although in the solid state $\phi_3C^+SnBr_5^-$ is undoubtedly a completely ionized species and in a solvent of high ionizing power the reaction would proceed far to the right, in poorly ionizing solvents both $K_1$ (ion pairs) and $K_2$ (free ions) can be quantitatively detected (54). As expected, the concentration of free ions is quite small, $K_2$ ranging from $5 \times 10^{-7}$ in benzene to $2 \times 10^{-4}$ in ethyl bromide. The value of $K_1$ is of course very much larger, but even in ethyl bromide, the most polar solvent used, the compound is only 8% ionized ($1/K_1 = 9.5$), which means that 92% of the originally ionized salt has returned to its covalent components.

A comprehensive study of $K_{eq}$ for trityl chloride with a large variety of Lewis acids has been carried out in acetonitrile (55) and is summarized in Table IV. Trityl chloride is not appreciably ionized in this solvent alone, but the addition of the appropriate halide acceptor (Lewis acid) can shift the equilibrium from virtually 100% ionization (e.g., $SbCl_5$, $K_{eq} = 10^5$) to barely detectable ionization ($SbCl_3$, $K_{eq} = 10^{-1}$). Again, it is of interest to

**TABLE IV**

**Equilibrium Constants for Ionization of Trityl Chloride in Acetonitrile with Various Lewis Acids at 25° (55)**

| Lewis Acid($MX_n$) | log $K_{eq}$[a] | $\Delta F°$, kcal/mole |
|---|---|---|
| $InCl_3$ | 5.2 | −7.1 |
| $SbCl_5$ | 5.1 | −7.0 |
| $GaCl_3$ | 4.8 | −6.6 |
| $TeCl_4$ | 4.67 | −6.4 |
| $SnCl_4$ | 4.30 | −5.9 |
| $NbCl_5$ | 3.54 | −4.9 |
| $TaCl_5$ | 3.2 | −4.4 |
| $TlCl_3$ | 2.59 | −3.6 |
| $TiCl_4$ | 1.89 | −2.6 |
| $ZnCl_2$ | 1.84 | −2.5 |
| $AlCl_3$ | 1.75 | −2.4 |
| $HgCl_2$ | 1.64 | −2.2 |
| $CdCl_2$ | 1.6 | −2.2 |
| $BCl_3$ | 1.32 | −1.8 |
| $BiCl_3$ | 0.94 | −1.3 |
| $SbCl_3$ | 0.70 | −0.97 |
| $PCl_5$ | 0.5 | −0.7 |
| $AsCl_3$ | −1.40 | 1.9 |

[a] $K_{eq} = \dfrac{[\phi_3C^+ MX_{n+1}^-]}{[\phi_3CCl][MX_n]}$, $\Delta F° = -RT \ln K_{eq}$.

compare the extent of carbonium ion formation in solution with the ease of isolation of the crystalline salt. The reaction of equimolar amounts of $BCl_3$ and trityl chloride in $CH_2Cl_2$, followed by subsequent precipitation of the insoluble yellow salt with cyclohexane, affords a 79% yield of the ionic salt $\phi_3C^+BCl_4^-$, as shown by its analysis and its chemistry (56). However, dissolving the salt in the better ionizing solvent acetonitrile should, according to $K_{eq}$ of Table IV, yield an equilibrium mixture containing at most 45% of ionized salt.

With the weaker Lewis acid antimony trichloride, no stable salts have been reported, but solution of trityl chloride in liquid $SbCl_3$ indicate, from both cryoscopic (57) and conductivity (58) measurements, that trityl chloride is a strong 1:1 electrolyte of the same order as $Me_4N^+Cl^-$ and that therefore the large excess of solvent—Lewis acid—has forced $K_{eq}$ toward complete formation of $\phi_3C^+SbCl_4^-$ (as free ions). In contrast, $\phi_2CHCl$ and $\phi CH_2Cl$ are both weakly conducting in this solvent, and the values of $K_{eq}$ for the formation of free ions are (58), respectively, only $3 \times 10^{-5}$ and $1 \times 10^{-7}$.

The most comprehensive study of the ionization of triarylmethyl compounds in the presence of a weak Lewis acid is that of Evans and coworkers (59–61), who have examined the thermodynamics of carbonium ion formation with a variety of substituted trityl chlorides in various solvents in the presence of $HgCl_2$. Some of these data have been incorporated in Table V, where the ionization equilibrium constants for

**TABLE V**

Ionization Equilibria and Free Energy Values for Trityl Chlorides in Nitromethane in the Absence and Presence of $HgCl_2$ at 20° [a]

| Trityl substituent | $K_1^b \times 10^4$ | $K_2^c$ | $\Delta F_1^{°\,b}$ | $\Delta F_2^{°\,c}$ | $\Delta F_1^° - \Delta F_2^°$ |
|---|---|---|---|---|---|
| None | 4.4 | 80.4 | 4.5 | −2.54 | 7.1 |
| Mono-p-Cl | 1.9 | 26.2 | 5.0 | −1.91 | 6.9 |
| Mono-o-Cl | 1.8 | 42.6 | 5.0 | −2.18 | 7.2 |
| Mono-p-$CH_3$ | 20.5 | 1,114 | 3.6 | −4.18 | 7.8 |
| Mono-o-$CH_3$ | 18.0 | 882 | 3.7 | −3.96 | 7.7 |
| Tri-p-Cl | 0.4 | 1.6 | 5.9 | −0.27 | 6.2 |

[a] Data from Ref. 60, converted from mole fraction units to moles/liter.

[b] $K_1 = \dfrac{[Ar_3C^+Cl^-]}{[Ar_3C-Cl]}$, $\Delta F_1^° = -RT \ln K_1$. Data from Refs. 41 and 63.

[c] $K_2 = \dfrac{[Ar_3C^+HgCl_3^-]}{[Ar_3C-Cl][HgCl_2]}$, $\Delta F_2^° = -RT \ln K_2$.

six different trityl chlorides in $CH_3NO_2$ have been compared in the absence ($K_1$) and in the presence ($K_2$) of $HgCl_2$.*

Comparing first the $\Delta F°$ of unsubstituted trityl chloride with $HgCl_2$ in nitromethane to that in acetonitrile (Table IV), we note that the latter solvent favors ionization by 0.4 kcal as expected on the basis of solvent ionizing power; owing to the immeasurably small value of $K_1$ in $CH_3CN$ (55), however, this effect is not verifiable in the absence of the Lewis acid. The effects of the substituents are approximately as expected based on their ability to stabilize a positive charge. It is of interest to observe that the $\Delta F_1° - \Delta F_2°$ values of Table V are reasonably constant for all the cations, in distinct contrast to the individual $\Delta F_1°$ and $\Delta F_2°$ values. This suggests that the $HgCl_2$ (and presumably other Lewis acids as well) have a constant effect on the free energy of ionization and that, therefore, the stability of the carbonium ion will be the prime factor in determining the extent of $\Delta F°$. It may or may not be theoretically significant that the most positive deviations of $\Delta F_1° - \Delta F_o^2$ occur with the most stable carbonium ion and the most negative with the least stable. Further work along these lines, using other Lewis acids and a much greater spread in carbonium ion stability, is indicated.

The effect of solvent on $K_{eq}$ in the presence of anion acceptors is, if anything, more complex than in the absence of acceptors. Using trityl chloride as a constant chloride ion donor, Baaz and Gutmann and their

### TABLE VI

**Influence of Solvent and Lewis Acid on Ionization Equilibrium Constant $K_{eq}$ of para-Methyltrityl Chloride**

| Solvent | Lewis acid | $K_{eq}$ | $\Delta F°_{(25°)}$ |
|---|---|---|---|
| $CH_3NO_2$ | None | $20 \times 10^{-4}$ | 3.6[a] |
| $CH_3NO_2$ | $HgCl_2$ | $1.11 \times 10^3$ | −4.2[b] |
| $CH_3COOH$ | None | $4.9 \times 10^{-5}$ | 5.8[c] |
| $CH_3COOH$ | $HgCl_2$ | $1.3 \times 10^{-2}$ | 1.5[d] |
| $CH_3COOH$ | $SnCl_4$ | 6.07 | −1.05[d] |
| $CH_3COOH$ | $FeCl_3$ | $2.41 \times 10^2$ | −3.2[d] |
| $CH_3COOH$ | $SbCl_5$ | $3.24 \times 10^2$ | −3.4[d] |

[a] Ref. 63.
[b] Ref. 60.
[c] Ref. 36.
[d] Ref. 61.

* Both equilibrium constants $K_1$ and $K_2$ refer only to ionization to ion pairs, although a minute but measurable dissociation of the ion pairs to free ions does indeed take place in the presence of $HgCl_2$ ($K_{diss} \cong 1 \times 10^{-6}$) (60).

co-workers have comprehensively investigated the effect of solvent on the halide-acceptor properties of a large number of Lewis acids; the same authors have summarized these and related studies concerning the role of Lewis acids as catalysts in Friedel-Crafts and similar reactions (62).

Table VI summarizes the available data on the ionization constants $K_{eq}$ for mono-$p$-methyltrityl chloride in $CH_3NO_2$ and in $CH_3COOH$ both with and without Lewis acids. It is rather disconcerting to note that, in contrast to the expected relative solvating abilities of acetic acid and nitromethane, the latter solvent is 2.2 kcal/mole more efficient in promoting ionization in the absence of catalyst, and this difference is increased to 5.7 kcal/mole in the presence of $HgCl_2$. On the other hand, an even more remarkable leveling effect is observed if we take the difference in $\Delta F°$ for the weakest and strongest Lewis acids common to Tables IV and VI:

| Lewis acid | $\Delta F°$ (Trityl chloride in $CH_3CN^a$) | $\Delta F°$ ($p$-$CH_3$–Trityl chloride in $CH_3COOH^b$) |
|---|---|---|
| $HgCl_2$ | −2.2 | 1.5 |
| $SbCl_5$ | −7.0 | −3.4 |
| | $\Delta\Delta F° = -4.8$ kcal/mole | $\Delta\Delta F° = -4.9$ kcal/mole |

[a] From Table IV.
[b] From Table VI.

The virtually identical values for $\Delta\Delta F°$ even though we are comparing different carbonium ions in different solvents, may indeed be only fortuitous and indicative of the desirability for further studies in this area.

Tin tetrachloride is a relatively good halide acceptor and presents the interesting complication that the resulting anion, the pentacoordinate $SnCl_5^-$ anion, is also a good halide acceptor, forming the stable hexacoordinate dinegative anion, $SnCl_6^{-2}$. Therefore, in solution, two equilibria will be established

$$Ar_3CCl + SnCl_4 \rightleftharpoons Ar_3C^+SnCl_5^- \quad (5)$$

$$Ar_3C^+SnCl_5^- + Ar_3CCl \rightleftharpoons (Ar_3C)_2^+SnCl_6^{-2} \quad (6)$$

The value of the equilibrium constants $K_{eq(5)}$ and $K_{eq(6)}$ will presumably be determined by the relative acidities of the three Lewis acids present,— $SnCl_4$, $Ar_3C^+$, and $SnCl_5^-$. Inasmuch as $SnCl_5^-$ is a much weaker Lewis acid than is $SnCl_4$, reaction 6 will be appreciable only with the comparatively stable carbonium ions such as tropylium, which forms $(C_7H_7)_2^+SnCl_6^{-2}$ as the only isolable product (64). With less stable carbonium ions such as trityl, $SnCl_5^-$ is not a strong enough Lewis acid to react with the unionized trityl chloride and reaction 6 takes place to only a minor extent, leaving

$\phi_3C^+SnCl_5^-$ as the product actually isolated (65a). With carbonium ions of intermediate stability, such as tri-$p$-tolyl carbonium ion, it is possible to drive the equilibrium in either direction: addition of excess $SnCl_4$ to a methylene chloride solution of $(p\text{-}CH_3C_6H_4)_3C\text{-}Cl$ affords only the pentachlorostannate salt. When this is treated with 1 mole of the triaryl chloride, only the hexachlorostannate salt of correct analysis is obtained.

$$Ar_3C^+SnCl_5^- + Ar_3CCl \longrightarrow (Ar_3C^+)_2SnCl_6^{-2} \downarrow$$
$$Ar = p\text{-}CH_3C_6H_4$$

If now the insoluble hexachlorostannate is slurried with $SnCl_4$ in $CH_2Cl_2$, its dissolution leads to the re-formation of the pentachlorostannate (65a):

$$(Ar_3C^+)_2SnCl_6^{-2} + SnCl_4 \longrightarrow 2Ar_3C^+SnCl_5$$
$$Ar = p\text{-}CH_3C_6H_4$$

It would be of interest to obtain data on the thermodynamic distribution of the three species in solution.*

No discussion on stabilizing anions would be complete without mention of the perchlorate ($ClO_4^-$) anion, the nonnucleophilic partner par excellence. Although not a Lewis acid in the usual sense, anhydrous $HClO_4$, or its silver salt, readily forms ionic perchlorates with all but the most unstable carbonium ions. Unlike anions such as $SnCl_5^-$ and $SbCl_6^-$, which can donate a halide to reform covalent material, there is no such possibility with $ClO_4^-$. However, the ionic perchlorate $R_3C^+ClO_4^-$ can rearrange to form a covalent perchlorate ester $R_3COClO_3$, and this apparently happens with $n$-alkyl perchlorates (66). Trityl perchlorate, known to be completely ionic in the crystal, is a strong electrolyte in liquid sulfur dioxide (67) and has recently been shown to be completely ionized in a number of halogenated aliphatic solvents (68). No studies have been reported on the extent of covalent perchlorate ester formation with carbonium ions of intermediate stability, and presumably this could be accomplished by the usual spectroscopic techniques.

A number of unusual anions have been used to stabilize trityl salts. The hydrogen dichloride anion ($HCl_2^-$) is, as expected, among the poorest of the nonnucleophilic anions and readily re-forms covalent triaryl chloride and HCl

$$Ar_3Cl + HCl \rightleftharpoons Ar_3C^+HCl_2^-$$

The equilibrium constant in solution for this reaction has not been reported, but with stable enough carbonium ions, a crystalline species can be isolated. Sharp (23) has prepared the $HCl_2^-$ salts of the tri-$p$-methoxy

---

* A comprehensive study on the stability and thermodynamic properties of various trityl penta- and hexachlorostannates has been recently reported (65b).

and 9-phenylxanthenyl carbonium ions and concluded from their infrared spectra that they were ionic. A similar conclusion for the former compound was reached by others on the basis of solid state absorption (31) and X-ray crystallographic data (16). The preparation of trityl tetrahaloborate salts containing mixed halogenoborate ions was attempted by reacting $\phi_3CX$ and $BY_3$ (where X or Y = F or Cl); but although the presence of the trityl ion was spectroscopically confirmed, it was not possible to show definitely the presence of mixed anions (26). On the other hand, a mixed hexacoordinate-antimonate anion has been reported by Paiska (69), who prepared trityhydroxypentachloroantimonate ($\phi_3C^+SbCl_5OH^-$), as a reddish solid of correct analysis, from trityl alcohol and $SbCl_5$.

Finally, and reserving the most unusual salt for last, a recent report (70) states that tritylether (or alcohol) reacts in toluene with the bifunctional Lewis acid, 1,2-bis(dichloroboryl) ethane, to form the following remarkable salt.

$$(C_6H_5)_3COR + (BF_2CH_2)_2 \longrightarrow (C_6H_5)_3C^+ \left[ RO \begin{array}{c} BF_2\text{—}CH_2 \\ | \\ BF_2\text{—}CH_2 \end{array} \right]^-$$

$$R = H \text{ or } CH_3$$

## III. THE EFFECT OF STRUCTURE ON ARYLCARBONIUM ION STABILITY

### A. General Background

Any discussion of the "stability" or "stabilization energy" of carbonium ions requires some elaboration of definitions and conventions. When a molecule is designated as either stable or unstable, we are commonly referring to its ease of isolation, tendency to decompose, its ability to exist, or a combination of these related, descriptive qualities. We are here interested only in the quantitative, thermodynamic stability of carbonium ions, defined as the difference in the energy required for the ionization of an ionogenic compound (R–X) as compared with that required for recombination of the ions ($R^+X^-$) to covalent species; i.e., $\Delta F°$ for equation 7

$$R\text{–}X \underset{}{\overset{K_{eq}}{\rightleftharpoons}} R^+X^- \qquad (7)$$

obtained by methods discussed subsequently from the equilibrium constant $K_{eq}$ for this reversible process:

$$K_{eq(7)} = \frac{[R^+X^-]}{[R\text{–}X]} \qquad \Delta F°_{(7)} = -RT \ln K_{eq(7)}$$

The $\Delta F°_{(7)}$ for ionization is a quantitative measure of the stability of any carbonium ion ($R^+$) relative to its covalent precursor.

Since we are actually interested in information on the effects of structural change on carbonium ion stability, we can utilize the difference in free energy of equation 7 as found for two different carbonium ions (i.e., $\Delta\Delta F°$) as a quantitative measure of the result of this structural dissimilarity. However, if the stabilization energy in such a comparison is indeed to reflect only internal electronic factors supporting positive charge in the ion, then all other factors that affect $\Delta F°$ must be constant or identical. This makes a common solvent and leaving group mandatory (Section II-B) and requires the assumption that contributions to $\Delta\Delta F°$ from the covalent species are similar enough to be neglected. For minor structural changes this is a reasonable premise, but it rapidly becomes invalid when conformational or steric differences are of increasing importance in the covalent compounds. In the latter case, $\Delta\Delta F°$ now includes inseparable contributions from changes in $\Delta F°$ in both the ions and their progenitors and is no longer a measure of internal electronic effects. Examples elaborating on the importance of such steric factors are given in Section III-C.

To best gauge the magnitude of a structural change on cation stability it is convenient to refer all $\Delta F°$ values to a common reference compound. The unsubstituted trityl cation is the obvious choice for such a standard (71) and such a comparison is designated as $\Delta F°_{R^+}$ (72). To obtain $\Delta F°_{R^+}$ for a carbonium ion whose $\Delta F°$ for ionization can be determined by any of the available techniques, it is only necessary to subtract this value from the corresponding value for trityl. This operation corresponds to the free energy for equation 8 where $R_0$ = trityl and $\Delta F°_{(8)} = \Delta F°_{R^+}$ for $R^+$. By definition, $\Delta F°_{R^+ \text{trityl}} = 0$, and ions more stable than trityl will have a

$$R^+Y^- + R_0X \rightleftharpoons RX + R_0^+Y^- \qquad (8)$$

positive $\Delta F°_{R^+}$, and vice versa. Although this is not in keeping with the usual thermodynamic convention in which a more negative $\Delta F°$ represents a more favored process, it has the distinct psychological advantage of associating a greater stability with a more positive number. Finally, we note that use of $\Delta F°_{R^+}$, in contrast to the dimensionless, often-used $pK_{R^+}$, allows the assignment of a numerical value in kilocalories to a particular structural change (substituent) and makes possible the direct comparison of carbonium ion stabilities measured by a variety of methods.

## B. Quantitative Determination of Carbonium Ion Stabilization Energies ($\Delta F°_{R^+}$)

### 1. Conductivity Measurements

This fundamental technique for directly measuring the free ions produced from an ionogenic compound, has been comprehensively reviewed

(4). The experimental equilibrium constant $K_{exp}$ is a composite of two constants, that for ionization $K_1$ and for dissociation $K_2$

$$R-X \xrightleftharpoons{K_1} R^+X^- \xrightleftharpoons{K_2} R^+ + X^-$$

$$K_{exp} = \frac{K_1 K_2}{K_1 + 1}$$

For closely related series with the same leaving group in the same solvent, $K_2$ should be reasonably constant and can be quantitatively estimated (4). Knowing $K_{exp}$, we can solve for $K_1$, the desired measure of the relative tendency of the compound to ionize. Table VII lists $K_1$, $\Delta F°$, and $\Delta F_{R^+}°$ for a series of substituted trityl chlorides in liquid sulfur dioxide. The variation in stability is generally in agreement with the expected electronic effect of the substituents.

The range of stabilities determined by conductivity methods is limited: the upper limit appears to be $\Delta F_{R^+}° = 3\text{-}4$, a value expected for 4,4',4''-t-butyltrityl chloride, which was regarded as too highly ionized for an accurate determination of $K_{exp}$ (4,73). The lower limit has not been explicitly set but appears to be as shown in Table VII. Change of solvent

TABLE VII

Stability Data from Conductivity Measurements of Triarylmethyl Chlorides in $SO_2$ at $0°$ [a]

| Substituent | $K_1 \times 10^2$ | $\Delta F_{0°}°$, kcal/mole | $\Delta F_{R^+}°$ |
|---|---|---|---|
| 4,4'-t-$C_4H_9$ | 320 | −0.6 | 2.9 |
| 4,4',4''-t-$C_4H_9$ | 87 | 1.0 | 2.2 |
| 2-$CH_3$ | 27 | 0.7 | 1.6 |
| 4-$CH_3$ | 23 | 0.8 | 1.5 |
| 4-t-$C_4H_9$ | 23 | 0.8 | 1.5 |
| 4,4'-$C_6H_5$ | 22 | 0.8 | 1.5 |
| 4-$C_6H_5$ | 5.6 | 1.6 | 0.7 |
| 3-t-$C_4H_9$ | 4.8 | 1.6 | 0.7 |
| 3-$CH_3$ | 3.4 | 1.8 | 0.5 |
| None | 1.46 | 2.3 | (0) |
| 3-$C_6H_5$ | 0.73 | 2.7 | −0.4 |
| 3,3'-$C_6H_5$ | 0.42 | 3.0 | −0.7 |
| 4-Cl | 0.38 | 3.0 | −0.7 |
| 2-Cl | 0.31 | 3.1 | −0.8 |
| 3,3',3''-$C_6H_5$ | 0.24 | 3.3 | −1.0 |
| 3-Cl | 0.47 | 4.2 | −1.9 |
| 4,4'-4''-Cl | 0.033 | 4.4 | −2.1 |

[a] Data from Ref. 4.

or leaving group could theoretically extend the range of stabilities toward higher or lower $\Delta F_{R^+}^{\circ}$ values, but it is unlikely that a common solvent, or leaving group, will become available for extension of the stability range beyond a span of $\sim 6$ kcal/mole. Nitrobenzene affords $K_1$ values at 25° about $10^{10}$ smaller than in sulfur dioxide (42) and thereby permits determination of conductivity data for the highly ionizable 4,4′,4″-methoxytrityl chloride ($K_1 = 4 \times 10^{-3}$), but not for the much weaker electrolyte trityl chloride; therefore, $\Delta F_{R^+}^{\circ}$ values are not available in this solvent.

## 2. Electronic Spectrophotometry

An important identifying feature of arylcarbonium ions is found in their intense absorption in the ultraviolet-visible region of the spectrum. These highly characteristic spectra (74), typically with extinction coefficients $> 10^4$, can afford accurate quantitative determination of even minute ($\sim 10^{-5} M$) concentrations of ion in the presence of relatively large amounts of their normally low-extinction, covalent precursors. The necessary $\lambda_{max}$ and $\varepsilon_{max}$ data are usually obtained either from the pure carbonium salt in an inert solvent or from the covalent carbinol in a medium sufficiently acidic to afford complete ionization (see Section III-B-3). It is standard practice to assume that both $\lambda_{max}$ and $\varepsilon$, once determined, will not vary with solvent or ion source, and it is common to find spectral data obtained from concentrated sulfuric acid being used in conjunction with acetonitrile or similar solvents. Fortunately, as documented in Table VIII for the trityl

**TABLE VIII**

Variation in $\lambda_{max}$ and $\varepsilon_{max}$ with Source of Ion and Solvent for Trityl Cation

| Source | Solvent | $\lambda_{max}$, $m\mu$ | $\log \varepsilon$ | Ref. |
|---|---|---|---|---|
| Carbinol | $> 20\%$ $H_2SO_4$ | 431 | 4.57 | 75 |
| Carbinol | $H_2SO_4$–aq. HOAc | 433 | 4.52 | 76 |
| $BI_3^-$ Salt | $H_2SO_4$ | 432 | 4.5 | 77 |
| $SnCl_5OH^-$ Salt | $CH_2Cl_2$ | 431 | 4.5 | 69 |
| $SbCl_6^-$ Salt | $CH_2Cl_2$ | 430 | 4.56 | 78 |
| $SbCl_6^-$ Salt | $H_2SO_4$ | 430 | 4.57 | 78 |
| $ClO_4^-$ Salt | $H_2SO_4$ | 433 | 4.6 | 34 |

cation generated from diverse sources in various solvents, this assumption appears to be reasonably justified.

Utilizing the spectral data as obtained for the pure ion, and having established that complications arising from either irreversible reactions with solvent or dissociation phenomena are absent, $K_{eq}$ (and $\Delta F^{\circ}$) can be obtained for the ionization of trityl halides from equation 7 when R = Ar$_3$C. Thus, after proving that in nitromethane a direct proportionality is observed between the spectrophotometrically determined concentrations of carbonium ion and unionized trityl halide, Evans and co-workers (63,41) concluded that only simple ion pairs were produced and obtained the extensive data listed in Table IX. A comparison of these results with those obtained by other techniques follows.

In general, this method suffers from the same deficiencies as the conductivity method, it is not possible to find a single solvent or leaving group with which to cover the entire stability range, or even a substantial part of

### TABLE IX
### Stability Data for Trityl Chlorides in Nitromethane by Spectroscopy[a]

| Substituents | $\Delta F°$, kcal/mole[b] | $\Delta F_R°+$ |
|---|---|---|
| 2,2′,2″-CH$_3$ | 1.4 | 3.1 |
| 4,4′,4″-CH$_3$ | 1.7 | 2.8 |
| 4,4′,4″-C$_6$H$_{11}$ | 1.7 | 2.8 |
| 4,4′,4″-$i$-C$_3$H$_7$ | 1.9 | 2.6 |
| 4,4′,4″-$t$-C$_4$H$_9$ | 2.0 | 2.5 |
| 4,4′-CH$_3$ | 2.6 | 1.9 |
| 4′-$t$-C$_4$H$_9$ | 2.7 | 1.8 |
| 2-Cl,4′,4″-CH$_3$ | 2.8 | 1.7 |
| 2-Br,4′4″-$t$-C$_4$H$_9$ | 2.9 | 1.6 |
| 2,5-di-CH$_3$ | 3.1 | 1.4 |
| 2-Cl,4′,4″-CH$_3$ | 3.1 | 1.4 |
| 4-C$_6$H$_{11}$ | 3.5 | 1.0 |
| 4-$i$-C$_3$H$_7$ | 3.6 | 0.9 |
| 4-CH$_3$ | 3.6 | 0.9 |
| 4-$t$-C$_4$H$_9$ | 3.7 | 0.8 |
| 2-CH$_3$ | 3.7 | 0.8 |
| 2-Br, 2-$t$-C$_4$H$_9$ | 4.0 | 0.5 |
| 3-CH$_3$ | 4.1 | 0.4 |
| None | 4.5 | (0) |
| 4-F | 4.5 | 0 |
| 2-Br | 4.7 | −0.2 |
| 4-Cl | 5.0 | −0.5 |
| 4-Br | 5.0 | −0.5 |
| 4,4′,4″-Cl | 5.9 | −1.4 |

[a] Data from Refs. 63 and 41.
[b] At 18–23°.

it. Furthermore, the polar nonnucleophilic solvents that are suitable for carbonium ion spectroscopy are notoriously difficult to free from trace impurities, a frustrating situation when determining a $K_{eq}$ involving concentrations of reactive ions in the $10^{-5}M$ range.

### 3. Ionization of Alcohols ($pK_{R^+}$)

The basicity of arylcarbinols was recognized long before the investigators acknowledged that the colored species produced in acid media were indeed carbonium ions. Cryoscopic data provided important evidence. In concentrated sulfuric acid, trityl and even diphenyl carbinols yielded a van't Hoff $i$-factor of 4, indicating complete dissociation (79):

$$Ar_3C-OH + H_2SO_4 \rightleftharpoons Ar_3C^+ + H_3O^+ + 2HSO_4^-$$

A very early attempt (80) to utilize arylcarbinol basicity for stability measurements by comparing the amounts of aqueous ethanol needed to quench the color generated from equimolar amounts of triarylcarbinols in acidic solution, presaged the sophisticated techniques available today.* These developments closely paralleled the evolution of Hammett's $H_0$ function and, considering the importance of this technique, a brief historical survey is given.

For the reversible hydration of a carbonium ion $R^+$ we may write equation 9 and, assuming that the concentration of $ROH_2^+$ is negligible,

$$R^+ + H_2O \rightleftharpoons ROH_2^+ \rightleftharpoons ROH + H^+ \tag{9}$$

as justified by Gold and Tye (82), we obtain equation 10, which can be

$$R^+ + H_2O \overset{K}{\rightleftharpoons} ROH + H^+ \tag{10}$$

readily transformed to equation 11, where pH and p$K$ represent the usual

$$pK = pH + \log \frac{[R^+]}{[ROH]} \tag{11}$$

criteria for Brønsted acidity measurements. Thus the p$K$ of water-soluble arylcarbinols that can be 50% ionized (i.e., $[R^+]/ROH] = 1$) in the normal pH range can be readily obtained by suitable titration or spectrophotometric methods. Lund (83) determined the p$K$'s for a series of polymethoxy-substituted triaryl carbinols in aqueous acetone; the 4,4′,4″-trimethoxy-trityl compound (p$K \cong 1$) probably being the lower limit for a trityl alcohol stable enough to be measured by normal acid-base methods.

For carbinols yielding less stable carbonium ions, the low pH's required are no longer measurable in aqueous solution and recourse must be made

---

* However, it is remarkable that as late as 1957, the Bayer-Villiger method still had its adherents (81).

to the $H_0$ acidity function techniques developed by Hammett for extending the pH scale to concentrated acid solutions (84). An early, moderately successful, attempt by Conant (85) was followed by the spectrophotometric investigations of low-stability trityl alcohols in nearly anhydrous sulfuric acid by Murray and Williams (86), who suggested that their equilibrium constant for ionization was related to Hammett's $H_0$. However, it remained for Gold and Hawes (75) to firmly establish this relationship and to define a new acidity function $J_0$, specifically adapted to the ionization of arylcarbinols in sulfuric acid. Using these $J_0$ values, Deno and co-workers (87) determined the thermodynamic p$K$ (designated as p$K_{R^+}$) for a series of trityl alcohols, but they later discovered that these values were seriously in error because of the breakdown of the $J_0$ values in acid more dilute than 80% (88). Therefore, a new acidity function, evaluated by Hammett $H_0$ techniques utilizing 18 arylcarbinols was determined and shown to be valid and reasonably independent of carbinol concentration in the 5–92% range of aqueous sulfuric acid.

Originally designated as $C_0$, Deno's popular carbonium ion-carbinol acidity function (88) is now commonly referred to as $H_R$ and defined as in equation 12. Using the $H_R$ value appropriate to the particular sulfuric

$$H_R \equiv pK_R + -\log\frac{[R^+]}{[ROH]} \tag{12}$$

acid concentration necessary to obtain a concentration of carbonium ion suitable for spectrophotometric evaluation, the p$K_{R^+}$ of a very wide range of ionizable carbinols can be obtained from equation 12. Conversion from p$K_{R^+}$ to standard $\Delta F°$ values is accomplished as in equation 13

$$\Delta F° = RT\, 2.303\, pK_{R^+} \tag{13}$$

and assuming a temperature of 25°, we obtain

$$\Delta F° = 1.365\, (pK_{R^+}) \tag{14}$$

Finally, from our definition of $\Delta F°_{R^+}$ and using the well-established value of $-6.6$ for the p$K_{R^+}$ of the trityl cation ($\Delta F° = 9.0$) we obtain

$$\Delta F°_{R^+}\ (\text{kcal/mole}) = 9.0 + (1.37)\, pK_{R^+} \tag{15}$$

which is useful for converting literature p$K_{R^+}$ values directly to $\Delta F°_{R^+}$.

The vast majority of thermodynamic carbonium ion stability data has been obtained from p$K_{R^+}$ measurements. A comprehensive compilation of all the currently available p$K_{R^+}$ data, together with the corresponding $\Delta F°$ and $\Delta F°_{R^+}$ values, is given in Tables X–XII for trityl, diphenylmethyl, and malachite green (4,4′-dimethylaminotrityl) carbonium ions, respectively. In all cases the temperature has been assumed to be 25° and when

## TABLE X
### Stability of Triphenylcarbonium Ions from p$K_{R^+}$ Measurements of Trityl Alcohols

| Substituents in | | | | | | |
|---|---|---|---|---|---|---|
| Ring 1 | Ring 2 | Ring 3 | p$K_{R^+}$ | $\Delta F^{\circ a}$ | $\Delta F_R^{\circ}{}_+$ | Ref. |
| 4-$(CH_3)_2$N | 4-$(CH_3)_2$N | 4-$(CH_3)_2$N | 9.36 | 12.8 | 21.8 | 89 |
| 4-$C_6H_5$NH | 4-$C_6H_5$NH | 4-$C_6H_5$NH | 7.78 | 10.6 | 19.6 | 90 |
| 4-$NH_2$ | 4-$NH_2$ | 4-$NH_2$ | 7.57 | 10.3 | 19.3 | 89 |
| 4-$(CH_3)_2$N | 4-$(CH_3)_2$N | — | $\begin{cases}6.90\\7.07\end{cases}$ | 9.4 | 18.4 | 89<br>91 |
| 4-$NH_2$ | 4-$NH_2$ | — | 5.38 | 7.3 | 16.3 | 89 |
| 4-$(CH_3)_2$N | — | — | $\begin{cases}3.54\\3.7\end{cases}$ | 4.8 | 13.8 | 92<br>94 |
|  |  |  | 4.75 | 6.5 | 15.5 | 89 |
| 4-$NH_2$ | — | — | 4.6 | 6.2 | 15.2 | 87 |
|  |  |  | 2.61 | 3.6 | 12.2 | 92 |
| 4-OH | 4-OH | 4-OH | 1.97 | 2.7 | 11.7 | 90 |
| 4-$OCH_3$ | 4-$OCH_3$ | 4-$OCH_3$ | 0.82 | 1.1 | 10.1 | 88,93 |
| 4-$OCH_3$ | 4-$OCH_3$ | — | $\begin{cases}-1.24\\-0.89\end{cases}$ | −1.5 | 7.5 | 88<br>93 |
| 4-$CH_3$CONH | 4-$CH_3$CONH | 4-$CH_3$CONH | −1.78 | −2.4 | 6.6 | 90 |
| 4-$C_6H_5$CONH | 4-$C_6H_5$CONH | 4-$C_6H_5$CONH | −2.30 | −3.1 | 5.9 | 90 |
| 2-$CH_3$ | 2-$CH_3$ | 2-$CH_3$ | −3.4 | −4.6 | 4.4 | 88 |
| 4-$OCH_3$ | — | — | $\begin{cases}-3.40\\-3.23\\-3.20\end{cases}$ | −4.5 | 4.5 | 88<br>76<br>93 |
| 4-$CH_3$ | 4-$CH_3$ | 4-$CH_3$ | −3.56 | −4.9 | 4.1 | 88 |
| 4-$CH_3$ | 4-$CH_3$ | — | −4.39 | −6.0 | 3.0 | 76 |
| 4-$SCH_3$ | 4-$SCH_3$ | — | −4.6 | −6.2 | 2.8 | 94 |
| 4-$SCH_3$ | — | — | −5.1 | −6.9 | 2.1 | 94 |
| 4-$CH_3$ | — | — | $\begin{cases}-5.3\\-5.41\end{cases}$ | −7.3 | 1.7 | 88<br>76 |
| 4-F | 4-F | 4-F | $\begin{cases}-6.00\\-6.05\end{cases}$ | −8.2 | 0.8 | 95<br>96 |
| 4-$t$-$C_4H_9$ | — | — | −6.1 | −8.3 | 0.7 | 88 |
| 4-F | 4-F | — | −6.21 | −8.6 | 0.4 | 95 |
| 3-$CH_3$ | 3-$CH_3$ | 3-$CH_3$ | −6.35 | −8.7 | 0.3 | 88 |
| 4-F | — | — | −6.42 | −8.8 | 0.2 | 95 |
| 4-$t$-$C_4H_9$ | 4-$t$-$C_4H_9$ | — | −6.6 | −9.0 | 0.0 | 88 |
| 4-$t$-$C_4H_9$ | 4-$t$-$C_4H_9$ | 4-$t$-$C_4H_9$ | −6.5 | −8.9 | 0.1 | 88 |
| 4-$iso$-$C_3H_7$ | 4-$iso$-$C_3H_7$ | 4-$iso$-$C_3H_7$ | −6.54 | −8.9 | 0.1 | 88 |
| None | None | None | $\begin{cases}-6.63\\-6.65\\-6.44\end{cases}$ | −9.0 | (0.0) | 88<br>76<br>93 |
| 2-$SCH_3$ | — | — | −6.8 | −9.3 | −0.3 | 94 |
| 4-Cl | 4-Cl | 4-Cl | $\begin{cases}-7.74\\-7.43\end{cases}$ | −10.3 | −1.3 | 88<br>93 |

**TABLE X** (*Continued*)

| Substituents in | | | $pK_{R^+}$ | $\Delta F^{\circ a}$ | $\Delta F^{\circ}_{R^+}$ | Ref. |
|---|---|---|---|---|---|---|
| Ring 1 | Ring 2 | Ring 3 | | | | |
| 3-F | — | — | −7.81 | −10.7 | −1.7 | 95 |
| 3-F | 3-F | — | −9.08 | −12.4 | −3.4 | 95 |
| 4-NO$_2$ | — | — | $\begin{cases} -9.15 \\ -9.44 \end{cases}$ | −12.8 | −3.8 | 88<br>93 |
| 3-F | 3-F | 3-F | 10.72 | −14.6 | −5.6 | 95 |
| 3-Cl | 3-Cl | 3-Cl | −11.03 | −15.0 | −6.0 | 88 |
| 4-[(CH$_3$)$_3$N]$^+$ | 4-[(CH$_3$)$_3$N]$^+$ | — | −12.1 | −16.5 | −7.5 | b |
| 4-NO$_2$ | 4-NO$_2$ | — | $\begin{cases} -12.90 \\ -13.45 \end{cases}$ | −17.9 | −8.9 | 88<br>93 |
| 4-[(CH$_3$)$_3$N]$^+$ | 4-[(CH$_3$)$_3$N]$^+$ | 4-[(CH$_3$)$_3$N]$^+$ | −15.2 | −20.6 | −11.6 | b |
| 4-NO$_2$ | 4-NO$_2$ | 4-NO$_2$ | $\begin{cases} -16.27 \\ -16.5 \\ (-18.08) \end{cases}$ | −22.3 | −13.3 | 88<br>b<br>93 |
| penta-F | penta-F | penta-F | ∼ −17.5 | −24.0 | −15.0 | 96 |

[a] For multiple values of $pK_{R^+}$ which are in reasonably good agreement, the $\Delta F^{\circ}$ values are averaged. The temperature is assumed to be 25°.
[b] Calculated from data in Ref. 86, using $H_R$ values of Ref. 88.

**TABLE XI**

**Stability of Diphenyl Carbonium Ions from Ionization of Alcohols ($pK_{R^+}$)**

| Substituents | $pK_{R^+}$ | $\Delta F^{\circ}_{R^+}$ | $\Delta\Delta F^{\circ a}$ | Ref. |
|---|---|---|---|---|
| 4,4′-N(CH$_3$)$_2$ | 5.61 | 16.6 | 25.9 | 89 |
| 4,4′-OCH$_3$ | −5.71, −5.60 | 1.3 | 10.6 | 88,93 |
| 4,4′-Mesityl | −6.6 | 0 | 9.3 | 88 |
| 4-OCH$_3$ | −7.9 | −1.8 | 7.5 | 97 |
| 4-OCH$_3$, 4′-CH$_3$ | −8.32 | −2.4 | 6.9 | 76 |
| 4,4′-OC$_6$H$_5$ | −9.85 | −4.5 | 4.8 | 90 |
| 4,4′-CH$_3$ | −10.4 | −5.2 | 4.1 | 88 |
| 4-CH$_3$ | −11.6 (−12.3) | −6.9 | 2.4 | 97,76 |
| 4-C$_2$H$_5$ | −11.6 | −6.9 | 2.4 | 97 |
| 4-$i$-C$_3$H$_7$ | −11.6 | −6.9 | 2.4 | 97 |
| 2,2′-CH$_3$ | −12.45 | −8.1 | 1.2 | 88 |
| 4,4′-F | −13.03 | −8.9 | 0.4 | 90 |
| None | −13.3, −13.56 | −9.3 | (0) | 88,76 |
| 4-Cl | −13.7 | −9.7 | −0.4 | 97 |
| 4,4′-Cl | −13.96 | −10.0 | −0.7 | 88 |
| 4,4′-Br | −14.16 | −10.3 | −1.0 | 90 |
| 4,4′-I | −14.26 | −10.5 | −1.2 | 90 |

[a] Relative to unsubstituted diphenylmethyl.

## TABLE XII
### Stability Data for Derivatives of Malachite Green

| R | $pK_{R^+}$ [a] | $\Delta F°_{R^+}$ | $\Delta\Delta F°$ [b] | Ref. |
|---|---|---|---|---|
| H | 7.07 | 18.6 | (0) | 91 |
| 4-$(CH_3)_3N^+$ | 6.15 | 17.4 | −1.2 | 91 |
| 4-$NO_2$ | 6.00 | 17.2 | −1.4 | 91 |
| 4-Cl | 6.80 | 18.3 | −0.3 | 91 |
| 4-$OCH_3$ | 7.87 | 19.7 | 1.1 | 91 |
| 4-OH | 8.06 | 19.9 | 1.3 | 91 |
| 2-Cl | 8.26 | 20.3 | 1.7 | 91 |
| 2-$NO_2$ | 7.5 | 19.2 | 0.6 | 91 |
| 4-$(CH_3)_2N$ | 9.46 | 21.9 | 3.3 | 98 |
| 2-OH | 7.80 | 19.7 | 1.1 | 99 |
| 3-OH | 6.95 | 18.5 | −0.1 | 100 |
| 3-$OCH_3$ | 6.90 | 18.4 | −0.2 | 100 |
| 2-$OCH_3$ | 7.83 | 19.7 | 1.1 | 101 |
| 2-F | 7.09 | 18.2 | −0.4 | 102 |
| 2-Br | 8.44 | 20.5 | 0.9 | 103 |
| 2-I | 9.05 | 21.3 | 2.7 | 103 |
| 2-Cl | 8.27 | 20.1 | 1.5 | 103 |
| 4-F | 7.15 | 18.8 | 0.2 | 104 |
| 4-Br | 6.82 | 18.3 | −0.3 | 104 |
| 4-I | 6.89 | 18.4 | −0.2 | 104 |
| 3-Cl | 6.53 | 17.9 | −0.7 | 105 |
| 3-Br | 6.58 | 18.0 | −0.6 | 105 |
| 3-I | 6.53 | 17.9 | −0.7 | 105 |
| 2-$CH_3$ | 9.07 | 21.4 | 2.8 | 106 |
| 3-$CH_3$ | 7.22 | 18.5 | −0.1 | 106 |
| 4-$CH_3$ | 7.32 | 19.0 | 0.4 | 106 |

[a] Calculated from $-\log[K_4/K_6]$ as given in references cited.
[b] Relative to malachite green.

multiple values of $pK_{R^+}$ were available, the derived $\Delta F°$ and $\Delta F°_{R^+}$ data are averaged when the data are in reasonable agreement, or the more reliable value is utilized.

The reliability of carbonium stabilization energies by $pK_{R^+}$ measurements in a particular solvent is limited by those factors which are inherent in all systems that depend on acidity function theory (107). The validity of equation 12 rests on the premise that the activity coefficients of the carbinol and carbonium ion vary in a manner similar to the standard bases used to establish the $H_R$ scale (93). This can be established experimentally for a given compound (88) or for a given solvent system (76) but, until verified, it remains only an assumption.

Quantitative evaluation of $[R^+]$ in equation 12 is invariably obtained by electronic spectrophotometry in the visible region and depends heavily on a knowledge of $\varepsilon_{max}$ of the pure carbonium ion and the assumption that Beer's law is valid over the range of concentrations used. This normally presents no difficulties for the more stable ions; but some species require very concentrated acid to convert the carbinol completely to ion, and these often undergo side reactions that necessitate an extrapolation to zero time for the estimation of the changing $\varepsilon_{max}$. All the side reactions in concentrated sulfuric acid solution can lead to destruction of either carbinol or carbonium ion; among them are sulfonation (108), oxidization, and ring closure; even if these artifacts are produced only in minute amounts, their spectral properties may seriously affect the calculation of $[R^+]$. Finally, we note that aqueous sulfuric acid is not a particularly good solvent for the high molecular weight carbinols and, in some cases, it has the further undesirable property of affording colloidal dispersions that appear, on casual inspection, to be clear solutions.

Tritylcarbinols that produce exceptionally stable ions ($pK_{R^+} > 2$) are almost invariably stabilized with strongly electron-donating groups; in this respect, the *p*-amino function is particularly important.* Although it is not necessary to use $H_R$ methods to measure the $pK_{R^+}$ of *p*-aminotrityl carbinols, there is a complication arising from the comparable basicity of the amino and hydroxyl functions in the monoamino carbinols, e.g., **9**. Competition of the amino and carbinol functions of **9** for a proton leads to equilibration with finite concentrations of both the carbonium ion **10** and the ammonium ion **11**; now the $pK_{R^+}$ of **9** is no longer a simple

---

* Historically, these amino-substituted trityl ions were an important source of dyestuffs and indicators and the early literature is replete with descriptive, rather than systematic, nomenclature for these compounds and their precursors. Some examples still in common use are: malachite green (4,4'-dimethylaminotrityl), crystal violet (4,4',4''-dimethylaminotrityl), brilliant green (4,4'-diethylaminotrityl), and Dobner's violet (4,4'-diaminotrityl).

function of pH and $K_a$ but involves an additional equilibrium constant $K_b$, which includes [**11**]. The large discrepancy in the $pK_{R^+}$ for both the 4-amino and the 4-dimethylaminotrityl carbinol in Table X reflects the fact that the lower, undoubtedly more reliable, values were obtained by recognition of this complication. Of peripheral interest is the observation that both **10** and **11** are capable of further protonation to yield the destabilized ammonium ion **12**; since this will occur only at very low pH, **12** is not involved in the equilibria discussed above. In fact, the difference in $\Delta F_{R^+}^\circ$ between 4-NH$_2$ and 4-NH$_3^+$ can be estimated to be $\sim 16$ kcal/mole in favor of the former, a graphic illustration of the relative ability of these two groups to stabilize a positive charge.

Substitution of a second $p$-amino function in trityl greatly increases the carbinol acidity (carbonium ion stability) and thereby minimizes the competitive acidity of the amino and hydroxyl groups. We would therefore expect a simple measurement of pH to be reasonably correct. An exact $pK_{R^+}$ value, however, must take into account all the complex equilibria involved. A comprehensive investigation of the protolysis reactions of

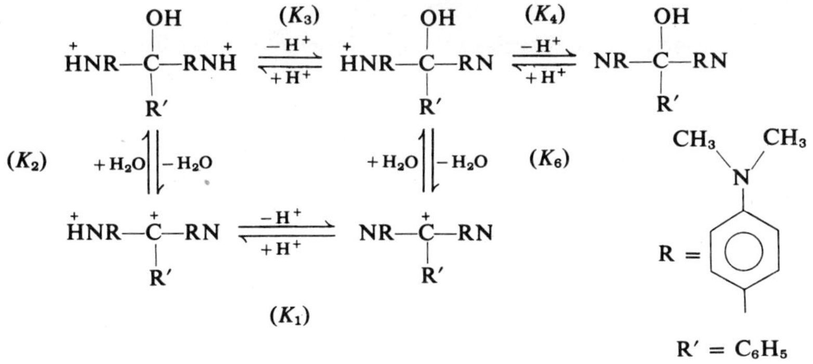

Fig. 1. Protolysis scheme for malachite green according to Cigen (91,98–105).

malachite green by Cigen (91) has yielded the scheme shown in Figure 1. The interrelated equilibrium constants for malachite green (Fig. 1) were evaluated by diverse techniques and are as follows: $K_1 = 28$, $K_2 = 22$, $K_3 = 2.7 \times 10^{-5}$, $K_4 = 5.8 \times 10^{-6}$, and $K_6 = 68$. From these data we can calculate the $pK_{R^+}$ of malachite green as being $-\log K_4/K_6 = 7.07$. Comparison of this value with the previous value of 6.90, obtained by simple ionization techniques that neglected all but direct carbinol–carbonium ion equilibration (89), corroborates the suggestion that complex protolysis equilibria are of major importance only for monoamino trityl carbinols. Cigen has extended his investigations to a series of malachite green derivatives (98–105) and his results have been used to calculate the stability data summarized in Table XII.

## 4. Emf Measurements

Jensen and Taft (71) used a novel approach to obtaining stability data. They determined the reversible electrode potentials ($E_0$, V) for the electrolytic reduction of triarylmethyl cations to the corresponding radicals (or dimers) in acetonitrile. The overall reaction is

$$Ar_3C^+ + e^- \rightleftharpoons \tfrac{1}{2}[Ar_3C\text{—}CAr_3]$$

and the actual cell

$$\frac{Pt/(Ar_3C\text{—})_2}{(Ar_3C^+)/Ag}$$

The stabilization energies ($\Delta F_{R^+}^\circ$) obtained by measuring the standard cell potential ($E_{eq}^\circ$) and converting this to a standard free energy are given in Table XIII. In general, these compare favorably with the corresponding values in Table X and this was regarded as compelling evidence for the

## TABLE XIII
### Stabilization Energies of Trityl Cations Obtained from Emf Data in Acetonitrile[a]

| Substituent | $-E_{eq}^{\circ}$, V | $\Delta F_{R^+}^{\circ}$ |
|---|---|---|
| H | (0) | (0) |
| 4-F | 0.006 | 0.1 |
| 3-OCH$_3$ | 0.022 | 0.5 |
| 4-C$_6$H$_5$ | 0.035 | 0.8 |
| 4-CH$_3$ | 0.086 | 2.0 |
| 4,4'-CH$_3$ | — | 3.2 |
| 4-SCH$_3$ | 0.155 | 3.6 |
| 4,4',4''-CH$_3$ | 0.203 | 4.7 |
| 4-OCH$_3$ | 0.219 | 5.1 |
| 4,4'-OCH$_3$ | 0.381 | 8.8 |
| 4,4',4''-OCH$_3$ | 0.518 | 12.0 |
| 4-N(CH$_3$)$_2$ | 0.660 | 15.2 |
| 4,4',4''-N(CH$_3$)$_2$ | 1.037 | 23.9 |

[a] Data from Ref. 71 and private communication from R. W. Taft.

validity of the $H_R$ function (71). The authors further point out that this agreement, despite the difference in solvent, suggests that only internal energy effects are influencing $\Delta F_{R^+}^{\circ}$; this point is further amplified below. In general, it is too early to speculate on the potential of the emf method,* nor is it obvious to what extent the recent discovery that trityl radicals do not always form simple dimers (109) will influence the results.

### 5. The Nmr Method

Considering the enormous versatility of nmr spectroscopy for the detection and characterization of charged species in general, and arylcarbonium ions in particular (30), it is not surprising that it has been adapted for $\Delta F_{R^+}^{\circ}$ determination (72). All the previous methods involve the determination of $\Delta F^{\circ}$ for a single covalent species in equilibrium with its carbonium ion; the nmr method, however, measures the free energy difference between two carbonium ions and their respective ionizable, or exchangable, covalent precursors. The general reaction is

$$R^+Y^- + R_0X \rightleftharpoons R_0^+Y^- + RX$$

and is identical to equation 8, where $\Delta F_{(8)}^{\circ}$ was obtained as a resultant of two independent equilibria, rather than the single process now under

* No pun intended. . . .

consideration. This is accomplished by determining the equilibrium distribution that results when equimolar amounts of a carbonium ion salt ($R^+Y^-$) and a suitable covalent precursor ($R_0X$) of a different carbonium ion ($R_0^+X^-$) are combined in an appropriate solvent. As before, when $R_0$ = trityl then $\Delta F°_{(8)} = \Delta F°_{R^+}$.

Inasmuch as limitations of nmr analysis require that the maximum difference in concentration of the two carbonium ions present at equilibrium be no greater than $\sim 19:1$, direct equilibration is limited to carbonium ions that differ in stability by no more than $\sim 3.5$ kcal/mole. Eventual relation to the $\Delta F°_{R^+}$ scale can be accomplished either sequentially or directly, if a carbonium ion (or precursor) of known $\Delta F°_{R^+}$ and comparable stability is available for equilibration with the unknown. This is illustrated by the following equilibrations (72,110)

$$Ar\phi_2C^+ + \phi_3C\text{---}Cl \rightleftharpoons Ar\phi_2C\text{---}Cl + \phi_3C^+ \qquad \Delta F°_{(16)} = 1.6 \text{ kcal} \quad (16)$$

$$Ar_2\phi C^+ + Ar\phi_2C\text{---}Cl \rightleftharpoons Ar_2\phi C\text{---}Cl + Ar\phi_2C^+ \qquad \Delta F°_{(17)} = 1.6 \quad (17)$$

$$Ar_3C^+ + Ar_2\phi C\text{---}Cl \rightleftharpoons Ar_3C\text{---}Cl + Ar_2\phi C^+ \qquad \Delta F°_{(18)} = 1.4 \quad (18)$$

$$Ar = p\text{-tolyl}$$

where $\Delta F_{(16)}$ yields directly $\Delta F°_{R^+}$ for the 4-methyltrityl, $\Delta F°_{(16)} + \Delta F°_{(17)}$ gives $\Delta F°_{R^+}$ for the 4,4'-dimethyltrityl and $\Delta F°_{(16)} + \Delta F°_{(17)} + \Delta F°_{(18)}$ affords $\Delta F°_{R^+}$ for the 4,4',4''-trimethyltrityl cations. Since $\Delta F°_{(16)} + \Delta F°_{(17)}$ is less than 3.5 kcal, it would also be possible to obtain $\Delta F°_{R^+}$ for 4,4'-dimethyltrityl by direct equilibration with tritylchloride; but this is not possible for the trimethyl compound, whose $\Delta F°_{R^+}$ of 4.6 kcal requires that less than 0.1% of trityl cation, undetectable by nmr, be present at equilibrium.

Nmr analysis of the equilibrium mixture also requires that at least one of the components have a single, sharp resonance line of sufficiently different chemical shift for the covalent and cationic species. This requirement is readily fulfilled by the presence of groups such as methyl, methoxyl, fluoro and dimethylamino. If both $R^+$ and $R_0^+$ have such groups, a built-in cross check for accuracy of $K_{eq}$ is available, but only one "indicator" is mandatory. The technique for measuring $[R^+]$ and $[R\text{--}X]$ varies with the stability of $R^+$ and the nature of X. When X = Cl or Br, the resulting rapid $S_N1$ exchange between each cation and its respective halide leads to time-averaged resonance lines whose chemical shift is proportional to the mole fraction of each component (72). This can be illustrated for the determination of $\Delta F°_{(17)}$. Individually, the chlorides have methyl lines at 139.8 (4-methyltrityl) and 138.3 Hz (4,4'-dimethyltrityl) and their respective carbonium ions appear at 164.9 and 163.8 Hz. An equimolar solution in $CH_2Cl_2$ of either chloride with the other carbonium ion salt leads to instant equilibration and the presence of two lines, one representing the time-averaged resonance of the mono-$CH_3$ cation and its chloride at 145.0 Hz (3-H singlet) and the other pair at 158.0 Hz (6-H singlet). From these data, we can calculate that the ratio of mono to dimethyl cation as 78:22 with $\Delta F°_{(17)} = 1.6$ kcal.

With ions of $\Delta F°_{R^+} > \sim 10$, the halides are partly or completely ionized and are therefore no longer suitable. This obstacle can be circumvented through the use of poorer leaving groups such as $OCH_3$, $N_3$, $OCOCH_3$, $OCOC_6H_5$, and OH (in the presence of a trace of acid) (72,110). These groups, although not sufficiently ionizable to lead to rapid $S_N1$ exchange on the nmr time scale, still afford equilibration at a reasonable rate, and analysis of the components can be obtained by conventional nmr integration techniques. The equivalence of the fast and slow exchange methods has been demonstrated (72,110) by obtaining similar $\Delta F°$ values with both chloride and hydroxyl leaving groups. The mechanism of the slow exchange process has not yet been established, but it probably involves $S_N2$ exchange (111).

One of the main advantages of the nmr method is its flexibility: $\Delta F°_{R^+}$ for a particular cation can be obtained with a variety of exchanging partners in different solvents and equilibrium can be approached from either direction with various leaving groups and nonnucleophilic anions.

Table XIV lists $\Delta F°_{R^+}$ values determined by the nmr method. The values for the amino-substituted trityls are 2–3 kcal higher than the corresponding values from $pK_{R^+}$, and the nmr results must be regarded as tentative pending the redetermination of $\Delta F°_{R^+}$ for the triphenylcyclopropenium cation, to which the values in question have been related. This points out one of the major weaknesses of this method: an inaccurate value assigned to a key compound used to bridge a gap in the series is systematically reflected in the value for all cations of higher stability. This does not affect the relative $\Delta F°_{R^+}$ values and may be corrected when an additional reference compound becomes available. Although the agreement with $pK_{R^+}$ data is generally good, a serious discrepancy is observed in $\Delta F°_{R^+}$ for the 4,4′,4″-t-butyltrityl cation as found by $pK_{R^+}$ (0.1, Table X) and nmr (4.4, Table XIV). Inasmuch as the agreement for the 4,4′,4″-methyl compound by $pK_{R^+}$, emf and nmr is excellent ($\Delta F°_{R^+} = 4.4 \pm 0.3$), and the methyl and t-butyl groups are electronically similar, the nmr value for the latter is considered to be more reliable. It follows that the reported value for the corresponding triisopropyl compound in Table X requires reinvestigation.

Thus far the nmr method has been extended down only to $\Delta F°_{R^+} = -1.8$, and it seems unlikely that this method will prove useful for cations of $\Delta F°_{R^+} < -5$, where difficulties in isolation of salts of such unstable cations become formidable. The $pK_{R^+}$ method, therefore, is the method of

## TABLE XIV
### $\Delta F_{R^+}^\circ$ by the Nmr Method[a]

| Compound | $\Delta F_{R^+}^{\circ\, b}$ |
|---|---|
| Trityl; 4,4'-(CH$_3$)$_2$N | 21.9 |
| Trityl; 4,4'-(CH$_3$)$_2$N, 4''-CF$_3$ | 20.5 |
| Trityl; 4-(CH$_3$)$_2$N, 4',4''-OCH$_3$ | 19.4 |
| Trityl; 4-(CH$_3$)$_2$N, 4'-OCH$_3$ | 17.8 |
| Trityl; 4-(CH$_3$)$_2$N, 4'-CH$_3$ | 16.8 |
| Trityl; 4-(CH$_3$)$_2$N | 16.3 |
| Trityl; 4-(CH$_3$)$_2$N; 4'-CF$_3$ | 14.6 |
| Cyclopropenium; triphenyl | 14.9 |
| Xanthyl; 9-(2-ethylphenyl) | 11.8 |
| Xanthyl; 9-(2-trifluoromethylphenyl) | 11.5 |
| Trityl; 4,4',4''-OCH$_3$ | 11.4 |
| Xanthyl; 9-(2-methylphenyl) | 10.9 |
| Xanthyl; 9-(4-methoxyphenyl) | 10.3 |
| Xanthyl; 9-(4-methylphenyl) | 8.4 |
| Xanthyl; 9-(3-methylphenyl) | 8.2 |
| Xanthyl; 9-phenyl | 8.0 |
| Trityl; 4,4'-OCH$_3$ | 8.2 |
| Trityl; 4,4'-CH$_3$, 4''-OCH$_3$ | 7.3 |
| Xanthyl; 9-(4-trifluoromethylphenyl) | 7.2 |
| Trityl; 4-CH$_3$, 4'-OCH$_3$ | 5.9 |
| Trityl; 4,4',4''-CH$_3$ | 4.6 |
| Trityl; 4-OCH$_3$ | 4.5 |
| Trityl; 4,4',4''-$t$-C$_4$H$_9$ | 4.4 |
| Trityl; 4,4'-CH$_3$ | 3.2 |
| Trityl; 2-C$_2$H$_5$ | 1.9 |
| Trityl; 2-CH$_3$ | 1.8 |
| Trityl; 4-CH$_3$ | 1.6 |
| Trityl; 3-CH$_3$ | 0.8 |
| Diphenylmethyl; 4,4'-OCH$_3$ | 0.5 |
| Trityl | (0) |
| Trityl; 3-CH$_3$, 4-Cl | −0.2 |
| Trityl; 3-CH$_3$, 4,4'-Cl | −0.8 |
| Cyclobutenium; 2-Cl, 1,2,3,4-tetraphenyl | −1.8 |

[a] From Refs. 72 and 110.
[b] Values for compounds of $\Delta F_R^\circ + > 14$ are tentative, subject to revision of triphenylcyclopropenium.

choice. However, for cations that undergo irreversible reactions in the presence of water, the aqueous media required by $pK_{R^+}$ is unsuitable, whereas the nmr method can be utilized with any anhydrous solvent polar enough to dissolve the cation. Some of the solvents that have been used successfully include methylene chloride, acetonitrile, and nitromethane. The cyclobutenyl cation **8** is an example of one that requires an anhydrous medium; in the aqueous acid needed to determine its $pK_{R^+}$, it is hydrolyzed irreversibly by the water present to form tetraphenylfuran (112), but in dry methylene chloride a $\Delta F_{R^+}^\circ = -1.8$ can be obtained by nmr exchange (72).

## 6. Miscellaneous Methods

A linear relation between the $^{19}F$ nmr shielding parameter, $\delta_F$, and the stabilization energies, $\Delta F_{R^+}^\circ$, of six trityl cations has been demonstrated by Taft and McKeever (113). Tests with cations of $\Delta F_{R^+}^\circ < 0$ or with any but $p$-substituted trityl compounds have not yet been performed, but the possibility that stability data can be obtained by a single $^{19}F$ nmr measurement is intriguing.

Hydride transfer reactions to carbonium ions is a well-established process (114). The feasibility of using hydride exchange to determine carbonium ion stabilities (i.e., eq. 8, X = H) was explored by Dauben and McDonough (115). Only limited success was achieved inasmuch as the rates of hydride exchange between carbonium ions of comparable stability proved to be prohibitively slow. However, it is noteworthy that, for cations of widely differing stability, hydride exchange can be an excellent preparative method. For example, cycloheptatriene is rapidly and quantitatively converted to the very stable tropylium cation when treated with trityl perchlorate (116)

$$\text{C}_7\text{H}_7\text{(H}_2\text{)} + \phi_3\text{C}^+\text{ClO}_4^- \rightleftharpoons \text{C}_7\text{H}_7^+ \text{ClO}_4^- + \phi_3\text{CH}$$

Table XV compares the $\Delta F_{R^+}^\circ$ values obtained by the five major methods for those trityl cations where at least three values were available. Except for the spectroscopic $\Delta F_{R^+}^\circ$ values, which are invariably lower, the agreement among the different methods is satisfactory, the differences reflecting general experimental inaccuracies and, as amplified later, leaving group steric effects. It appears that it would be optimistic to consider differences of less than 0.5 kcal to be truly significant, but the general agreement, despite the enormous differences in reaction conditions, clearly supports the view that $\Delta F_{R^+}^\circ$ by equilibration procedures is a true measure of internal carbonium ion stabilization energy.

### TABLE XV
### A Comparison of $\Delta F_{R^+}^\circ$ Values for Trityl Compounds Obtained by Various Methods

| Substituted trityl | Spectra[a] | Conductance[b] | $pK_{R^+}$[c] | Emf[d] | Nmr[e] |
|---|---|---|---|---|---|
| 2-CH$_3$ | 0.8 | 1.6 | — | — | 1.8 |
| 4-CH$_3$ | 0.9 | 1.7 | 1.7 | 2.0 | 1.6 |
| 3-CH$_3$ | 0.4 | 0.5 | — | — | 0.8 |
| 4,4'-CH$_3$ | 1.9 | — | 3.0 | 3.2 | 3.2 |
| 4,4'4''-CH$_3$ | 2.8 | — | 4.1 | 4.7 | 4.6 |
| 4,4',4''-Cl | −1.4 | −2.1 | −1.3 | — | — |
| 4-F | 0.0 | — | 0.2 | 0.1 | — |
| 4-OCH$_3$ | — | — | 4.5 | 5.1 | 4.5 |
| 4,4'-OCH$_3$ | — | — | 7.5 | 8.8 | 8.2 |
| 4,4',4''-OCH$_3$ | — | — | 10.1 | 12.0 | 11.4 |

[a] Table IX.
[b] Table VII.
[c] Table X.
[d] Table XIII.
[e] Table XIV.

## C. The Influence of Steric Factors on $\Delta F_{R^+}^\circ$

The introduction of steric or bond angle strain in a molecule invariably raises its ground state energy relative to its less strained counterpart. Whether such a perturbation will lower or raise $\Delta F^\circ$ for the ionization equilibrium (eq. 1), will depend on whether such destabilization is more important in the covalent or cationic component. In general, the lower steric requirements of the $sp^2$, trivalent carbonium ion with its 120° bond angles will result in less destabilization than in its $sp^3$-hybridized precursor. Therefore, increased steric crowding at the cationic center will increase the tendency toward ionization, with an accompanying increase in cation stability. On the other hand, electronic delocalization of charge in the ion requires coplanarity that is sensitive to steric factors in the ion, but not its precursor. It is seldom possible to rigorously separate the relative contributions from these steric influences, although the change in $\Delta F_{R^+}^\circ$ can detect which effect predominates.

Some indication of the influence of steric relief by ionization is available from a study of the effects of leaving group on carbonium ion stability (110). For a family of ionizable compounds with similar steric requirements (i.e., *m*- or *p*-substituted trityl) the nonbonded steric interactions of the leaving group X are constant in the covalent component and absent in the ion. Consequently, $\Delta F_{R^+}^\circ$ should be independent of the size of X. Such is not the case for compounds with steric requirements different from those of trityl (i.e., xanthyl or cyclopropenium). Here we might expect differences in $\Delta F_{R^+}^\circ$ reflecting the relative ground state destabilization of the covalent species, and these should increase with the size of the leaving group. The $\Delta F_{R^+}^\circ$ data of Table XVI offer some support for these conclusions. Except

**TABLE XVI**
Effect of Leaving Group on $\Delta F_{R^+}^\circ$

| Cation | Leaving group | | | | |
|---|---|---|---|---|---|
| | Cl[a] | OH[a] | OCH$_3$[a] | $\phi_3$C·[b] | OH[c] |
| Tri-4-methoxytrityl | — | 11.4 | 11.3 | 11.9 | 10.2 |
| Di-4-methoxytrityl | — | 8.2 | 8.2 | 8.6 | 7.4 |
| Tri-4-methyltrityl | 4.6 | — | 4.6 | — | 4.2 |
| 4-Methoxytrityl | 4.5 | — | — | 5.0 | 4.4 |
| 4,4′-Methyltrityl | 3.2 | 3.4 | — | 3.3 | 3.0 |
| 4-Methyltrityl | 1.6 | 1.8 | — | 1.8 | 1.8 |
| Triphenylcyclopropenium | — | 14.4 | 12.9 | — | 13.3 |
| 9-(4-Methoxyphenyl)xanthyl | — | 10.3 | 9.1 | 8.6 | 11.1 |
| 9-Phenylxanthyl | — | 9.1 | 8.0 | 7.4 | 10.5 |
| 9-(4-Trifluoromethylphenyl)xanthyl | — | 7.2 | 5.9 | 5.5 | — |

[a] Nmr method.
[b] Emf method.
[c] p$K_{R^+}$ method.

for the higher value p$K_{R^+}$ data, which are thought to be influenced by differing solvation effects, the $\Delta F_{R^+}^\circ$ values for the trityl compounds are reasonably independent of leaving-group size. In contrast, the $\Delta F_{R^+}^\circ$ values for triphenylcyclopropenium and for the 9-arylxanthyls differ by 1–2 kcal, the $\Delta F_{R^+}^\circ$ values decreasing as the leaving group becomes larger. Direct acid-catalyzed equilibration of covalent trityl and 9-arylxanthyl compounds by nmr (eq. 19) has been used (110) to quantitatively assess the

$$\text{Tr—OR} + \text{Xan—OR}' \underset{}{\overset{H^+}{\rightleftharpoons}} \text{Tr—OR}' + \text{Xan—OR} \tag{19}$$
$$\text{Tr} = \text{trityl}$$
$$\text{R or R}' = \text{H, CH}_3, \text{C}_2\text{H}_5, \text{C}_3\text{H}_7$$
$$\text{Xan} = \text{9-phenylxanthyl}$$

effect of steric size of R on $\Delta F_{(19)}^\circ$. The finding that $\Delta F_{(19)}^\circ$ decreases from +0.9 for R = OH to −0.9 kcal/mole for R = OC$_3$H$_7$ demonstrates that the steric requirement of trityl is larger than for 9-arylxanthyl, a conclusion also required by the data of Table XVI.

Bulky *ortho* substituents may lower or raise $\Delta F_R^\circ{}^+$, depending on whether steric destabilization of the ion or its precursor predominates. By assuming that electronic effects are similar in the *ortho* and *para* positions, we can obtain the net steric influence from the difference ($\Delta\Delta F^\circ$) between $\Delta F_R^\circ{}^+$ (*para*) and $\Delta F_R^\circ{}^+$ (*ortho*), as summarized in Table XVII. The small

**TABLE XVII**
**The Effect of *Ortho* Substituents on $\Delta F_R^\circ{}^+$**

| Type of compound | Substituent[a] | $\Delta F_R^\circ{}^+$ (*para*) | $\Delta F_R^\circ{}^+$ (*ortho*) | $\Delta\Delta F^\circ$(*o-p*) |
|---|---|---|---|---|
| Trityl | CH$_3$ | 1.6 | 1.8 | 0.2 |
| Trityl | Tri-CH$_3$ | 4.1 | 4.4 | 0.3 |
| Trityl | Mono-CH$_2$OCH$_3$ | ~ −0.5[b] | 0 | ~0.5 |
| Trityl | Di-CH$_2$OCH$_3$ | ~ −0.7[b] | 0 | ~0.7 |
| Trityl | Tri-CH$_2$OCH$_3$ | ~ −0.8[b] | 0.1 | ~0.9 |
| Trityl | Mono-SCH$_3$ | 2.0 | −0.3 | −2.3 |
| Trityl | Di-SCH$_3$ | 2.7 | −1.0 | −2.8 |
| Malachite green[c] | NO$_2$ | 17.2 | 19.2 | 2.0 |
| Malachite green[c] | Cl | 18.3 | 20.1 | 1.8 |
| Malachite green[c] | OH | 19.9 | 19.7 | −0.2 |
| Malachite green[c] | OCH$_3$ | 19.7 | 19.7 | 0.0 |
| Malachite green[c] | F | 18.8 | 18.2 | −0.6 |
| Malachite green[c] | I | 18.4 | 21.3 | 2.9 |
| Malachite green[c] | Br | 18.3 | 20.5 | 2.2 |
| Malachite green[c] | CH$_3$ | 19.0 | 21.4 | 2.4 |
| 9-Phenylxanthyl[d] | CH$_3$ | 8.4 | 10.9 | 2.5 |
| 9-Phenylxanthyl[d] | CF$_3$ | 7.2 | 11.5 | 4.3 |

[a] Substituents in different rings.
[b] Estimated from $\sigma$ data.
[c] Substituents on unsubstituted ring.
[d] Substituents on 9-aryl ring.

positive values for $\Delta\Delta F^\circ_{(o-p)}$ of methyl- and methoxymethyl-trityl suggests that steric destabilization of the covalent and ionic components are quite similar, with the former predominating. With the larger –SCH$_3$ group, however, destabilization of the ion predominates. With the malachite greens the trend is reversed; small groups (OH and F) destabilize the ion while large groups destabilize the carbinol, the increase in $\Delta F_R^\circ{}^+$ being approximately proportional to the steric bulk of the *ortho* substituent. Inasmuch as steric interactions must be similar in both the covalent trityls and malachite greens, it would appear that a conformational difference in the ions must account for the observed differing steric influences. It is likely that the ring containing only the *ortho* substituent is twisted more out of plane in the malachite greens than in the symmetrical trityl ions (cf. Section IV) and that *ortho* substituents will therefore be less effective in destabilizing the ion in the former than in the latter.

The 9-arylxanthyl compounds in Table XVII offer a particularly graphic demonstration of how steric interactions can raise $\Delta F_R^\circ{}_+$ by destabilizing the covalent precursor. Steric interactions between the 1,8-protons of the coplanar xanthyl array and o-substituents of the 9-aryl ring force the latter to adopt a conformation perpendicular to the xanthyl plane in both the carbinol (**13**, X = OH) and the carbonium ion (**14**). Once constrained to such orthogonality, an *ortho* substituent introduces little or no additional

(13)    (14)

steric crowding in the ion but does destabilize the carbinol. The result is an increase in ion stability and in $\Delta F_R^\circ{}_+$, as documented in Table XVII. In fact, steric crowding in the carbinol (**13**) is so severe that rotation about the 9-aryl bond is hindered to the point that the resulting barrier can be detected and evaluated by variable temperature nmr methods (117). The observation that this barrier is significantly decreased by increasing the size of the 9-X group is best rationalized by steric destabilization of the ground state, in complete agreement with our present conclusion. The consequence of a similar barrier in the cation is discussed in Section IV-B.

It is instructive to contrast the effect of structural change on cation stability when steric factors allow or forbid coplanarity with the trigonal plane. Thus if we make the reasonable assumption that two $p$-OCH$_3$ groups in the diarylcarbonium ion (**15a**) is electronically equivalent to the

$\Delta F_R^\circ{}_+$

(15a) R = H   −1.1
(15b) R = Ph   8.2

$\Delta F_R^\circ{}_+$

(16a) R = H   7.8 (90)
(16b) R = Ph   8.0

oxygen bridge of the xanthyl cation **16a**, we see that the greater coplanarity of the latter, due to the absence of *o*-proton interference, accounts for a net gain of almost 9 kcal of stabilization energy. In contrast, the further substitution of phenyl at the central carbon, as in **15b** and **16b** leads to a completely opposing effect: the additional phenyl in **15b** is reasonably coplanar, compared to the perpendicular 9-phenyl of **16b**, and results in a gain of stability of 9.3 kcal for the former and only 0.2 kcal for the latter. For exactly analogous reasons the sesquixanthyl cation (**17**, $\Delta F_R^\circ{}^+ = 21.8$) is 9.4 kcal more stable than its electronically similar counterpart, the tetramethoxy xanthyl cation **18** (118).

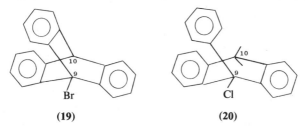

(17)  (18)

Bond angle strain can also be of major influence on cation stability. The classic example is Bartlett's observation (119) that 9-bromotrypticene (**19**) gives no evidence for carbonium ion formation even though the potential cationic center is $\alpha$ to three phenyl groups as in the trityl cation. Failure for **19** to ionize rests in the prohibitively large strain energy needed to force the C-9 bridgehead carbon into the trigonal coplanar arrangement needed

for charge delocalization to the phenyls. Some bond angle strain must still be present in 9-chloro-9-phenyl-10,10-dimethyldihydroanthracene (**20**), formally obtained from **19** by cleavage of a C-10 bond. Here the boat shape of the central ring places no restriction on coplanarity at C-9, but does present a barrier to coplanarity of the carbonium site with the benzo rings. However, this loss of stabilization is more than compensated by the absence of the *ortho*-proton interactions in the benzo groups and the ion from **20** is reported (120) to have $\Delta F_R^\circ{}^+ = 2.1$. Finally, we note that the xanthyl cation **16b** probably does not suffer from bond angle destabilization owing to a boat shaped central ring; the delocalizing central oxygen is undoubtedly $sp^2$ hybridized, unlike the $sp^3$ C-10 of **20**, and the entire xanthyl moiety therefore can be planar.

### D. The Influence of Electronic Factors on $\Delta F_R^\circ{}^+$

Our major interest here will be limited to the effect of *m*- and *p*-aryl substituents on $\Delta F_R^\circ{}^+$. This is most economically accomplished by means of the well-known Hammett equation 20, which correlates reaction rates or equilibria for an amazingly extensive series of reactions involving *m*- and

$$\log \frac{K}{K_0} = \rho\sigma \tag{20}$$

*p*-substituted aromatic compounds (121). For carbonium intermediates, or carboniumlike transition states, electron-donating substituents in direct conjugation with the positive site afford anomalous results. To overcome this deficiency Brown and co-workers (122) developed a new set of electrophilic substituent constants ($\sigma^+$) based on the solvolysis of cumyl halides. These are generally not significantly different from the usual Hammett $\sigma$ for *meta* or electron-attracting *para*-substituents, but they are considerably more negative (stabilizing) for donating *para* groups, as anticipated from the carboniumlike transition state for cumyl halide solvolysis. Therefore, a linear relation between $\sigma^+$ and $\Delta F_R^\circ{}^+$ is a reasonable expectation. Such a relationship, for a limited number of $pK_R^+$ data has indeed been found (90) and has been extended by Ritchie (123) and Hine (124) to include a series of malachite green derivatives. Similar plots of $\sigma^+$ versus $\Delta F_R^\circ{}^+$ values, containing significantly more data, appear in Figures 2 and 3 for trityl and diphenylmethyl cations, respectively.

Curve *A* of Figure 2 is remarkable in that it affords a good linear plot for a span of over 30 kcal of stabilization energy; to the extent that $\sigma^+$ reflects only internal, electronic effects of a structural change (and there is little doubt of this), $\Delta F_R^\circ{}^+$ is acting accordingly. The slopes of the curves ($\rho$ of eq. 20) are also of interest inasmuch as they reflect the overall susceptibility of a particular family of compounds to electronic effects; the more negative the $\rho$, the greater the electronic demand at the cationic center. For the trityl compounds of curve *A* (Fig. 2), $\rho$ is $-4.5$, identical with that found by Brown for cumyl halide solvolysis (122), suggesting that the two systems are equally sensitive to substituent effects. It follows that for a family of intrinsically less stable carbonium ions, such as the diphenylmethyls, the $\rho$ should be more negative, whereas for the more stable malachite greens, $\rho$ should be less negative than for trityl. This is indeed the case: $\rho$ for the former (Fig. 3) is $-8$, and for the latter $\rho$ is $-1.4$ (124).

oxygen bridge of the xanthyl cation **16a**, we see that the greater coplanarity of the latter, due to the absence of *o*-proton interference, accounts for a net gain of almost 9 kcal of stabilization energy. In contrast, the further substitution of phenyl at the central carbon, as in **15b** and **16b** leads to a completely opposing effect: the additional phenyl in **15b** is reasonably coplanar, compared to the perpendicular 9-phenyl of **16b**, and results in a gain of stability of 9.3 kcal for the former and only 0.2 kcal for the latter. For exactly analogous reasons the sesquixanthyl cation (**17**, $\Delta F_R^\circ{}^+ =$ 21.8) is 9.4 kcal more stable than its electronically similar counterpart, the tetramethoxy xanthyl cation **18** (118).

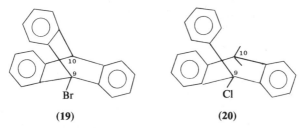

(17)          (18)

Bond angle strain can also be of major influence on cation stability. The classic example is Bartlett's observation (119) that 9-bromotrypticene (**19**) gives no evidence for carbonium ion formation even though the potential cationic center is α to three phenyl groups as in the trityl cation. Failure for **19** to ionize rests in the prohibitively large strain energy needed to force the C-9 bridgehead carbon into the trigonal coplanar arrangement needed

(19)          (20)

for charge delocalization to the phenyls. Some bond angle strain must still be present in 9-chloro-9-phenyl-10,10-dimethyldihydroanthracene (**20**), formally obtained from **19** by cleavage of a C-10 bond. Here the boat shape of the central ring places no restriction on coplanarity at C-9, but does present a barrier to coplanarity of the carbonium site with the benzo rings. However, this loss of stabilization is more than compensated by the absence of the *ortho*-proton interactions in the benzo groups and the ion from **20** is reported (120) to have $\Delta F_R^\circ{}^+ = 2.1$. Finally, we note that the xanthyl cation **16b** probably does not suffer from bond angle destabilization owing to a boat shaped central ring; the delocalizing central oxygen is undoubtedly $sp^2$ hybridized, unlike the $sp^3$ C-10 of **20**, and the entire xanthyl moiety therefore can be planar.

### D. The Influence of Electronic Factors on $\Delta F_{R^+}^{\circ}$

Our major interest here will be limited to the effect of *m*- and *p*-aryl substituents on $\Delta F_{R^+}^{\circ}$. This is most economically accomplished by means of the well-known Hammett equation 20, which correlates reaction rates or equilibria for an amazingly extensive series of reactions involving *m*- and

$$\log \frac{K}{K_0} = \rho\sigma \tag{20}$$

*p*-substituted aromatic compounds (121). For carbonium intermediates, or carboniumlike transition states, electron-donating substituents in direct conjugation with the positive site afford anomalous results. To overcome this deficiency Brown and co-workers (122) developed a new set of electrophilic substituent constants ($\sigma^+$) based on the solvolysis of cumyl halides. These are generally not significantly different from the usual Hammett $\sigma$ for *meta* or electron-attracting *para*-substituents, but they are considerably more negative (stabilizing) for donating *para* groups, as anticipated from the carboniumlike transition state for cumyl halide solvolysis. Therefore, a linear relation between $\sigma^+$ and $\Delta F_{R^+}^{\circ}$ is a reasonable expectation. Such a relationship, for a limited number of $pK_{R^+}$ data has indeed been found (90) and has been extended by Ritchie (123) and Hine (124) to include a series of malachite green derivatives. Similar plots of $\sigma^+$ versus $\Delta F_{R^+}^{\circ}$ values, containing significantly more data, appear in Figures 2 and 3 for trityl and diphenylmethyl cations, respectively.

Curve *A* of Figure 2 is remarkable in that it affords a good linear plot for a span of over 30 kcal of stabilization energy; to the extent that $\sigma^+$ reflects only internal, electronic effects of a structural change (and there is little doubt of this), $\Delta F_{R^+}^{\circ}$ is acting accordingly. The slopes of the curves ($\rho$ of eq. 20) are also of interest inasmuch as they reflect the overall susceptibility of a particular family of compounds to electronic effects; the more negative the $\rho$, the greater the electronic demand at the cationic center. For the trityl compounds of curve *A* (Fig. 2), $\rho$ is $-4.5$, identical with that found by Brown for cumyl halide solvolysis (122), suggesting that the two systems are equally sensitive to substituent effects. It follows that for a family of intrinsically less stable carbonium ions, such as the diphenylmethyls, the $\rho$ should be more negative, whereas for the more stable malachite greens, $\rho$ should be less negative than for trityl. This is indeed the case: $\rho$ for the former (Fig. 3) is $-8$, and for the latter $\rho$ is $-1.4$ (124).

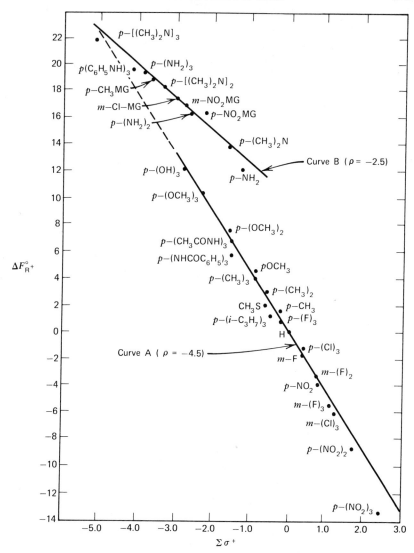

Fig. 2. Plot of $\Sigma\sigma^+$ versus $\Delta F^\circ_{R^+}$ for trityl cations; MG = malachite green.

Curve *B* of Figure 2 plots trityl compounds, all with $\Delta F^\circ_{R^+} > 12$ and all substituted with one, two or three *p*-amino groups. For the two most stable members of this series, it is futile to try to ascertain which curve they best fit; but no doubt exists that as a class, the *p*-amino trityls, including the malachite greens, are electronically less demanding ($\rho$ of curve $B \simeq -2.5$) than nonamino-substituted trityls, even of comparable stability. The

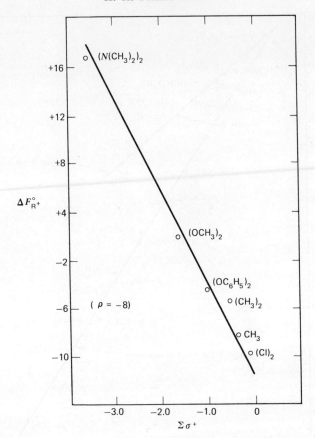

Fig. 3. Plot of $\Sigma\sigma^+$ versus $\Delta F_{R^+}^\circ$ for p-substituted diphenylmethyl cations.

reason for this, especially for the mono-*para* amino compounds, is not particularly obvious. Despite the complications mentioned previously, it is highly unlikely that errors either in $\sigma^+$ or $pK_{R^+}$ can account for these results. Although the "saturation effect" to be discussed is a possible explanation for the polyamino derivatives, it is inapplicable to the monoaminotrityls, and their large deviation from curve A remains unanswered.

Some insight concerning the ability of a *p*-amino group to delocalize charge and, specifically the contribution from the iminium resonance form to charge delocalization in the 3-methyl-4-dimethylaminotrityl cation (**21**) has been obtained from a recent nmr study (125) of the rotational barrier about the aryl–nitrogen bond. This barrier, a direct measure of the extent of participation by the nitrogen of **21**, could be detected by the non-

(21)

equivalence of the dimethylamino methyl groups at temperatures low enough to lead to slow rotation on the nmr time scale. The value of $\Delta F^{\ddagger}_{25°}$ for rotation was found to be 14 kcal/mole; it was also sensitive to *para* substituents in the unsubstituted ring, increasing with electron-attracting and decreasing with electron-donor substituents. A plot of the barriers versus $\Delta F^{\circ}_{R^+}$ values showed a reasonable correlation, suggesting that for this system the kinetic and thermodynamic properties were equally sensitive indicators of total electronic charge delocalization in the ion. This result may be compared to the findings of Taft and collaborators (126), who also found a linear relation between the rates of ionization and the p$K_{R^+}$ of amino-substituted trityl alcohols. Interestingly, a *different* linear relation exists for the nondimethylamino trityls, and Taft (126) concluded that important and severe structural limitations exist in linear relations involving cation rates and equilibria. On the basis of the previous discussion, it seems likely that here again we have an example of the uniqueness of the *p*-amino trityls.

The term "saturation effect" has been used by Hine (124) and by Taft (127) to describe the often observed phenomenon in which the effect of multiple substitution on stabilization energies is not strictly additive. Thus the fact that the net stabilization for donating substituents is less than the sum of their potential individual contributions is partly responsible for the smaller $\rho$ of curve B (Fig. 2), although the effect is apparently too weak to detect in curve A.

A clearer insight into the nature and magnitude of the saturation effect with respect to substituents of varying electronic capabilities can be obtained from Table XVIII, which lists the stabilization energy gained by consecutive substitution ($\Delta\Delta F^{\circ}$) of groups whose $\sigma^+$ varies from $-1.7$ for the strongly donating dimethylamino to 0.8 for the strongly electron attracting (destabilizing) nitro. Using the stabilization energy of the monosubstituted trityl compounds as a reference point, we can calculate the ratio of $\Delta F^{\circ}_{R^+}$ found for the di- and trisubstituted compounds to that expected on the assumption that contributions to $\Delta F^{\circ}_{R^+}$ are additive, i.e., that $\Delta F^{\circ}_{R^+}$ for identical di- and trisubstitution should be, respectively,

## TABLE XVIII
### Effect of Multiple Substitution on Stability of Trityl Cations

| $\sigma^+$ | $\Delta\Delta F°$ Due to | $\Delta F_R°^+$ Mono | $\Delta F_R°^+$-Di Theoretical | $\Delta F_R°^+$-Di Found | SR[a] | $\Delta F_R°^+$-Tri Theoretical | $\Delta F_R°^+$-Tri Found | SR[a] |
|---|---|---|---|---|---|---|---|---|
| −1.7 | 4-(CH$_3$)$_2$N | 13.8 | 27.6 | 18.4 | 0.67 | 41.4 | 21.8 | 0.53 |
| −1.3 | 4-NH$_2$ | 12.2 | 24.4 | 16.3 | 0.67 | 36.6 | 19.3 | 0.53 |
| −0.8 | 4-OCH$_3$ | 4.5 | 9.0 | 7.5 | 0.83 | 13.5 | 10.1 | 0.75 |
| −0.3 | 4-CH$_3$ | 1.7 | 3.4 | 3.0 | 0.88 | 5.1 | 4.1 | 0.80 |
| −0.07 | 4-F | 0.2 | 0.4 | 0.4 | 1.0 | 0.6 | 0.8[d] | 1.3 |
| 0.35 | 3-F | −1.7 | −3.4 | −3.4 | 1.0 | −5.1 | −5.6 | 1.1 |
| 0.8 | 4-NO$_2$ | −3.8 | −7.6 | −8.9 | 1.1 | −11.4 | −13.3 | 1.2 |
| 0.8[b] | 4-(CH$_3$)$_3$N$^+$ | −3.8[c] | −7.6 | −7.5 | 1.0 | −11.4 | −11.6 | 1.0 |

[a] "Saturation ratio": ratio of $\Delta F_R°^+$ found to theoretical, assuming additivity.
[b] Calculated from Figure 2 curve $A$ (see text).
[c] Estimated from $\sigma^+$.
[d] Probably 0.8 ± 0.2 with SR = 1.0 ± 0.3.

twice and three times that for the mono. This ratio, which is here designated as a saturation ratio (SR), is equal to 1.0 if substituent effects are strictly additive and becomes progressively larger or smaller than unity, the greater is the deviation from additivity.

The SRs of Table XVIII clearly shows that the saturation effect increases both with degree of substitution and with electron-donor ability of the substituent. Thus the more stable the carbonium ion, the less is the stabilizing effect of added donor groups.* For the donating substituents of Table XVIII, the relation of SR with $\sigma^+$ is roughly linear (Fig. 4). For

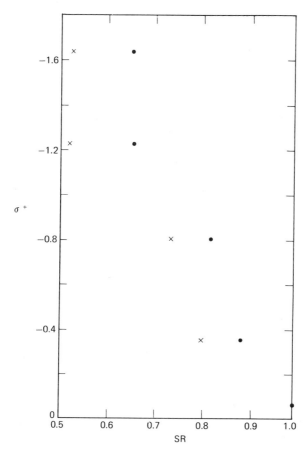

Fig. 4. Plot of SR for p-di- and trisubstituted trityls of Table XVIII; O = Di, X = Tri.

* To use an anthropomorphic analogy, substituents, like people, will donate only a minimum amount and will increase their donation only on demand!

destabilizing substituents (positive $\sigma^+$), however, the effect of polysubstitution on $\Delta F_{R^+}^\circ$ is apparently additive and, considering the probable experimental uncertainty in $pK_{R^+}$ for the $p$-$NO_2$ and $p$-$(CH_3)_3N^+$ data,* the apparent deviations from SR = 1.0 for these groups is not considered to be significant. Final documentation for the saturation effect is available from a comparison of the effect of a given substituent on trityl (Table X) as compared to malachite green (Table XII); in every case, whether stabilizing or destabilizing, the effect is smaller with the malachite greens.

## IV. THE CONFORMATION OF TRITYL AND RELATED CATIONS

### A. Possible Structures and their Relative Merits

The conformational requirements of the trityl cation—a problem that has intrigued workers in the field since it was first pointed out by Lewis and Calvin (130)—arise as a direct consequence of the steric repulsions encountered in arranging three coplanar phenyl groups about a trigonal central carbon atom. It is instructive to examine first the potential precursors of the cation, the $sp^3$-hybridized molecules triphenylmethane and triphenylmethyl bromide.

The symmetrical disposition of the three phenyls about the central tetrahedral carbon leads to a molecule with a threefold axis of symmetry in which, in order to avoid inter-ring steric repulsions, the phenyl groups must twist by an angle $\theta$, where $\theta$ is defined as equal to zero when a plane perpendicular to the phenyl plane lies in the plane determined by the threefold axis and the central carbon bond. For triphenylmethane, an electron-diffraction study (131) shows that $\theta = 45°$ and an X-ray crystallographic investigation of triphenylmethyl bromide (132) indicates $\theta = 50°$.

---

* Okamato and Brown (128) have reported a $\sigma^+$ value of 0.4 for the 4-$(CH_3)_3N^+$ group and have considered explanations for this anomalously low value. Correlation of the $\Delta F_{R^+}^\circ$ values for 4,4'-$(CH_3)_3N^+$ and 4,4',4''-$(CH_3)_3N^+$ with curve A of Figure 2 gives a $\sigma^+ = 0.8$, in good agreement with the $\sigma$ for $(CH_3)_3N^+$ of 0.7–1.0 (129).

Loss of hydride and bromide, respectively, from triphenylmethane and trityl bromide yields the trityl cation and a corresponding change in hybridization of the central carbon from $sp^3$ to $sp^2$. In this conversion we have reduced considerably the steric crowding in the resulting planar, trigonal cation relative to its tetrahedral precursor, but not to the point where all three phenyl groups can simultaneously lie in the plane defined by the central trigonal bonds without significant interhydrogen repulsions. Thus assuming trigonal bond angles of 120°, a completely coplanar model (**22**) for the trityl cation would force the *ortho* hydrogens of adjacent rings to within 0.6 Å of each other, a value well below the normal van der Waals distance of about 2.2 Å. A planar model for the trityl cation therefore need not be considered; to overcome the steric repulsions the phenyl groups must rotate out of the plane of the central trigonal carbon. Inasmuch as maximum delocalization of the positive charge requires complete coplanarity, the extent of rotation would be expected to be minimal, consistent with the relief of the steric interactions. Furthermore, the phenyls may rotate in a symmetrical or nonsymmetrical manner, resulting in conformations that may or may not differ sufficiently in energy to be noninterconvertible and, therefore, amenable to detection by modern techniques.

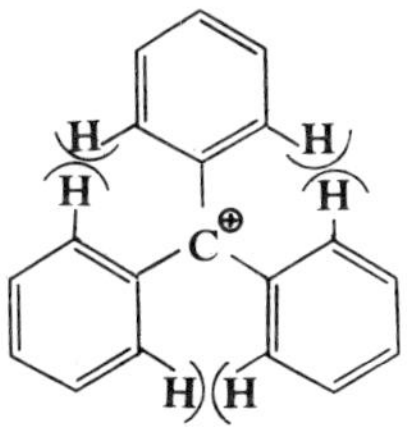

(**22**) Planar

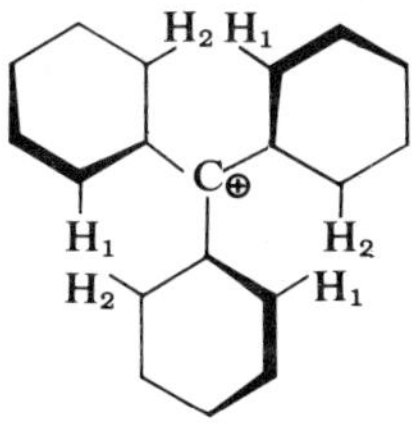

(**23**) Symmetrical propeller

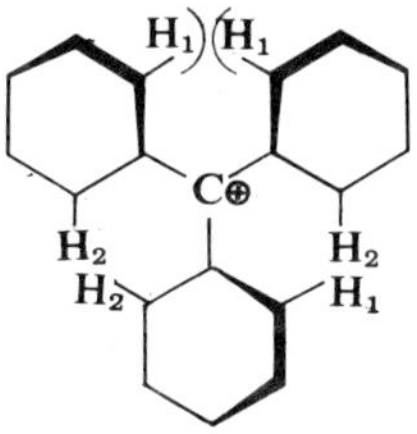

(**24**) Unsymmetrical propeller

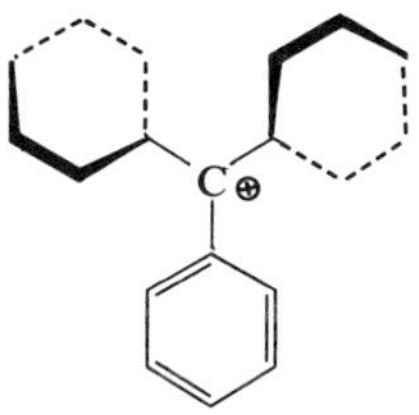

(**25**) Plane propeller

Rotation of the three phenyls about each of the central carbon–phenyl bonds by the same amount and in the same direction leads to the three-bladed propeller conformation (23) in which the *ortho* protons labeled $H_1$ are above and those labeled $H_2$ are below the plane of the central trigonal plane. This was first suggested by Lewis et al. (133) and structure 23 was assigned to the band appearing at 5900 Å in the ethanol solution of crystal violet [tris-(*p*-dimethylaminophenyl)carbonium ion] at very low temperatures. At higher temperatures a shoulder appeared at lower wavelength and this was assigned to the unsymmetric screw helix (unsymmetrical propeller) structure (24), arising from the twisting of one of the phenyl rings by the same amount but in an opposite direction to the other two. The energy difference between conformers 23 and 24 was calculated to be approximately 0.5 kcal in favor of the symmetrical propeller model 23, in which the steric interactions are considerably less than in 24 where two pairs of protons on adjacent rings lie in the same plane rather than alternating above and below the central plane, as in 23. For the *ortho* protons of structure 24 to maintain a minimum van der Waals contact distance, considerably greater rotation of the phenyls out of the trigonal plane must occur, with a concurrent decrease in resonance stabilization, but with no obvious compensatory factors. For this reason, and since no subsequent investigation has provided any evidence for rotamer 24, it is unlikely that this conformation contributes appreciably to the structure of the triaryl carbonium ion. A similar conclusion, based on additional ultraviolet evidence, has been reached by Barker (134), who suggests that the shoulder at 5900 Å can best be assigned to the presence of aggregated molecules.

One further unsymmetrical conformation for the triaryl carbonium ion merits consideration. Structure 25, in which two of the three rings are symmetrically twisted sufficiently out of the trigonal plane to allow the third ring to remain in the plane, was proposed by Newman and Deno (79) to explain certain similarities encountered in the electronic spectra of similarly substituted diaryl and triaryl carbonium ions. This argument assumes that shifts in the ultraviolet absorption maxima are a direct measure of the stabilizing influences on the ions, a conclusion unjustified by the experimental facts. Furthermore, the Newman-Deno structure (25) suggests that a monoaryl carbonium ion would have an electronic absorption spectrum similar to that of an analogously substituted di- and triaryl species, but this again was not supported by the subsequent work of Deno et al. (87), who concluded that it was not possible to present a consistent picture based only on electronic spectra. However, evidence for structure 25 was now deduced from the observation that the tri-*o*-tolyl and the tri-*p*-tolyl carbinols had similar $pK_{R^+}$ values, even though the cation generated from the *ortho*-substituted carbinol must suffer from considerably more steric inhibition to coplanarity than its *para*-substituted analog (79,87). This observation is consistent with structure **25**, since the coplanarity of but a single ring at the expense of the other two (although all three rings become equivalent on a time average) would greatly de-emphasize the role of *ortho* substituents in destabilizing the ion by steric inhibition of resonance. The Deno-Newman plane-propeller structure (**25**) has been discussed and criticized (135,136) but notwithstanding the bulk of experimental evidence supporting the symmetrical propeller **23**,* it remains extremely difficult to unequivocally rule out **25**, especially for trityl cations in which one of the rings contains a strongly donating *para* substituent.

A possible approach to evaluating the relative merits of structures **23** and **25** would be to estimate the total resonance energy $E_{tot}$ retained by each when maximum coplanarity, consistent with steric repulsions, is realized. This requires an exact knowledge of the relation between $E_{res}$ and $\theta$, the angle of twist of the phenyl rings from the trigonal plane. Although a rigorous solution to this problem is not presently available, a method of approximating the relationship has been suggested by Pauling and Corey (138) and by Dewar (139), who conclude that $E_{res}$ will vary with coplanarity according to $E_{res} = \text{const.} \times \cos^2 \theta$. A plot of this relationship (Fig. 5) reveals that the resonance energy does not change significantly for small values of $\theta$ (up to ca. 25%), is still 50% of its original value when $\theta = 45°$,

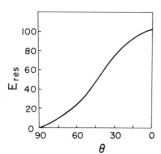

Fig. 5. Retention of resonance energy $E_{tot}$ with angle of twist of phenyl ring from trigonal plane $\theta$ in an arylcarbonium ion.

* An apparent exception to this is the recent study by DeGroot et al. (137) of the photoinduced triplet of the trityl cation obtained by analysis of the paramagnetic resonance spectrum of irradiated glassy solutions of deuterated trityl alcohol in acid solvents. These authors conclude that the absence of trigonal symmetry of the spectral parameters rules out the symmetrical propeller conformation, but they propose no alternative. Regardless of the merit of this conclusion, it is clear that they are not observing the ground state conformation of the trityl cation and, therefore, that their results are not strictly comparable to other results discussed in this section.

and then drops off to zero when the phenyl plane is perpendicular to the trigonal plane ($\theta = 90°$). Using this approximate relationship, we can estimate the relative merits of the symmetrical propeller form **23** and the unsymmetrical form **25**. In the former, we assume that each phenyl group contributes equally to stabilization and that therefore the total energy is equal to $3E_{res}(\cos^2 \theta)$. For structure **25** we assume that two rings now maintain an angle of $2\theta$ to allow the third ring to be coplanar and that, therefore, the total energy is $E_{res} + 2E_{res}(\cos^2 2\theta)$. For small angles of $\theta$ we can substitute the approximation resulting from a Taylor series expansion for $\cos^2 x$ ($\cos x = 1 - x^2/2$ and $\cos^2 x = 1 - x^2$) and thus obtain

for structure **23** $\qquad E_{tot} = 3E_{res} - 3E_{res}\theta^2$

for structure **25** $\qquad E_{tot} = 3E_{res} - 8_{res}\theta^2$

Therefore, for small values of $\theta$, the difference in $E_{tot}$ for the two forms is $5E_{res}\theta^2$ in favor of the symmetrical propeller structure **23**.

By a similar approach we can calculate the difference in energy for the symmetrically twisted **26** and the one-ring coplanar **27** model of the diphenyl carbonium ion. Again assuming that the angle of twist for both

(26)    (27)

phenyls in **26** is $\theta$ and for the noncoplanar phenyl in **27** is $2\theta$, we find that **26** is more stable than **27** by $2E_{res}\theta^2$. Ingraham (140) has used a similar argument in estimating the difference in resonance energy for the asymmetrical (one-ring coplanar) and symmetrical conformations for *cis*-stilbene. He concludes that for all angles of $\theta$ from 0 to 45°, the symmetrical structure is the preferred one. Ingraham (140) points out that this treatment becomes meaningless for values of $\theta > 45°$, since structure **27** cannot reduce the steric repulsions any further after the 90° structure—$\theta = 45°$ in **26** and $2\theta = 90°$ in **27**—is realized.

It is important to note that, in contrast to the trityl ion, an additional means of relieving *ortho* repulsions is possible for the diphenyl carbonium ion by spreading of the angle formed by the central carbon and the two phenyl ring bonds. Increase in this angle beyond the normal value of 120° would undoubtedly decrease charge delocalization $E_{res}$ but exactly to what extent is not known. It seems likely that the ion would adopt a compromise conformation involving both symmetrical phenyl twisting and angle spread. This is supported by an X-ray examination of 3,3'-dibromobenzophenone (141), not an unreasonable model for the diphenyl carbonium ion, in which the normally trigonal angle of 120° is now 126.2° and $\theta = 22.4°$. Inasmuch as a similar angle spread cannot be a factor in the stabilization of triaryl carbonium ions, it is all the more perilous to draw conclusions concerning the relative stabilities of tri- and diaryl carbonium ions from properties in which angle spread can be a contributing factor.

A number of interesting observations arise from an inspection of accurately scaled molecular models of the triphenylmethyl cation. Specifically, we can estimate the distance between the *ortho* positions of adjacent rings as a function of $\theta$. This is illustrated graphically in Figure 6 for the symmetrical propeller form **23**. The results are equally applicable to the

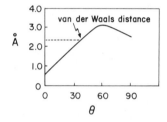

Fig. 6. Distance $A$ between nearest neighbor *ortho* protons as a function of angle of twist of phenyl groups $\theta$ for the symmetrical propeller model of the trityl cation.

diarylmethyl cations, provided angle spread is absent. The maximum angle of $\theta$ which is effective in relieving inter-ring steric repulsions is approximately 60° and will accommodate substituents that require a minimum van der Waals distance of 3.2 Å. However, at this extreme angle of $\theta$, only 25% of the total $E_{res}$ is retained, making such a structure highly unlikely. At angles of $\theta > 60°$, the *ortho*-steric interactions between nearest hydrogens actually increase and are approximately the same at 45° as at 90°. Assuming a van der Waals radius of 1.2 Å per hydrogen, we can estimate from Figure 6 that the minimum value for $\theta$ for the trityl cation is approximately 35° and that this angle of twist will not be affected by *meta* or *para* substituents.

Introduction of *ortho* substituents will greatly increase $\theta$, with a corresponding decrease in resonance stabilization. Thus the sum of the van der Waals radii for methyl and hydrogen (3.2 Å) suggests that in the symmetrical 2,2',2''-trimethyltriphenyl carbonium ion each phenyl group is 60° out of the trigonal plane, leading to a retention of only 25% of the available resonance energy, as compared with approximately 65% in the 4,4′,4″-trimethyl isomer. The similar $\Delta F_R^\circ+$ of these isomers must be due then to an unexpectedly large steric assistance to ionization in the tri-*ortho*-substituted compound, as discussed previously.

Recent structural studies, although thus far limited to relatively simple trityl salts, have provided unequivocal proof for the correctness of the symmetrical propeller structure **23**, both in the solid state and in solution. An analysis of the solid phase infrared spectra of trityl salts in which the anion is derived from complex fluoro acids ($BF_4^-$, $PF_6^-$, $AsF_6^-$, $SO_3F^-$, $TaF_6^-$) agreed best with a structure possessing $D_3$ symmetry and therefore excluded all conformers but **23** (27). Final confirmation for the symmetrical propeller structure, at least in the crystal, is now available from a detailed three-dimensional X-ray diffraction study of trityl perchlorate (15). Disorder in the perchlorate anion complicated the analysis somewhat, but the molecule is clearly propeller shaped of $D_3$ symmetry, each phenyl twisted from the plane of the $sp^2$ central carbon by $31.8 \pm 0.6°$, corresponding to an angle of 54.3° between the planes of adjacent aromatic rings.

Results substantially in agreement with these data are reported for the tri-*p*-methoxyphenyl carbonium ion (16), although random distribution of the anion ($HCl_2^- \cdot 4H_2O$) made accurate refinement impossible. A preliminary report (142) on the three-dimensional X-ray structure of tris-*p*-aminotrityl perchlorate (*para*-rosaniline) indicates that although the ion is propeller shaped, one of the phenyls has $\theta = 29°$ and the remaining two phenyls have $\theta = 34°$. It is not obvious why a *p*-amino group should affect the structure in this manner; perhaps it is because of packing in the crystal, or it may reflect a fundamental structural difference between substituted and unsubstituted trityl cations.

Inasmuch as solvation effects on charged species are fundamental in influencing their stability and structure, it is most desirable to confirm the solid state structures with equally good conformational evidence obtained from solution studies. This is now available from nmr investigations (30). Early work using specifically deuterated and methoxy-substituted trityl cations led to the conclusions that all three phenyl rings were equivalent within the nmr time scale of $\sim 10^{-2}$ sec and that the best fit of calculated to experimentally observed spectra is obtained on the basis of a symmetrical propeller model with an angle of twist of about 30° for each ring (143,144). Finally, evidence in support of the symmetrical propeller as obtained from relatively sophisticated variable temperature $^{19}F$ nmr techniques, has been reported recently in two separate investigations (95,145). These results, which are important relative to the existence of optical activity in the trityl cation, are amplified in the following section.

## B. Chirality in Arylcarbonium Ions

It has been firmly established that optically active reactants that proceed to products via carbonium ion formation at the optically active site lead, as a consequence of the symmetry properties of the planar $sp^2$-hybridized intermediate, to racemic products (146,147). With very short lived carbonium ion intermediates, it is possible to obtain a dissymmetrically solvated species, and this can presumably lead to some retention of optical activity. However, with an intermediate as stable as a triaryl carbonium ion, the intermediate should be long lived enough to be symmetrically solvated, which would be expected to lead to complete racemization.

Surprisingly, this seemingly straightforward conclusion is not borne out by the experimental results of Wallis (148). Wallis successfully resolved phenyl-$p$-biphenyl-$\alpha$-naphthylthioglycolic acid **28** into its $d$ and $l$ enantiomers and, in his early work, investigated retention of activity in the product formed by solvolysis of the highly colored carbonium ion derived from **28**. With concentrated sulfuric or perchloric acids **28** formed the expected carbonium ion intermediate **28a**, which on hydrolysis yielded racemic **29**. In the presence of zinc or ferric chloride in chloroform, however, the apparently similar intermediate, as judged by the color, gave optically unchanged starting material. Wallis's (148) conclusion that the latter

$$(d \text{ or } l \ \mathbf{28}) \xrightarrow{H_2SO_4 \text{ or } HClO_4} [p\text{-}\phi\text{-}\phi\text{-}C^{\oplus}] \ HSO_4^- \text{ or } ClO_4^- \ (\alpha\text{-napthyl}) \ (\mathbf{28a}) \xrightarrow{H_2O} p\text{-}\phi\text{-}\phi\text{-}C\text{-}OH \ (\alpha\text{-napthyl}) \ (\text{racemic } \mathbf{29})$$

$$(d \text{ or } l \ \mathbf{28}) \xrightarrow{ZnCl_2 \text{ or } FeCl_3 \text{ in } CHCl_3} [\text{deep-violet solution}] \xrightarrow{H_2O} d \text{ or } l \ \mathbf{29}$$

reactions proved the existence of optical activity in the cation was thoroughly demolished by Gomberg (149), who showed that only a trace amount of carbonium ion was formed in the reaction of **28** with zinc or ferric chloride, the major product being unchanged $d$ or $l$ **28**.*

---

* It is of historical interest that Wallis's goal in this early work was to decide between the benzenoid and quinonoid structure for the triaryl carbonium ion. Although in the light of current knowledge none of his results conclusively favors either structure, Wallis interpreted the data to support the benzenoid carbonium structure and Gomberg (149) found the evidence equally convincing for his "quinocarbonium ion salt" structure. It is also evident from these results that the formation of a deeply colored solution per se is insufficient evidence for the presence of significant amounts of a postulated carbonium ion intermediate.

The isolation of racemic products alone from **28** upon hydrolysis of the preformed carbonium ion **28a** is in contrast to the later work of Wallis and Adams (150) in which they demonstrated retention of optical activity (or incomplete racemization) when **28** is treated with silver nitrate in a solvolyzing medium. Reacting optically pure dextrorotatory acid **28a** in the presence of water gave levorotatory alcohol **29a**, whereas levorotatory acid **28b** afforded the enantiomer **29b**. Exactly parallel retention of activity was

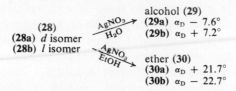

alcohol (29)
(29a) $\alpha_D - 7.6°$
(29b) $\alpha_D + 7.2°$

ether (30)
(30a) $\alpha_D + 21.7°$
(30b) $\alpha_D - 22.7°$

(28)
(28a) *d* isomer
(28b) *l* isomer

obtained in the formation of the ether **30**. On the other hand, an attempt to repeat the foregoing transformations under identical experimental conditions with 12-phenyl-12-β-benzoxanthenethioglycolic acid **31** led only to completely racemized alcohol or ether, respectively (150).

$$\xrightarrow[\text{H}_2\text{O or EtOH}]{\text{AgNO}_3}$$ racemic alcohol or ether

(31) (*d* or *l*)

Except in the recent work of Murr, to be discussed, the significance of Wallis's demonstration of retention of optical activity in a triaryl carbonium ion intermediate seems to have been completely ignored or discounted.* The simplest explanation for Wallis's results—that dissymmetric solvation of the carbonium ion intermediate is responsible for the retention of activity—is not compatible with the known stability of the triaryl carbonium ion, and we must therefore look elsewhere for the answer. A possible explanation lies in the inherent dissymmetry of the symmetrical propeller structure of the trityl carbonium ion. The ion, which has $D_3$ symmetry can exist in both left-handed and right-handed propeller conformations which, being nonsuperimposable mirror images, correspond to *d* and *l* enantiomers. It should therefore be theoretically possible to retain optical activity when passing from an optically active covalent triarylmethyl compound to a trigonal triaryl carbonium intermediate, inasmuch as loss of asymmetry at the central carbon is accompanied by the simultaneous

---

* Eliel (146) discusses only Wallis's earlier work (148) as evidence that a planar, stable carbonium ion intermediate must lead to racemization, but he does not comment on Wallis's latter findings (150), although a reference to this work is given.

Left handed propeller   Right-handed propeller

generation of a dissymmetric species. This is clearly evident in the X-ray structure of trityl perchlorate, which has a noncentrosymmetric space group (15) and must contain in each single crystal either a $d$ or an $l$ cation; therefore, with suitable crystals, it is theoretically capable of being mechanically separated (à la Pasteur) (151) into optical enantiomers. On the other hand, tri-$p$-aminotrityl perchlorate has a centrosymmetric space group ($p^2$ $1/c$) which contains within each unit cell two $d$ and two $l$ propellers (15).

Whether chirality in a triaryl carbonium ion will be experimentally demonstrable will depend on the rate of racemization, and this in turn depends on the activation energy for the interconversion of the left-handed to the right-handed propeller, i.e., the rotational barrier about the delocalized phenyl–$C^+$ bond. The mechanics of such a process as analyzed by Kurland (95) may involve either a ring flip, in which the rotating phenyl passes through a plane perpendicular to the trigonal plane, or a ring twist, in which a ring inverts without passing through this plane, or a combination of both processes. Inasmuch as a ring flip is accompanied by loss of the resonance stabilization from the phenyls that become perpendicular in the transition state and a ring twist involves overcoming the steric interactions of neighboring *ortho* protons, studies of the energetics and mechanism for this racemization process provide valuable information on the relative importance of steric and electronic effects in polyarylcarbonium ions.

Substituents in two or more of the trityl phenyl rings, with at least one ring being asymmetrically substituted (i.e., *ortho* or *meta*) results in the formation of geometric isomers (diastereomers) as well as enantiomers. This is illustrated in Figure 7, where conformers $a$ and $c$ (or $b$ and $d$) are equal-energy *syn* and *anti* diastereomers, respectively, whereas $a$ and $b$ (or $c$ and $d$) are enantiomeric. Interconversion of diastereomeric propellers can only occur if the *meta*-R phenyl group undergoes a ring flip, as in process 1 or 2 of Figure 7, the other two rings "reversing pitch" by either a two-ring flip or twist process in order to maintain the propeller conformation. Thus geometrical isomerization must always be accompanied by optical isomerization, if we assume that $D_3$ symmetry is inviolable. On the other

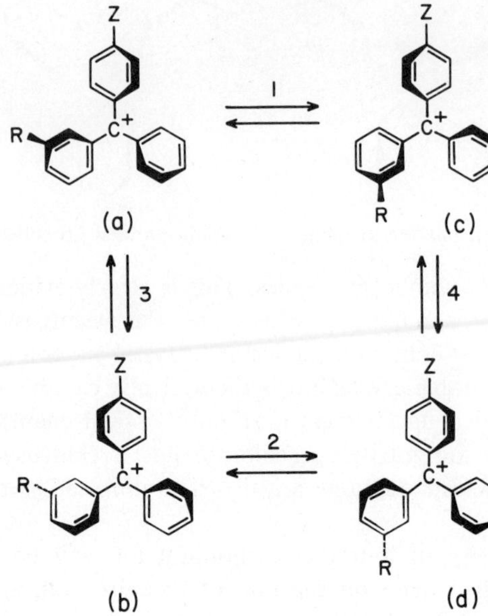

Fig. 7. Interconversion of trityl enantiomers and diastereomers.

hand, it is possible to interconvert enantiomers (racemize) for any one diastereomer by process 3 or 4 of Figure 7. This occurs only when the *m*-R phenyl group undergoes a ring twist, the other two rings again reversing pitch as before. As pointed out by Kurland (95), the energy required to force three phenyl rings into coplanarity in the transition state is prohibitively high and we can therefore, a priori, eliminate a three-ring twist mechanism. Similarly, it seems likely that the energy necessary to overcome the steric interactions for two rings to become coplanar would be greater than that needed for two rings to become perpendicular and, therefore, that pathways involving ring flips (1 and 2 of Fig. 7) should be preferred to those involving ring twists (3 and 4 of Fig. 7).

The presence and magnitude of the rotational barrier in trityl cations which is responsible for the interconversion of diastereomers was elegantly demonstrated by the nmr study of polyfluoro-substituted trityl cations (95). The variable temperature, proton-decoupled, $^{19}$F nmr spectra of 3-F, 4-F, 4,4'-di-F and 4,4',4"-tri-F trityl cations showed no temperature dependence, as expected from the symmetry properties of these ions. However, the less symmetrical, poly-*m*-substituted ions, 3,3'-di-F, and 3,3',3"-tri-F cations showed a strong temperature dependence that could clearly be attributed to the generation of diastereomers at temperatures low enough to overcome the aryl rotational barrier. The three diastereomers, containing equal populations of the four magnetically different fluorines are illustrated for the 3,3'-di-F cation (32). Line shape analysis of the nmr data yielded an activation energy of ~13 kcal at −43°, a reasonable value for the rotational barrier induced by charge delocalization to the aryl rings.

(32)

Although without question their work is a pioneering and valuable addition to arylcarbonium ion chemistry, accomplished despite formidable experimental difficulties, the conclusions of Schuster et al. (95) are open to refinement in two aspects. Their assertion that the demonstration of nonequivalent fluorines in 32 (or in the analogous 3,3',3''-trifluoro ion) proves the correctness of the propeller structure is suppositious. The presence of such diastereomers clearly establishes the presence of the rotational barrier but is noncommittal regarding the propeller conformation. (In fact, in the absence of additional steric information to the contrary, these results are compatible with a completely coplanar structure.) From a detailed analysis of the possible pathways for the exhange of fluorines in 32, in conjuction with computer matching of the spectra, a one-ring flip process was eliminated, leaving the two- or three-ring flip alternatives. Inasmuch as only in the two-ring flip transition state does the molecule retain the resonance stabilization of the remaining (now coplanar), ring, their choice of the three-ring flip as the most favored process is surprising. The very recent results to be discussed suggest that this conclusion may also be either unjustified or, less likely, peculiar to the polyfluorotrityl cations.

An alternative approach takes advantage of the capability of nmr to detect the chirality inherent in the trityl propeller conformation by using a suitable diastereotopic probe. Rakshys et al. (145) synthesized and examined the variable-temperature $^{19}$F nmr spectra of $m$-CF$_2$H-substituted trityl cations, whose fluorines are potentially nonequivalent when chirality in the cation is maintained by slow propeller racemization. The limiting $^{19}$F spectra for the $m$-CF$_2$H trityl cation is reproduced in Figure 8; the change of the spectra from an A$_2$X doublet at 25° to an ABX octet below −30° can be attributed unambiguously to the trityl propeller. Line shape

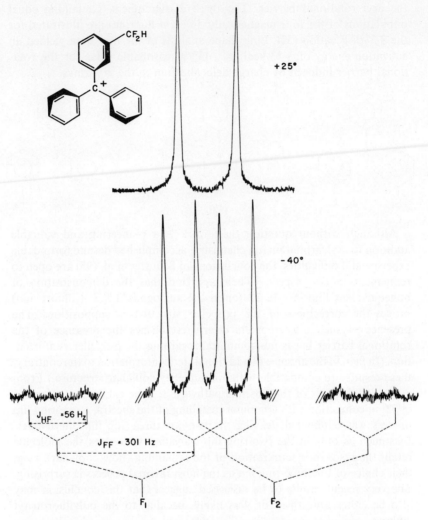

Fig. 8. The $^{19}$F nmr spectrum (94.1 MHz) of $m$-CF$_2$H-trityl$^+$ SbCl$_5$OH$^-$ in acetonitrile (145).

analysis yielded the activation data of Table XIX, the value of 12.7 kcal/mole for $\Delta G^{\ddagger}_{25°}$ of racemization being in reasonable agreement with the rotational barrier for interconversion of the diastereomeric fluorines of the 3,3′,3″-trifluorotrityl cation (95). Analogous data obtained with the –CF$_2$H probe in a *para* position (Table XIX) confirm the trends found with the *meta* compounds.

## TABLE XIX
### Activation Parameters for Substituted-Trityl Cation Propeller Interconversion[a]

$BF_4^-$ or $SbCl_5OH^-$

| —$CF_2H$ | X | $\Delta F^{\ddagger}_{25°}$ | $\Delta S^{\ddagger}_{25°}$ | $T_c^°$ | $\sqrt{F_1} - \sqrt{F_2}$, Hz[b] |
|---|---|---|---|---|---|
| meta | H | 12.7 | −6.2 | 0 | 140 |
| meta | $CH_3$ | 11.6 | −0.7 | −14, −48[c] | 127, 14[d] |
| meta | $OCH_3$ | — | — | ∼ −80 | — |
| para | H | 12.3 | 13.3 | −2 | 62 |
| para | $CH_3$ | (10.4)[e] | — | −30 | 47 |
| para | $OCH_3$ | (8.7)[e] | — | −85 | — |

[a] Ref. 145.
[b] Chemical shift between diastereotopic fluorines at 94.1 MHz at −40°. Chemical shift increases with decreasing temperature with rate 1.0 and 0.3 Hz/deg. for the $m$-$CF_2H$ and $p$-$CF_2H$ derivatives, respectively. This temperature dependence was taken into account in the lineshape calculations.
[c] Coalescence temperature for diastereomer interconversion process.
[d] Chemical shift between diastereomers.
[e] Calculated from rate observed at $T_c$ and $\Delta S^{\ddagger}$ calculated for $p$-$CF_2H$, $p'$-H compound.

The mechanism of propeller interconversion was tested by noting the effect of cation-stabilizing substituents on the rotational barriers (145). Introduction of a $p'$ substituent in the $m$-$CF_2H$ trityl cation leads to an additional complication: two equal-energy diastereomers can now be present, and each can exist as an enantiomeric pair (i.e., $a + b$ and $c + d$ of Fig. 7 with R = $CF_2H$, Z = $CH_3$ or $OCH_3$). The low-temperature, lower-limit nmr spectrum now consists of two, rather than one, octet, one each for the *syn* and *anti* diastereomers (inner eight lines shown in Fig. 9, along with calculated representative spectra). These more complex spectra were helpful in eliminating a zero- or one-ring flip mechanism but could not distinguish between the two- and three-ring flip possibilities. However, the observation that electron-donating groups lower the barrier (Table XIX) is consistent only with a two-ring flip mechanism; if all three rings become perpendicular in the transition state a *para*-donor sub-

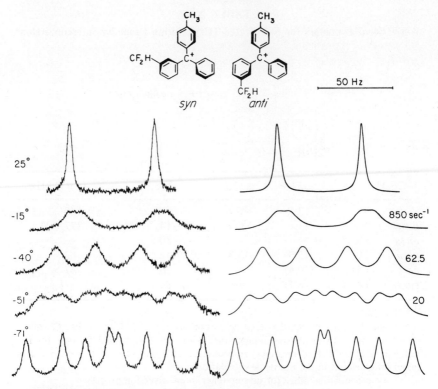

Fig. 9. Experimental and calculated variable-temperature $^{19}$F nmr spectra of $m$-CF$_2$H, $p'$-CH$_3$-trityl$^+$ BF$_4^-$ in propionitrile. Only the inner eight lines are shown.

stituent must stabilize the ground more than the transition state and yield a net increase in the barrier, in contrast to the results observed. It is possible that this disagreement with the previous conclusion (95) arises from the assumption that the probability of an individual ring flip is substituent independent. This assumption is probably not justified, but we cannot unequivocally rule out the tenuous possibility that fundamentally different pathways for isomer interconversion occur in the fluoro (95) and the difluoromethyl (145) trityl cations.

The previously discussed results of Wallis are not inconsistent with the foregoing discussion. The intermediate cation **28a** would not be expected to maintain its dissymmetry at normal temperatures and is instantly racemized. Therefore, when **28a** was generated from optically active **28** and then solvolyzed, it yielded only inactive **29**. However, when **28a** was generated as a short-lived dissymmetric intermediate in the presence of a nucleophile, it is possible that solvolysis was more rapid than racemization, yielding optically active alcohol **29** or ether **30**.

After lying fallow for 40 years, Wallis's results (150) have been recently confirmed and elegantly extended by Murr (152,153), who realized their potential significance for the study of the stereochemistry of $S_N1$ substitution processes involving free ions and ion pairs. Murr repeated Wallis's work and found his conclusions to be substantially correct. By the route shown, Murr (152) isolated the pure dextrorotatory isomer of the optically active phenyl-$p$-biphenyl-$\alpha$-naphthyl carbinol (29), whose rotation of +17° indicated that Wallis's silver-assisted solvolysis had an

$dl$-ROH  $\longrightarrow$ ROCH$_2$COOH $\longrightarrow$ [$l$-ROCH$_2$COOH]
(m.p. 161–162°)

$\longrightarrow$ $l$-ROCH$_2$CONH$_2$ $\xrightarrow[\text{2. MeO, MeOH}]{\text{1. }t\text{-BuOCl}}$ $d$-ROH
(m.p. 180–181°, $\alpha_D$ + 17°)

$$R = p\text{-}\phi\text{-}\phi\underset{\alpha\text{-naphthyl}}{\overset{\phi}{C}}\text{—}$$

optical yield of 60%. Treatment of Wallis's $d$-ROH with ethyl iodide and silver carbonate yielded $l$-ROEt and recovered, unracemized $d$-ROH, indicating that the $d$-ROH and $l$-ROEt have the same configuration. Murr (152) concluded that Wallis's stereospecific nucleophilic substitution reactions took the same stereochemical course and probably occurred with net retention of configuration.

Further investigations (153) utilizing $l$-phenyl-$p$-biphenyl-$\alpha$-naphthyl methyl benzoate 33 revealed that solvolysis in 95% aqueous acetone and in 80% dioxane was subject to a large common ion effect (diagnostic for the $S_N1$ mechanism) and produced a $d$-ROH in 50% optical yield. These results are consistent with a scheme in which a preferred orientation of the bulky naphthyl group in $l$-ROBz 33 is maintained on ionization, yielding the dissymmetric cation 34, which is attacked by solvent on the less sterically hindered face to yield $d$-ROH. The optical yield reflects either

($l$)  
(33)  
Ar = $p$-biphenyl

$\xrightleftharpoons[k_{-1}]{k_1}$  

$\downarrow k_3$ (retention)  
$d$-ROH  
(34)  

$\xrightleftharpoons[k_{-2}]{k_2}$  

$\downarrow k_4$ (inversion)  
$l$-ROH  
(35)

the relative magnitude of $k_3$ and $k_2$; or if $k_2$ is very much slower than $k_3$, then it measures the relative rates of attack on the front and back sides of the carbonium ion. A final possibility is that the optical yield may be a measure of the amounts of enantiomers **34** and **35** resulting from the dissociation of an initially produced ion pair. Further results in this area should provide definitive information on the nature of the intermediates involved in solvolytic processes proceeding via stable carbonium ions.

We can now understand why the xanthene thioglycolic acid **31** ($d$ or $l$) yielded only racemic products under conditions identical to those which yielded optically active products with its nonoxygen bridged analog **28** ($d$ or $l$). Ionization of **31** yields a carbonium ion intermediate **31a**, in which the oxygen bridge makes a propeller conformation no longer possible and optical isomers are precluded by the plane of symmetry that passes through

the planar upper half of the molecule and bisects the 12-phenyl ring, perpendicular to this plane. In general, we may predict that the isolation of optical isomers of *ortho*-substituted xanthylium cations will closely parallel the requirements found for the optically active biphenyls, both being examples of atropisomerism. Thus for the xanthylium cation **36** or the biphenyl molecule **37**, the upper half of each molecule is in the plane of the paper and the lower half is perpendicular to this plane. If $R = R'$ or $R'' = R'''$, there is present a plane of symmetry and both are therefore nonresolvable. On the other hand, if $R \neq R'$ and $R'' \neq R'''$ then both molecules become dissymmetric and their resolution into stable enantiomers will depend on how effectively the steric bulk of the *ortho* groups

(36)        (37)

prevents rotation about the central carbon–phenyl bond, which leads to the interconversion of the *d* and *l* forms.

The factors affecting racemization have been analyzed for the biphenyl derivatives in terms of a number of molecular parameters (154) and analogous factors should apply to the xanthylium cation. A study of models of **36** and **37** indicates that steric interference between the *ortho* position of the upper and lower rings is considerably greater in the cation **36**, leading to the conclusion that once resolved, the enantiomers of **36** will racemize at a slower rate than the corresponding biphenyl derivatives! An attempt to resolve such an asymmetric 9-arylxanthyl compound and to convert it to its optically stable carbonium ion has been attempted in these laboratories (155), but without success. Similarly, incorporation of potentially diastereotopic probes in the asymmetric carbinol precursors **38** and **39** exhibited the expected nonequivalence of the methyls of the isopropyl

(38)  (39)

group in **38** and the methylene protons of the ethyl group in **39**, but this diastereotopism was not observed in their respective carbonium ions (117). This was attributed to a reduction in the anisochronous character of these groups, rather than to the absence of chirality due to rapid rotation. The author is quite confident that the isolation of a completely optically stable xanthylium, or equivalent, carbonium ion is experimentally feasible and will be accomplished eventually.

Interestingly enough, an analogous steric barrier has been recently detected by nmr in suitably substituted trityl cations (145). Referring to the previous discussion on the mechanism of racemization of trityl cations (Fig. 7), further analysis indicates that an additional element of chirality, directly analogous to the atropisomerism of the chiral biphenyls and xanthylium cations, is also present in suitably substituted tri- (and di-) aryl carbonium ions. This chirality is completely independent of that arising from the propeller conformation and is present in the *p, m'*-disubstituted ions of Figure 7 so long as processes 3 or 4, in which the asymmetrically substituted rings becomes coplanar in the transition state, is slow. The resulting nonsuperimposable mirror images (enantiomers) are illustrated in Figure 10. From Figure 10 it is evident that this biphenyl-type enantiomeric interconversion will occur only when the ring containing

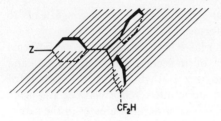

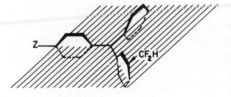

Fig. 10. Enantiomers arising from atropisomerism in an *m, p'*-substituted trityl cation.

the diastereotopic substituent rotates through the trigonal plane. Thus this chirality is completely independent of the conformation of the remaining aryl rings, whose rapid conformational contortions have no effect on the biphenyl-type chirality of the molecule. (In fact, if we rotate the *p*- and unsubstituted ring into the plane and add an *ortho*-oxygen bridge, we return to our chiral xanthyl cation.)

From the foregoing discussion it is clear that the two sources of chirality in suitably substituted trityl (or diaryl) carbonium ions may be individually detected with the same diastereotopic probe, provided the energy for their respective enantiomer interconversion is sufficiently high and different. Introduction of large *ortho*- substituents in the probe ring, or strongly donating substituents in those rings lacking the probe, will raise the barrier for process 3–4 while simultaneously lowering the barrier for process 1–2. Such a case has been described recently (94). Although the nonequivalence of the diastereotopic protons of **40** has been attributed to the presence of

(40)

propeller chirality (94), it is observed at the relatively high temperature of 30° and diastereomers are absent; these circumstances suggest that the dissymmetry present is of the biphenyl type and that propeller interconversion by the two-ring flip mechanism is fast.

With a moderately donating *para* substituent, it should be possible, by means of our $m'$-probe, to observe the consequences of both types of chirality in an accessible temperature range. Here the change in the spectrum with increasing temperature should reflect the simultaneous disappearance of diastereomers and propeller dissymmetry, as process 1–2 becomes facile, yielding a spectrum characteristic of the biphenyl-type induced dissymmetry, a change tantamount to an epimerization. At still higher temperatures, process 3–4 becomes fast and the nonequivalence in our probe group vanishes. A spectrum indicative of just such a consecutive epimerization and racemization has indeed been observed in the variable temperature nmr study of the $m$-$CF_2H$, $p'$-$OCH_3$-trityl cation (156).

It would appear that the polyarylcarbonium ions offer a unique opportunity to study aryl rotational barriers in a molecule whose ground state is neither coplanar (as in the amides) nor orthogonal (as in the biphenyls) and in which a fortunate combination of steric and electronic factors makes both the coplanar and orthogonal pathways for isomer interconversion of sufficient energy for individual nmr observation. As nmr techniques become increasingly sophisticated, we can expect future developments in the conformational dynamics of arylcarbonium ions to be both rapid and exciting.

## V. ACKNOWLEDGMENTS

The author welcomes this opportunity to express his gratitude to his colleagues, S. V. McKinley, J. W. Rakshys, Jr., A. E. Young, and V. R. Sandel for their invaluable contributions to the arylcarbonium ion studies originating in this laboratory.

EDITOR'S NOTE ADDED IN PROOF. A systematic approach to the analysis of stereoisomerism and isomerization of arylcarbonium ions due to restricted rotation of aryl groups attached to the carbonium ion center has recently been developed [F. Strohbusch, *Tetrahedron*, **28** 1915 (1972); D. Gust and K. Mislow, *J. Am. Chem. Soc.*, in press]. Since the interconversion networks resulting from stereoisomerization by way of the flip mechanisms may be complex, a systematic analysis is required for a thorough treatment of these multiple pathways and their associated energetics. The necessarily exhaustive study is rendered less arduous by the use of graphical representations. Such topological analyses have been applied to compounds discussed in several of the studies cited in this chapter, and complement the discussion contained therein. The reader is referred to the cited papers for further details.

**REFERENCES**

1. J. F. Norris and W. W. Sanders, *J. Am. Chem. Soc.*, **25**, 54 (1901); F. Kehrmann and F. Wentzel, *Ber.*, **34**, 3815 (1901); M. Gomberg, *J. Am. Chem. Soc.*, **25**, 328 (1901).
2. H. Burton, *Quart. Rev. (London)*, **6**, 302 (1952); D. Bethel and V. Gold, *ibid.*, **12**, 173 (1958); J. E. Leffler, *The Reactive Intermediates of Organic Chemistry*, Interscience, New York, 1956, Chapters V and VI; N. C. Deno, *Progr. Phys. Org. Chem.*, **2**, 177 (1964).
3. A. Streitwieser, *Solvolytic Displacement Reactions*, McGraw-Hill, New York, 1962, pp. 38ff; C. A. Bunton, *Nucleophilic Substitution at a Saturated Carbon Atom*, Elsevier, New York, 1963; Y. Pocker, in *Reaction Kinetics*, G. Porter, Ed.
4. N. N. Lichtin, *Progr. Phys. Org. Chem.*, **1**, 75 (1963).
5. M. J. S. Dewar, *Electronic Theory of Organic Chemistry*, Oxford University Press, London, 1949, pp. 305ff; *Recent Advances in the Chemistry of Coloring Matters*, London, The Chemical Society, 1956, pp. 64ff; C. C. Barker, in *Steric Effects in Conjugated Systems*, G. W. Gray, Ed., Academic Press, London, 1950, Chapter IV.
6. G. A. Olah and M. W. Meyer, in *Friedel-Crafts and Related Reactions*, Vol. I, Interscience, New York, 1963, pp. 646ff.
7. E. D. Hughes, C. K. Ingold, S. F. Mok, S. Patai, and Y. Pocker, *J. Chem. Soc.*, **1957**, 1265.
8. S. Winstein, B. Appel, R. Baker, and A. Diaz, *Organic Reaction Mechanisms*, The Chemical Society, London, 1965, p. 109.
9. J. Schmidlin, *Das Triphenylmethyl*, F. Enke, Stuttgart, 1914.
10. P. Pfeiffer, *Organische Molekulverbindungen*, F. Enke, Stuttgart, 1927.
11. G. R. Cowie, H. J. M. Fitches, and G. Kohnstan, *J. Chem. Soc.*, **1963**, 1585; P. Casapieri and E. R. Swart, *J. Chem. Soc.*, **1961**, 4343.
12. H. Voltz, *Angew. Chem.*, **75**, 21 (1963); J. Holmes and R. Pettit, *J. Org. Chem.*, **28**, 1695 (1963).
13. M. Gomberg, *Ber.*, **40**, 1853 (1907); L. C. Anderson, *J. Am. Chem. Soc.*, **52**, 4567 (1930).
14. W. Dilthey, *Z. Angew. Chem.*, **37**, 313 (1924); A. Hantzsch, *Ber.*, **43**, 339 (1910).
15. A. H. Gomes de Mesquita, C. H. MacGillavry, and K. Eriks, *Acta Crystallogr.*, **18**, 437 (1965).
16. P. Anderson and B. Klewe, *Acta Chem. Scand.*, **19**, 791 (1965).
17. R. Breslow, *J. Am. Chem. Soc.*, **79**, 5318 (1957); R. Breslow and C. Yuan, *ibid.*, **85**, 5991 (1958).
18. M. Sundaralingam and L. H. Jensen, *J. Am. Chem. Soc.*, **85**, 3302 (1963); *ibid.*, **88**, 198 (1966).
19. R. Breslow and H. W. Chang, *J. Am. Chem. Soc.*, **83**, 2369 (1961).

20. R. F. Bryan, *J. Am. Chem. Soc.*, **86**, 733 (1964).
21. R. F. Bryan, private communication.
22. D. W. A. Sharp and N. Sheppard, *J. Chem. Soc.*, **1957**, 674.
23. D. W. A. Sharp, *J. Am. Chem. Soc.*, **1958**, 2558.
24. S. D. Ross, *Spectrochim. Acta*, **18**, 225 (1962).
25. Unpublished work from this laboratory.
26. R. D. Kemmitt, R. S. Milner, and D. W. A. Sharp, *J. Chem. Soc.*, **1963**, 111.
27. D. W. A. Sharp, *J. Chem. Soc.*, **1957**, 4804.
28. R. E. Weston, A. Tsukamoto, and N. N. Lichtin, *Spectrochim. Acta*, **22**, 433 (1966).
29. L. J. Bellamy, *The Infrared Spectra of Complex Molecules*, 2nd ed., Methuen, London, 1958, p. 71.
30. G. Fraenkel and D. G. Farnum, in *Carbonium Ions*, Vol. I, G. Olah and P. v. R. Schleyer, Eds., Interscience, New York, Chapter 7, 1968.
31. B. L. Evans and A. D. Yoffe, *J. Chem. Phys.*, **30**, 1437 (1959).
32. R. F. Fuoss, *J. Chem. Educ.*, **32**, 527 (1955).
33. G. Briedleib, W. Ruttiger, and W. Jung, *Angew. Chem., Int. Ed.*, **2**, 545 (1963).
34. W. B. Smith and P. S. Rao, *J. Org. Chem.*, **26**, 254 (1961).
35. A. G. Evans, I. H. McEwan, A. Price, and J. H. Thomas, *J. Chem. Soc.*, **1955**, 3098.
36. A. G. Evans, A. Price, and J. H. Thomas, *Trans. Faraday Soc.*, **1955**, 481.
37. W. E. B. Arthur, A. G. Evans, and E. Whittle, *J. Chem. Soc.*, **1959**, 1940.
38. A. L. Gatzke and R. Stewart, *Can. J. Chem.*, **39**, 1849 (1961).
39. A. G. Evans, A. Price, and J. H. Thomas, *Trans. Faraday Soc.*, **52**, 332 (1956).
40. E. M. Arnett, *Progr. Phys. Org. Chem.*, **1**, 324 (1963).
41. A. G. Evans, J. A. G. Jones, and G. O. Osborne, *J. Chem. Soc.*, **1954**, 3803 (and earlier papers in this series).
42. E. Price and N. N. Lichtin, *Tetrahedron Lett.*, **10** (18) (1960).
43. Y. Pocker, *J. Chem. Soc.*, **1958**, 240.
44. G. Olah, *Chem. & Eng. News*, **1967**, 77.
45. C. G. Swain and C. R. Scott, *J. Am. Chem. Soc.*, **75**, 246 (1953).
46. C. G. Swain, T. E. C. Knee, and A. MacLachlan, *J. Am. Chem. Soc.*, **82**, 6101 (1960).
47. A. J. Parker, *Chem. Rev.*, **69**, 1 (1969).
48. E. R. Thornton, *Solvolysis Mechanisms*, Ronald Press, New York, 1964, pp. 161–163.
49. C. G. Swain, C. R. Scott, and K. H. Lohmann, *J. Am. Chem. Soc.*, **75**, 136 (1953).
50. S. J. Cristol and J. E. Leffler, *J. Am. Chem. Soc.*, **76**, 4468 (1954); R. T. Marrow and R. H. Boschan, *ibid.*, **76**, 4622 (1954).
51. R. G. R. Bacon and J. Kochling, *J. Chem. Soc.*, **1964**, 5609.
52. R. J. Gillespie, in *Friedel-Crafts and Related Reactions*, Vol. I, Interscience, New York, 1963, Chapter III.
53. J. W. Bayles, J. L. Cotter, and A. G. Evans, *J. Chem. Soc.*, **1955**, 3104.
54. F. Fairbrother and B. Wright, *J. Chem. Soc.*, **1949**, 1058.
55. (a) M. Baaz, V. Gutman, and O. Kunze, *Monatsch.*, **93**, 1142 (1962); (b) *ibid.*, **93**, 1162 (1962).
56. K. M. Harmon, A. B. Harmon, and F. E. Cummings, *J. Am. Chem. Soc.*, **86**, 5511 (1964).
57. R. Porter and E. C. Baughan, *J. Chem. Soc.*, **1958**, 744.

58. A. G. Davies and E. C. Baughan, *J. Chem. Soc.*, **1961**, 1711.
59. J. W. Bayles, A. G. Evans, and J. R. Jones, *J. Chem. Soc.*, **1955**, 206.
60. J. W. Bayles, A. G. Evans, and J. R. Jones, *J. Chem. Soc.*, **1957**, 1020.
61. J. L. Cotter and A. G. Evans, *J. Chem. Soc.*, **1959**, 2988.
62. M. Baaz and V. Gutmann, in *Friedel-Crafts and Related Reactions*, Vol. I, Interscience, New York, 1963, Chapt. V.
63. A. G. Evans, J. A. G. Jones, and G. O. Osborne, *Trans. Faraday Soc.*, **50**, 16, 470 (1954).
64. D. Bryce-Smith and N. A. Perkins, *Chem. Ind. (London)*, **1959**, 1022.
65. (a) A. E. Young, private communication; (b) K. M. Harmon, L. L. Hesse, L. P. Kleman, C. W. Kocher, S. V. McKinley, and A. E. Young, *Inorg. Chem.*, **8**, 1054 (1969).
66. J. Radell, J. W. Connolly, and A. J. Raymond, *J. Am. Chem. Soc.*, **83**, 3958 (1961).
67. N. N. Lichtin and P. Pappas, *Trans. N.Y. Acad. Sci.*, **20**, 143 (1957).
68. W. R. Longworth and C. P. Mason, *J. Chem. Soc.*, (A) **1966**, 1164.
69. W. M. Paiska, *Tetrahedron*, **22** 557 (1966).
70. D. F. Shriver and M. J. Biallas, *J. Am. Chem. Soc.*, **89**, 1078 (1967).
71. E. D. Jensen and R. W. Taft, *J. Am. Chem. Soc.*, **86**, 116 (1964).
72. A. E. Young, R. R. Sandell, and H. H. Freedman, *J. Am. Chem. Soc.*, **88**, 4352 (1966).
73. N. N. Lichtin and H. P. Leftin, *J. Phys. Chem.*, **60**, 164 (1956).
74. For examples see S. A. Olah, C. U. Pittman, Jr., and M. C. R. Symons, in *Carbonium Ions*, Vol. I, G. Olah and P. v. R. Schleyer, Eds., Interscience, New York, 1968, Chapter 5.
75. V. Gold and B. W. V. Hawes, *J. Chem. Soc.*, **1951**, 2102.
76. W. N. White and C. A. Stout, *J. Org. Chem.*, **27**, 2915 (1962).
77. K. M. Harmon and F. E. Cummings, *J. Am. Chem. Soc.*, **87**, 539 (1965).
78. B. A. Timimi, *Chem. Ind. (London)*, **1967**, 2148.
79. M. S. Newman and N. C. Deno, *J. Am. Chem. Soc.*, **73**, 3644 (1951).
80. A. Bayer and D. Villiger, *Ber.*, **35**, 1189 (1902).
81. S. T. Bowden and D. T. Zalichi, *J. Chem. Soc.*, **1957**, 4240.
82. V. Gold and F. L. Tye, *J. Chem. Soc.*, **1952**, 2172.
83. H. Lund, *J. Am. Chem. Soc.*, **49**, 1346 (1927).
84. L. P. Hammett, *Physical Organic Chemistry*, McGraw-Hill, New York, 1940, Chapter IX.
85. J. B. Conant and T. H. Werner, *J. Am. Chem. Soc.*, **52**, 4436 (1930).
86. M. A. Murray and G. Williams, *J. Chem. Soc.*, **1950**, 3322.
87. N. C. Deno, J. Jaruzelski, and A. Schriesheim, *J. Org. Chem.*, **19**, 155 (1954).
88. N. C. Deno, J. Jaruzelski, and A. Schriesheim, *J. Am. Chem. Soc.*, **77**, 3044 (1955).
89. R. J. Goldacre and J. N. Phillips, *J. Chem. Soc.*, **1949**, 1724.
90. N. C. Deno and W. L. Evans, *J. Am. Chem. Soc.*, **79**, 5804 (1957).
91. R. Cigen, thesis, University of Lund, 1956.
92. R. Renaud, *Compt. Rend.*, **260**, 6933 (1965).
93. E. M. Arnett and R. D. Bushick, *J. Am. Chem. Soc.*, **86**, 1564 (1964).
94. R. Breslow, S. Garratt, L. Kaplan, and D. LaFollette, *J. Am. Chem. Soc.*, **90**, 4051, 4056 (1968).
95. I. I. Schuster, A. K. Colter, and R. J. Kurland, *J. Am. Chem. Soc.*, **90**, 4679 (1968).

96. R. Filler, et al., *J. Am. Chem. Soc.*, **89**, 1026 (1967).
97. N. C. Deno and A. Schriesheim, *J. Am. Chem. Soc.*, **77**, 3051 (1955).
98. R. Cigen, *Acta Chem. Scand.*, **12**, 1456 (1958).
99. R. Cigen, *Acta Chem. Scand.*, **16**, 192 (1962).
100. R. Cigen, *Acta Chem. Scand.*, **15**, 1892 (1961).
101. R. Cigen, *Acta Chem. Scand.*, **15**, 1905 (1961).
102. R. Cigen and C. G. Ekström, *Acta Chem. Scand.*, **17**, 1189 (1963).
103. R. Cigen and C. G. Ekström, *Acta Chem. Scand.*, **17**, 1843 (1963).
104. R. Cigen and C. G. Ekström, *Acta Chem. Scand.*, **17**, 2083 (1963).
105. R. Cigen and C. G. Ekström, *Acta Chem. Scand.*, **18**, 157 (1964).
106. C. G. Ekström, *Acta Chem. Scand.*, **19**, 1381 (1965).
107. E. M. Arnett, *Progr. Phys. Org. Chem.*, **1**, 223 (1963).
108. G. Branch and H. Walba, *J. Am. Chem. Soc.*, **76**, 1564 (1954).
109. K. Lankamp, W. T. Nauta, and C. MacLean, *Tetrahedron Lett.*, **1968**, 249.
110. S. V. McKinley, J. W. Rakshys, Jr., A. E. Young, and H. H. Freedman, *J. Am. Chem. Soc.*, **93**, 4715 (1971).
111. A. E. Young, H. H. Freedman, and V. R. Sandel, preprints of papers presented at the 150th National Meeting of the American Chemical Society, Atlantic City, N.J., September, 1965, p. 775.
112. H. H. Freedman and A. M. Frantz, *J. Am. Chem. Soc.*, **84**, 4165 (1962).
113. R. W. Taft and L. D. McKeever, *J. Am. Chem. Soc.*, **87**, 2489 (1965).
114. N. C. Deno, H. J. Peterson, and G. S. Saines, *Chem. Rev.*, **60**, 7 (1960).
115. L. M. McDonough, Ph.D. thesis, University of Washington (H. J. Dauben, supervisor), 1960.
116. H. J. Dauben et al., *J. Am. Chem. Soc.*, **79**, 4557 (1957).
117. S. V. McKinley, P. A. Grieco, A. E. Young, and H. H. Freedman, *J. Am. Chem. Soc.*, **92**, 5900 (1970).
118. J. C. Martin and R. G. Smith, *J. Am. Chem. Soc.*, **86**, 2252 (1964).
119. P. D. Bartlett and E. S. Lewis, *J. Am. Chem. Soc.*, **72**, 1005 (1950).
120. N. N. Lichtin and P. D. Bartlett, *J. Am. Chem. Soc.*, **73**, 5530 (1951).
121. C. D. Ritchie and W. F. Sager, *Progr. Phys. Org. Chem.*, **2**, 323 (1964); S. Ehrensen, *ibid.*, **2**, 195 (1964).
122. H. C. Brown and Y. Okamoto, *J. Am. Chem. Soc.*, **80**, 4979 (1958).
123. C. D. Ritchie, W. F. Sager, and E. S. Lewis, *J. Am. Chem. Soc.*, **84**, 2349 (1962).
124. J. Hine, *Physical Organic Chemistry*, 2nd ed., McGraw-Hill, New York, 1962, p. 101.
125. J. W. Rakshys, Jr., S. V. McKinley, and H. H. Freedman, *Chem. Commun.*, **1969**, 1180.
126. R. A. Diffenbach, Y. Sano, and R. W. Taft, *J. Am. Chem. Soc.*, **88**, 4747 (1966).
127. L. D. McKeever and R. W. Taft, *J. Am. Chem. Soc.*, **88**, 4544 (1966).
128. Y. Okamoto and H. C. Brown, *J. Am. Chem. Soc.*, **80**, 4976 (1958).
129. H. H. Jaffe, *Chem. Rev.*, **53**, 191 (1953).
130. G. N. Lewis and M. Calvin, *Chem. Rev.*, **25**, 273 (1939).
131. P. Anderson, *Acta Chem. Scand.*, **19**, 622 (1965).
132. C. Stora and N. Poyer, *Bull. Soc. Chim. France*, **1966**, 841.
133. G. N. Lewis, T. T. Magel, and D. Lipkin, *J. Am. Chem. Soc.*, **64**, 1774 (1942).
134. C. C. Barker, M. H. Bride, and A. Stamp, *J. Chem. Soc.*, **1959**, 3957.
135. V. Gold, *J. Chem. Soc.*, **1956**, 3944.
136. N. N. Lichtin and M. T. Vignale, *J. Am. Chem. Soc.*, **79**, 579 (1957).

137. M. S. deGroot, I. A. M. Hesselmann, and J. H. van der Waals, *Mol. Physics*, **10**, 241 (1966).
138. L. Pauling and R. B. Corey, *J. Am. Chem. Soc.*, **74**, 3964 (1952).
139. M. J. S. Dewar, *J. Am. Chem. Soc.*, **74, 3341** (1952).
140. L. L. Ingraham, in *Steric Effects in Organic Chemistry*, Wiley, New York, 1956, Chapter 11.
141. S. Rameshan and K. Vankatesan, *Experientia*, **1958**, 237.
142. K. Erics, report submitted to the Petroleum Research Fund, PRF/230-A, September, 1963.
143. R. Dehl, W. R. Vaughan, and R. S. Berry, *J. Org. Chem.*, **24**, 1616 (1959).
144. R. Dehl, W. R. Vaughan, and R. S. Berry, *J. Chem. Phys.*, **34**, 1460 (1961).
145. J. W. Rakshys, Jr., S. V. McKinley, and H. H. Freedman, *J. Am. Chem. Soc.*, **92**, 3518 (1970); **93**, 6522 (1971).
146. E. L. Eliel, *Stereochemistry of Carbon Compounds*, McGraw-Hill, New York, 1962, pp. 372–376.
147. Reference 124, p. 101.
148. E. S. Wallis, *J. Am. Chem. Soc.*, **53**, 2253 (1931).
149. M. Gomberg and W. E. Gordon, *J. Am. Chem. Soc.*, **57**, 119 (1935).
150. E. S. Wallis and F. H. Adams, *J. Am. Chem. Soc.*, **55**, 3338 (1933).
151. L. Pasteur, *Ann. Chem. Phys.*, **24** (3), 442 (1848).
152. B. L. Murr, *J. Am. Chem. Soc.*, **85**, 2866 (1963).
153. B. L. Murr and C. Santiago, *J. Am. Chem. Soc.*, **88**, 1826 (1966).
154. F. H. Westheimer, in Reference 140, Chapter 12.
155. J. W. Rakshys, Jr., and H. H. Freedman, unpublished.
156. J. W. Rakshys, Jr., S. V. McKinley, and H. H. Freedman, *J. Am. Chem. Soc.*, **93**, 6522 (1971).

CHAPTER 29

# Cycloheptatrienylium (Tropenylium) Ions

KENNETH M. HARMON
*Oakland University, Rochester, Michigan*

| | |
|---|---|
| I. Introduction | 1580 |
|    A. Nomenclature | 1580 |
|    B. Arrangement of the Chapter | 1581 |
|    C. Background | 1582 |
| II. General Preparations | 1582 |
|    A. Primary Ionogenic Reactions | 1582 |
|       1. Bromination–Dehydrobromination | 1583 |
|       2. Oxidation of Cycloheptatriene | 1584 |
|       3. Hydride Transfer to Carbonium Ions; The Dauben Reaction | 1586 |
|       4. Hydride Transfer to Metal Halides | 1588 |
|       5. Miscellaneous Preparations | 1590 |
|    B. Preparations via Tropenylium Ion | 1590 |
| III. Unsubstituted Tropenylium Ion and its Salts | 1591 |
|    A. Properties of the Ion | 1591 |
|       1. Structure | 1591 |
|       2. Infrared and Raman Spectra | 1592 |
|       3. Ultraviolet Spectrum | 1593 |
|       4. Charge-Transfer Complexes and Spectra | 1594 |
|       5. Nmr Spectra and Diamagnetic Susceptibility | 1596 |
|       6. Conductivity and Polarographic Reduction | 1596 |
|    B. Known Tropenylium Ion Salts | 1597 |
|       1. Halides | 1601 |
|       2. Polyhalides | 1601 |
|       3. Hydrogen Dihalides | 1602 |
|       4. Oxygen Acid Anions | 1602 |
|       5. Halometallates | 1602 |
|       6. Miscellaneous | 1603 |
|    C. Reactions of Tropenylium Ion | 1603 |
|       1. Reduction by Hydride | 1603 |
|       2. Reaction with Electrophilic Reagents | 1605 |
|       3. Reaction with Nucleophilic Reagents | 1605 |
|       4. Reaction as an Electrophile | 1607 |
|       5. Alkylation of Carbonyl and Activated Methylene Compounds | 1607 |
|       6. Reduction to Tropenyl Radical; Coupling Reactions | 1609 |
|       7. Miscellaneous Reactions | 1609 |
| IV. Substituted Tropenylium Ions | 1610 |
|    A. Preparations of Substituted Tropenylium Ions | 1610 |
|       1. Substituted Cycloheptatrienes | 1610 |
|       2. Substituted Tropenylium Ions | 1616 |

B. Physical Properties of Substituted Tropenylium Ions . . . 1617
      1. Structure . . . . . . . . . . 1617
      2. Acidities . . . . . . . . . . 1617
      3. Infrared Spectra . . . . . . . . . 1618
      4. Electronic Spectra . . . . . . . . . 1619
      5. Nmr Spectra and Diamagnetic Susceptibility . . . . 1622
      6. Polarographic Reduction . . . . . . . 1624
      7. Miscellaneous Measurements and Free Energy Correlations . 1626
   C. Reactions and Properties of Representative Substituted Tropenylium
      Ions . . . . . . . . . . . 1627
      1. Alkyltropenylium Ions . . . . . . . . 1627
      2. Aryltropenylium Ions . . . . . . . . 1628
      3. Halotropenylium Ions . . . . . . . . 1628
      4. Hydroxytropenylium Ions . . . . . . . 1629
      5. Alkoxytropenylium Ions . . . . . . . . 1630
      6. Ousene-Type Compounds . . . . . . . 1630
V. Conclusion . . . . . . . . . . . 1631
References . . . . . . . . . . . . 1632

## I. INTRODUCTION

Many readers of this chapter are aware that it was to have been written by Professor Hyp J. Dauben, Jr., of the University of Washington. Upon Professor Dauben's tragic death in April, 1968, I was asked by the editors to take over this task; as a former student and colleague of Professor Dauben I am honored by the request and wish to dedicate this work to his memory as an inspiring teacher and warm friend.

Professor Dauben's death left much of his work on tropenylium ions unpublished. During the next year or so I will be working with his former students to prepare manuscripts on this material, which should appear in American Chemical Society publications. In preparing this chapter I have referred to much of this work, and have included the University Microfilms numbers for doctoral dissertations. Inquiries about unpublished work may be directed to me or to the student involved.

### A. Nomenclature

The cycloheptatrienylium cation has been variously called tropylium (1), tropenium (2), and tropenylium (3) ion. Tropylium has the right of precedence; however, I find tropenylium both more descriptive and more useful in naming derivatives and use it in this chapter. The three names are accepted equally by American journals and are now cross-referenced in *Chemical Abstracts*, although this was not always true in former years. Examples of the derivative nomenclature to be used in this chapter are given in Table I.

## TABLE I
### Nomenclature of Cycloheptatriene Derivatives

| Formula | Structure | Nomenclature Used herein (3,80) | Other | Ref. |
|---|---|---|---|---|
| $C_7H_8$ | | Cycloheptatriene | Tropilidene | 1 |
| $C_7H_7-$ | | Tropenyl group | Tropyl group | 1 |
| $C_7H_7{}^+$ | | Tropenylium ion | Tropylium ion<br>Tropenium ion | 1<br>2 |
| $C_7H_6-{}^+$ | | Tropenyliumyl group | Tropylium group<br>Tropenylium group | 1<br>227 |

### B. Arrangement of the Chapter

Section II of this chapter discusses general methods of preparation of tropenylium species with particular application to the unsubstituted ion. These preparations are divided into primary ionogenic reactions with cycloheptatrienes and secondary methods of synthesis utilizing tropenylium ion itself as an intermediate in the preparations. Section III covers the properties and reactions of unsubstituted tropenylium ion and its salts, and Section IV deals with the preparations, properties, and reactions of substituted tropenylium ions and their salts, and some observations on substituent effects.

The intended tone of this work is practical rather than theoretical, with emphasis on the chemical and physical properties of tropenylium salts as laboratory materials and consideration of techniques of preparing and studying these salts that have been developed in Professor Dauben's and our own laboratories over the last 15 years. In addition, I hope to provide a survey of the field and a reference guide to the current literature. Of necessity some subjects must be slighted, and I have not included discussions of three extremely interesting but somewhat oblique subjects: detailed molecular orbital calculations of the resonance energy and spectral properties of the ions (4–34), mass spectral studies of the rearrangements of benzene derivatives and other species to tropenylium ion in the gas phase (35–49), and compounds in which a tropenylium ring is fused into a larger ring system (8,20,25,31,50–63).

## C. Background

Tropenylium bromide was inadvertently prepared by Merling (64) in 1891; however, neither the expectations nor techniques of the time were compatible with this material, and it was not identified. The stability of a seven-carbon, six-$\pi$-electron cyclic cationic system was predicted from molecular orbital theory by Hückel (4,5) and foreshadowed in the properties of tropone, tropolone, and their derivatives (65); however, the first synthesis of the parent cation was not achieved until 1954 when Doering and Knox prepared and characterized the bromide salt (1). Dauben and Pearson (66), in 1954, independently recognized tropenylium ion as the product of the acid cleavage of 7-succinimidylcycloheptatriene, and in 1955 they announced the synthesis of tropenylium salts by hydride exchange with triphenylcarbonium ion (2,67). Interest in tropenylium ion and its derivatives grew rapidly and has remained at a healthy although not exorbitant level through the last decade. Several prior reviews are available (65,68–73).

## II. GENERAL PREPARATIONS

### A. Primary Ionogenic Reactions

This section discusses preparative reactions in which tropenylium ions are generated directly from covalent substrates, which are usually 1,3,5-cycloheptatriene or its derivatives. 1,3,5-Cycloheptatriene (hereafter called cycloheptatriene) is commercially available from a number of sources. It can be readily freed of impurities arising from air oxidation by fractionation under nitrogen; the distillate should be stored in an inert atmosphere in the cold. A minute amount of yellow impurity often co-distills (74), which renders the distillate pale yellow. Although this material does not affect tropenylium salt preparations, it can be removed, if desired, by chromatography on neutral alumina. Most samples of cycloheptatriene contain varying amounts (1–10%) of toluene, which is extremely difficult to remove by fractional distillation. This does not interfere in any way with tropenylium ion preparations, but allowance for it must be made in calculating yield data. In most instances the preparations of substrates for salt formation are adequately discussed in the primary references and are not considered here.

## 1. Bromination–Dehydrobromination

The original method of Merling (64) was adapted by Doering and Knox (1) for the first preparation of a tropenylium ion salt. Their improved

$$\text{C}_7\text{H}_8 + \text{Br}_2 \longrightarrow \text{dibromide} \xrightarrow{\Delta} \text{tropenylium}^+, \text{Br}^- + \text{HBr}$$

technique (75) consists of slow addition of 1 mole eq of bromine to a cold carbon tetrachloride solution of cycloheptatriene, followed by removal of solvent and pyrolysis of the dibromide product to give a 60% yield of tropenylium bromide. The process requires several days and the product needs recrystallization; however, the method is easily adapted to the preparation of large quantities of salt.

The bromination–dehydrobromination technique has been used to prepare some substituted tropenylium ions. Johnson and Tisler (76–78) prepared carboxytropenylium bromide in low yield from 7-carboxycycloheptatriene by this method, and Doering and Krauch (69) have reported the preparation of alkyl and aryl tropenylium ions. Cairncross (79) prepared phenyltropenylium ion in low yield via bromination, and Dauben and Rhoades (3,80) have extended the method to aminotropenylium (tropenylidenimmonium) ions. Doering and Knox (75) have shown that the dibromide product of tropenyl methyl ether can be quantitatively pyrolyzed to hydroxytropenylium bromide; this latter salt is a convenient source of tropone (75).

$$\text{dibromide-OCH}_3 \xrightarrow{\Delta} \text{tropenylium}^+\text{-OH}, \text{Br}^- + \text{CH}_3\text{Br}$$

The intermediate dibromide product of cycloheptatriene is not normally isolated during the preparation of tropenylium bromide; it has been characterized, however, and is believed to consist primarily of the 1,6-isomer on the basis of its ultraviolet spectrum (1,67,75). Several workers have investigated the effect of basic or ionizing solvents on 1,6-dibromo-2,4-cycloheptadiene. Pearson (67) found that solution in liquid sulfur dioxide converted the dibromide to tropenylium bromide in 50–60% yields, and Kitahara and Funamizu (81) report even higher conversion in this solvent. Direct bromination of cycloheptatriene in sulfur dioxide, however, does not give satisfactory yields of salt (67). Dauben and Pearson

(67) have found that bromination of cycloheptatriene in dimethylformamide containing added lithium chloride gives fair yields of tropenylium ion and that bromination in acetonitrile proceeds smoothly to give 80–90% yields of an equimolar mixture of tropenylium bromide and 1,6-dibromo-2,4-cycloheptadiene. It was shown that the bromide arose from the activated complex of the bromination step, rather than by decomposition of preformed dibromide.

Harmon (82) has shown that an equilibrium exists between iodine and cycloheptatriene with tropenylium iodide and hydrogen iodide; if nonpolar

$$C_7H_8 + I_2 \rightleftharpoons C_7H_7^+, I^- + HI$$

solvents (in which tropenylium salts are insoluble) or complexing agents (to complex iodide and shift the equilibrium) are used, excellent yields of tropenylium iodide or complex iodide salts can be obtained. It is not known whether this reaction proceeds through a hydride transfer process, as has been suggested for the reaction of triarylcarbonium ions with hydrogen iodide (83) or through an addition-elimination sequence; however, a colorless, iodine-containing material is isolated from the equilibrium mixture, which may be an addition diiodide.

## 2. Oxidation of Cycloheptatriene

All conversions of cycloheptatriene to tropenylium ion are, of course, oxidations; this section considers specifically the use of common chemical oxidizing agents or electrochemical methods to carry out this process.

Direct chemical oxidation of cycloheptatriene affords an apparently attractive route to tropenylium ion, and several workers have investigated such processes; however, results are usually unsatisfactory. Pearson (67) found spectroscopic evidence for ion formation when cycloheptatriene was treated with sulfur trioxide or lead dioxide but was unable to isolate salts. Kursanov and Vol'pin (84,85) have isolated tropenylium ion by precipitation as the perchlorate or chloroplatinate salt from reactions of cycloheptatriene with sulfuric acid, nitric acid, chromium trioxide in glacial acetic acid, selenium dioxide, and sulfuryl chloride; but the yields do not exceed 25%, and extensive polymerization accompanies the preparations.

The low yields and concurrent polymerization in the above-mentioned reactions make it difficult to determine to what extent tropenylium ion arises from direct oxidation. Thus the reaction of cycloheptatriene with sulfuric acid to form tropenylium ion has been described as an oxidation (67,84,85) but is probably best represented as acid-catalyzed disproportionation in which protonation of cycloheptatriene is followed either by cationic polymerization or hydride transfer to yield cycloheptadiene and tropenylium ion. Hydridic termination of the polymerization would also produce the ion. Thus Honnen (86) has shown that slow extraction of cycloheptatriene from pentane into concentrated sulfuric acid yields about 20% tropenylium ion, whereas high-boiling hydrocarbon residues can be isolated from the pentane layer, and ter Borg et al. (87) have shown that perchloric acid in glacial acetic acid reacts with cycloheptatriene to afford low yields of ion and a polymer with an average molecular weight of 450. Phosphoric acid—a nonoxidizing acid—gave similar results. Since most of the attempted chemical oxidations of cycloheptatriene involve acidic conditions or reagents that *are* acids or else yield acids with moisture, it is possible that acid-catalyzed disproportionation contributes significantly to the yields.

A second problem in the chemical oxidation is that many of the reagents either react with tropenylium ion itself or lead to more highly oxidized seven-membered products. For example, tropenylium ion is oxidized by chromic acid or silver oxide to benzaldehyde (75) and reacts with hydrogen peroxide to give primarily benzene and carbon monoxide (88). The oxidation of cycloheptatriene with chromium trioxide has been shown to involve a tropenylium ion intermediate; however, the main product is benzoic acid (89). Selenium dioxide has been reported to oxidize cycloheptatriene to tropone in good yield (90),* and tropolone is an important product of oxidation by potassium permanganate (91).

Thus a combination of polymerization, further reaction, and concurrent formation of reactive or sensitive side products has obviated attempts to use conventional oxidants to prepare tropenylium ion. Autooxidation, however, has been carried out successfully; ter Borg et al. (87) have shown that cycloheptatriene in glacial acetic acid is oxidized under acid conditions by a stream of oxygen to give satisfactory yields of isolable salts. This reaction is catalyzed by ferric ion.

In contrast to the problems attendant on chemical oxidation, Geske (92) found that cycloheptatriene is oxidized smoothly and quantitatively at the rotating platinum electrode in a two-electron step to tropenylium ion; this procedure can be adapted to preparative scale (93).

---

* We have found it nearly impossible to remove all the selenium from the product of this reaction.

### 3. Hydride Transfer to Carbonium Ions; The Dauben Reaction

The Dauben reaction, i.e., hydride transfer from cycloheptatriene or a substituted cycloheptatriene to triphenylcarbonium ion (2,67,94) is by far the most widely used and generally applicable synthetic method for the

$$C_7H_8 + (C_6H_5)_3C^+, A^- \longrightarrow C_7H_7^+, A^- + (C_6H_5)_3CH$$

synthesis of tropenylium ions. The method is successful with a wide range of substituent groups, anions, and solvents, and generally gives nearly quantitative yields of high-purity salts, from which the by-product triphenylmethane is easily separated. For large-scale preparation of tropenylium fluoroborate, the method of Kursanov (84) as modified by Conrow (95) is preferred; however, we find hydride exchange to be both more rapid and convenient for preparation of up to 0.2 mole of this commonly used salt. Experimental directions for representative hydride exchanges have been given by Harmon (96,97).

Triphenylmethyl halides are available commercially from a variety of sources, or they may be prepared by the method of Bachmann (98). Triphenylcarbonium ion salts of anions derived from strong acids may be conveniently prepared by the method of Dauben et al. (94), which is adapted from that of Hofmann and Kirmreuther (99); it consists of slow addition of the acid to a chilled solution of triphenylmethanol in acetic or propanoic anhydride. This method has been used to prepare triphenylcarbonium perchlorate and fluoroborate (94), hexafluoroantimonate and hexafluoroarsenate (100), picrysulfonate (101), pentachloro- and pentabromostannate(IV) (102), and iodide (103,104), among others.

Best results are obtained if the original directions (94) are modified as follows. After initial addition of a few drops of acid to the carbinol solution at room temperature (to prevent later crystallization of carbinol) the solution is chilled to ice-bath temperature, stirred vigorously, and the acid added *very slowly* in a dropwise manner, to prevent the slightest temperature increase. The solvent is then decanted while cold and the product washed with ethyl acetate and pentane and dried *in vacuo*. Triphenylcarbonium hexafluorophosphate, -arsenate, and -antimonate are now commercially available (105). Triphenylcarbonium ion salts of complex metal halide anions can be conveniently prepared *in situ* by addition of anhydrous metal halide to triphenylmethyl halide in the solution (usually methylene chloride) in which the exchange reaction is to be run (102–104, 106–108).

In theory the Dauben reaction can be carried out in any solvent in which the reactants are reasonably soluble and which does not react harmfully with carbonium ions; the one important restriction is that the triphenylmethyl species be ionized. This restriction does not present any problem with triphenylcarbonium ion salts of anions of low nucleophilicity, such as perchlorate, fluoroborate, or hexafluoroantimonate, since these are ionized even in the solid; however, triphenylmethyl halides only react to an appreciable extent in strongly ionizing solvents. For practical purposes, liquid sulfur dioxide is the solvent of choice for the halides, since reaction in this solvent is quantitative (67,96,97) and the solvent is easily removed. Reaction of the halides in acetonitrile is slow and incomplete and there is little if any reaction in halocarbons (86). Salts of anucleophilic anions, on the other hand, react readily in any polar solvent in which carbonium ions are stable, such as acetonitrile, nitroalkanes, 1,2-dichloroethane, or methylene chloride. Acetonitrile is the most widely used solvent. The dependence of the rate and extent of hydride transfer from cycloheptatriene on the degree of ionization of the triphenylmethyl species has been used to demonstrate the ionic nature of a number of triphenylmethyl halide–metal halide complexes (102,104,106,108).

The Dauben reaction has been used to prepare a wide range of substituted tropenylium ions from substituted cycloheptatrienes; a partial list of substituents includes alkyl (2,109,110), aryl (111,112), vinyl (113–115), allyl (116), substituted ethynyl (113,117), halogens (2,96), alkoxy (2,110, 118), thioalkoxy (96,117), carboxy (96), cyano (119), carboranyl (100), and polyethylenyl (120). A more complete listing of substituents and references thereto is given in Section IV. The presence of a substituent on either the 1-, 2-, or 3-position of cycloheptatriene does not normally interfere with the exchange reaction, although electron-withdrawing groups may slow the rate of reaction considerably (96,100,119). When the substituent is on the 7-position the reaction may proceed abnormally. Labile groups such as methoxy (2,96) or alkylamino (3,80) are preferentially abstracted by triphenylcarbonium ion, and either steric bulk (109) or electronic effects (100,119) can slow the reaction to an impractical rate.

In theory any carbonium ion sufficiently less stable than tropenylium ion could be used to abstract hydride from cycloheptatriene; in practice the triphenylcarbonium ion has proved so satisfactory that little work has been done in a preparative sense with other acceptors. Successful hydride transfer with cycloheptatriene has been used to demonstrate the carbonium ion nature of a number of species, including derivatives of cyclopentadienylium cation (121,122) mono- (123) and diarylcarbonium ions (124–126) and aryl-substituted polycarbonium ions (127–130). *t*-Butyl halides react with cycloheptatriene in the presence of strong Lewis acids (74,86, 106,108,131), but hydride abstraction by *t*-butyl cation has not been demonstrated (Section II-A-4). The primary fate of the *t*-butyl cation under such conditions appears to be addition to the olefinic system. Parnes and co-workers (132) have shown that methylcyclohexane can be isolated from the reaction of 1-methylcyclohexene with cycloheptatriene in trifluoroacetic acid; however, the yield is very low and the predominant reaction is polymerization of cycloheptatriene.

### 4. Hydride Transfer to Metal Halides

In 1957 Kursanov and Vol'pin (84,85) reported that the action of phosphorus pentachloride on cycloheptatriene gave quantitative yields of

$$C_7H_8 + 2PCl_5 \longrightarrow C_7H_7^+, PCl_6^- + PCl_3 + HCl$$

tropenylium ion. This reaction is an excellent source of large quantities of the ion; specific experimental directions are given by Conrow (95) for the fluoroborate, and Harmon et al. (133) have utilized this process in an improved preparation of ditropenyl ether, a basic starting material for the synthesis of tropone. The initial product in the phosphorus pentachloride–cycloheptatriene reaction was first reported to be the chloride (84) but was later shown to be a complex salt containing the hexachlorophosphate(V) ion (134) and excess tropenylium chloride (135); we have also discovered that the filtered and apparently dry solid contains variable and significant quantities of the reaction solvent carbon tetrachloride. The white reaction product is difficult to store and is not stable to the atmosphere; it is normally converted to other salts by solution in alcohol and/or water and precipitation as tropenylium fluoroborate, perchlorate, or chloroplatinate (84,85,95).

The mechanism of the phosphorus pentachloride reaction has not been investigated in detail; however, the quantitative yield and absence of any polymerization rule out (see below) addition of $Cl^+$ (136) to the olefinic system, and the most likely reaction course is hydride abstraction by $PCl_4^+$ (137,138) followed by dissociation of $HPCl_4$ to $HCl$ and $PCl_3$.

$$C_7H_8 + PCl_4^+ \longrightarrow C_7H_7^+ + HPCl_4$$

$$HPCl_4 \longrightarrow PCl_3 + HCl$$

Holmes and Pettit (124) have shown that antimony pentachloride quantitatively converts a variety of hydrocarbons, including cycloheptatriene, to carbonium ion hexachloroantimonates, and that antimony trichloride is formed in these reactions; it is likely that the phosphorus pentachloride reaction follows a similar path. This reaction has been adapted to the preparation of bromo- (139), methyl- (84), and alkoxytropenylium ions (140).

Harmon et al. (96,103,104,106–108) have examined in detail the reaction of boron halides with cycloheptatriene. In cocatalyst-free hydrocarbon solvent, boron fluoride fails to react with cycloheptatriene and, although boron chloride reacts cleanly to give tropenylium chloroborate, the reaction is very slow. Equimolar quantities of either boron bromide or iodide, however, react smoothly with cycloheptatriene in cyclohexane to produce 75% yields (100% based on available halogen) of the tetrahaloborates. The stoichiometry of the reaction and the complete absence of any polymerization argue strongly for direct hydride transfer from carbon to boron (104,108) followed by disproportionation of $HBX_3^-$.

$$C_7H_8 + BX_3 \longrightarrow C_7H_7^+ + HBX_3^-$$
$$HBX_3^- + BX_3 \longrightarrow HBX_2 + BX_4^-$$

An extremely reactive substance remains in the mother liquor which we presumed (108) to be a highly reduced boron species, perhaps containing alkenyl groups from hydroboration reactions. Parnes et al. have subsequently demonstrated that this substance contains boron–hydrogen bonds (141), which lends significant support to the hydride transfer mechanism. The reaction of boron halides with cycloheptatriene in methylene chloride is much more rapid than in cyclohexane; in the case of the bromide (106) this reaction can be adapted to a rapid, large-scale preparation of tropenylium bromide, since the bromoborate salt can be easily converted to the bromide either by sublimation or by cautious solution in methanol followed by precipitation with ether.

Bruce-Smith and Perkins (74,131) have investigated the reaction of boron, aluminum, ferric, stannic, and titanium chlorides with cycloheptatriene in the presence of *t*-butyl chloride cocatalyst and find that the corresponding complex metal chloride salts of tropenylium ion are produced. They attribute these reactions to hydride transfer to metal halide; however, the low yields and copious concurrent polymerization of cycloheptatriene argue against hydride transfer as the principal source of ion. We find (108) that cycloheptatriene is completely inert to stannic chloride in the absence of cocatalyst; but if alkyl halide, hydrogen halide, or water is present, a low yield of tropenylium ion is formed along with much dark polymeric material. It appears that any reaction in which an electrophile adds to cycloheptatriene can give ion and will give polymer (86,87,108,132,141,142). Further evidence for the role of the *t*-butyl cation in such reactions is given by Honnen (86), who found that *t*-butyl fluoride is a necessary cocatalyst for the reaction of boron fluoride with cycloheptatriene or perinaphthene and that in the latter reaction, equal quantities of perinaphthenium ion and 7-*t*-butylperinaphthane are isolated.

An interesting reaction, which most likely proceeds by hydride abstraction by metal halide, has been reported by Kistner et al. (143). They find that cycloheptatriene reacts with platinum(IV) halides or hexahaloplatinic

$$4C_7H_8 + 3PtX_4 \longrightarrow (C_7H_7^+)_2PtX_6^{2-} + 2C_7H_8 \cdot PtX_2 + 2HX$$

acids to yield tropenylium hexahaloplatinate(IV) salts and the cycloheptatriene adduct of platinum(II) halide in a 1:2 ratio. This reaction is believed to proceed by hydride abstraction with concurrent reduction of platinum(IV) followed by rapid complexing of the platinum(II) species by cycloheptatriene.

### 5. Miscellaneous Preparations

A number of miscellaneous methods of producing tropenylium ion have been reported; some of these are of limited applicability, and others appear to be of general synthetic interest. Tropenylium halides have been prepared by the action of monohalocarbenes on benzene (144,145), by chlorination of cycloheptatriene (146), by hydrogen bromide cleavage of a cycloheptatriene–ethyl azodicarboxylate adduct (147), and by irradiation of cycloheptatriene–$t$-butyl bromide mixtures with gamma rays (148). Dewar et al. (149–152) have prepared a variety of tropenylium species from the action of acids or oxidizing agents on 2,4,6-cycloheptatrienecarboxylic acid and its derivatives and have also demonstrated that the ion is formed in small amounts in the oxidation of cyclooctatetraene (153–155). A 60% yield of tropenylium ion has been reported from the acid-catalyzed fragmentation of "$t$-butyl-norcaradienepercarboxylate" (156), and the action of phosgene (157) and methyllithium (158) on tropone has been shown to give substituted tropenylium ions.

Two methods that have not been widely used for preparative purposes appear to have considerable potential. Nesmeyanov and Golovnya (159, 160) have shown that azobenzene accepts hydride from cycloheptatriene in the presence of acids to give good yields of tropenylium ion, and Reid

$$C_7H_8 + C_6H_5N{=}NC_6H_5 + H^+A^- \longrightarrow C_7H_7^+, A^- + C_6H_5NHNHC_6H_5$$

et al. (161) report the preparation of a number of salts of the ion by the action of benzoquinone on cycloheptatriene. Similarly, Seto and Ogura (162,163) have shown that $p$-benzoquinone dibenzenesulfonimide with boron trifluoride etherate converts cycloheptatriene to tropenylium ion in quantitative yield.

### B. Preparations via Tropenylium Ion

Since simple tropenylium ion salts such as the bromide (2,75,97) and fluoroborate (2,95,97) are readily prepared in large quantities, they are attractive starting materials for the preparation of more esoteric salts.

This section considers the preparation of unsubstituted salts; conversion of the ion into substituted tropenylium ion salts is discussed in Section IV-A.

The simplest method for interconversion of tropenylium salts is anion interchange based on mass action or solubility. The action of hydrogen halide on tropenylium bromide has been used to prepare the chloride (1) and is the method of choice for the iodide (75); colorless hydriodic acid should be used in the latter preparation to avoid triiodide formation. A wide variety of insoluble tropenylium salts can be prepared by addition of the desired anion to a solution of the bromide or fluoroborate in appropriate solvent; familiar examples are the chloroplatinate (84,164) and tetraphenylborate (165) salts.

Tropenylium halides react with a variety of complexing agents to produce salts of complex anions such as tribromide (97,166), triiodide (2,82,161), mixed polyhalide (166), and halometallate (74,82,106,108,167,168) ions.

Tropenyl ethers are easily cleaved by acid to tropenylium ion and alcohol (1), and this reaction may be used to convert tropenyl methyl ether—which is conveniently prepared from the bromide or fluoroborate (1)—to any salt for which the corresponding acid is available. One

$$\text{C}_7\text{H}_7\text{-OCH}_3 + \text{H}^+\text{A}^- \longrightarrow \text{C}_7\text{H}_7^+, \text{A}^- + \text{CH}_3\text{OH}$$

advantage of this method is that the ether is soluble in a variety of nonpolar solvents from which salts separate quantitatively on treatment with acid. For example, Harmon and Davis (169) have prepared tropenylium hydrogen dichloride and hydrogen dibromide by addition of tropenyl methyl ether to saturated solutions of hydrogen halide in diethyl ether, and Harmon and Thompson (100) have used ether cleavage by hydrogen bromide to prepare 1-methyl-2-tropenyliumyl-1,2-dicarba-*closo*-dodecaborane(10) bromide which is difficult, if not impossible, to prepare by other methods.

Tropenyl methyl ether should be handled in a fume hood, since it causes severe headaches in some workers on brief exposure.

## III. UNSUBSTITUTED TROPENYLIUM ION AND ITS SALTS

### A. Properties of the Ion

#### 1. Structure

Theory predicts (4,5) that the $C_7H_7^+$ cyclic cation should be a planar, regular heptagon possessing considerable resonance energy. Calculations based on molecular orbital (6), valence bond (24), or extended Hückel theory (26) techniques agree that the ion should have significantly more resonance stabilization than benzene. This stability is reflected by a p$K_R+$ value (170) for tropenylium ion of 4.75 (1), more than 10 orders of magnitude greater than that of triphenylcarbonium ion.

There have been few X-ray crystallographic investigations of tropenylium ion salts. Gould (171) carried out preliminary studies on tropenylium chloride, but the instability of the material and the complexity of the crystal prevented detailed analysis. Kitaigorodskii and co-workers (172, 173) report detailed studies on tropenylium iodide and perchlorate. Free rotation of the rings in these salts prevents direct observation of carbon positions; however, the ring is planar and the ion assumes the symmetry of a cylinder, from the radius of which a mean carbon–carbon bond distance of 1.47 Å can be calculated. The ions pack in columns of alternating cation and anion; in the iodide the anion may be slightly displaced toward one of the two sandwiching cations, as a result of a weak charge-transfer bond. In 1,2-dihydroxytropenylium chloride (tropolone hydrochloride) the rings do not rotate, and Sasada et al. (174) have shown that the cationic ring is a planar, slightly irregular heptagon. The variations in bond length are not reflected in the mirror plane containing C-5 and splitting the C-1–C-2 bond, and thus do not seem to depend on distance from the substituents. The distortions appear small enough to arise from packing effects. Similar results are found for the tropolone anion (175). The $D_{7h}$ symmetry of the tropenylium ion has been shown clearly by infrared studies (see below).

The chemical equivalence of the seven-carbon atoms in tropenylium ion has been shown by Vol'pin and co-workers (176,177), who reacted

$$C_6H_5MgBr + C_7H_7^+ \longrightarrow C_6H_5-\text{(cycloheptatrienyl)} \xrightarrow{[O]} C_6H_5CO_2H$$

$^{14}$C-labeled tropenylium ion with phenyl magnesium bromide and oxidized the resulting 7-phenylcycloheptatriene to benzoic acid. The benzoic acid contained one-seventh of the theoretical activity at the carboxyl group.

## 2. Infrared and Raman Spectra

The infrared spectrum of the tropenylium ion is extremely simple, and at normal concentrations in a potassium bromide pellet or Nujol mull consists of three absorptions: carbon–hydrogen stretch at 3020 cm$^{-1}$, carbon–carbon stretch at 1479 cm$^{-1}$, and a doublet at 678 and 651 cm$^{-1}$, of which one peak is assigned to carbon–hydrogen out-of-plane bending.

In concentrated spectra a variety of smaller peaks are seen, particularly in the region of 1000–1300 cm$^{-1}$. These are not consistent in several reported spectra (1,119) and with our experience, and it is not clear which are definitely associated with the ion and which with possible contaminants.

The infrared and Raman spectra of tropenylium ion in solution (hydrobromic acid) have been analyzed in detail by Fateley et al. (178–180). The ion shows the three infrared absorptions and seven Raman lines, with a fourth infrared band inferred in the region of 228 cm$^{-1}$; there are no coincidences between Raman and infrared frequencies. These spectra are in excellent agreement with those calculated for a planar, regular heptagon with $D_{7h}$ symmetry (180–182) and are not in accord with any other likely structure for tropenylium ion. There is great similarity between the spectra of benzene and tropenylium ion.

### 3. Ultraviolet Spectrum

Tropenylium ion shows two absorption bands in the ultraviolet and does not absorb in the visible region at all. Doering and Knox (1) reported that the spectrum of tropenylium bromide in aqueous acid showed a long tail into the visible region which accounted for the yellow color of the ion. It is now clear (2,97) that the ion is completely colorless and that solutions of its salts are also colorless unless the anion is colored or anion-to-cation charge transfer occurs; nevertheless, the "long tail" is frequently mentioned in theoretical articles, and a few workers have been moved to perform calculations upon it.

The most useful solvent for the ultraviolet spectral analysis of tropenylium ion salts is reagent grade 96% sulfuric acid (Fig. 1), in which the ion shows $\lambda_{max}$ 213 m$\mu$ (sh), 217 ($\varepsilon$ 43,000), 221 (sh), 268 (sh), 274 (4350), and 279.5 (3890), and $\lambda_{min}$ 244 m$\mu$ ($\varepsilon$ 420) (2,96,97). Other useful solvents [$\lambda_{max}$ ($\varepsilon$ $C_7H_7^+$)] include dilute aqueous acid [$\lambda_{max}$ 275 m$\mu$ ($\varepsilon$ 4360) and 280 (sh) (1)], acetonitrile [$\lambda_{max}$ 275 m$\mu$ ($\varepsilon$ 4250) and 280 (sh) (96)], and methylene chloride [$\lambda_{max}$ 271.5 m$\mu$ (sh), 278 ($\varepsilon$ 4270), and 283 (sh) (97)]. The intense short-wavelength band is often obscured by end absorption in organic solvents; however, the long-wavelength band serves for analytical purposes, and the intensity difference between the bands makes observation of both in a single spectrum difficult in any case.

Quantitative ultraviolet spectral analysis in 96% sulfuric acid is the method of choice for estimation of tropenylium ion; we use this technique routinely for analyzing new salts and checking the purity of known materials. With good technique in solution preparation, values are within the limits of accuracy expected of combustion elemental analyses. TD pipets calibrated for water may be used with sulfuric acid if they have a capacity of 5 ml or greater and a 25-sec draining time is allowed. Pipets

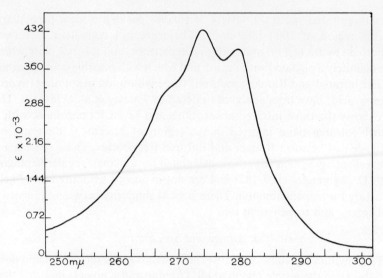

Fig. 1. Long-wavelength ultraviolet absorption band of tropenylium ion in 96% sulfuric acid.

10 ml or larger are preferred. Large fluctuations in environmental temperature should be avoided, as sulfuric acid has a large thermal coefficient of expansion.

The theoretical ultraviolet spectrum of tropenylium ion has been calculated by a number of workers (9,11,14,17,19,23,25,27,30–32). Such calculations give fair to excellent agreement—depending on the method employed—with the experimentally determined wavelengths, and they agree in predicting that the ion should show two doubly degenerate absorptions in the ultraviolet region: an allowed, intense band near 200 mμ, and a weaker, symmetry-forbidden band in the 270–280 mμ region.

## 4. Charge-Transfer Complexes and Spectra

Observation of the colors of solid tropenylium halides (chloride white, bromide yellow, iodide red) led Kosower (183) and Doering (75) to postulate charge-transfer interactions in these salts, and Kitaigorodskii (172,173) reported that the crystal structure of tropenylium iodide could be interpreted as showing a weak transfer bond between ions. Harmon et al. (97,106,184) have recorded the charge-transfer bands of the halides in methylene chloride (Fig. 2) and demonstrated that they in fact arise from anion-to-cation electron transfer. In tropenylium hydrogen dihalides (169,185) the charge-transfer bands lie at higher energies than in the

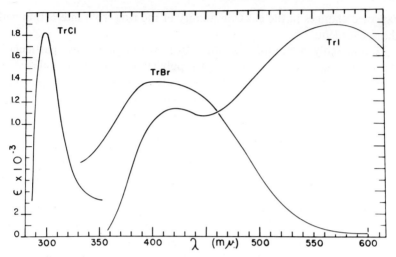

Fig. 2. Charge-transfer absorption bands of tropenylium halides in methylene chloride; cf. Ref. 97.

corresponding halides, since hydrogen bonding to halide increases the ionization potential. Kosower (186) has measured the solvent sensitivity of the tropenylium iodide charge-transfer band; the Z-value (187) plot for this absorption shows a positive slope, as expected for charge-transfer between ions. Harmon and Cummins (188) report the charge-transfer band of solid tropenylium bromide to be 430 mμ, which is a bathochromic shift of 27 mμ compared to the value in methylene chloride solution.

Feldman and Winstein (189) found that tropenylium ion gives 1:1 charge-transfer complexes with a variety of aromatic hydrocarbons; the absorption bands of these complexes were in good agreement with calculated values. The energies of the charge-transfer bands correlate well with the ionization potentials of the donors (190). Since the electron affinity of tropenylium ion is known (191), such spectra can be used to calculate the electron affinities of a variety of cationic donors (192,193). Bhat and Rao (194) have examined the spectral and electrical properties of a variety of solid complexes between tropenylium ion and aromatic donors. These complexes in general exhibit some electrical conductivity but are not good semiconductors; the charge-transfer excitation appears to give localized ion pairs and not delocalized electrons. Dauben and Wilson (195) report the charge-transfer bands of complexes of pyrene with hydroxy-, methoxy-, *t*-butyl-, bromo-, chloro-, cyano-, and unsubstituted tropenylium ions; electron-donating substituents raise the energy of the transition and electron-withdrawing substituents lower it. This reflects the expected changes in the electron affinity of the ring from interaction with the substituent.

Undoubtedly a number of other tropenylium ion salts, particularly those of complex metal halide anions, polyhalides, or pseudo-halides will show charge-transfer spectra. One interesting example is the tropenylium salt of the 1,2,3,4,5-pentacarbomethoxycyclopentadienyl anion (196), which shows a charge-transfer band at 470 nm in methylene chloride.

### 5. Nmr Spectra and Diamagnetic Susceptibility

Tropenylium ion shows a single, sharp $^1$H nmr signal for the seven equivalent protons at 0.70 ppm ($\tau$, TMS) in acetonitrile (100), and 1.32 ppm ($\tau$, TMS) in deuteriosulfuric acid. The significant downfield shift from benzene, which is not accounted for by ring current considerations (197), appears to mainly result from the interaction of the charge in the carbon $p$ orbital with the electrons in the carbon–hydrogen bond (198). This charge effect has been confirmed by examination of $^{13}$C nmr spectra (199). The tropenylium ion does, however, possess a considerable aromatic ring current. Vol'pin and co-workers (167) measured the diamagnetic susceptibility of tropenylium salts and concluded that the exaltation, although appreciable, was less than that of benzene. More recently, Laity (200) has extended the approach of Dauben et al. (201) to include ions and has demonstrated that when the effect of charge is accounted for, the tropenylium ion shows a greater diamagnetic susceptibility exaltation (and presumably a greater ring current) than benzene.

### 6. Conductivity and Polarographic Reduction

Gadecki (202) studied the conductivity of tropenylium bromide in water and acetonitrile, and Lichtin (203) has reported conductivity studies in liquid sulfur dioxide. In water the bromide acts as a typical strong electrolyte, when allowance is made for the presence of a hydrolysis equilibrium, and in acetonitrile a dissociation constant of $2.37 \times 10^{-3}$, may be calculated by the method of Shedlovsky (204). Since the ultraviolet spectrum of tropenylium bromide in acetonitrile shows complete ionization, the association in such solvents must be electrostatic.

Gadecki (202) also carried out preliminary polarographic reduction studies of tropenylium ion in water and demonstrated that a one-electron process was taking place. Harmon (96) and Honnen (86) found $E_{1/2}$ for tropenylium ion in acetonitrile to be $-0.75$ V measured against a mercury pool and showed that the one-electron reduction was irreversible.

Zhdanov and co-workers (165,205,206) investigated the reduction of the ion in unbuffered aqueous solution and found a complex series of waves. In dilute solutions a single principal wave at $-0.3$ V was found, whereas in concentrated solutions a second wave—attributed to adsorption—occurred at $-0.7$ V; several further waves at more negative voltages were observed in both cases. They postulated that the irreversible character of the reduction arose from dimerization of the reduced species and suggested, from a comparison of polarization and electrocapillary curves, that the dimerization reaction occurs in the adsorbed state. In a later paper (207) Hopin and Zhdanov report that the dimerization is between free radicals, and that the lifetime of the radicals is less than 1 msec.

Zuman and co-workers (208–210) have carried out a detailed study of the reduction of tropenylium ion in buffered solutions. Again the ion shows a single, one-electron, irreversible reduction wave at $-0.3$ V in dilute solutions. The height of this wave decreases with increasing pH in the form of a dissociation curve; the limiting current at high pH is controlled by the rate of formation of tropenylium ion from tropenyl alcohol, which itself is reduced at very negative potentials. The half-wave potential was found to be somewhat concentration dependent, and this was interpreted in terms of a dimerization reaction following the reduction. At higher concentrations they found three waves due to adsorption and showed that the adsorbed species had an area of 29 $Å^2$ compared with a calculated value of 22 $Å^2$ for tropenylium ion. Thus it appears that the adsorbed species is tropenylium ion, contrary to the views of the Zhdanov group. Gerovich and Polyanevaskaya (211) have studied the adsorption of tropenylium ion on the mercury-solution boundary and find that the ion adsorbs—presumably by $\pi$-bond interaction—even on a positive mercury surface. This adsorption goes through a minimum at $+0.37$ V and then increases, as a layer of tropenylium ions begins to form on adsorbed iodide ions.

## B. Known Tropenylium Ion Salts

Nearly 50 different salts of the tropenylium ion have been reported. Some of these salts are familiar, widely used materials; others may have been prepared only once for some specific purpose. Table II lists all reasonably well-characterized salts described in the literature reviewed in the preparation of this chapter. Table II has been constructed without editorializing; many—if not most—of the salts have been rigorously characterized, although the proof of identity of others may well require further analytical or instrumental work. The reader interested in a particular species is advised to examine the original reference with judgment.

Table II is intended as a guide to synthetic methods; no attempt has been made to give a complete list of references for the commoner salts. In each case, the reference listed first is my choice of the best method of preparation; obviously, individual preference or expediency may render an alternative method more satisfactory.

## TABLE II
### Known Tropenylium Ion Salts

| Anion name | Anion formula | Melting point, °C | Color | Sensitive[a] | Ref. |
|---|---|---|---|---|---|
| Halides | | | | | |
| Chloride | $Cl^-$ | 121 (dec.) | White | $H_2O$, light | 185,1 |
| Bromide | $Br^-$ | 204 | Yellow | $H_2O$ | 97,75 |
| Iodide | $I^-$ | 127 | Scarlet | $O_2$ | 75,97,161 |
| Polyhalides | | | | | |
| Dichlorobromide | $BrCl_2^-$ | 79–80 | Orange | $H_2O$ | 166 |
| Tetrachloroiodide | $ICl_4^-$ | 195–196 | Yellow | — | 166 |
| Tribromide | $Br_3^-$ | 118 | Yellow | Stable | 97,166 |
| Dibromoiodide | $IBr_2^-$ | 142–143 | Red | — | 166 |
| Bromodiiodide | $I_2Br^-$ | 179–180 | Orange | — | 166 |
| Triiodide | $I_3^-$ | 134.5 | Dark red | $O_2$ | 82,161 |
| Heptaiodide | $I_7^-$ | 69–70 | Grey-black | — | 166 |
| Hydrogen dihalides | | | | | |
| Hydrogen dichloride | $HCl_2^-$ | dec. 100 | White | $H_2O$, light | 185 |
| Hydrogen dibromide | $HBr_2^-$ | dec. | Yellow | $H_2O$ | 185 |
| Oxygen acid anions | | | | | |
| Tetraoxalate | $H_3(C_2O_4)_2^-$ | 140 (dec.) | White | — | 161 |
| Picrate | $C_6H_2N_3O_7^-$ | 120 (dec.) | Yellow | — | 161,150 |
| Pyrophosphate | $P_2O_7^{4-}$ | 125 (dec.) | White | — | 161 |
| Bisulfate | $HSO_4^-$ | 175 (dec.) | White | — | 161 |
| Toluene-p-sulfonate | $C_7H_7SO_3^-$ | 125 (dec.) | Buff | — | 161 |
| Perchlorate | $ClO_4^-$ | 270 (expl.) | White | Caution[b] | 2,96,165 |

Halometallates[c]

| | | | | | |
|---|---|---|---|---|---|
| Fluoroborate | $BF_4^-$ | ca. 210 | White | Stable | 95,97,94 |
| Chloroborate | $BCl_4^-$ | 119–120 | White | $H_2O$ | 108,74 |
| Bromoborate | $BBr_4^-$ | 173 (dec.) | White | $H_2O$ | 106 |
| Iodoborate | $BI_4^-$ | 223 | Yellow | $H_2O$, $O_2$, light | 104 |
| Tetrachloroaluminate | $AlCl_4^-$ | 208 (dec.) | White | $H_2O$ | 74 |
| Hexachlorophosphate | $PCl_6^-$ | — | White | $H_2O$ | 95,84,133 |
| Pentachlorotitanate(IV) | $TiCl_5^-$ | — | White | $H_2O$ | 74 |
| Tetrachloroferrate(III) | $FeCl_4^-$ | 228 | | Stable | 74 |
| Tetrabromopalladiate(II) | $PdBr_4^{2-}$ | 214–216 | Red-brown | — | 168 |
| Tetraiodocadmate(II) | $CdI_4^{2-}$ | 176 | Yellow | Stable | 212 |
| Decaiodotricadmate(II) | $(CdI_4 \cdot \frac{1}{2}CdI_2)^{2-}$ | 175 | Orange | Stable | 212 |
| Hexachlorostannate(IV) | $SnCl_6^{2-}$ | 257 | White | Stable | 102,131 |
| Tetrachlorodibromostannate(IV) | $SnCl_4Br_2^{2-}$ | 246 | Yellow | Stable | 74 |
| Tetrabromodichlorostannate(IV) | $SnBr_4Cl_2^{2-}$ | 242 | Yellow | Stable | 74 |
| Hexabromostannate(IV) | $SnBr_6^{2-}$ | 253 (dec.) | Yellow | Stable | 102,74 |
| Hexachloroantimonate | $SbCl_6^-$ | 190 | Yellow | Stable | 124 |
| Hexachloroplatinate(IV) | $PtCl_6^{2-}$ | 175 (dec.) | Orange | Stable | 165,164,143 |
| Tetrabromoplatinate(II) | $PtBr_4^{2-}$ | 180 (dec.) | Purple | — | 168 |
| Hexabromoplatinate(IV) | $PtBr_6^{2-}$ | 191–194 | Orange | Stable | 143 |
| Tribromomercurate(II) | $HgBr_3^-$ | 150.5 | Yellow | Stable | 96 |
| Tetrabromomercurate(II) | $HgBr_4^{2-}$ | 186 (dec.) | Yellow | Stable | 86 |
| Triiodomercurate(II) | $HgI_3^-$ | 167[d] | Orange | Stable | 82,213 |
| Tetraiodomercurate(II) | $HgI_4^{2-}$ | 175–175.5 | Red-orange | Stable | 167,213 |
| Pentaiododimercurate(II) | $Hg_2I_5^-$ | 151 | Yellow | Stable | 82 |
| Pentabromodiplumbate(II) | $Pb_2Br_5^-$ | — | Yellow | Stable | 214 |

*(continued)*

**TABLE II** (*Continued*)

| Anion name | Anion formula | Melting point, °C | Color | Sensitive[a] | Ref. |
|---|---|---|---|---|---|
| Miscellaneous | | | | | |
| Tetraphenylborate | $(C_6H_5)_4B$ | 120–121.5 | Orange | Stable | 165 |
| 1,2-Bis(difluoroboryl)ethane–methoxide | $CH_2-\bar{B}F_2\overset{+}{\diagdown}O-CH_3$ $CH_2-\bar{B}F_2\diagup$ | — | White | — | 215 |
| 1,2,3,4,5-Pentacarbomethoxycyclopentadienide | $(CCO_2CH_3)_5$ | 157–159 | Red | Stable | 196 |

[a] If blank, not indicated in original reference.
[b] Tropenylium perchlorate is a violent explosive when ignited or subjected to shock or grinding; it is otherwise completely stable to the atmosphere. It can detonate on contact with metal hydrides.
[c] Arranged by increasing atomic number of metal atom.
[d] A yellow, dendroidal polymorph melting at 157° has been reported (96).

## 1. Halides

The tropenylium halides vary considerably in their stability. Tropenylium chloride is extraordinarily hygroscopic and is destroyed by heating or brief exposure to light (185). The bromide is also hygroscopic, but there is a threshold below which the salt will not take up water. The bromide appears to be stable to ordinary light. It can be sublimed readily; however, sublimation of even completely pure bromide always leaves a residue containing several as yet unidentified colored products that are similar to the thermal or light-catalyzed decomposition products of the chloride. The nature of the gas phase species in the sublimation has not been established. Tropenylium iodide does not appear to be hygroscopic but darkens on exposure to the air with an increase in iodine content.

Harmon and Cummings (97,184) have shown that solutions of tropenylium halides in nonaqueous solvents are decomposed by traces of water. The chloride and bromide mainly disproportionate to tropone and cycloheptatriene, and some benzaldehyde also forms in the bromide case. With the iodide, benzaldehyde is the major product; this arises through an oxidative rearrangement of tropenyl alcohol by molecular iodine released by air oxidation of hydrogen iodide (96,97,216). Since tropenylium ion is relatively stable in dilute aqueous solution, and since other salts such as the fluoroborate are almost completely unaffected by exposure to water in nonaqueous solvents, the halide ions must play a part in these decompositions. The nature of this role has not been established.

The colors of the tropenylium halides have been shown to arise from charge-transfer interactions (97,184).

Heat of hydrogenation data for tropenylium bromide and chloride (217) have been interpreted in terms of partial covalency for the latter salt. Reevaluation is necessary in light of the ready disproportionation of the chloride to covalent products in the presence of traces of moisture (97).

## 2. Polyhalides

Tropenylium polyhalides are brightly colored salts. The predominant color is that of the anion; charge-transfer bands presumably occur but are masked by the strong anion absorptions. Tropenylium tribromide was first reported by Dewar and Pettit (150), who assigned a covalent structure to the compound; Dauben and Harmon (96,97,216) later identified it as a salt. The tribromide gives benzaldehyde on basic hydrolysis through oxidative rearrangement of tropenyl alcohol by bromine. Tropenylium iodide complexes readily with varying amounts of iodine (161,166) and it is difficult to prepare stoichiometric triiodide. The best method is that of Hesse (82).

### 3. Hydrogen Dihalides

Tropenylium hydrogen dichloride and hydrogen dibromide are unusually stable salts of these interesting, hydrogen-bonded anions. Equilibrium vapor pressure studies of tropenylium, pyridinium, and alkylammonium hydrogen dibromides have revealed (185) that the tropenylium ion is much more effective at stabilizing this anion than can be accounted for by cation size alone. This stabilization arises only from enthalpy effects; the entropy of formation of the tropenylium complex salt is less favorable than in the other cases. The dissociation vapor pressure of tropenylium hydrogen dichloride cannot be measured because the product chloride is unstable to heat. Anion-to-cation charge transfer occurs in both salts (169,185).

Tropenylium hydrogen diiodide cannot be prepared, since the tropenylium ion is reduced by hydrogen iodide (82,218).

### 4. Oxygen Acid Anions

The perchlorate is the only commonly encountered salt of those listed in this section of Table II. Tropenylium perchlorate is a colorless, nonhygroscopic, air-stable salt that is easily prepared in high purity (2). The perchlorate must be handled with extreme caution. I have seen the detonation of 0.2 mg shatter a Vicor combustion train, and accounts of serious explosions of larger quantities are in the literature (219). The perchlorate is easily ignited and has been reported to be more shock sensitive than lead azide (220).

Honnen (86) suffered a series of explosions on attempted reduction of tropenylium perchlorate with lithium aluminum hydride; the initial explosion occurred when ether solvent evaporated and allowed the reactants to dry on the walls of the flask. The reduction of methyltropenylium perchlorate with lithium aluminum hydride reported by Conrow (221) must be assumed to be unsafe.

### 5. Halometallates

A large number of tropenylium ion salts of complex metal halide anions have been prepared. Only a few representative species are discussed.

Tropenylium fluoroborate is the most widely used salt of this cation. The fluoroborate is easily prepared in quantity (95,97,94) and keeps indefinitely without precautions. It is sparingly soluble in most solvents, which makes the bromide a preferred reactant if solution volume is a problem. Tropenylium fluoroborate is not affected by ordinary light; however, van Tamelen et al. (222) demonstrated that irradiation with a high-pressure mercury lamp converts the salt to a variety of bicyclic products. The anion is presumably not involved in the photochemical reaction, and the observations should apply to other tropenylium salts.

A variety of anucleophilic anions form extremely stable, nonhygroscopic salts with the tropenylium ion. Examples of these are hexachlorostannate(IV), hexabromostannate(IV), hexachloroplatinate(IV), and hexachloroantimonate ions. Addition of a concentrated solution of stannic chloride pentahydrate in $12N$ hydrochloric acid to hydrochloric acid solutions of tropenylium species affords nearly quantitative precipitation of tropenylium hexachlorostannate(IV); similarly, hexachloroplatinic acid can be used to quantitatively precipitate the ion from aqueous solution. A less expensive reagent, lead bromide and potassium bromide, has been reported (214) to precipitate tropenylium ion quantitatively from water as the pentabromodiplumbate salt.

Tropenylium iodide forms a series of complex iodide salts with group 2b metal iodides that exhibit interesting polymorphism. For several years our preparations (86,97) of tropenylium triiodomercurate(II) gave yellow dendroidal crystals that melted at 157°; however, after Honnen (86) and Harmon (82) had each inadvertently prepared a higher-melting form, neither was able to again prepare the low-melting species. Harmon and Hesse (82,212) have examined a series of complex iodide salts of mercury, cadmium, and zinc; all exist in at least two crystal forms with identical analyses and melting points.

### 6. Miscellaneous

Among the miscellaneous salts, the tetraphenylborate is particularly useful; its quantitative precipitation from water can be used for estimation of tropenylium ion.

## C. Reactions of Tropenylium Ion

The manifold reactions of the tropenylium ion are arbitrarily divided here into (1) reduction by hydride, (2) reaction with electrophilic reagents, (3) reaction with nucleophilic reagents, (4) reaction as an electrophile, (5) alkylation of carbonyl compounds, (6) reduction to tropenyl radical and coupling reactions, and (7) miscellaneous reactions. When the ion reacts with a nucleophilic reagent it is, of course, functioning as an electrophile; the division between (3) and (4) is based on a subjective opinion of what is substrate and what is attacking reagent.

### 1. Reduction by Hydride

Tropenylium ion is smoothly reduced to cycloheptatriene by metal hydrides such as lithium aluminum hydride or sodium borohydride (223); such reductions should not be carried out on perchlorate salts because explosions can occur (Section III-B-4). Other nonmetal hydrides such as trialkylsilanes (223,224) convert the ion to cycloheptatriene. Excess anhydrous hydrogen iodide reduces tropenylium ion (82,218) in what is presumed to be a hydridic process (83).

Reversible hydride transfer between cycloheptatriene-$d_8$ and tropenylium ion has been demonstrated by isotopic scrambling (225). Tropenylium ion abstracts hydride readily from cycloheptatrienes bearing electron-donating substituents, which indicates that such substituents stabilize the cationic

$$\text{C}_7\text{H}_7\text{–Y} + \text{C}_7\text{H}_7^+ \longrightarrow \text{C}_7\text{H}_6^+\text{–Y} + \text{C}_7\text{H}_8$$

aromatic system. Conrow (226) has shown that passage of 7-methylcycloheptatriene vapor through a tropenylium fluoroborate–nitromethylpimelonitrile column yields only cycloheptatriene in the effluent gas; he has also performed equilibrium studies (221) of hydride transfer in solution to prove that the methyltropenylium ion is about 3.7 kcal/mole more stable than the unsubstituted ion. When the substituent is capable of $+T$ electron donation to the ring, this effect is enhanced; cycloheptatrienes substituted with amine groups (3,80) or anionic borane cages (227–229) give rapid and quantitative hydride transfer to tropenylium ion to yield the corresponding substituted tropenylium ion salts.

The most extensively studied reactions of this type are those of tropenyl alcohol and tropenyl ethers. Tropenylium ion abstracts hydride from tropenyl alcohol, which accounts for the presence of cycloheptatriene and tropone (or hydroxytropenylium ion) as hydrolytic decomposition products of a number of tropenylium salts (97,184). Catalytic amounts of tropenylium ion initiate the disproportionation of ditropenyl ether to tropone and cycloheptatriene; this reaction can be utilized as a source of

$$\text{(C}_7\text{H}_7\text{)–O–(C}_7\text{H}_7\text{)} \xrightarrow{\text{C}_7\text{H}_7^+} \text{C}_7\text{H}_6\text{=O} + \text{C}_7\text{H}_8$$

tropone (133,230–232). Primary and tertiary alkyl tropenyl ethers donate hydride from the cycloheptatrienyl ring to tropenylium ion to form tropone, cycloheptatriene, and products derived from the corresponding alkyl carbonium ion (233).* With secondary alkyl or benzyl tropenyl ethers, hydride abstraction from the methyne or benzyl carbon of the

---

* Solutions of tropenyl methyl ether stored over molecular sieve become contaminated with tropone via this acid-catalyzed disproportionation; magnesium sulfate is preferred for drying this reagent.

substituent group competes with abstraction from the ring; benzaldehyde or a ketone and tropenylium ion are formed. This is in accord with the observation that tropenylium ion will abstract hydride from diisopropyl ether (234,235) but not from bis-primary alkyl ethers.

Grishin and Yasnikov (236) have examined the kinetics of the reduction of tropenylium ion by $N$-substituted 1,4-dihydronicotinamides. The ability of the ion to oxidize this model compound suggests that it might show

$$\underset{R}{\text{[dihydronicotinamide]}} + C_7H_7{}^+ \longrightarrow \underset{R}{\text{[nicotinamide}^+\text{]}} + C_7H_8$$

similar activity toward the reduced form of nicotinamide adenine dinucleotide (NADH) in biological systems.

## 2. Reaction with Electrophilic Reagents

As expected, the tropenylium ion is nearly inert to electrophilic reagents; e.g., Vol'pin et al. (237) found less than 1% incorporation of deuterium after treatment of tropenylium salts with deuterium bromide and aluminum chloride for extended periods, and we find that the nmr spectra of tropenylium and substituted tropenylium ions show no change on standing in 100% $D_2SO_4$. Pearson (67) found that reaction with concentrated nitric acid at 100° for 24 hr destroyed the ion, but a substituted tropenylium ion was not formed. Harmon (96) showed that tropenylium ion could be recovered unchanged after reaction with fuming sulfuric acid at 100° for seven days or with bromine in concentrated sulfuric acid at 210° for five days. Reaction with fuming sulfuric acid at 210° for six days converted the ion into a new species with an ultraviolet spectrum similar to that of a tropenylium ion substituted (100) with an electron-withdrawing group. Further studies have not been carried out.

## 3. Reaction with Nucleophilic Reagents

The tropenylium ion, although remarkably stable in comparison with nonaromatic hydrocarbon ions, is still susceptible to attack by a variety of nucleophilic reagents. Vol'pin et al. conclude that anions will form ionic salts with tropenylium ion when the conjugate acid of the anion has $K_a$ greater than $1.2 \times 10^{-4}$ and will form covalent cycloheptatriene derivatives when the corresponding $K_a$ is less than $1.75 \times 10^{-5}$ (238,239). The ultraviolet absorption of the tropenylium ion in 96% sulfuric acid (2,97) or $0.1F$ hydrochloric acid (1) is identical; however, in water at pH 7 the absorption is somewhat reduced, and at spectral concentrations (about

$10^{-4}$ mole/liter) in ethanol the cation spectrum is replaced by that of a covalent 7-substituted cyclophetatriene at 256 m$\mu$ (67). Doering and Knox (1,75) have shown that tropenylium ion can be smoothly titrated with dilute hydroxide ion to a covalent alcohol and that the process is reversible. They calculate that the tropenylium ion has a $K_a$ of $1.8 \times 10^{-5}$.

$$C_7H_7{}^+ \underset{H^+}{\overset{OH^-}{\rightleftharpoons}} C_7H_7OH$$

Tropenyl alcohol has not been isolated; in basic solutions the insoluble ditropenyl ether is formed (1,131,133,135). Similarly, 7-amino- or 7-sulfhydrylcycloheptatrienes cannot be prepared; reaction of tropenylium ion with ammonia or hydrogen sulfide yields di- or tritropenylamine and ditropenyl sulfide (75). Alkyl amines (3,75,80,240,241) and mercaptans (242) react normally to give tropenylalkylamines, ditropenylalkylamines, tropenyldialkylamines, tropenyltrialkylammonium ions, and tropenylalkyl sulfides. Ditropenyl selenide has been prepared in a manner similar to the sulfide (243).

Derivatives of 7-aminocycloheptatriene can be prepared by the reaction of tropenylium ion with amides in aqueous or pyridine solution (75). The

$$C_7H_7{}^+ + C_6H_5CONH_2 \longrightarrow C_7H_7NHCOC_6H_5 + H^+$$

tropenylamides are easily cleaved by acid to regenerate tropenylium ion (66,67) and, since they are easily crystallizable, stable compounds, they should be a convenient and easily stored source of the ion.

Strong nucleophiles such as alkoxide (1),* cyanide (75,119,244), azide (214), cyclopentadienide (202,244), and isocyanate (245) anions readily attack tropenylium ion to form the corresponding 7-substituted cycloheptatrienes. Tropenyl alkyl ethers are readily cleaved by acid to give tropenylium salts (Section II-B), and 7-cyanocycloheptatriene is a useful intermediate in the synthesis of cyanotropenylium ion (119) and the Büchner acids (75). Tropenyl azide adds to substituted alkynes to give triazoles (246). A compound previously reported to be an ionic tropenylium isocyanate (149) has been shown to be ditropenylurea (245).

Grignard and lithium reagents react with tropenylium ion to give 7-alkyl- and 7-arylcycloheptatrienes (69,241,244); however, the homogeneous reaction of tropenyl methyl ether with Grignard reagents (61,109,

C₇H₇—OCH₃ —RMgX→ C₇H₇—R

* See footnote, page 1604.

116,247) is the preferred route to such compounds. Murray and Kaplan (248,249) have prepared 1,4-bis- and 1,3,5-tris-tropenylbenzene by reaction of tropenylium bromide with the Grignard or lithium reagents derived from the corresponding bromobenzenes, and Harmon et al. have prepared a variety of tropenylcarboranes by reaction of tropenyl methyl ether with mono- and dilithiocarboranes (100,250). Triethylaluminum reacts with tropenylium ion to give 7-ethylcycloheptatriene (251).

### 4. Reaction as an Electrophile

Tropenylium ion attacks vinyl ethers or esters in aqueous solution to yield 2-tropenylaldehydes (252–255); the reaction is believed to proceed by electrophilic attack of tropenylium ion on the double bond, followed by

$$C_7H_7^+ + CH_2=CHOCH_3 \longrightarrow C_7H_7CH_2CHO$$

hydrolysis of the resulting carbonium ion (255). A hemiacetal intermediate has been postulated (253) but not isolated. Tropenylium ion also adds to other activated olefins such as cyclopentadiene (252) or tetramethoxyethylene (256).

Tropenylium ion does not act as an electrophile toward benzene and apparently attacks anisole only to a limited extent (257); however, the ion readily effects substitution for hydrogen on $N,N$-dialkylanilines (258),

$$C_7H_7^+ + \text{[aryl-OH]} \longrightarrow C_7H_7-\text{[aryl-OH]} + H^+$$

phenols (252,259–262), or phenoxide ions (259). A variety of nonbenzenoid aromatic systems such as tropolone (263), azulene (264–266), and polyhedral borane anions (227–229) are also attacked by tropenylium ion in facile aromatic electrophilic substitution reactions.

A discussion of the products of these reactions is deferred to Section IV-C-6.

### 5. Alkylation of Carbonyl and Activated Methylene Compounds

Tropenylium ion reacts with simple aldehydes (252,267) and ketones (252,268) in aqueous or alcohol solution to yield α-tropenyl carbonyl

$$C_7H_7^+ + CH_3COCH_3 \longrightarrow C_7H_7CH_2COCH_3 + H^+$$

compounds. The rate of the reaction with isobutyraldehyde is inversely proportional to hydrogen ion concentration (269), which suggests that enolate ions are involved. The α-tropenyl carbonyl compounds are easily cleaved by acid (270). Tropenylium ion does not normally react with

$$2C_7H_7^+ + CH_3CN + (C_2H_5)_3N \longrightarrow (C_7H_7)_2CH_2CH=N(C_2H_5)_2^+$$

acetonitrile; when trialkylamines are present, however, an unexpected alkylation reaction leads to the formation of dialkylimmonium salts (241). Nitroalkanes are also alkylated by tropenylium ion (270).

A considerable amount of work has been done on the tropenylation of compounds with a doubly activated methylene group. Such reactions can be carried out under strongly basic, mildly basic, mildly acidic, or strongly acidic conditions, with a concurrent change in mechanism from nucleophilic attack by enolate anion on tropenylium ion to electrophilic addition of tropenylium ion to the enol form of the methylene compound. Kinetic studies of the reaction of tropenylium ion with malonic acid or acetoacetic ester in dilute aqueous acid (269) reveal that the rate is independent of tropenylium ion, first order with respect to dicarbonyl compound, and directly proportional to hydrogen ion concentration. Table III summarizes the alkylation reactions that have been reported.

## TABLE III
### Reactions of Tropenylium Ion with Activated Methylenes

| Methylene, $X-CH_2-Y$ | | Compound[a] | Methods[b] | Ref. |
|---|---|---|---|---|
| X | Y | | | |
| $-COCH_3$ | $-COCH_3$ | M | B, D | 270,271[c] |
| $-COCH_3$ | $-CO_2C_2H_5$ | M | B, C | 268–270 |
| $-CO_2H$ | $-CO_2H$ | M | B, C | 268–270 |
| $-CO_2CH_3$ | $-CO_2CH_3$ | M, D | A | 273 |
| $-CN$ | $-CO_2H$ | M | B | 270 |
| $-CN$ | $-CO_2CH_3$ | M, D | A | 274 |
| $-CN$ | $-CO_2C_2H_5$ | M, D | A, C | 268,274,275 |
| $-CN$ | $-CONH_2$ | M, D | B, C | 275 |
| $-CN$ | $-CN$ | M, D | B | 275 |
| $-NO_2$ | $-CO_2C_2H_5$ | M | C | 268 |
| Miscellaneous | | | | |
| 2-Carboethoxycyclopentanone | | M | B | 276 |
| Phthalimide | | M | A | 276 |
| Succinimide | | M | A | 67 |
| Indanone | | M | D | 272 |
| 1,3-Indanedione | | M, D | B | 276 |

[a] Mono- (M) or disubstitution (D) by tropenyl group(s).

[b] A, Reaction with preformed enolate ion, usually generated with sodium or sodium hydride in ethanol or tetrahydrofuran; B, reaction in pyridine solvent; C, reaction in slightly acid aqueous or alcohol solution; D, reaction in acetonitrile–water–fluoroboric acid solution.

[c] This reference describes the preparation of 3-tropenyl-2,4-pentane dione in quantitative yield, via the method developed by Bertelli (272) for indanone.

The mono- and ditropenyl substituted active methylene compounds are stable, colorless oils or crystalline solids which may be readily characterized by analysis and nmr spectra. Decarboxylation of acids leads to a wide variety of monofunctional tropenyl compounds of considerable synthetic interest (270). The most characteristic reaction of tropenyl-substituted active methylene compounds is facile cleavage by a wide variety of acidic reagents to regenerate the methylene starting material and tropenylium ion (270,275,277).

Tropenylium ion reacts with aluminum acetylacetonate to give 3-tropenyl-2,4-pentanedione and aluminum ion (278).

### 6. Reduction to Tropenyl Radical; Coupling Reactions

Electrochemical reduction of tropenylium ion gives high yields of ditropenyl (93), which is presumably formed by coupling of the known

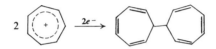

(279) tropenyl radical. A number of chemical reducing agents also give reductive coupling with tropenylium ion; among these are zinc (75), transition metal complexes (164,280,281), and the cyclooctatetraene dianion (282). Tropenylium bromide has been reduced to the radical itself by reaction with sodium-potassium alloy in dimethoxyethane solution at −80° (283).

Ledwith and co-workers (284) have shown that tropenylium ion is an effective initiator for the cationic polymerization of vinyl ethers or vinyl-substituted aromatic or heterocyclic compounds. They rule out addition of tropenylium ion to the double bond as the initiation event and propose an initial charge-transfer complex formation followed by electron transfer to yield the cation radical. Rapid termination of the radical site allows a

$$C_7H_7^+ + CH_2=CH-OR \rightleftharpoons \text{complex} \longrightarrow C_7H_7\cdot + \cdot CH_2CH=\overset{+}{O}R$$

normal cationic polymerization to proceed. More recently (285) these workers have examined the charge-transfer processes involved in the polymerizations in detail, and have also shown that tropenylium ion can oxidize $N$-methylcarbazole and 1,4-bis-dimethylaminobenzene to stable free radicals. Apparently no attempt was made to isolate ditropenyl from these reactions.

### 7. Miscellaneous Reactions

Tropenylium ion has been reported to react with dimethyl sulfoxide to give reasonable yields of tropone and dimethyl sulfide (286). We have found that a variety of tropenylium salts react with this solvent, which makes it unsuitable for crystallizations.

Vol'pin et al. (287) have demonstrated the mutual interconversion of the tropenylium and benzene aromatic systems by converting the ion to benzene by reaction with hydrogen peroxide and benzene to the ion by treatment with monohalocarbene.

Beechey and Knight (288) have shown that tropenylium ion reversibly inhibits respiration in rat heart mitochondria by interference with the energy transfer systems associated with the electron transport system.

## IV. SUBSTITUTED TROPENYLIUM IONS

About 150 mono- and polysubstituted monocyclic tropenylium ions have been reported, and for each of these ions a number of different methods of preparation and a variety of counterions may exist. This chapter can only give a broad overlook at the general preparations and properties of substituted tropenylium ion salts; the reader seeking specific information on a given compound should turn to the original literature.

Table IV lists the substituted tropenylium ions reported to date; in some cases very similar compounds are grouped under a single entry.

### A. Preparations of Substituted Tropenylium Ions

#### 1. Substituted Cycloheptatrienes

Substituted cycloheptatriene starting materials for the preparation of substituted salts are generally prepared in one of four ways: (*a*) ring enlargement of the corresponding substituted benzene with diazomethane, (*b*) attack by a nucleophile on tropenylium ion to yield a 7-substituted cycloheptatriene, (*c*) thermal rearrangement of a 7-substituted cycloheptatriene [prepared as in (*b*)] to an *x*-substituted cycloheptatriene (*x*- designates either the 1-, 2-, or 3-substituted isomer, or a mixture of several of these isomers), and (*d*) electrophilic attack by tropenylium ion on an aromatic or olefinic substrate. Miscellaneous methods are also available.

The photochemical ring enlargement of benzenes with diazomethane was used by Doering and Knox (69,300) and Dauben and Harmon (2,96) for the synthesis of a variety of substituted cycloheptatrienes; however, this method suffers from low yield, difficult separation of products, and a finite fire hazard. The cuprous chloride catalyzed reaction of diazomethane with substituted benzenes developed by Muller and co-workers (110,301, 302) is much easier to carry out, and excellent yields have been obtained with a wide variety of substituted benzenes.

## TABLE IV
### Substituted Tropenylium Ions

| Substituent | Starting Material[a] | Preparation[b] | Spectrum[c] | Anion notes[d] | Ref. |
|---|---|---|---|---|---|
| **Alkyl and arylalkyl** | | | | | |
| Methyl | A, B | A, C | I | not $BF_4^-$, $X^-$ | 2,84,85,96,109,110,117 |
| 1,2-Dimethyl | A | A | I | not $X^-$ | 110 |
| 1,3,5-Trimethyl | E | A | ? | not $X^-$ | 79 |
| Ethyl | B | A | I | not $X^-$ | 109,117 |
| n-Dodecyl | B | A | ? | not $X^-$ | 289 |
| Polyethylenyl | B–E | A | ? | not $X^-$ | 120 |
| Allyl | B | A | III | not $BF_4^-$ | 116 |
| Benzyl | B | A | ? | $SbCl_6^-$ best | 109 |
| i-Propyl | B | A | I | not $X^-$ | 109 |
| Benzhydryl | B, C | A | ? | not $X^-$ | 109 |
| t-Butyl | A, C | A, B | I | all | 69,109,117 |
| **Alkenyl and arylalkenyl** | | | | | |
| Vinyl | B | A | III | — | 289 |
| Styryl and substituted styryl | B, E[e] | A, D | III | — | 113 (24 ions) |
| 2-Arylvinyl | D | A | III | — | 114 (7 ions) |
| Aryl and diaryl conjugated polyenes | B, D | A | III | — | 115 (16 ions) |
| 2-Indenyl | D | A | III | — | 272 |
| t-Butylethynyl | B | A | III | — | 117 |
| Phenylethynyl | B | A | III | — | 113 |
| p-Bromophenylethynyl | B | A | III | — | 113 |

(*continued*)

TABLE IV (*Continued*)

| Substituent | Starting Material[a] | Preparation[b] | Spectrum[c] | Anion notes[d] | Ref. |
|---|---|---|---|---|---|
| Aryl | | | | | |
| Phenyl | A–C | A, B | III | all | 69,79,109,111,112,117 |
| 1,2-Diphenyl | B | A | ? | — | 79 |
| 1,3-Diphenyl | B | A | III | — | 79 |
| 1,4-Diphenyl | B, D | A | III | — | 79,113 |
| 1,2,3,4-Tetraphenyl | E | A | III | all | 290 |
| Hexaphenyl | E | F | III | all | 290 |
| Heptaphenyl | E | B | ? | all | 291 |
| 3-Tropenyliumylphenyl | B | A | III | all | 249 |
| 4-Tropenyliumylphenyl | B | A | III | all | 248 |
| 3,5-bis-Tropenyliumylphenyl | B | A | III | all | 249 |
| Monosubstituted phenyl[f] | B–E | A, D | III | — | 112,258,261,262,292, 293 (28 ions) |
| Alkylhydroxyphenyl | D | A, D | III | — | 231,261 (3 ions) |
| Dimethoxyphenyl | D | B, C | III | — | 262 (4 ions) |
| 4-Dimethylamino-1-naphthyl | D | A | III | — | 258 |
| 2-Naphthyl | B | A | III | — | 112 |
| 1-Methyl-2-phenyl | E | A | III | — | 292 |
| 1-Methyl-3-phenyl | E | A | III | — | 292 |
| Carbonyl and acid derivatives | | | | | |
| Carboxy | E | A, B | I | all | 77,78,96,117 |
| Carboethoxy | E | A | I | — | 117 |
| Benzoyl | E | A | I | — | 117 |
| Cyano | C | A | I | — | 117,119 |
| (2-*t*-Butylperacetyl) | D | A | I | — | 294 |

| | | | | | |
|---|---|---|---|---|---|
| Halogen | | | | | |
| Fluoro | A | A | I | not X⁻ | 96,117 |
| Chloro | A | A | II | not X⁻ | 2,96,110,117,157 |
| Bromo | A | A–C, E | II–III | not X⁻ | 2,69,85,96,117,139 |
| Iodo | E | E[g] | III | I⁻ | 96 |
| Heptachloro | E | F | ? | $MCl_n^-$ | 295 |
| Hydroxy and alkoxy | | | | | |
| Hydroxy | E | E, F[h] | II | need H-bond to anion | 133 and Refs. therein |
| 1 2-Dihydroxy | E | F[i] | II | X⁻ good | 91 |
| Methoxy | A, C | A–C | II | not X⁻ | 2,96,118,140 |
| 1,2-Dimethoxy | C, E | C, F | II | — | 140 |
| 1,4-Dimethoxy | A | A | II | — | 117 |
| 1,3,5-Trimethoxy | A | A | ? | — | 117 |
| Ethoxy | C, E | C, F | II | — | 296 |
| Sulfur | | | | | |
| Thiomethoxy | A, C | A | III | not X⁻ ? | 96,117 |
| Amino and alkylamino | | | | | |
| Amino | B | F[j] | II | all | 3,80 |
| Methylamino | B | A,[k] F[j] | II | all | 3,80,297 |
| n-Propylamino | B | A[k] | II | all | 297 |
| Tropenylamino | B | A[k] | II | — | 3,80 |
| Dimethylamino | B | A[k] | II | all | 3,80 |
| Diethylamino | B | A[k] | II | all | 297 |
| Piperdino | B | B | II | — | 240 |
| Polyhedral borane ions and carboranes (named as ousene-type compounds (228)) | | | | | |
| [7.10²]Hemiousenide ion | D | D | III | Cs⁺, R₄N⁺ | 227,229 |
| [7.12]Hemiousenide ion | D | D | III | Cs⁺, R₄N⁺ | 228,229 |
| [7.7.10²·⁷]Ousene | D | D | III | neutral | 298 |

*(continued)*

TABLE IV (*Continued*)

| Substituent | Starting Material[a] | Preparation[b] | Spectrum[c] | Anion notes[d] | Ref. |
|---|---|---|---|---|---|
| [7.12$^1$]-1,2-Dicarbahemiousenium ion | C | A | I | X$^-$ covalent | 250 |
| [7.12$^1$]-1,7-Dicarbahemiousenium ion | C | A | I | X$^-$ covalent | 250 |
| 1-Methyl[7.12$^2$]-1,2-dicarba-hemiousenium ion | C | A | I | X$^-$ covalent | 100 |
| [7.7.12$^{1,2}$]-1,2-Dicarbaousenium ion | C | A | I | SnCl$_6^{2-}$ best | 250 |
| Cationic | | | | | |
| Tropenyliumyl | E[l] | C | II | PCl$_6^-$ best | 299 |

[a] Preparations of cycloheptatriene starting materials include (A) ring enlargement of substituted benzene with diazomethane, (B) attack of nucleophile on tropenylium ion to form a 7-substituted cycloheptatriene, (C) process B followed by thermal rearrangement to x-substituted cycloheptatriene, (D) electrophilic attack of tropenylium ion on aromatic or olefinic substrate, and (E) miscellaneous (see Section IV-A-1).

[b] Preparations of substituted tropenylium ions include (A) hydride transfer to triphenylcarbonium ion, (B) bromination–dehydrobromination, (C) hydride transfer to phosphorus pentachloride, (D) electrophilic substitution by tropenylium ion followed by *in situ* transfer of hydride to tropenylium ion, (E) nucleophilic exchange of substituent, and (F) miscellaneous (see Section IV-A-2).

[c] See Section IV-B-4.

[d] If not otherwise noted fluoroborate, hexahaloantimonate, or perchlorate (*caution, explosive*) salts satisfactory. "All" means halide as well as anucleophilic anions.

[e] Condensation of aldehydes with methyltropenylium ion.

[f] Substituents on phenyl include 4-*t*-butyl, 4-phenyl, 3-phenyl, 4-fluoro, 3-fluoro, 3-chloro, 3-bromo, 3-trifluoromethyl, 4-benzoyl, 4-thiomethoxy, and 4-benzamido (112), 4-methyl, 3-methyl, 4-chloro, and 4-bromo (112,292), 2-chloro, 2-bromo, 4-nitro, 2-methyl, 4-carbomethoxy, and 2-carbomethoxy (292), 4-hydroxy (112,261,293), 4-methoxy and 3-methoxy (112,262,292), 2-methoxy (262), and 4-dimethylamino (112,258).

[g] Reaction of chlorotropenylium ion with iodide ion in acetonitrile.

[h] Prepared by action of water on halotropenylium ion (E) or protonation of tropone (F).

[i] Protonation of tropolone.

[j] Solvolytic detropenylation of corresponding ditropenylamine.

[k] Hydride transfer to tropenylium ion.

[l] Ditropenyl (1).

Ring enlargement remains the method of choice for halocycloheptatrienes and for many di-, tri-, or polysubstituted cycloheptatrienes; however, most other substituted cycloheptatrienes can be prepared more easily by other methods.

The reaction of tropenylium ion with nucleophiles yields 7-substituted cycloheptatrienes. Many of these, such as most 7-alkyl or 7-arylcycloheptatrienes obtained by the reaction of tropenylium ion or tropenyl methyl ether with Grignard or lithium reagents (69,109,247), can be converted directly to substituted ions; however, when bond energy, steric, or inductive factors interfere (see Section IV-A-2) the 7-substituted cycloheptatrienes must be rearranged to $x$-substituted species.

The intramolecular rearrangement of 7-substituted cycloheptatrienes to $x$-substituted cycloheptatrienes has been investigated in detail (70,100,109, 119,303) for a number of substituents, including alkyl, aryl, carboranyl, cyano, alkoxyl, and deuterium. The cycloheptatriene is normally heated neat at 100–170° for 45–200 min; specific conditions for optimum yield vary from compound to compound, and the isomer(s) produced are also a function of substituent. For example, 7-carboranylcycloheptatrienes give quantitative yields of the 3-isomers when heated at 165° for 45 min (100,250), and 7-dimethylaminocycloheptatriene gives exclusively the 1-isomer under similar conditions (3,80). Any of the $x$-isomers or mixtures thereof are satisfactory for salt preparation in cases where the 7-isomers cannot be used.

Reactive aromatic and olefinic molecules such as phenols (260,261), dimethoxybenzenes (262), indene (272), and *closo*-borane anions (227–229) undergo electrophilic substitution by tropenylium ion to yield 7-substituted cycloheptatrienes. In many cases the cycloheptatriene is not isolated; instead, rapid hydride abstraction by a second tropenylium ion occurs to yield cycloheptatriene and the stabilized substituted tropenylium ion salt.

$$2C_7H_7^+ + B_{10}H_{10}^{2-} \longrightarrow C_7H_6B_{10}H_9^- + C_7H_8$$

A variety of miscellaneous methods have been used for individual starting materials. Cycloheptatriene carboxylic acids and their derivatives have been prepared by the photochemical ring enlargement of benzene (304) and substituted benzenes (79) with ethyldiazoacetate; however, a preferred method (117) of preparation of the acids is acid-catalyzed hydrolysis of $x$-cyanocycloheptatriene (119). $x$-Carboxycycloheptatrienes can be converted by conventional means to a variety of derivatives.

## 2. Substituted Tropenylium Ions

Substituted tropenylium ions can be prepared by most methods applicable to the unsubstituted ion, including (a) hydride transfer to triphenylcarbonium ion, (b) bromination–dehydrobromination, and (c) hydride transfer to phosphorus pentachloride (see Sections II-A-1,3,4). In addition, (d) electrophilic attack by tropenylium ion followed by *in situ* hydride transfer to tropenylium ion and (e) nucleophilic interchange of substituents offer convenient routes to certain substituted ions.

Hydride abstraction by triphenylcarbonium ion is by far the most widely used method for the synthesis of substituted tropenylium ions; the reaction has been discussed in detail (Section II-A-3) for the unsubstituted case, and the general directions also apply to most substituted ions.

The hexafluoro- and hexachloroantimonate anions appear most generally satisfactory for the preparation of salts of the substituted cations. Fluoroborates are generally stable; however, some substituted tropenylium ion fluoroborates are oils at room temperature, and many are too soluble in organic solvents to be easily recrystallized. The perchlorates are normally highly crystalline, but present an explosion hazard, and halide ions are reactive with many species. The hexahalometallate salts are usually more crystalline, less hygroscopic, and have longer shelf life than other salts. In some cases the hexafluoroarsenate ion is superior to the antimony analog, since it gives somewhat more soluble salts.

Certain limitations must be considered in applying the Dauben hydride transfer reaction to the synthesis of substituted tropenylium ions. A labile substituent in the 7-position on cycloheptatriene, such as —OCH$_3$, will be abstracted preferentially instead of hydride by triphenylcarbonium ion (2,96), and bulky substituents such as *t*-butyl (109) or electron-withdrawing substituents such as cyano (119) or carboranyl (100) slow the reaction to an impractical rate and maximize unwanted side reactions. In such cases thermal rearrangement to an *x*-substituted cycloheptatriene usually provides a satisfactory starting material.

The bromination–dehydrobromination reaction originally used by Doering and Knox (1) for the synthesis of tropenylium bromide has been used to prepare a variety of stable tropenylium ions with substituents such as carboxyl (77,78), *t*-butyl (69), bromo (69), and phenyl (69). The method is limited by the ability of the cycloheptatriene to withstand the reaction conditions and by the stability of the resulting bromide salts.

Hydride abstraction by phosphorus pentachloride (see Section II-A-4) is a good general method for forming tropenylium ions if the substituted cycloheptatriene is stable to the reagent otherwise. Since the hygroscopic PCl$_6{}^-$ anion is generally not satisfactory, and since conversion of the hexachlorophosphate salts to other salts involves solution in water or alcohol, the method is limited to the synthesis of hydrolytically stable substituted tropenium ions.

## B. Physical Properties of Substituted Tropenylium Ions

### 1. Structure

Very little X-ray crystallographic work has been done on substituted tropenylium ions, in spite of the excellent crystallinity and stability of many of the salts. In 1,2-dihydroxytropenylium ion (174) the ring is a slightly irregular heptagon (see Section III-A-1), and in diazulenium azulene perchlorate (305)—which contains what are essentially two 1,2-dialkyltropenylium ions—the cationic rings are planar heptagons with an average bond length of 1.382 Å and an average bond angle of 128.4° (306). In the [7.10²]hemiousenide ion [2-tropenyliumyl-*closo*-nonahydrodecaborate anion (229)] the average bond length in the seven-membered ring is 1.399 Å and the average bond angle is 128.6° (306).

Bertelli (307) has deduced from a detailed analysis of the nmr coupling constants of the ring hydrogens in hydroxy-, methoxy-, and indenyltropenylium ions that electron-donating substituents perturb the symmetry of the seven-membered ring and introduce a degree of bond alternation.

### 2. Acidities

The acidity, i.e., the propensity of the ion to react with the nucleophile water, has been measured for a number of substituted tropenylium ions,

$$\text{Tr}^+\text{-R} + H_2O \rightleftharpoons \text{Tr(R)-OH} + H^+$$

including carboxy- (77,78), phenyl- and substituted phenyl- (112), and a variety of arylalkenyltropenylium ions (113–115). In general, electron-donating substituents lower the acidity and electron-withdrawing substituents increase the acidity. Jutz (112) found that the acidities of substituted phenyltropenylium ions correlate reasonably well with the Hammett σ values of the substituents.

At first consideration it might appear that the acidities of a series of substituted ions would reflect the ability of substituents to "stabilize" the tropenylium ion, i.e., a more "stable" ion would react to a lesser extent with a given nucleophile. However, since the reaction with water is under thermodynamic control, the influence of the substituent on both the substituted cycloheptatriene and on the corresponding ion must be considered. For example, the p$K_a$ of phenyltropenylium is 3.88 compared to a value of 4.02 for tropenylium (112), which suggests that the phenyl group acts as an electron-withdrawing group and destabilizes the ion. However, Cairncross (79) has shown by equilibration studies that the phenyl group actually stabilizes the tropenylium nucleus and that the observed acidity effect occurs because the phenyl group stabilizes the starting cycloheptatriene more than the product ion.

Parnes has given an excellent review of the acidities of tropenylium ions (70).

### 3. Infrared Spectra

Relatively few infrared spectra of substituted tropenylium ions have been reproduced in the literature, and little detailed analysis has been made of such spectra.

The infrared spectra of substituted tropenylium ions are similar to those of substituted benzenes, as would be expected (178–180). The carbon–hydrogen stretching bands of the ring hydrogens in the 3000-cm$^{-1}$ region are usually weak, and may be difficult to observe. The carbon–hydrogen stretching frequencies of hydrogens on the $\beta$ carbon of an alkyl substituent are not affected by the cationic ring; however, those of hydrogens on an $\alpha$ carbon appear to be shifted and broadened (109). This may reflect the markedly acid nature of such hydrogens (see Section IV-C-1). Medz (109) has assigned strong bands near 750 cm$^{-1}$ in the spectra of alkyltropenylium ions to out-of-plane, in-phase carbon–hydrogen bending frequencies, and he notes that these shift to higher energies with increasing bulk in the substituent.

In the 1200–1600 cm$^{-1}$ region the spectrum of a substituted tropenylium ion is quite sensitive to the degree of electronic conjugation with the substituent. The major C=C stretching mode in the 1400–1500 cm$^{-1}$ region is strongly enhanced and split by the presence of an electron-donating substituent; similar exaltation of C=C stretching frequencies through conjugation between substituent and aromatic ring is found in benzene derivatives (308–310). In addition, the carbon–hydrogen in-plane deformation bands near 1200 cm$^{-1}$ (assigned by analogy with monosubstituted benzenes (311) and tropone (312)), which are forbidden in the unsubstituted ion (178–180) and weak or nonexistent in ions with electron-withdrawing substituents, become progressively enhanced as the degree of conjugation with the substituent increases. Harmon (229) has given a discussion and representative spectra.

The spectra of some individual ions are discussed in Section IV-C. Among the compounds whose infrared spectra have been reproduced in full are phenyl- (112), diphenyl- (79), alkyl and arylalkyl- (109), cyano- (119), hydroxy- (188,313,314), methoxy- (188), chloro- (229), —$B_{10}H_9^{2-}$ (229), —$B_{12}H_{11}^{2-}$ (229), and amino- and alkylamino- (80) tropenylium ions.

## 4. Electronic Spectra

The unsubstituted tropenylium ion shows (2,97) an intense absorption at 217 mμ and a much weaker, symmetry-forbidden absorption at 274 mμ; both these bands are doubly degenerate, and they are designated the $^1E_{1u}$ and $^1E_{3u}$ bands, respectively (11).

In Table IV, I have arbitrarily divided the reported spectra of substituted tropenylium ions into three classes based on the appearance of the spectra; these classes reflect the extent to which the substituent perturbs and interacts with the π-electron system of the parent ring. Figure 3 presents a representative example of each spectral type.

Type I spectra, which consist of an intense absorption in the 220–230 mμ region (ε about 40,000) and a weaker absorption in the 280–300 mμ region (ε 3000–6000) closely resemble the spectrum of tropenylium ion itself; the most significant change is the red shift and (usually) the intensity in the long-wavelength band. This type of spectrum is shown by substituents that do not conjugate with the ring (e.g., fluoro, alkyl, carbonyl, carboxyl and

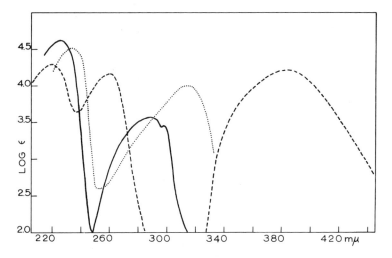

Fig. 3. Representative electronic spectra of substituted tropenylium ions in 96% sulfuric acid: solid curve, methyltropenylium ion (type I); dotted curve, methoxytropenylium ion (type II); and dashed curve, thiomethoxytropenylium ion (type III).

derivatives, cyano, and carboranyl) and presumably result from a symmetry change from $D_{7h}$ to $C_{2v}$ with concurrent increase in the intensity of the previously symmetry-forbidden $^1E_{3u}$ band.

Substituents capable of charge-transfer conjugation with the cationic ring perturb the spectrum of the tropenylium ion to a much greater extent. Interaction with the substituent breaks down the degeneracy of both transitions and makes the $^1E_{3u}$ transition allowed along the axis of the ring–substituent bond (22).

I have classified as type II a group of generally similar spectra, including those of hydroxy, alkoxy, chloro, and bromotropenylium ions. These spectra exhibit an intense band ($\varepsilon$ 30,000–40,000) in the 230–250 m$\mu$ region and a second strong band ($\varepsilon$ about 10,000) in the 300–330 m$\mu$ region. Theory predicts (22) a third strong absorption at shorter wavelengths; this band is seen in the spectrum of bromotropenylium ion (Table V). The type

### TABLE V
Spectra of Halotropenylium Ions[a,b]

| $-x$ | $-\lambda$, m$\mu$ ($\varepsilon$) | | |
|---|---|---|---|
| —F  | —            | 221 (43,100) | 283 (3680)  |
| —Cl | —            | 237 (40,300) | 310 (8800)  |
| —Br | 210 (18,800) | 247 (28,000) | 323 (11,000) |
| —I  | 216 (23,200) | 263 (9300)   | 378 (5060)  |

[a] In 96% sulfuric acid.
[b] References 2,96.

II spectra represent a transition between the type I, where there is no electronic interaction between substituent and ring, and the type III, where actual electron transfer occurs. Increased polarizability of the substituent in type II compounds results in a red shift of all bands, a decrease in the intensity of the shorter-wavelength transition, and an increase in the intensity of the long-wavelength band. The spectrum of hydroxytropenylium ion has been discussed in detail by Hosoya et al. (22). Although the spectra of amino- and alkylaminotropenylium ions (3,80,297) in general resemble type II spectra, the structure of the bands and the intensity ratios suggest that they are better interpreted as the spectra of trienimines, as tropone is interpreted as a trienone (22).

Type III spectra are characteristic of compounds in which there is a well-developed charge-transfer excitation of an electron from the substituent to the ring (33,229,292), such as iodo-, alkenyl-, aryl-, and thiomethoxytropenylium ions, and the ousene-type compounds (229) in which tropenylium ion is substituted for hydrogen on a polyhedral borane anionic cage. The spectra of all these ions have two bands that are nearly invariant from ion to ion; these are an intense absorption ($\varepsilon = 25{,}000$–$40{,}000$) at 220–225 mµ and a second absorption ($\varepsilon = $ ca. $10{,}000$) at 260–263 mµ; in aryltropenylium ions this second band is found at slightly higher wavelengths. These bands are assigned to transitions within the ring (33,292). In addition, there is a band in the visible region, and the position and intensity of the visible band vary widely from substituent to substituent. This band has been assigned to a substituent-to-ring charge-transfer excitation on the basis of theoretical calculations and through observation of the effects that substituents on the benzene ring of aryl- and arylalkenyltropenylium ions have on the position of the band (33,292). This assignment is confirmed by observation of solvent effects on the long-wavelength band of the hemiousenide ions, where positive slopes of the Kosower Z-value plots (187) show neutralization of internal dipole in the excited state (229).

The spectra of halotropenylium ions show the transition through these three types of spectrum as the ability of the substituent to enter into charge-transfer interaction with the ring increases. The spectrum of fluorotropenylium is almost identical to that of the parent unsubstituted ion. Chlorotropenylium is a typical type II spectrum as expected (22) for a substituent that will polarize the ring but not transfer an electron in a charge-transfer excitation. A strong red shift and the appearance of a third absorption band make the spectrum of bromotropenylium ion intermediate between type II and type III, and the iodotropenylium ion shows a pure type III spectrum. These spectra are listed in Table V.

The long-wavelength band of the iodotropenylium ion has a shoulder at 335 mµ; the shape of the structured band is identical to that shown by iodide ion change-transfer spectra (97,315,316) which arises from the differing electron affinities of the $P_{3/2}$ and $P_{1/2}$ states of the iodine atom. The energy gap ($3400 \text{ cm}^{-1}$) is, however, much less than in the case of free iodide ion ($6400$–$7200 \text{ cm}^{-1}$).

The spectrum of allyltropenylium ion (116) is anomolous; although apparently well characterized as the allyl-substituted ion, the material shows a type III spectrum as would be expected for the 2-methylvinyltropenylium ion.

Relatively little work has been done on the spectra of disubstituted tropenylium ions. In general, polyalkyl-, polyhydroxy-, or polyalkoxytropenylium ions show spectra similar to the monosubstituted compounds. This appears to be true, also, for polyaryltropenylium ions; with an increase in the number of aryl substituents, however, the aryl ring absorptions mask the bands of the tropenylium ring. We have found that the spectra of 2-halotropones in 96% sulfuric acid (1-hydroxy-2-halotropenylium ions) are very similar to the spectra of the corresponding halotropenylium ions with small (2–10 mμ) shifts in the positions of absorption. Two tropenylium ions substituted on an aromatic moiety such as a benzene ring, a carborane cage, or a borane anion appear to have little spectral interaction.

### 5. Nmr Spectra and Diamagnetic Susceptibility

Relatively little has been published on the nmr spectra of substituted tropenylium ions; spectra are occasionally published attendant to synthetic work, but by far the largest systematic collection of spectral data appears to be in the unpublished work of Professor Dauben and his students (109,117,119,317). Table VI lists some selected representative $^1$H nmr spectra of substituted ions.

### TABLE VI
### Nmr Spectra of Selected Substituted Tropenylium Ions[a,b]

| Substituent | Ring protons | Substituent protons | Ref. |
|---|---|---|---|
| $CH_3$— | 0.88 (s; 6H) | 6.75 (s; 3H) | 109,117 |
| $(CH_3)_3C$— | 0.66 (s; 6H) | 8.32 (s; 9H) | 109 |
| NC— | 0.46 (ss) | — | 119 |
| $HO_2C$— | 0.41–0.61 (m) | — | 117 |
| $1\text{-}C_2B_{10}H_{11}$—[c] | 0.68 (ss; 6H) | 6.25 (s; 1H)[d] | 250 |
| HO— | 1.3–1.9 (m) | — | 117 |
| $CH_3O$— | 0.84–1.62 (m; 6H) | 5.52 (s; 3H) | 117 |
| $CH_3S$— | 1.24–1.48 (m; 6H) | 7.00 (s; 3H) | 117 |
| $B_{12}H_{11}$— | 0.59 (s; 2H) | [d] | 229 |
|  | 1.47–1.62 (m; 4H) | — | — |
| $2\text{-}B_{10}H_9$— | 1.53, 1.69 (d; 2H) | [d] | 229 |
|  | 2.03–2.61 (m; 4H) | — | — |
| F— | 0.58–0.75 (m) | — | 117 |
| Cl— | 0.64–0.86 (ss) | — | 117 |
| Br—[c] | 0.64, 0.74 (d; 2H) | — | 229 |
|  | 0.90–1.15 (m; 4H) | — | — |
| I—[e] | 1.10, 1.30 (d; 2H) | — | 317 |
|  | 1.70–2.00 (m; 2H) | — | — |
|  | 2.00–2.40 (m; 2H) | — | — |
| $C_6H_5$— | 0.47–0.93 (m; 6H) | 1.87–2.43 (m; 5H) | 117 |

[a] Chemical shifts are in ppm relative to tetramethylsilane = 10 (τ); abbreviations are: (s) singlet, (ss) structured singlet, (m) multiplet.
[b] Spectra are in nitromethane unless otherwise indicated.
[c] Acetonitrile solvent; values are about 0.3 ppm upfield from nitromethane.
[d] Protons on boron give broad (up to 10 ppm) absorption; not resolvable.
[e] 96% $D_2SO_4$ solvent; values about 0.6 ppm upfield from nitromethane.

In general, electron-withdrawing substituents shift the absorptions of the ring protons to lower field, and electron-donating substituents shift them to higher field. The spectra of monosubstituted ions range in appearance from a sharp singlet for all six protons to a trio of complex absorptions, each representing two protons. The most common pattern is that of a two-proton doublet and a four-proton multiplet; with $-I, +T$ substituents such as bromo, iodo, thiomethoxy, and the borane anion cages, the two-proton doublet of the hydrogens adjacent to the substituent is at a lower field than the remaining hydrogens. Bertelli (307) has analyzed the nmr spectra of hydroxy- and methoxytropenylium ions at high resolution; in these cases the doublet is at higher field. From his consideration of coupling constants, Bertelli (307) concludes that there is some degree of bond alteration in these two ions.

There is a reasonable correlation between the extent of the broadening of the region of absorption of the ring protons and the degree of charge-transfer electronic interaction between the substituent and ring; those ions that show significant splitting between various types of ring hydrogen also show type III electronic spectra (compare Tables IV and VI). Schuster et al. (292) report the nmr spectra of a series of substituted phenyltropenylium ions. They find that the presence of electron-donating substituents on the phenyl ring results in a significant broadening and an increase in the complexity of the nmr spectrum of the tropenylium ring protons, whereas ions with electron-withdrawing substituents on the phenyl ring give narrow ring proton absorptions. An increase in electron-donating ability of the substituent on phenyl in substituted phenyltropenylium ions has previously been correlated with an increase in internal charge-transfer interactions (112).

In a particularly interesting example, Schuster et al. (292) examined the nmr spectra of 4-methylphenyltropenylium ion and 2-methylphenyltropenylium ion. The former shows the expected splitting of the tropenylium ring protons into a two-proton doublet and a four-proton multiplet, and the phenyl protons show a pair of doublets in accord with an AA'-BB' system. The second compound, in which the rings are forced near a 90° relationship to each other by steric effects, shows only a sharp singlet nmr absorption for both the tropenylium and the phenyl ring protons, and the long-wavelength band of the electronic spectrum is moved to shorter wavelengths and decreased in intensity. The authors (292) ascribe these changes in the nmr spectrum to steric effects; however, it appears more reasonable to suppose that the prime effect is electronic, since a comparison of the nmr spectra of tropenylium ions substituted by an o-carborane cage or a $-B_{12}H_{11}{}^{2-}$ cage—where the steric requirements are nearly identical—show a structured singlet in the former case and a broad separation of absorptions in the latter (Table VI), in accord with the differing degree of electronic interaction (100,229,250).

Laity (200,201) has examined the diamagnetic susceptibility of a series of substituted tropenylium ions (Table VII). The exaltation of the mean molar diamagnetic susceptibility $\Lambda$ of a compound over the value estimated on the assumption of no electronic delocalization is taken as a criterion of cyclic, aromatic delocalization of electrons (318,319), and the similarity of the values for most of the substituted ions in Table VII to

**TABLE VII**
**Diamagnetic Susceptibility Exaltations of Substituted Tropenylium Ions**

| Substituent(s) | $\Lambda$, $10^{-6}$ cm$^3$/mole |
|---|---|
| Hydrogen | 16.0 |
| Methyl | 14.8 |
| $t$-Butyl | 14.7 |
| Phenyl | 15.5 |
| Benzyl | 14.5 |
| $t$-Butylethynyl | 12.8 |
| Hydroxy | 14.9 |
| Methoxy | 13.9 |
| 1,4-Dimethoxy | 11.8 |
| 1,3,5-Trimethoxy | 15.2 |
| Chloro | 10.8 |
| Amino | 6.5 |
| Dimethylamino | 4.0 |

that of tropenylium ion itself suggests that most substituents do not disturb the $\pi$-electron system of the ring to a great degree. The main exceptions are the amino- and dimethylaminotropenylium ions, which have a greatly suppressed exaltation; this agrees with the similarity of the electronic spectrum to that of tropone (Section IV-B-4) and supports the immonium ion structure proposed (3,80) for these ions.

Since the exaltation of the chlorotropenylium ion is noticeably lowered, an experimental determination of the exaltations of bromo- and iodotropenylium ions would be of interest.

### 6. Polarographic Reduction

Harmon and Dauben (96) made a preliminary study of the polarographic reduction potentials of a series of substituted tropenylium ions in aceto-

## TABLE VIII
## Polarographic Reduction Potentials of Substituted Tropenylium Ions[a]

| Substituent | $\Delta E_{1/2}$ from Ref. 96[b,c] | $\Delta E_{1/2}$ from Ref. 117[c,d] |
|---|---|---|
| $CH_3S$— | −0.23 | — |
| HO— | −0.20 | −0.203 |
| $CH_3O$— | — | −0.154 |
| $(CH_3)_3C$— | — | −0.106 |
| $CH_3$— | −0.06 | −0.069 |
| H— | — | — |
| $C_6H_5$— | — | 0.014 |
| Br— | 0.23 | 0.247 |
| Cl— | 0.26 | — |
| $HO_2C$— | 0.34 | 0.343 |
| I— | 0.44 | — |

[a] In acetonitrile at the dropping mercury electrode.
[b] Two-electrode system with mercury pool reference anode.
[c] Values for $\Delta E_{1/2}$, V, relative to $E_{1/2}$ for tropenylium ion.
[d] Three-electrode system with silver–silver ion reference anode.

nitrile solvent (Table VIII) at the dropping mercury electrode (dme) using a two-electrode system and a mercury pool reference; all these reductions were one-electron processes. The choice of reference electrode was unfortunate, since the mercury pool anode potential is dependent on the concentration of mercurous ions at its surface and is difficult to stabilize; nevertheless, the data showed reasonable trends, with electron-donating substituents shifting $E_{1/2}$ to more negative values, and electron-withdrawing substituents shifting it to more positive values. The half-wave potentials for the $-I, +T$ halogen substituents are more positive than for tropenylium ion itself, in contrast to the nmr (Table VI) and electronic (Table V) spectra, which indicate significant substituent-to-ring electron donation in the $\pi$ system.

Dauben and Gresham (117) have carried out carefully controlled studies of the reduction potentials of substituted tropenylium ions at the dme in acetonitrile solvent (Table VIII) with a three-electrode system and silver–silver ion reference. According to preliminary analysis of their data, the polarographic waves for these reductions in acetonitrile show little if any deviation from reversibility, although irreversibility arising from dimerization of radicals has been demonstrated for the reduction wave of tropenylium species in aqueous solution (208–210,229) and in acetonitrile with a different electrode system (96).

The question of irreversibility is important if data of this type are to be used for free energy correlations (Section IV-B-7) since, strictly speaking, it is the rate of the electrode process and not the half-wave potential which is related to the free energy of the electrode reaction for irreversible processes; however, for similar compounds under similar conditions, the half-wave potentials adequately reflect the electron affinities (320,321).

A detailed study of the polarographic reduction of the $C_7H_6B_{10}H_9^-$ and $C_7H_6B_{12}H_{11}^-$ hemiousenide ions (229) accompanied by separate millicoulometric experiments demonstrated conclusively that these ions are reduced irreversibly in $1N$ potassium chloride in a one-electron process, and allowed calculation of diffusion coefficients for the ions. The borane cages shift $E_{1/2}$ for these ions to quite negative values compared to the unsubstituted tropenylium ion.

### 7. Miscellaneous Measurements and Free Energy Correlations

Jutz and Voithenleitner (112) have measured the $pK_a$ values of phenyltropenylium ions with a wide variety of substituents on phenyl (Section IV-B-2) and find a reasonable correlation between these $pK_a$ values and Jaffe's $\sigma$ constants (322). When conjugated alkenyl or alkynyl groups separate the tropenylium and phenyl rings, the correlation is not as good (113-115).

Preliminary studies by Dauben and Wilson (195), in connection with electron-transfer reactions, have shown that the polarographic reduction potentials and the energies of charge-transfer complex absorption bands appear to be linearly related for a series of carbonium ions. Both of these measurements are theoretically related to electron affinity (320), which is the simplest parameter to use for the comparison of a series of like carbonium ions (320).

Dauben and Gresham (117) have examined the half-wave reduction potentials in acetonitrile (Section IV-B-6), the energies of charge-transfer absorption bands with hexamethylbenzene in nitropropane and 1,2,4,5-tetramethoxybenzene in nitroethane, and the equilibrium constants of formation of charge-transfer complexes with hexamethylbenzene donor in butyronitrile for a series of 14 substituted tropenylium ions. Their results indicate that these quantities are all linearly related and, more important, all three quantities give excellent linear agreement with the Hammett $\sigma_{para}$ constants of the substituents.

The close correlation of the electron affinities of substituted tropenylium ions with the Hammett $\sigma_{para}$ constant is unexpected, since most reactions involving interactions with a present or developing positive charge are better fitted to a $\sigma^+_{para}$ plot (322), and in tropenylium ions the substituent is already attached to a carbonium ion. The striking similarity between benzene and tropenylium ion, which is shown by their electronic (1,2,11) and infrared (178,180) spectra and $^1$H nmr chemical shifts attributable to ring currents (198), is reinforced by their similar interaction with substituents. Species such as alkyl, cyclohexadienyl, and mono-, di-, or triarylmethyl cations, although varying widely in stability, share much in common; the Hückel aromatic tropenylium ion is altogether different. It is a reasonable shift in semantics—and, perhaps, in viewpoint—to say that tropenylium ion is not a carbonium ion in the same sense that benzene is not an alkene.

### C. Reactions and Properties of Representative Substituted Tropenylium Ions

Substituted tropenylium ions in general undergo the same types of reactions as the unsubstituted ion; however, there are certain reactions unique to the substituted species. Most of these reactions can be classified into two main types: nucleophilic attack at the ring carbon bearing the substituent

$$\text{(Tr}^+\text{)}-X + Y^- \rightleftharpoons \text{(Tr}^+\text{)}-Y + X^-$$

or nucleophilic attack on a $\beta$ atom of the substituent (96).

$$\text{(Tr}^+\text{)}-A-B + N: \longrightarrow \text{(Tr)}=A + B-N$$

Sometimes a combination of the two is found, as in the hydrolysis of chlorotropenylium ion salts to tropone (96).

$$\text{(Tr}^+\text{)}-Cl \xrightarrow{H_2O} \text{(Tr}^+\text{)}-OH + HCl \xrightleftharpoons{H_2O} \text{(Tr)}=O + H_3O^+$$

### 1. Alkyltropenylium Ions

Alkyltropenylium ion salts are colorless, and are, in general, more soluble in organic solvents than the corresponding unsubstituted salts. The fluoroborate salts are often low-melting and extremely soluble; for separation and stability purposes hexafluoroantimonate salts are preferred (109).

Hydrogens attached to the $\alpha$ carbon of an alkyl substituent are extremely acidic and are removed by any mild nucleophile (e.g., water or halide ion); thus such alkyltropenylium ions are sensitive to the atmosphere, and it is not possible to prepare their halide salts. Doering and Wiley (323) found that methyltropenylium bromide could be prepared by addition of dry hydrogen bromide to heptafulvene at $-80°$, but the compound lost hydrogen bromide on heating and a resin resulted. Harmon (96) found similar results when the bromide was prepared by hydride exchange in liquid sulfur dioxide at $-78°$. Isopropyl- and benzhydryltropenylium ions are particularly sensitive (109).

This property has been used to synthetic advantage, as in the condensation of alkyltropenylium ions with aldehydes (113–115) or ethylorthoformate (296) to give alkenyltropenylium ions.

$t$-Butyltropenylium ion, with no hydrogens on the $\alpha$ carbon, appears to be completely stable to the atmosphere (109), and a bromide salt can be prepared (69).

## 2. Aryltropenylium Ions

Aryltropenylium ions are resistant to nucleophilic attack of both types and form stable salts with a wide variety of anions. These ions are generally colored from internal charge-transfer interactions, and the color becomes deeper with an increase in the number of phenyl substituents. The phenyl group appears to act as an electron-withdrawing group when substituted on the tropenylium nucleus, and the electrophilic character of the ring is markedly increased in polyphenyltropenylium ions (290,291). Hexaphenyltropenylium halides show a borderline covalency (290,324), and heptaphenyltropenylium ion is readily reduced with potassium to the corresponding anion (325).

Correlation of theory with experimental data suggests that the rings are at $45°$ to each other in phenyltropenylium ion (292).

## 3. Halotropenylium Ions

Fluoro-, chloro-, and bromotropenylium ion salts of anucleophilic anions are colorless (2,96). Bromotropenylium bromide and iodotropenylium iodide are orange and red, respectively, because of anion-to-cation charge-transfer (96,97). Although the iodotropenylium ion has only been isolated as the iodide, the electronic spectrum in 96% sulfuric acid suggests that the ion itself would be light yellow from substituent-to-ring charge-transfer absorption (Table V).

The halogen substituents are readily displaced by nucleophilic reagents, and all these ions are rapidly converted to hydroxytropenylium ion or tropone by contact with wet solvents, alcohols, or atmospheric moisture (2,69,85,96,139). It is perhaps surprising that the ease of hydrolysis increases in going from iodine to fluorine; fluorotropenylium ion is so sensitive to moisture that its preparation can be carried out only under the most rigorously dry conditions. Halide ion can readily displace halogen substituent; both chlorotropenylium bromide and bromotropenylium chloride equilibrate to mixtures of the two substituted ions, and iodotropenylium ion is made by displacement of chloride from chlorotropenylium ion by iodide ion (96).

Treatment of chlorotropenylium ion with strong nucleophiles such as mercaptide ion (96) or diethyl malonate anion (273) can lead to new substituted tropenylium ions, but substitution also occurs at carbons other than that bearing the halogen substituent, and the reaction has been of little synthetic importance. Chlorotropenylium ion reacts with nitrite ion under anhydrous conditions to yield free tropone and a yellow compound believed to be nitrosyl chloride (96).

## 4. Hydroxytropenylium Ions

Hydroxytropenylium ion salts are usually prepared by protonation of tropone; we have recently published convenient preparative details for a variety of salts (133). Cation hydroxyl-to-anion hydrogen bonding may be a necessary requirement for salt formation; such hydrogen bonding is present in all salts examined to date—including the perchlorate (314)—and we have not been successful in preparing a salt of an anion that cannot accept hydrogen bonding.

Harmon et al. (188,313,314) have carried out an intensive study of the physical properties of hydroxytropenylium ion salts. Hydroxytropenylium bromide and iodide form stable, crystalline monohydrates with strong hydrogen bonds between the cation hydroxyl and hydrate water; equilibrium vapor pressure measurements show the strength of these hydrogen bonds to be at least 14 kcal/mole. Infrared spectra show secondary hydrogen bonding from water hydroxyl to anion in these hydrates and hydrogen bonding from cation hydroxyl to anion in the anhydrous chloride, bromide, iodide, and perchlorate salts.

Hydroxytropenylium bromide shows a sharp, reversible pale-to-brilliant yellow chromogenic transition at 125°. Infrared studies show that the cation hydroxyl-to-anion hydrogen bonds weaken in a smooth and reversible manner on heating, with no discontinuity at 125°. However, electronic spectra of the solid show that the anion-to-cation charge-transfer absorption undergoes an abrupt bathochromic shift and change in line shape at 125°; this reaction is ascribed to a crystal lattice reorientation, since the infrared studies reveal no significant change in donor–acceptor properties at this point. The lattice change is accompanied by an endotherm of 580 cal/mole.

With more complete infrared data available, it has been possible to assign the oxygen–hydrogen and carbon–oxygen absorptions of the hydroxytropenylium ion, thus clarifying some previous (326) ambiguities.

## 5. Alkoxytropenylium Ions

Alkoxytropenylium ion perchlorate or halometallate salts are generally stable, colorless, highly crystalline materials; however, the ions are readily hydrolyzed by water to hydroxytropenylium ion or tropone and should be protected from the atmosphere. The highly colored halide salts are less stable, and on heating they undergo nucleophilic attack by halide ion on the $\beta$ atom of the substituent to yield tropone and alkyl halide (2,96). This reaction is presumably operative in the tropone syntheses developed by Doering with Detert (327) and Knox (75) based on bromination and subsequent pyrolysis of cycloheptatrienyl ethers.

Slow addition of methoxide ion to methoxytropenylium ion in the cold yields 7,7-dimethoxycycloheptatriene, the dimethylketal of tropone (119,328).

## 6. Ousene-Type Compounds

The parent type of the ousenes is a *closo*-borane anion cage substituted by two fully aromatic, carbocyclic cationic rings to form a neutral molecule. In the ousene nomenclature (229) the numbers in brackets refer to the number of carbons in the rings and borons in the cage, with the boron number last; superscript numbers on the boron number give the position of the rings on the cage. The prefix *hemi-* is used if a single ring is present. The presence of a positive or negative charge, or the substitution of a carborane cage for a borane anion are represented in the normal ways. Examples of some ousene-type compounds and their names are given in Figure 4.

The $[7.10^2]$- and $[7.12]$hemiousenide ions (227–229) are highly colored species with excellent thermal and chemical stability; the cesium salts of these ions can be heated to 400° in air without change and are unaffected by air, indefinitely stable in water, and attacked only slowly by concentrated acids. Electronic, infrared, and $^1$H and $^{11}$B nmr spectra, Z-value correlations, and polarographic reduction potentials show extensive electron donation from cage to ring. The substituent-to-ring charge-transfer absorptions are very solvent sensitive. A single crystal X-ray study by Sly (306) has confirmed the position of substitution in the $[7.10^2]$hemiousenide ion.

A $[7.7.10^{x,y}]$ousene has been prepared (298) and characterized by elemental analysis and solid state infrared spectra; the extreme insolubility of this deep purple, crystalline material makes study by other techniques difficult.

The $[7.12^1]$-1,2- and $[7.12^1]$-1,7-dicarbahemiousenium ions and the $[7.7.12^{1,7}]$-1,7-dicarbaousenium ion are typical of tropenylium ions with

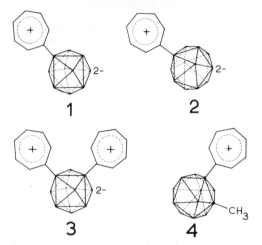

Fig. 4. Some representative ousene-type compounds: (1) $[7.10^2]$hemiousenide ion; (2) [7.12]hemiousenide ion; (3) $[7.7.10^{2,3}]$ousene; (4) 1-methyl$[7.12^2]$-1,2-dicarbahemiousenium ion. Cf. Ref. 229.

an electron-withdrawing substituent; little or no electronic interaction exists between cage and ring (100,250). Halometallate salts of these ions are colorless, stable to the atmosphere, and insoluble in nonpolar solvents; however, the bromides are covalent and dissolve easily in hydrocarbons or ether.

## V. CONCLUSION

No work of this type can be complete; however, if I have overlooked or underdiscussed anyone's work it has not been by intention. In general, I have chosen to write in greater detail on work with which I am familiar and can speak of with more assuredness.

I wish to acknowledge the invaluable support of my wife, Ann B. Harmon, who carried out most of the literature studies for this work and has assisted with the preparation of the manuscript.

Thanks are also due to the Department of Chemistry at the University of Washington for support of this writing while I was a visiting professor there in 1968–1969 and again in the summer of 1970, and to the Petroleum Research Fund of the American Chemical Society, whose funds supported portions of this work under grants to Harvey Mudd College and Oakland University.

## REFERENCES

1. W. v. E. Doering and L. H. Knox, *J. Am. Chem. Soc.*, **76**, 3203 (1954).
2. H. J. Dauben, Jr., F. A. Gadecki, K. M. Harmon, and D. L. Pearson, *J. Am. Chem. Soc.*, **79**, 4557 (1957).
3. H. J. Dauben, Jr., and D. F. Rhoades, *J. Am. Chem. Soc.*, **89**, 6764 (1967).
4. E. Hückel, *Z. Physik*, **70**, 204 (1931).
5. E. Hückel, *Z. Elektrochem.*, **43**, 752, 827 (1937).
6. J. D. Roberts, A. Streitwieser, and C. M. Regan, *J. Am. Chem. Soc.*, **74**, 4579 (1952).
7. J. L. Franklin and F. H. Field, *J. Am. Chem. Soc.*, **75**, 2819 (1953).
8. E. Heilbronner and A. Eschenmoser, *Helv. Chim. Acta*, **36**, 1101 (1953).
9. A. Julg and B. Pullman, *J. Chim. Phys.*, **52**, 481 (1955).
10. Wei-Chuwan Lin, *Chin. Chem. Soc. (Taipei)*, **2**, 37 (1955).
11. J. N. Murrell and H. C. Longuet-Higgins, *J. Chem. Phys.*, **23**, 2347 (1955).
12. H. Siebert, *Elektronentheorie Homöopolaren Bindung, Hauptjahrestag, Chem. Ges. Deut. Demokrat. Rep.*, **1955**, 7; *Chem. Abstr.*, **53**, 3803g (1959).
13. K. Aida, *Sci. Rept. Res. Inst. Tohoku Univ. Ser. A*, **8**, 361 (1956).
14. H. C. Longuet-Higgins and K. L. McEwen, *J. Chem. Phys.*, **26**, 719 (1957).
15. H. C. Longuet-Higgins, in *Theoretical Organic Chemistry; Kekulé Symposium*, Butterworths, London, 1958.
16. R. B. Turner, in *Theoretical Organic Chemistry; Kekulé Symposium*, Butterworths, London, 1958.
17. N. S. Ham, *J. Chem. Phys.*, **32**, 1445 (1960).
18. A. Streitwieser, Jr., *J. Am. Chem. Soc.*, **82**, 4123 (1960).
19. N. Bouman, *J. Chem. Phys.*, **35**, 1661 (1961).
20. E. Heilbronner, *Chimia*, **15**, 35 (1961).
21. A. D. Liehr, *Z. Naturforsch., A*, **16**, 641 (1961).
22. H. Hosoya, J. Tanaka, and S. Nagakura, *Tetrahedron*, **18**, 859 (1962).
23. A. Julg, *J. Chim. Phys.*, **59**, 367 (1962).
24. H. Fischer and J. N. Murrell, *Theor. Chim. Acta*, **1**, 463 (1963).
25. E. Heilbronner and J. N. Murrell, *Mol. Phys.*, **6**, 1 (1963).
26. R. Hoffmann, *J. Chem. Phys.*, **39**, 1397 (1963).
27. A. Julg, *Tetrahedron*, **19**, *Suppl.* 2, 25 (1963).
28. D. Peters, *Tetrahedron*, **19**, *Suppl.* 2, 143 (1963).
29. R. Ettinger, *Tetrahedron*, **20**, 1579 (1964).
30. J. Koutecky, P. Hochman, and J. Michl, *J. Chem. Phys.*, **40**, 2439 (1964).
31. J. Feitelson, *J. Chem. Phys.*, **43**, 2511 (1965).
32. A. Julg, *J. Chim. Phys.*, **62**, 1372 (1965).
33. G. Hohlneicher, R. Kiessling, H. C. Jutz, and P. A. Straub, *Ber. Bunsenges. Phys. Chem.*, **70**, 60 (1966).
34. G. V. Boyd and N. Singer, *Tetrahedron*, **22**, 547 (1966).
35. M. M. Bursey and F. W. McLafferty, in *Carbonium Ions*, Vol. I, G. Olah and P. v. R. Schleyer, Eds., Interscience, New York, 1968, p. 257.
36. S. Meyerson and P. N. Rylander, *J. Chem. Phys.*, **27**, 901 (1957).
37. P. N. Rylander, S. Meyerson, and H. M. Grubb, *J. Am. Chem. Soc.*, **79**, 842 (1957).
38. V. Hanus, *Nature*, **184**, *Suppl.* 23, 1796 (1959).
39. S. Meyerson, P. N. Rylander, Jr., E. L. Eliel, and J. D. McCollum, *J. Am. Chem. Soc.*, **81**, 2606 (1959).

40. V. Hanus and Z. Dolejsek, *Kernenergie*, **3**, 836 (1960).
41. J. D. McCollum and P. N. Rylander, *J. Am. Chem. Soc.*, **83**, 1401 (1961).
42. B. A. Thrush and J. J. Zwolenik, *Proc. Chem. Soc.*, **1962**, 339.
43. H. Wincel and Z. Kecki, *Nukleonika*, **8**, 215 (1963).
44. F. Meyer and A. G. Harrison, *J. Am. Chem. Soc.*, **86**, 4757 (1964).
45. K. R. Jennings and J. H. Futrell, *J. Chem. Phys.*, **44**, 4315 (1966).
46. P. Brown, *J. Am. Chem. Soc.*, **90**, 2694 (1968).
47. G. G. Meisels and D. R. Arnold, *J. Phys. Chem.*, **72**, 3061 (1968).
48. N. M. M. Nibbering and T. J. de Boer, *Tetrahedron*, **24**, 1435 (1968).
49. K. L. Rinehart, Jr., A. C. Buchholz, G. E. Van Lear, and H. L. Cantrill, *J. Am. Chem. Soc.*, **90**, 2983 (1968).
50. H. Fernholz, *Justus Liebigs Ann. Chem.*, **568**, 63 (1950).
51. A. Eschenmoser and H. H. Rennhard, *Helv. Chim. Acta*, **36**, 290 (1953).
52. W. H. Schaeppi, R. W. Schmid, E. Heilbronner, and A. Eschenmoser, *Helv. Chim. Acta*, **38**, 1874 (1955).
53. W. Simon, G. Naville, H. Sulser, and E. Heilbronner, *Helv. Chim. Acta*, **39**, 1107 (1956).
54. G. Berti, *J. Org. Chem.*, **22**, 230 (1957).
55. H. H. Rennhard, G. DiModica, W. Simon, E. Heilbronner, and A. Eschenmoser, *Helv. Chim. Acta*, **40**, 957 (1957).
56. D. Meuche, H. Strauss, and E. Heilbronner, *Helv. Chim. Acta*, **41**, 414 (1958).
57. M. J. S. Dewar and C. R. Ganellin, *J. Chem. Soc.*, **1959**, 3139.
58. D. Meuche, W. Simon, and E. Heilbronner, *Helv. Chim. Acta*, **42**, 452 (1959).
59. G. Naville, H. Strauss, and E. Heilbronner, *Helv. Chim. Acta*, **43**, 1221 (1960).
60. W. Tochtermann and K. Stecher, *Tetrahedron Lett.*, **1967**, 3847.
61. D. J. Bertelli and W. J. Rossiter, *Tetrahedron*, **24**, 609 (1968).
62. J. J. Looker, *J. Org. Chem.*, **33**, 1304 (1968).
63. W. Tochtermann, G. Schnabel, and A. Mannschreck, *Justus Liebigs Ann. Chem.*, **711**, 88 (1968).
64. G. Merling, *Chem. Ber.*, **24**, 3108 (1891).
65. T. Nozoe, "Non-Benzenoid Aromatic Compounds," Interscience, New York, 1959, pp. 339–464.
66. H. J. Dauben, Jr., and D. L. Pearson, paper presented at the 126th National Meeting of the American Chemical Society, New York, September, 1954, *Abstr.*, p. 18-O.
67. D. L. Pearson, Ph.D. dissertation, University of Washington, Seattle, 1955; *Diss. Abstr.*, **15**, 978 (1955); No. 55-1311, University Microfilms, Inc., Ann Arbor, Mich.
68. W. v. E. Doering, in *Theoretical Organic Chemistry; Kekulé Symposium*, Butterworths, London, 1958, pp. 35–48.
69. W. v. E. Doering and H. Krauch, *Angew. Chem.*, **68**, 661 (1956).
70. G. D. Kolomnikova and Z. N. Parnes, *Russ. Chem. Rev.*, **36**, 735 (1967).
71. D. N. Kursanov, M. E. Vol'pin, and Z. N. Parnes, *Khim. Nauka i Prom.*, **3**, 159 (1958).
72. T. Nozoe, *Progr. Org. Chem.*, **5**, 132 (1961).
73. M. E. Vol'pin, *Russ. Chem. Rev.*, 129 (1960).
74. D. Bryce-Smith and N. A. Perkins, *J. Chem. Soc.*, **1961**, 2320.
75. W. v. E. Doering and L. H. Knox, *J. Am. Chem. Soc.*, **79**, 352 (1957).
76. M. Tisler, *Univ. Ljubljani Teh. Fak. Acta Tech., Ser. Chim.*, **1956**, 2; *Chem. Abstr.*, **53**, 18927i (1959).

77. A. W. Johnson and M. Tisler, *Chem. Ind. (London)*, **1954**, 1427.
78. A. W. Johnson, A. Langemann, and M. Tisler, *J. Chem. Soc.*, **1955**, 1622.
79. A. Cairncross, Ph.D. dissertation, Yale University, 1963; *Diss. Abstr.*, **25**, 825 (1964); No. 64-7136, University Microfilms, Inc., Ann Arbor, Mich.
80. D. H. Rhoades, Ph.D. dissertation, University of Washington, Seattle, 1969.
81. K. Kitahara and M. Funamizu, paper presented at the Annual Meeting of the Chemistry Society, Japan; Tokyo, April 1959, *Abstr.*, p. 230.
82. K. M. Harmon, A. B. Harmon, S. D. Alderman, P. A. Gebauer, and L. Hesse, *J. Org. Chem.*, **32**, 2012 (1967).
83. N. C. Deno, N. Friedman, J. D. Hodge, F. P. MacKay, and G. Saines, *J. Am. Chem. Soc.*, **84**, 4713 (1962).
84. D. N. Kursanov and M. E. Vol'pin, *Dokl. Akad. Nauk SSSR*, **113**, 339 (1957).
85. D. N. Kursanov, *Uch. Zap., Khar'Kovsk. Gos. Univ.*, **110**, *Tr. Khim. Fak. i Nauchn.-Issled. Inst. Khim.* (17), 7 (1961), *Chem. Abstr.*, **58**, 4399a (1963).
86. L. R. Honnen, Ph.D. dissertation, University of Washington, Seattle, 1962; *Diss. Abstr.*, **24**, 972 (1963); No. 63-4420, University Microfilms, Inc., Ann Arbor, Mich.
87. A. P. ter Borg, R. van Helden, and A. F. Bickel, *Rec. Trav. Chim. Pays-Bas*, **81**, 164, 177 (1962).
88. M. E. Vol'pin and D. N. Kursanov, *Dokl. Akad. Nauk SSSR*, **126**, 780 (1959).
89. G. Juppe and A. P. Wolf, *Chem. Ber.*, **94**, 2328 (1961).
90. P. Radlick, *J. Org. Chem.*, **29**, 960 (1964).
91. W. v. E. Doering and L. H. Knox, *J. Am. Chem. Soc.*, **73**, 828 (1951).
92. D. H. Geske, *J. Am. Chem. Soc.*, **81**, 4145 (1959).
93. J. Mizuguchi, Y. Uetani, T. Sato, T. Matsubayashi, and T. Kashiwaya, *Denki Kagaku*, **34**, 124 (1966).
94. H. J. Dauben, Jr., L. R. Honnen, and K. M. Harmon, *J. Org. Chem.*, **25**, 1442 (1960).
95. K. Conrow, *Org. Syn.*, **43**, 101 (1963).
96. K. M. Harmon, Ph.D. dissertation, University of Washington, Seattle, 1958; *Diss. Abstr.*, **19**, 1563 (1958); No. 58-7362, University Microfilms, Inc., Ann Arbor, Mich.
97. K. M. Harmon, F. E. Cummings, D. A. Davis, and D. J. Diestler, *J. Am. Chem. Soc.*, **84**, 3349 (1962).
98. W. E. Bachmann, *Organic Syntheses*, Coll. Vol. III, Wiley, New York, 1955, p. 841.
99. K. A. Hofmann and H. Kirmreuther, *Chem. Ber.*, **42**, 4856 (1909).
100. K. M. Harmon, A. B. Harmon, and B. C. Thompson, *J. Am. Chem. Soc.*, **89**, 5309 (1967).
101. H. J. Dauben, Jr., and N. T. Owen, private communication.
102. K. M. Harmon, L. Hesse, L. P. Klemann, C. W. Kocher, S. V. McKinley, and A. E. Young, *Inorg. Chem.*, **8**, 1054 (1969).
103. K. M. Harmon and F. E. Cummings, *J. Am. Chem. Soc.*, **84**, 1751 (1962).
104. K. M. Harmon and F. E. Cummings, *J. Am. Chem. Soc.*, **87**, 539 (1965).
105. From Ozark-Mahoning Co., 1870 South Boulder, Tulsa, Okla. 74119.
106. K. M. Harmon and A. B. Harmon, *J. Am. Chem. Soc.*, **83**, 865 (1961).
107. K. M. Harmon, A. B. Harmon, and F. E. Cummings, *J. Am. Chem. Soc.*, **83**, 3912 (1961).
108. K. M. Harmon, A. B. Harmon, and F. E. Cummings, *J. Am. Chem. Soc.*, **86**, 5511 (1964).

109. R. B. Medz, Ph.D. dissertation, University of Washington, Seattle, 1964; *Diss. Abstr.*, **25**, 2766 (1964); No. 64-11,186 University Microfilms, Inc., Ann Arbor, Mich.
110. E. Mueller, H. Kessler, H. Fricke, and W. Kiedaisch, *Justus Liebigs Ann. Chem.*, **675**, 63 (1964).
111. J. W. Wilt and D. Piszkiewicz, *Chem. Ind. (London)*, **1963**, 1761.
112. C. Jutz and F. Voithenleitner, *Chem. Ber.*, **97**, 29 (1964).
113. C. Jutz and F. Voithenleitner, *Chem. Ber.*, **97**, 1337 (1964).
114. C. Jutz, *Chem. Ber.*, **97**, 1349 (1964).
115. C. Jutz and F. Voithenleitner, *Chem. Ber.*, **97**, 1590 (1964).
116. D. J. Bertelli, C. Golino, and D. L. Dreyer, *J. Am. Chem. Soc.*, **86**, 3329 (1964).
117. H. J. Dauben, Jr., and W. R. Gresham, unpublished results.
118. E. Weth and A. S. Dreiding, *Proc. Chem. Soc.*, **1964**, 59.
119. T. J. Pratt, Ph.D. dissertation, University of Washington, Seattle, 1964; *Diss. Abstr.*, **25**, 4962 (1965); No. 65-1898 University Microfilms, Inc., Ann Arbor, Mich.
120. G. Manecke and M. Schoeneshoefer, *Makromol. Chem.*, **112**, 293 (1968).
121. H. Volz, *Tetrahedron Lett.*, **1964**, 1899.
122. H. Volz, *Tetrahedron Lett.*, **1963**, 1965.
123. H. Volz and M. J. Volz de Lecea, *Tetrahedron Lett.*, **1965**, 3413.
124. J. Holmes and R. Pettit, *J. Org. Chem.*, **28**, 1695 (1963).
125. H. Volz, *Angew. Chem.*, **75**, 921 (1963).
126. H. Volz and H. W. Schnell, *Angew. Chem.*, **77**, 864 (1965).
127. H. Volz and M. J. Volz de Lecea, *Tetrahedron Lett.*, **1964**, 1871.
128. H. Volz and M. J. Volz de Lecea, *Tetrahedron Lett.*, **1966**, 4675.
129. H. Volz and M. J. Volz de Lecea, *Tetrahedron Lett.*, **1966**, 4683.
130. H. Volz and W. D. Mayer, *Tetrahedron Lett.*, **1966**, 5249.
131. D. Bryce-Smith and N. A. Perkins, *Chem. Ind. (London)*, **1959**, 1022.
132. Z. N. Parnes, V. I. Zdanovich, E. E. Kugucheva, G. I. Basova, and D. N. Kursanov, *Dokl. Akad. Nauk SSSR*, **166**, 122 (1966).
133. K. M. Harmon, A. B. Harmon, T. T. Coburn, and J. M. Fisk, *J. Org. Chem.*, **33**, 2567 (1968).
134. I. R. Beattie and M. Webster, *J. Chem. Soc.*, **1963**, 38.
135. D. Bryce-Smith and N. A. Perkins, *J. Chem. Soc.*, **1962**, 1339.
136. W. Bracke, W. J. Cheng, J. M. Pearson, and M. Szwarc, *J. Am. Chem. Soc.*, **91**, 203 (1969).
137. D. Clark, H. M. Powell, and A. F. Wells, *J. Chem. Soc.*, **1942**, 642.
138. D. S. Payne, *J. Chem. Soc.*, **1953**, 1053.
139. M. E. Vol'pin, I. S. Akhrem, and D. N. Kursanov, *Izv. Akad. Nauk SSSR, Otd. Khim. Nauk*, **1957**, 760.
140. T. Nozoe and K. Takahashi, *Bull. Chem. Soc. Japan*, **38**, 665 (1965).
141. Z. N. Parnes, M. I. Kalinkin, and D. N. Kursanov, *Dokl. Akad. Nauk SSSR*, **165**, 1093 (1965).
142. J. A. Blair, G. P. McLaughlin, and J. Paslawski, *Chem. Commun.*, **1967**, 12.
143. C. R. Kistner, J. R. Doyle, N. C. Baenziger, J. H. Hutchinson, and P. Kasper, *Inorg. Chem.*, **3**, 1525 (1964).
144. M. E. Vol'pin, V. G. Dulova, and D. N. Kursanov, *Dokl. Akad. Nauk SSSR*, **128**, 951 (1959).
145. G. L. Closs and L. E. Closs, *Tetrahedron Lett.*, **1960**, 38.
146. K. H. Buechel and A. Conte, *Z. Naturforsch., B*, **21**, 1111 (1966).

147. J. M. Cinnamon and K. Weiss, *J. Org. Chem.*, **26**, 2644 (1961).
148. D. H. Martin and F. Williams, *J. Am. Chem. Soc.*, **85**, 1014 (1963).
149. M. J. S. Dewar and R. Pettit, *Chem. Ind. (London)*, **1955**, 199.
150. M. J. S. Dewar and R. Pettit, *J. Chem. Soc.*, **1956**, 2021, 2026.
151. M. J. S. Dewar, C. R. Ganellin, and R. Pettit, *J. Chem. Soc.*, **1958**, 55.
152. M. J. S. Dewar and C. R. Ganellin, *J. Chem. Soc.*, **1959**, 2438.
153. C. R. Ganellin and R. Pettit, *Chem. Ber.*, **90**, 2951 (1957).
154. C. R. Ganellin and R. Pettit, *J. Am. Chem. Soc.*, **79**, 1767 (1957).
155. C. R. Ganellin and R. Pettit, *J. Chem. Soc.*, **1958**, 576.
156. C. Ruechardt and H. Schwarzer, *Angew. Chem.*, **74**, 251 (1962).
157. B. Föhlisch, P. Bürgle, and D. Krockenberger, *Angew. Chem.*, **77**, 1019 (1965).
158. G. L. Closs and L. E. Closs, *J. Am. Chem. Soc.*, **83**, 599 (1961).
159. A. N. Nesmeyanov and R. V. Golovnya, *Dokl. Akad. Nauk SSSR*, **133**, 1337 (1960).
160. A. N. Nesmeyanov and R. V. Golovnya, *Zh. Obshch. Khim.*, **31**, 1067 (1961).
161. D. H. Reid, M. Fraser, B. B. Molloy, H. A. S. Payne, and R. G. Sutherland, *Tetrahedron Lett.*, **1961**, 530.
162. S. Seto and K. Ogura, paper presented at the Tohoku Local Meeting of the Chemical Society, Japan; Akita, 1959, *Abstr.*, p. 29.
163. S. Seto and K. Ogura, *Tohoku Daigaku Hisuiyoeki Kagaku Kenkyusho Hokoku*, **9**, 111 (1960); *Chem. Abstr.*, **55**, 24602d (1961).
164. E. W. Abel, M. A. Bennett, R. Burton, and G. Wilkinson, *J. Chem. Soc.*, **1958**, 4559.
165. M. E. Vol'pin, S. I. Zhdanova, and D. N. Kursanov, *Proc. Acad. Sci. USSR, Sect. Chem.*, **112**, 33 (1957).
166. K. H. Buechel and A. Conte, *Z. Naturforsch.*, *B*, **21**, 1110 (1966).
167. V. I. Belova, M. E. Vol'pin, and Ya. K. Syrkin, *Zh. Obshch. Khim.*, **29**, 693 (1959).
168. A. E. Kemppainen and E. L. Compere, Jr., *J. Inorg. Nucl. Chem.*, **29**, 588 (1967).
169. K. M. Harmon and S. Davis, *J. Am. Chem. Soc.*, **84**, 4359 (1962).
170. N. C. Deno, J. Jaruzelski, and A. Schriesheim, *J. Am. Chem. Soc.*, **77**, 3044 (1955).
171. E. S. Gould, *Acta Crystallogr.*, **8**, 657 (1955).
172. A. I. Kitaigorodskii, T. L. Khotsyanova, and Yu. T. Struchkov, *Acta Crystallogr.*, **10**, 797 (1957).
173. A. I. Kitaigorodskii, Yu. T. Struchkov, T. L. Khotsyanova, M. E. Vol'pin, and D. N. Kursanov, *Izv. Akad. Nauk SSSR, Otd. Khim. Nauk*, **1960**, 39.
174. Y. Sasada, K. Osaki, and I. Nitta, *Acta Crystallogr.*, **7**, 113 (1954).
175. Y. Sasada and I. Nitta, *Acta Crystallogr.*, **9**, 205 (1956).
176. M. E. Vol'pin, D. N. Kursanov, M. M. Shemyakin, V. J. Maimind, and L. A. Neyman, *Chem. Ind. (London)*, **1958**, 1261.
177. M. E. Vol'pin, D. N. Kursanov, M. M. Shemyakin, V. I. Maimind, and L. A. Neyman, *Zh. Obshch. Khim.*, **29**, 3711 (1959).
178. W. G. Fateley and E. R. Lippincott, *J. Am. Chem. Soc.*, **77**, 249 (1955).
179. R. D. Nelson, W. G. Fateley, and E. R. Lippincott, *J. Am. Chem. Soc.*, **78**, 4870 (1956).
180. W. G. Fateley, G. Curnutte, and E. R. Lippincott, *J. Chem. Phys.*, **26**, 1471 (1957).
181. K. Aida, *Sci. Rept. Res. Inst. Tohoku Univ.*, **A8**, 361 (1956).

182. F. A. Savin, *Opt. i Spektrosk.*, **15**, 42 (1963).
183. E. M. Kosower and P. E. Klinedinst, Jr., *J. Am. Chem. Soc.*, **78**, 3493 (1956).
184. K. M. Harmon, F. E. Cummings, D. A. Davis, and D. J. Diestler, *J. Am. Chem. Soc.*, **84**, 120 (1962).
185. K. M. Harmon, S. D. Alderman, K. E. Benker, D. J. Diestler, and P. A. Gebauer, *J. Am. Chem. Soc.*, **87**, 1700 (1965).
186. E. M. Kosower, *J. Org. Chem.*, **29**, 956 (1964).
187. E. M. Kosower, *J. Am. Chem. Soc.*, **80**, 3253 (1958).
188. K. M. Harmon, S. M. Cummins, and T. T. Coburn, *J. Phys. Chem.*, **73**, 2939 (1969).
189. M. Feldman and S. Winstein, *J. Am. Chem. Soc.*, **83**, 3338 (1961).
190. M. Nepras and R. Zahradnik, *Collect. Czech. Chem. Commun.*, **29**, 1545 (1964).
191. B. A. Thrush and J. J. Zwolenik, *Proc. Chem. Soc.*, **1962**, 339.
192. M. Feldman and S. Winstein, *Tetrahedron Lett.*, **1962**, 853.
193. M. Feldman and S. Winstein, *Theor. Chim. Acta*, **10**, 86 (1968).
194. S. N. Bhat and C. N. R. Rao, *J. Chem. Phys.*, **47**, 1863 (1967).
195. H. J. Dauben, Jr., and J. D. Wilson, *Chem. Commun.*, **1968**, 1629.
196. E. Le Goff and R. B. LaCount, *J. Am. Chem. Soc.*, **85**, 1354 (1963).
197. J. R. Leto, F. A. Cotton, and J. S. Waugh, *Nature*, **180**, 978 (1957).
198. G. Fraenkel, R. E. Carter, A. McLachlan, and J. H. Richards, *J. Am. Chem. Soc.*, **82**, 5846 (1960).
199. H. Spiesecke and W. G. Schneider, *Tetrahedron Lett.*, **1961**, 468.
200. J. L. Laity, Ph.D. dissertation, University of Washington, Seattle, 1969; *Diss. Abstr.*, **30**, 565-B (1969); No. 69-13577, University Microfilms, Inc., Ann Arbor, Mich.
201. H. J. Dauben, Jr., J. D. Wilson, and J. L. Laity, *J. Am. Chem. Soc.*, **90**, 811 (1968).
202. F. A. Gadecki, Ph.D. dissertation, University of Washington, Seattle, 1957; *Diss. Abstr.*, **18**, 1253 (1958); No. 58-1081, University Microfilms, Inc., Ann Arbor, Mich.
203. N. N. Lichtin and P. Pappas, *Trans. N.Y. Acad. Sci.*, **20**, 143 (1957).
204. N. N. Lichtin, private communication.
205. S. I. Zhdanov and A. N. Frumkin, *Dokl. Akad. Nauk SSSR*, **122**, 412 (1958); *Chem. Abstr.*, **55**, 1239h (1961).
206. S. I. Zhdanov, *Z. Phys. Chem. (Leipzig)*, **1958**, 235.
207. A. M. Hopin and S. I. Zhdanov, *Elektrokhimiya*, **4**, 228 (1968).
208. P. Zuman, J. Chodkowski, H. Potesilova, and F. Santavy, *Nature*, **182**, 1535 (1958).
209. P. Zuman, J. Chodowski, and F. Santavy, *Collect. Czech. Chem. Commun.*, **26**, 380 (1961).
210. P. Zuman and J. Chodkowski, *Collect. Czech. Chem. Commun.*, **27**, 759 (1962).
211. M. A. Gerovich and N. S. Polyanovaskaya, *Nauch. Dokl. Vyssh. Shk., Khim. i. Khim. Tekhnol.*, **1958**, 651; *Chem. Abstr.*, **53**, 5814 (1959).
212. K. M. Harmon, A. B. Harmon, S. D. Alderman, and L. Hesse, paper presented at the 147th National Meeting of the American Chemical Society, Philadelphia, Pa., April 9, 1964; *Abstr.*
213. J. Degani and R. Fochi, *Boll. Sci. Fac. Chim. Ind. Bologna*, **22**, 7 (1964).
214. C. E. Wulfman, C. F. Yarnell, and D. S. Wulfman, *Chem. Ind. (London)*, **1960**, 1440.
215. M. J. Biallas and D. F. Shriver, *J. Am. Chem. Soc.*, **88**, 375 (1966).

216. H. J. Dauben, Jr., and K. M. Harmon, paper presented at the Pacific Southwest Regional Meeting of the American Chemical Society, Redlands, Calif., October 25, 1958; *Abstr.*
217. R. B. Turner, H. Prinzbach, and W. v. E. Doering, *J. Am. Chem. Soc.*, **82**, 3451 (1960).
218. K. M. Harmon and P. A. Gebauer, *Inorg. Chem.*, **2**, 1319 (1963).
219. P. G. Ferrine and A. Marxer, *Angew. Chem. Int. Ed. Engl.*, **1**, 405 (1962).
220. Letter from U.S. Army Testing Laboratory to H. J. Dauben, Jr.
221. K. Conrow, *J. Am. Chem. Soc.*, **83**, 2343 (1961).
222. E. E. v. Tamelen, T. M. Cole, R. Greeley, and H. Schumacher, *J. Am. Chem. Soc.*, **90**, 1372 (1968).
223. Z. N. Parnes, M. E. Vol'pin, and D. N. Kursanov, *Tetrahedron Lett.*, **1960**, 20.
224. A. E. Borisov, A. N. Abramova, and Z. N. Parnes, *Izv. Akad. Nauk SSSR, Ser. Khim.*, **1964**, 941.
225. Z. N. Parnes, M. E. Vol'pin, and D. N. Kursanov, *Izv. Akad. Nauk SSSR, Otd. Khim. Nauk*, **1960**, 763.
226. K. Conrow and L. L. Reasor, *J. Org. Chem.*, **30**, 4368 (1965).
227. K. M. Harmon, A. B. Harmon, and A. A. MacDonald, *J. Am. Chem. Soc.*, **86**, 5036 (1964).
228. A. B. Harmon and K. M. Harmon, *J. Am. Chem. Soc.*, **88**, 4093 (1966).
229. K. M. Harmon, A. B. Harmon, and A. A. MacDonald, *J. Am. Chem. Soc.*, **91**, 323 (1969).
230. A. P. ter Borg, R. v. Helden, A. F. Bickel, W. Renold, and A. S. Dreiding, *Helv. Chim. Acta*, **43**, 457 (1960).
231. M. E. Vol'pin, Z. N. Parnes, and D. N. Kursanov, *Izv. Akad. Nauk SSSR, Otd. Khim. Nauk*, **1960**, 950.
232. T. Ikemi, T. Nozoe, and H. Sugiyama, *Chem. Ind. (London)*, **1960**, 932.
233. Z. N. Parnes, G. D. Mur, R. V. Kudryavtsev, and D. N. Kursanov, *Dokl. Akad. Nauk SSSR*, **155**, 1371 (1964).
234. Z. N. Parnes, G. D. Mur, and D. N. Kursanov, *Dokl. Akad. Nauk SSSR*, **159**, 857 (1964).
235. D. N. Kursanov, Z. N. Parnes, and G. D. Kolomnikova, *Zh. Org. Khim.*, **3**, 1060 (1967).
236. O. M. Grishin and A. A. Yasnikov, *Ukr. Khim. Zh.*, **34**, 70 (1968).
237. M. E. Vol'pin, K. I. Zhdanova, D. N. Kursanov, V. N. Setkina, and A. I. Shatenshtein, *Izv. Akad. Nauk SSSR, Otd. Khim. Nauk*, **1959**, 754.
238. M. E. Vol'pin, I. S. Akhrem, and D. N. Kursanov, *Khim. Nauk i Prom.*, **2**, 656 (1957).
239. M. E. Vol'pin, I. S. Akhrem, and D. N. Kursanov, *Zh. Obshch. Khim.*, **28**, 330 (1958).
240. C. Jutz, *Chem. Ber.*, **97**, 2050 (1964).
241. S. G. McGeachin, *Can. J. Chem.*, **47**, 151 (1969).
242. J. Degani and R. Fochi, *Boll. Sci. Fac. Chim. Ind. Bologna*, **20**, 139 (1962).
243. A. E. Kemppainen and E. L. Compere, Jr., *J. Chem. Eng. Data*, **11**, 588 (1966).
244. I. S. Akhrem and D. N. Kursanov, *Bull. Akad. Sci. USSR, Otd. Khim.*, **1957**, 1905.
245. W. v. E. Doering and L. E. Helgen, *J. Chem. Soc.*, **1961**, 482.
246. J. J. Looker, *J. Org. Chem.*, **30**, 638 (1965).
247. A. G. Harrison, L. R. Honnen, H. J. Dauben, Jr., and F. P. Lossing, *J. Am. Chem. Soc.*, **82**, 5593 (1960).

248. R. W. Murray and M. L. Kaplan, *Tetrahedron Lett.*, **1965**, 2903.
249. R. W. Murray and M. L. Kaplan, *Tetrahedron Lett.*, **1967**, 1307.
250. K. M. Harmon, A. B. Harmon, B. C. Thompson, D. R. Ryan, and T. T. Coburn, *Inorg. Chem.*, in preparation.
251. D. B. Miller, *J. Org. Chem.*, **31**, 908 (1966).
252. M. E. Vol'pin, I. S. Akhrem, and D. N. Kursanov, *Izv. Akad. Nauk SSSR, Otd. Khim. Nauk*, **1957**, 1501.
253. D. N. Kursanov, M. E. Vol'pin, and I. S. Akhrem, *Dokl. Akad. Nauk SSSR*, **120**, 531 (1958).
254. M. E. Vol'pin, I. S. Akhrem, and D. N. Kursanov, *Tr. Konf. po Vopr. Str. i Reaks. Sposobnosti Atsetalei, Akad. Nauk King., SSR, Inst. Organ. Khim.*, **1961**, 96; *Chem. Abstr.*, **60**, 13180g (1964).
255. M. E. Vol'pin, I. S. Akhrem, and D. N. Kursanov, *Zh. Obshch. Khim.*, **30**, 159 (1960).
256. R. W. Hoffmann, *Angew. Chem. Int. Ed. Engl.*, **4**, 977 (1965).
257. D. Bryce-Smith and N. A. Perkins, *J. Chem. Soc.*, **1962**, 5295.
258. J. J. Looker, *J. Org. Chem.*, **30**, 4180 (1965).
259. T. Nozoe, S. Ito, and T. Tezuka, *Chem. Ind. (London)*, **1960**, 1088.
260. R. v. Helden, A. P. ter Borg, and A. F. Bickel, *Rec. Trav. Chim. Pays-Bas*, **81**, 599 (1962).
261. P. Bladon, P. L. Pauson, G. R. Proctor, and W. J. Rodger, *J. Chem. Soc., C*, **1966**, 926.
262. K. Takahashi, *Bull. Chem. Soc. Japan*, **40**, 1462 (1967).
263. T. Nozoe, T. Tezuka, and T. Mukai, *Bull. Chem. Soc. Japan*, **36**, 1470 (1963).
264. K. Hafner, A. Stephan, and C. Bernhard, *Justus Liebigs Ann. Chem.*, **650**, 42 (1961).
265. A. G. Anderson, Jr., and L. L. Replogle, *J. Org. Chem.*, **28**, 262 (1963).
266. T. Nozoe, T. Toda, T. Asao, and Y. Yamanouchi, *Bull. Chem. Soc. Japan*, **41**, 2935 (1968).
267. M. E. Vol'pin, I. S. Akhrem, and D. N. Kursanov, *Zh. Obshch. Khim.*, **29**, 2855 (1959).
268. M. E. Vol'pin, I. S. Akhrem, and D. N. Kursanov, *Zh. Obshch. Khim.*, **30**, 1187 (1960).
269. M. E. Vol'pin, I. S. Akhrem, E. A. Terent'eva, and D. N. Kursanov, *Izv. Akad. Nauk SSSR, Otd. Khim. Nauk*, **1963**, 802.
270. K. Conrow, *J. Am. Chem. Soc.*, **81**, 5461 (1959).
271. K. M. Harmon and G. O. Spessard, unpublished results.
272. D. J. Bertelli, P. O. Crews, and S. Griffin, *Tetrahedron*, **24**, 1945 (1968).
273. H. J. Dauben, Jr., and E. J. Gauglitz, Jr., unpublished results in M.S. thesis of E. J. G., University of Washington, Seattle, 1958.
274. F. Korte, K. H. Büchel, and F. F. Wiese, *Justus Liebigs Ann. Chem.*, **664**, 114 (1963).
275. T. Mukai and T. Miyashi, *Bull. Chem. Soc. Japan*, **38**, 469 (1965).
276. N. W. Jordan and I. W. Elliot, *J. Org. Chem.*, **27**, 1445 (1962).
277. M. E. Vol'pin and I. S. Akhrem, *Dokl. Akad. Nauk SSSR*, **161**, 597 (1965).
278. K. M. Harmon and B. G. Cook, unpublished results.
279. G. Vincow, M. L. Morrell, W. V. Volland, H. J. Dauben, Jr., and F. R. Hunter, *J. Am. Chem. Soc.*, **87**, 3527 (1965), and references therein.
280. R. P. M. Werner and S. A. Manastyrskyj, *J. Am. Chem. Soc.*, **83**, 2023 (1961).
281. J. Kwiatek and J. K. Seyler, *J. Organomet. Chem.*, **3**, 421 (1965).

282. R. W. Murray and M. L. Kaplan, *J. Org. Chem.*, **31**, 962 (1966).
283. J. dos Santos-Veiga, *Mol. Phys.*, **5**, 637 (1962).
284. C. E. H. Bawn, C. Fitzsimmons, and A. Ledwith, *Proc. Chem. Soc.*, **1964**, 391.
285. A. Ledwith and M. Sambhi, *Chem. Commun.*, **1965**, 64.
286. Y. Kitahara and M. Funamizu, Japanese Patent 11122 (1962); *Chem. Abstr.*, **59**, 10012b (1963).
287. M. E. Vol'pin, D. N. Kursanov, and V. G. Dulova, *Tetrahedron*, **8**, 33 (1960).
288. R. B. Beechey and I. G. Knight, *Nature*, **212**, 938 (1966).
289. G. A. Gladkovskii, S. S. Skorokhodov, S. G. Slyvina, and A. S. Khachaturov, *Izv. Akad. Nauk SSSR*, **1963**, 1273.
290. M. A. Battiste and T. J. Barton, *Tetrahedron Lett.*, **1968**, 2951.
291. M. A. Battiste, *J. Am. Chem. Soc.*, **83**, 4101 (1961).
292. P. Schuster, D. Vedrilla, and O. E. Polansky, *Monatsh. Chem.*, **100**, 1 (1969).
293. T. Nozoe, Japanese Patent 17674 (1964); *Chem. Abstr.*, **62**, 5234a (1965).
294. G. R. Jurch, Jr., and T. G. Traylor, *J. Am. Chem. Soc.*, **88**, 5228 (1966).
295. R. West and K. Kusuda, *J. Am. Chem. Soc.*, **90**, 7354 (1968).
296. K. Hafner, H. W. Riedel, and M. Danielisz, *Angew. Chem.*, **75**, 344 (1963).
297. N. L. Bauld and Y. S. Rim, *J. Am. Chem. Soc.*, **89**, 6763 (1967).
298. A. B. Harmon and K. M. Harmon, unpublished results.
299. I. S. Akhrem, E. I. Fedin, B. A. Kvasov, and M. E. Vol'pin, *Tetrahedron Lett.*, **1967**, 5265.
300. W. v. E. Doering and L. H. Knox, *J. Am. Chem. Soc.*, **75**, 297 (1953).
301. E. Mueller, H. Fricke, and H. Kessler, *Tetrahedron Lett.*, **1963**, 1501.
302. E. Mueller and H. Fricke, *Justus Liebigs Ann. Chem.*, **661**, 38 (1963).
303. R. W. Murray and M. L. Kaplan, *J. Am. Chem. Soc.*, **88**, 3527 (1966).
304. G. O. Schenk and H. Ziegler, *Justus Liebigs Ann. Chem.*, **584**, 221 (1953).
305. P. C. Myhre and R. D. Andersen, *Tetrahedron Lett.*, **1965**, 1497.
306. W. G. Sly, unpublished results.
307. D. J. Bertelli, T. G. Andrews, Jr., and P. O. Crews, *J. Am. Chem. Soc.*, **91**, 5286 (1969).
308. D. W. A. Sharp and N. Sheppard, *J. Chem. Soc.*, **1957**, 674.
309. H. Gotz, E. Heilbronner, A. R. Katritzky, and R. A. Jones, *Helv. Chim. Acta*, **444**, 387 (1961).
310. L. J. Bellamy, *The Infrared Spectra of Complex Molecules*, Wiley, New York, 1958, pp. 71ff.
311. D. H. Whiffen, *J. Chem. Soc.*, **1956**, 1350.
312. Y. Ikegami, *Bull. Chem. Soc. Japan*, **35**, 967 (1962).
313. K. M. Harmon and T. T. Coburn, *J. Am. Chem. Soc.*, **87**, 2499 (1965).
314. K. M. Harmon and T. T. Coburn, *J. Phys. Chem.*, **72**, 2950 (1968).
315. R. Platzman and J. Franck, *Z. Physik*, **138**, 411 (1954).
316. E. Rabinowitch, *Rev. Mod. Phys.*, **14**, 112 (1942).
317. H. J. Dauben, Jr., and K. M. Harmon, unpublished results.
318. J. Hoarau, *Ann. Chim. (Paris)*, **1**, 544 (1956).
319. C. K. Ingold, *Structure and Mechanism in Organic Chemistry*, Cornell University Press, Ithaca, N.Y., 1953, pp. 187–196.
320. A. Streitwieser, *Molecular Orbital Theory for Organic Chemists*, Wiley, New York, 1961, pp. 173ff.
321. S. Wawzonek, *Anal. Chem.*, **28**, 638 (1956).
322. H. H. Jaffe, *Chem. Rev.*, **53**, 191 (1953).
323. W. v. E. Doering and D. W. Wiley, *Tetrahedron*, **11**, 183 (1960).

324. M. A. Battiste and T. J. Barton, paper presented at the 152nd National Meeting of the American Chemical Society, New York, September 13, 1966; *Abstr.*, p. S51.
325. R. Breslow and H. W. Chang, *J. Am. Chem. Soc.*, **87**, 2200 (1965).
326. B. E. Zaitsev, Yu. D. Koreshkov, M. E. Vol'pin, and Yu. N. Sheinker, *Dokl. Akad. Nauk SSSR*, **139**, 1107 (1961).
327. W. v. E. Doering and F. L. Detert, *J. Am. Chem. Soc.*, **73**, 876 (1951).
328. R. W. Hoffmann and J. Schneider, *Tetrahedron Lett.*, **1967**, 4347.

CHAPTER 30

# Azacarbonium Ions

Francis L. Scott and Richard N. Butler
*Department of Chemistry, University College, Cork, Ireland*

I. Aminocarbonium Ions $[>N\stackrel{+}{=}C<]$ . . . . . 1644

    A. As Transient Species, Postulated but not Physically Detected . . 1645
        1. Under Kinetic Conditions . . . . . . . 1645
            a. Mannich Reactions . . . . . . . . 1645
            b. Fragmentation Reactions . . . . . . . 1649
            c. Aminoacylium Ions . . . . . . . . 1652
        2. Aminocarbonium Ions in Synthetic Reactions . . . . 1652
            a. Amine Oxidations . . . . . . . . 1653
            b. Reactions of $\alpha$-Dialkylamino Compounds . . . . 1655
    B. Aminocarbonium Ions Detected under Equilibrium Conditions . 1656
        1. General . . . . . . . . . . 1656
        2. Enamine Studies . . . . . . . . . 1657
        3. Imine Protonations . . . . . . . . 1660
        4. Mass Spectra . . . . . . . . . 1661
    C. Aminocarbonium Ions as Crystalline Salts . . . . . 1661
        1. $\alpha$-Aminohalides . . . . . . . . . 1661
        2. Immonium Salts . . . . . . . . . 1663
        3. Enamine Salts . . . . . . . . . 1665
        4. Vilsmeier-Haack Complexes . . . . . . . 1666
        5. Polyamino Compounds . . . . . . . . 1669
II. Amidocarbonium Ions . . . . . . . . . 1670

III. Nitrilium Ions $[RC\stackrel{+}{=}NR]$ . . . . . . 1672

    A. From Nitriles: Involving CN Bond-Order Change from 3 to 2 . 1673
        1. With Electrophilic Carbon Moieties . . . . . . 1673
            a. By Direct Alkylation . . . . . . . . 1673
            b. The Ritter Reaction . . . . . . . . 1674
            c. By Arylation . . . . . . . . . 1676
        2. By Protonation . . . . . . . . . 1676
        3. By Reaction with Metallic Cations . . . . . . 1678
    B. Involving CN Bond-Order Change from 2 to 3 . . . . 1679
        1. From Imidic Halides . . . . . . . . 1679
        2. From Oximes; as Intermediates in the Beckmann Rearrangement . 1680

IV. Aminonitrilium Ions [RC≡N—NHR′]⁺ . . . . . . 1682
   A. From Haloazines . . . . . . . . . 1683
      1. Ketazines . . . . . . . . . . 1683
      2. Aldazines . . . . . . . . . . 1684
   B. From Hydrazidic Halides . . . . . . . . 1685
      1. N-Arylhydrazidic Halides: ArC(Br)=NNHAr′ . . . . 1685
      2. N-Tetrazolylhydrazidic Halides . . . . . . 1686
      3. Nitrilimines . . . . . . . . . . 1688
References . . . . . . . . . . . . 1689

The delocalization of carbonium ion character onto a neighboring nitrogen (or oxygen) atom can stabilize the carbonium ion center, thus making it more easily formed and enhancing the reactivity of its precursors. In this chapter our aim is to try and see where the intervention of such species has been demonstrated or can legitimately be suspected.

We have organized the data under two broad headings, presenting first the material pertaining to charge delocalization of the type

$$\diagup\!\!\!\!\!\diagdown\!\!\!\!\text{N}\!-\!\overset{+}{\text{CH}}-\;\longleftrightarrow\;\diagup\!\!\!\!\!\diagdown\!\!\!\!\overset{+}{\text{N}}\!=\!\text{CH}-$$

i.e., where the bond order of the carbon–nitrogen bond increases from one to two as a result of the delocalization. Under this heading we have examined some of the chemistry of amino- and amidocarbonium ions and also some aminoacylium species. The aminocarbonium section is the most diffuse, and in discussing this topic we progress from reactions in which such species have been postulated but not detected, to those studies in which such cations have been detected but not isolated, to the final stage in which such materials have been isolated and studied as crystalline salts.

The second category of delocalization we consider is that wherein the carbon–nitrogen bond order increases from two to three by a charge distribution of the type

$$-\text{N}\!=\!\overset{+}{\text{C}}-\;\longleftrightarrow\;-\overset{+}{\text{N}}\!\equiv\!\text{C}-$$

In this category we examined some of the properties of both nitrilium and aminonitrilium ions.

## I. AMINOCARBONIUM IONS    $[\!\!>\!\!\text{N}\!\!\doteq\!\!\text{C}\!\!<\!]$

As we mentioned in the organization of this section, we progress from the postulated to the detected and thence to the isolated forms of aminocarbonium ions.

## A. As Transient Species, Postulated but not Physically Detected

### 1. Under Kinetic Conditions

**a. Mannich Reactions.** One of the major synthetic reactions in which aminocarbonium ions as transient species may be involved is $\alpha$-aminoalkylation (1) or the Mannich reaction (2). This reaction involves the combination of an aldehyde (usually formaldehyde) with ammonia or a primary and secondary amine and with a compound containing an "activated" or acidic hydrogen. It may be illustrated by the following sequence:

$$R_2NH + CH_2O + R'H \longrightarrow R_2N-CH_2R' + H_2O$$

The mechanism of this reaction has been the subject of considerable discussion. Bodendorf and Koralewski (3) concluded from their experiments that neither the condensation of the formaldehyde with the amine nor with the active hydrogen compound to yield the corresponding methylols "showed the true course of the reaction." The suggestion that an aminocarbonium ion was involved was presented by Lieberman and Wagner (4) without definite experimental evidence, however, other than persuasive arguments. They went further and suggested that the final irreversible step in a whole series of equilibria was the combination of the carbonium ion and a carbanion derived from the acidic component to yield the Mannich base.

Alexander and Underhill (5) shortly afterward presented kinetic data on the Mannich reaction using dimethylamine, formaldehyde, and ethylmalonic acid (as substrates) in aqueous acidic solution ([HX] 0.10–0.39$M$). Under their conditions the reaction obeyed third-order kinetics with no primary salt effect. This result is incompatible with the suggestion of Lieberman and Wagner.

A more complete study of the mechanism of the reaction was made more recently (6). Kinetic data on the reaction between cyclohexanone, dimethylamine, and formaldehyde were obtained. Different mechanisms were clearly indicated to be operative in acidic and basic media, with complex combinations of mechanisms taking place at intermediate pH values.

The following mechanism for the reaction in acidic solution, which fitted their data and those of Alexander (5), was suggested by Cummings and Shelton; their innovation was that they incorporated into their kinetic scheme the recognition that the amine involved was present in equilibrium with its ammonium salt.

$$Me_2\overset{+}{N}H_2 + A^- \underset{}{\overset{K_1}{\rightleftharpoons}} Me_2NH + HA \tag{1}$$

$$Me_2NH + CH_2O \underset{}{\overset{K_2}{\rightleftharpoons}} Me_2NCH_2OH \quad (2)$$

$$Me_2NCH_2OH + HA \underset{}{\overset{K_3}{\rightleftharpoons}} Me_2\overset{+}{N}CH_2 + H_2O + A^- \quad (3)$$

$$\text{C}_6\text{H}_{10}{=}O \underset{}{\overset{k_4}{\rightleftharpoons}} \text{C}_6\text{H}_9{-}OH \quad (4)$$

$$\text{C}_6\text{H}_9{-}OH + Me_2\overset{+}{N}CH_2 \xrightarrow{k_1} \underset{CH_2NMe_2}{\text{C}_6\text{H}_9}{=}\overset{+}{O}H \quad (5)$$

$$\underset{CH_2NMe_2}{\text{C}_6\text{H}_9}{=}\overset{+}{O}H + A^- \underset{}{\overset{K_5}{\rightleftharpoons}} \underset{CH_2NMe_2}{\text{C}_6\text{H}_9}{=}O \quad (6)$$

From the foregoing equations, with equation (5) being irreversible and rate determining, we have

$$\text{rate} = \frac{dx}{dt} = k_1 K_4 K_3 K_2 K_1 [Me_2\overset{+}{N}H_2][CH_2O][C_6H_{10}O] \quad (7)$$

$$= k[Me_2\overset{+}{N}H_2][CH_2O][C_6H_{10}O] \quad (8)$$

Thus Cummings and Shelton conclude that the mechanism pictured suggests that the rate is independent of acid concentration, which is what is observed between pH 1.02 and 4.19. They note that in the case of the study performed by Alexander and Underhill an acid much stronger than cyclohexanone was used as one of the substrates. Therefore, if the reactive form is in fact the un-ionized acid [as is shown in equations used (5)], then a rate term involving the acid dissociation would have to be introduced, and this would make the rate equation once again dependent on the concentration of hydroxonium ion. Cummings and Shelton propose the following mechanism for the reaction under basic conditions:

$$\underset{(1)}{Me_2NH} + \underset{(2)}{CH_2O} \underset{}{\overset{K_2}{\rightleftharpoons}} \underset{(4)}{Me_2NCH_2OH} \quad (9)$$

$$\underset{(3)}{\text{C}_6\text{H}_{10}}{=}O + OH^- \underset{}{\overset{K_6}{\rightleftharpoons}} \text{C}_6\text{H}_9{-}\overset{\ominus}{O} + H_2O \quad (10)$$

$$\underset{(5)}{\bigcirc\!\!\!\!-O^{\ominus}} + \underset{(4)}{\underset{NMe_2}{\overset{H\;\;H}{\underset{|}{C}}}\!\!-OH} \;\overset{K_7}{\rightleftharpoons}\; \underset{(6)}{\overset{O}{\bigcirc}\!\!\!\!-\underset{NMe_2}{\overset{H\;\;H}{\underset{|}{C}}}\!\!-OH} \quad (11)$$

$$\downarrow k_2$$

$$\underset{(7)}{\overset{O}{\bigcirc}\!\!\!\!-CH_2NMe_2 + OH^-}$$

$$\text{rate} = \frac{dx}{dt} = k_2(6) = k_2 K_7 (4)(5)$$
$$= k_2 K_7 K_6 K_2 (1)(2)(3)(OH^-)$$
$$= k[Me_2NH][CH_2O][C_6H_{10}O][OH^-]$$

These equations predict that, at any given pH at which this mechanism predominates, the reaction should exhibit third-order kinetics and, with increasing pH, the rate should increase.

The precursor to any aminocarbonium ion is not necessarily the aminomethylol compound used for convenience here. In fact, Lieberman and Wagner (4) reported that in general methylene bisamines $(R_2N)_2CH_2$ could be substituted for the amine plus formaldehyde in the Mannich reaction. In a more recent study (7a) of the Mannich reaction involving 2,4-dimethylphenol, morpholine, and formaldehyde under basic conditions (i.e., pH ca. 10), Burckhalter and his co-workers claim to have identified such a species as a "rate-determining intermediate" [however, see also the comments thereon of Fernandez (7b)]. They concluded that under such basic conditions the rate-determining step is the reaction between "free phenol and free amine" (amine being $N,N'$-methylenebismorpholine). They also state "preliminary results suggest that a different mechanism operates in acidic media."

We might conclude at the end of these various studies that if an aminocarbonium ion is involved in the Mannich reaction, it most likely will occur at low pH and *in aqueous solution*. That this point about medium is important has been brought into focus by the work of Fernandez (8).

Very recently, Fernandez examined the kinetics of the Mannich reaction of methylenebispiperidine with nitromethane, nitroethane, and 2-nitropropane in the solvents dioxane and DMF, varying in effect the medium dielectric constant over the range 5 to 28 and examining also the influence of small amounts of added water on the reaction rate. The following are the main features of the data obtained.

Reactivity (toward $(\underset{2}{\overset{}{C_6H_{10}}N})CH_2$) decreased in the order $CH_3CH_2NO_2 > (CH_3)_2CHNO_2 > CH_3NO_2$. Reactivity increased with increase in dielectric constant, and whereas for the nitroethane and 2-nitropropane reactions this increase in rate with dielectric constant was smooth, in the nitromethane reaction the rate increased sharply beyond an $\varepsilon$ value of 20, suggesting a change in mechanism. In dioxane, the $\Delta H^{\ddagger}$ decreased in the order of acid strength of the nitroalkane as follows.* $CH_3NO_2$ [23], $CH_3CH_2NO_2$ [16], $(CH_3)_2CHNO_2$ [9]. In dioxane–DMF mixtures this regularity in $\Delta H^{\ddagger}$ was not observed. The entropies of activation were also significant, although such significance is not without ambiguity (9). Thus for the solvent change dioxane to 67% DMF–dioxane, the $\Delta S^{\ddagger}$ changes were as follows: $CH_3NO_2$, $-14$ to $-29$ eu; $CH_3CH_2NO_2$, $-31$ to $-21$ eu; and $(CH_3)_2CHNO_2$, $-54$ to $-28$ eu. The nitromethane data were again out of line with the others and suggested a buildup of charge in the transition state.

For all three reactions in media of low dielectric constant, and for the nitroethane and 2-nitropropane reactions in media of moderately high ($\geqslant 25$) dielectric constant a cyclic transition state of type **A** was suggested

$$RCH_2NO_2 \rightleftharpoons RCH=NO_2H$$

$$RCH=NO_2H + CH_2(NR'_2)_2 \rightleftharpoons RCH \underset{\underset{NR'_2}{\overset{CH_2}{|}}}{\overset{\overset{O}{\underset{\|}{N}}\overset{O-H}{\diagup}}{\diagdown}} NR'_2$$

(A)

For the nitromethane reaction in media of moderately high dielectric constant ($\geqslant 25$), Fernandez envisages a linear polar transition state of type **B**

$$CH_2=N\underset{OH}{\overset{O}{\diagup}} \cdots \underset{R_2}{\overset{NR_2}{N}}-CH_2 \quad CH_2=N\underset{O}{\overset{OH}{\diagup}}$$

(B)

* The value of $\Delta H^{\ddagger}$ appears in brackets, expressed in kilocalories.

Although nitromethane and nitroethane failed to undergo a second alkylation with methylenebispiperidine in anhydrous media, a second alkylation did occur when water was present or when ethoxymethylpiperidine was used as reactant. These observations, along with rate studies employing 2-nitropropane and methylenebispiperidine in the presence of small amounts of water, point strongly to the $N$-hydroxymethylpiperidine (⟨NCH$_2$OH⟩) as a key intermediate in these Mannich reactions. Perhaps, in general, we could agree with Fernandez that "the $N$-hydroxymethylamine reacts [in the Mannich reaction] at a much faster rate than the methylenebisamine with which it is in equilibrium."

The cleavage of such methylenebisamines depends also on their structure. Thus formamidine formation rather than cleavage occurred with a cyanosubstituted compound (10), e.g.,

$$\langle N-CH-N \rangle \rightleftharpoons \langle N-CH=\overset{+}{N} \rangle \; CN^-$$
$$\quad\;\; |$$
$$\quad\; CN$$

Recent studies have added further kinetic evidence (10a) for the presence of an aminocarbonium ion in the Mannich reaction, and Masui et al. (10b) have interpreted a new polarographic reduction wave observed for aqueous solutions of formaldehyde and secondary amines as a direct observation of the aminomethylcarbonium ion. For interesting recent discussions of the Mannich reaction see (10c) and (10d).

**b. Fragmentation Reactions (11,11a,149).** Azacarbonium ions may be generated from certain $\gamma$-aminohalides by means of the fragmentation process:

$$\underset{/}{\overset{\backslash}{N}}-\underset{|}{\overset{|}{C}}-\underset{|}{\overset{|}{C}}-\underset{|}{\overset{|}{C}}-X \longrightarrow \left[ \underset{/}{\overset{\backslash}{\overset{+}{N}}}=C\underset{\backslash}{\overset{/}{}} \right] + \underset{/}{\overset{\backslash}{C}}=C\underset{\backslash}{\overset{/}{}} + X^-$$

Grob and his co-workers have established the principles underlying this process using kinetic and stereochemical tools. The products of solvolysis of appropriate systems in 80% aqueous ethanol were studied and the appropriate first-order rate constants were measured. These were compared with the rate constants for solvolysis of a sterically analogous alkyl halide without an amino group, a so-called homomorph. Whether the nitrogen atom influenced ionization in the rate-determining step was ascertained by observing whether rate enhancement of the amine solvolysis occurred relative to the homomorph.

In the case of γ-aminochlorides with a primary or secondary leaving group, ring-closure (involving anchimeric assistance by nitrogen) to form the azetidinium salt was the reaction observed.

$$Me_2N:\begin{matrix}\\ \diagdown \\ \diagup\end{matrix}\begin{matrix}R\\ |\\ CH-Cl\\ |\\ CH_2-CR_2\end{matrix} \longrightarrow \begin{matrix}Me\\ |\\ Me-N^+-\square\end{matrix}$$

A completely different mechanism goes into operation when the leaving group is tertiary, as in α,α-dimethyl-γ-aminopropyl chlorides. These compounds solvolyze at almost the same rate as their homomorphs, and the products involve considerable fragmentation. Grob called this route the unaccelerated two-step mechanism of fragmentation

$$Me_2N-CH_2-CH_2-\underset{Me}{\overset{Me}{\underset{|}{\overset{|}{C}}}}-Cl \xrightarrow[SOH]{k_1} \left[Me_2NCH_2CH_2-\underset{Me}{\overset{Me}{\underset{|}{\overset{|}{C^+}}}}\right]$$

$$\xrightarrow{k_t} [Me_2N^+=CH_2] + CH_2=CMe_2$$

Processes other than fragmentation (43%) were involved as well, namely, solvolysis (25%) and ring closure (3%). Grob also encountered an accelerated two-step mechanism

$$Me_2N-CH_2-\underset{Me}{\overset{Me}{\underset{|}{\overset{|}{C}}}}-\underset{Me}{\overset{Me}{\underset{|}{\overset{|}{C}}}}-Cl \xrightarrow{k_1} Me_2N^+\cdots\cdots\cdots\underset{\underset{Me}{\overset{|}{C}}-Me}{\overset{Me_2}{\underset{|}{C}}}\cdots Cl^- \longrightarrow$$

$$Me_2\overset{+}{N}=CH_2 + Me_2C=CMe_2$$
$$72\%$$

Elimination also occurred in this case to the extent of 20%. With geminal methyl groups in the γ position, fragmentation was enhanced (80%, elimination being 20%), and rate enhancement (relative to the homomorph) again occurred.

Finally, a synchronous or one-step fragmentation process was detailed. The stereochemical requirements are strict, and both bonds undergoing cleavage (i.e., $C_\alpha X$ and $C_\beta C_\gamma$ bonds) as well as the electron pair on the nitrogen atom must be antiparallel to each other or must enter that way. Tables I and II illustrate how the operation of steric effects proves decisive.

Perhaps one of the most striking results was encountered with 4-bromoquinuclidine (Table II): the products involved only fragmentation and yet the rate increase relative to its homomorph was 53,000. The rate of the

## TABLE I
### Fragmentation Rates and Products

| Halide | Product | $10^6 k$ | $k_{rel.}$ |
|---|---|---|---|
| (1) 3-β-Tropanyl chloride | $\xrightarrow{100\%}$ N-methyl pyrrolinium–CH$_2$–CH=CH$_2$ | 300 | $2.5 \times 10^4$ |
| (2) 3-β-Nortropanyl chloride | $\xrightarrow{100\%}$ pyrrolinium–CH$_2$–CH=CH$_2$ | 16.7 | 1400 |
| (3) 3-α-Tropanyl chloride | 3-α-Tropanol (74%)<br>3-β-Tropanol (6%)<br>Tropidine (20%) | 7.72 | 645 |
| (4) cyclohexyl chloride | | 0.012 | 1 |

## TABLE II
### Bridgehead Halides

| Halide | Product | $k\ (40°)$ | $k_{rel.}$ |
|---|---|---|---|
| bicyclo Br | bicyclo OR | calc. $8.7 \times 10^{-10}$ | 1 |
| quinuclidinyl Br | =CH$_2$ / =CH$_2$ iminium | $4.6 \times 10^{-5}$ | $5.3 \times 10^4$ |
| $t$-BuBr | | calc. $2.4 \times 10^{-3}$ | $27 \times 10^6$ |

heterocyclic compound approaches that of *t*-butyl bromide in spite of its bridgehead bromine atom. Grob has extended these investigations to certain Bechmann rearrangements also, in particular the fragmentations of the *syn* and *anti* forms of α-amino ketoxime derivatives, thus

$$\underset{\overset{|}{C_6H_5}}{C_6H_{10}\overset{+}{N}-CH_2-C=N-OCOC_6H_5} \longrightarrow C_6H_{10}\overset{+}{N}=CH_2 + C_6H_5CN$$

**c. Aminoacylium Ions.** There is a third category of compounds, namely, α-aminoacyl derivatives, whose rates of solvolysis suggest that azacationic character may be present. The kind of stabilization involved here is illustrated by

$$R_2N-C^+=O \longleftrightarrow R_2N^+=C=O$$

$$R_2NSO_2Cl \longrightarrow [R_2N-\overset{+}{S}O_2 \longleftrightarrow R_2\overset{+}{N}=SO_2]\ Cl^-$$

$$(R_2N)_2PCl \longrightarrow \left[ R_2N-\underset{+}{\overset{\overset{O}{\|}}{P}}-NR_2 \longleftrightarrow R_2\overset{+}{N}-\overset{\overset{O}{\|}}{P}-NR_2 \right] Cl^-$$

Hall has obtained evidence, using kinetic and product data, for all three types of acylium ions just listed and has recently summarized his findings (12). Toward mercuric perchlorate in water, dimethylcarbamyl chloride, dimethylsulfamyl chloride, tetramethyl diamido phosphorochloridate, and benzoyl chloride reacted rapidly to form the corresponding acylium ions. Benzenesulfonyl chloride, diethylphosphorochloridate and *n*-butyl chloroformate were completely inert under these conditions. While the dimethylcarbamyl cation (Me$_2\overset{+}{N}$CO) was trapped in low yields by azide ion, attempts to trap the corresponding sulfur and phosphorus cations by a variety of nucleophilic reagents were unsuccessful. The existence and trapping of the parent carbamylium ion (NH$_2$CO$^+$) has been demonstrated by Holmann et al. (13) in a kinetic and tracer study of the reactions of carbamoyl phosphate (NH$_2$COOPO$_3$H$_2$) in aqueous solution. Protonation of amides may also lead to species with aminocarbonium ion character (14a).

Klages (14) prepared some crystalline salts from antimony pentachloride and the appropriate carbamyl halide: Me$_2$NCO$^+$ SbCl$_6^-$, m.p. 136–140°; Et$_2$NCO$^+$ SbCl$_6^-$, m.p. 96–100°, and he confirmed their saltlike character by conductivity measurements in liquid sulfur dioxide.

### 2. *Aminocarbonium Ions in Synthetic Reactions*

Because this review is confined within the limits described, we exclude from our discussion the great variety of synthetic reactions discussed by Hellmann and Optiz (1) and by Blicke (2), in which azacarbonium ions may be involved. Similarly, we do not elaborate on the fields (e.g., certain alkaloidal reactions) that may entail fragmentation reactions other than outlining the basic work of Grob. This limitation is both to conserve space inasmuch as these topics have been covered elsewhere and also to focus our viewpoint on the species rather than its ubiquity. There are several mechanistic and synthetic areas left, however, which we feel it desirable to include—in particular, amine oxidations and reactions of α-dialkylamino compounds.

**a. Amine Oxidations.** Since it has been suggested that the oxidation of tertiary aliphatic amines with quinones (15) or with peroxides (16) involves azacarbonium ion species, we may write

$$\ce{>N-CH_2-CH_3} \xrightarrow{\text{Oxid}} [\ce{>N+=CH-CH_3}] \xrightarrow{-H^+} \ce{>NCH=CH_2}$$

$$\ce{>N-CH_2C_6H_5} \xrightarrow{\text{Oxid}} [\ce{>N+=CHC_6H_5}] \xrightarrow{H_2O} \ce{>NH} + C_6H_5CHO$$

Equally, in the so-called Polonovskii (17) reaction (namely, the reaction of amine oxides and acid anhydrides), the $N$-acyloxyammonium salts formed initially (18) can cleave easily to form azacarbonium ions (19).

$$\underset{\underset{CH_3}{|}}{R_2 N^+ OCOR} \xrightarrow{B^-} [R_2N^+=CH_2] + BH + RCOO^-$$

Other reactions that may involve such species include the reaction of aliphatic amines with $t$-butyl hydroperoxide (20) and the oxidation of tertiary amines using mercuric acetate (21).

Nitrosative cleavage of tertiary amines, a reaction that has aroused considerable discussion, is another area in which the presence of azacarbonium ion intermediates has been proposed. Very recently, Smith and Loeppky (22) described a considerable amount of evidence in support of the following path:

$$R_2N-CHR'_2 \xrightarrow{HNO_2} R_2\overset{+}{N}\underset{NO}{\overset{CHR'_2}{<}} \longrightarrow R_2\overset{+}{N}=CR'_2 + HNO$$

$$R_2\overset{+}{N}=CR'_2 \xrightarrow{H_2O} R'_2-C=O + R_2N^+H_2 \longrightarrow R_2N-NO$$

$$2HNO \longrightarrow H_2N_2O_2 \longrightarrow H_2O + N_2O$$

When the amine nitrosation was carried out in nonaqueous media with nitrosyl chloride and fluoroborate, the absence of water prevented the completion of the foregoing sequence of reactions, and some of the intermediate species could be isolated. For example, nitrosyl fluoroborate suspended in chloroform reacted with tribenzylamine at $-10°$ to give products that were isolated as a mixture of colorless solids by addition of petroleum ether. The infrared $\left(\diagup C=\overset{+}{N}\diagdown \text{ at } 1650 \text{ cm}^{-1}\right)$ and nmr spectra (singlet at $\tau = 0.50$, $\diagup\overset{+}{N}=\underset{|}{C}-H$, cis and trans $\diagup CH_2$ $\tau$, 4.78 and 4.61) of this mixture showed the presence of N,N-dibenzylbenzaldimmonium fluoroborate (8) (ca. 40%). On standing in moist air, this material slowly changed to dibenzylammonium fluoroborate, as evidenced by the

$$\begin{bmatrix} C_6H_5-CH_2 & & C_6H_5 \\ & \diagdown N^+=C \diagup & \\ C_6H_5-CH_2 & & H \end{bmatrix} BF_4^-$$
(8)

gradual disappearance of the $\diagup C=\overset{+}{N}\diagdown$ band (1650 cm$^{-1}$) and appearance of one at 3400 cm$^{-1}$ (N$^+$–H) in the infrared spectrum of the solid mixture.

It has also been theorized that hydride transfer from tertiary amines involves an aminocarbonium species. This reaction, with triphenylcarbonium ion as oxidant, has been investigated by Damico and Broaddus (23). These workers have found that if an amine contains only α-hydrogens the corresponding "ternary iminium salt" can be isolated. When β-hydrogens are present, this intermediate reacts with the amine producing an

$$\overset{R}{\underset{R'}{\diagdown}}N^+=CHR'' \longleftrightarrow \overset{R}{\underset{R'}{\diagdown}}N-\overset{+}{C}HR''$$

enamine and an amine hydroanion salt. In the case of the latter reaction, the involvement of the azacarbonium ion was detected indirectly, the overall reaction being pictured as follows:

$(CH_3)_2N(CH_2)_nCH_3 + Ph_3C^+ \ An^- \longrightarrow$
(A)         (B)
$(CH_3)_2N^+=CH-(CH_2)_{n-1}CH_3 + Ph_3CH$
(C)

$A + C \rightleftharpoons (CH_3)_2N^+H(CH_2)_nCH_3 \ An^- + (CH_3)_2NCH=CH-(CH_2)_{n-2}CH_3$
               (D)                                (E)

$E + B \longrightarrow Ph_3CH + (CH_3)_2N^+=CH-CH=CH-(CH_2)_{n-3}CH_3 \ An^-$
                                    (F)

$F + A \longrightarrow D + (CH_3)_2N-CH=CH-CH=CH-(CH_2)_{n-4}CH_3$
                                                            $\downarrow$
                                             undergoes further reactions

Finally, in another recent study (24) centering on the cleavage of amines with chlorine dioxide, similar azacationic species have been suggested. The overall reaction involved was, e.g.,

$$(CH_3CH_2)_3N + 2ClO_2 + H_2O \longrightarrow (CH_3CH_2)_2N^+H_2\ ClO_2^- + HClO_2 + CH_3CHO$$

From their kinetic studies of the reaction, Rosenblatt et al. favor a rate-determining step involving electron transfer from the amine with the resultant formation of an aminium cation radical. A fast proton abstraction from this intermediate results in the formation of a species with a three-electron $\pi$ bond. The overall mechanism postulated by these workers is depicted by the following sequence:

$$ClO_2 + RCH_2N\begin{smallmatrix}R'\\ \\R'\end{smallmatrix} \underset{}{\overset{slow}{\rightleftarrows}} ClO_2^- + RCH_2\overset{\cdot}{N}{}^+\begin{smallmatrix}R'\\ \\R'\end{smallmatrix}$$
$$(A)$$

$$A \xrightarrow{fast} H^+ + RCH\overset{\cdot}{\cdots}N\begin{smallmatrix}R'\\ \\R'\end{smallmatrix}$$
$$(B)$$

$$B \xrightarrow[ClO_2]{fast} RCH=\overset{+}{N}\begin{smallmatrix}R'\\ \\R'\end{smallmatrix} + ClO_2^-$$

$$RCH=\overset{+}{N}\begin{smallmatrix}R'\\ \\R'\end{smallmatrix} + H_2O \longrightarrow RCHO + \begin{smallmatrix}R'\\ \\R'\end{smallmatrix}\!\!NH_2{}^+$$

In more recent studies Masui et al. (24a) and Andrieux and Saveant (24b) have quoted further evidence for the presence of aminocarbonium ions in the anodic oxidation of a range of aliphatic amines. Oxidation of substituted tetrazenes may also involve aminocarbonium ion intermediates (24c).

**b. Reactions of α-Dialkylamino Compounds.** In addition to the sources already discussed, the species $R_2NCH_2{}^+$ may be generated in a number of other interesting ways (25). Thus Böhme (26,27) has found that α-dialkylaminoesters can generate such azacations

$$R_2NCH_2OAc + HClO_4 \longrightarrow [R_2N\text{---}CH_2{}^+]\ ClO_4^- + HOAc$$

Even α-aminoethers show enhanced reactivity compared to acetals, and this has been ascribed (28) to their facile formation of azacarbonium ions.

As we have already mentioned, similar reactive species can be generated from methylenebisamines (29), and these may be involved in the reactions between α-aminonitriles and transition metals (30) and Grignard reagents (31).

## B. Aminocarbonium Ions Detected under Equilibrium Conditions

### 1. General

In the reactions just described, azacations appeared as transient species generally undetectable save through their influence on reaction kinetics or on the synthetic course of reactions. In this section, data are examined which come closer to isolating such cations for study purposes; i.e., on a time scale they become detectable species or intermediates.

Stewart and Bradley (32), in some interesting early work, investigated the cleavage of α-dialkylaminomethylsulfonic acids and identified therein, chemically and within limits, azacarbonium ions. They found that such substituted sulfonic acids, prepared *in situ*, reacted reasonably readily ($t_{1/2}$, 8 to 10 min) with periodide ion thus

$$R_2N^+HCH_2SO_3^- + I_3^- + H_2O \xrightarrow{H^+} [R_2N^+{=}CH_2] + SO_2^{2+} + 3I^- + 3H^+$$

The resulting solution containing the immonium ion reacted instantaneously with sodium bisulfite (regenerating the sulfonic acid). Formaldehyde was not the reaction product of the periodide reaction inasmuch as it reacts slowly with bisulfite at the pH used, and the presence of dialkylammonium salts did not accelerate the formaldehyde reaction. Interestingly, the substituted aminosulfonic acid reacted at a zero-order rate in iodine, suggesting that the investigators were measuring a rate-determining transformation (elimination of sulfite) of the substituted sulfonic acid. They were able to produce the cycle of reactions: slow iodine titration, sulfite addition, slow iodine titration, at will. In one experiment, after the production of the immonium species, excess iodine trapped it as $ICH_2NR_2I_4$. They were also able to produce an unsaturated reactive species by treating α-dialkylaminoethers ($ROCH_2NR_2'$) with acid. In one such experiment, with *R*-isobutyl, *R'*-ethyl, they isolated a salt for which they suggested the structure $CH_2{=}N^+(C_2H_5)_2$ $Cl^- \cdot 6H_2O$.

Equally interesting is their work (33) on the rate of cleavage of α-dialkylaminosulphonic acids with iodine. In these reactions the roles of the amine moiety or of the acid level in the solution are not simple, and the studies certainly merit reinvestigation. Stewart and Bradley stated that in the equilibrium

$$R_2N^+HCH_2SO_3^- \longrightarrow [R_2N^+{=}CH_2] + HSO_3^-$$

the concentration of the free sulfite ion is less than 1%.

Much more recently, Le Henaff (34), using three analytical techniques (bisulfite treatment, cryoscopic studies, and iodine oxidations with $K_2HgI_4$), examined the mixtures obtained when primary and secondary amines are reacted with formaldehyde under equilibrium conditions at various pH levels. With secondary amines he identified such species as $R_2N^+HCH_2OH$, $R_2NCH_2OH$, and $R_2NCH_2NR_2$. With primary amines the existence of species such as $RN^+H_2CH_2OH$, $(RN{=}CH_2)_3H^+$, $RNHCH_2OH$, and $(RN{=}CH_2)_3$ was evaluated. From our present viewpoint, the only evidence he encountered of azacarbonium ions was very slight. In a concentrated solution of $R_2N^+HCH_2OH$ he found absorption in the ultraviolet at 210 m$\mu$ corresponding to the $CH_2{=}\overset{+}{N}\!\!<$ chromophore.

Very recently, Fernandez (35) examined the cryoscopic behavior of some methylenediamines $(R_2N)_2CH_2$ in concentrated sulfuric acid. These, in contrast with comparable compounds in which the amino functions are separated by more than one methylene group, display van't Hoff $i$ factors of approximately 4. The simplest explanation for these results was the equilibrium

$$R_2NCH_2NR_2 + 2H_2SO_4 \longrightarrow [R_2NHCH_2NHR_2]^{2+} + 2HSO_4^-$$
$$\Updownarrow$$
$$R_2NH_2^+ + 2HSO_4^- + [R_2N^+{=}CH_2] \longleftarrow R_2N^+HCH_2NR_2 + HSO_4^- + H_2SO_4$$

Finally, perhaps one of the most promising approaches to the study of azacarbonium ions in solution has been that of Skell (36). Skell and de Luis examined the nmr spectra of solutions of methylenebisamines in concentrated sulfuric acid media. They found evidence for the following equilibria:

$$Me_2NCH_2NMe_2 \xrightleftharpoons{H^+} Me_2NCH_2^+ + Me_2NH_2^+$$

Deno initially (36) cited $\delta$ values of 3.71 ($\tau$ 6.29) and 7.75 ($\tau$ 2.25) for the methyl and methylene proton signals in the azacarbonium ion, but elsewhere (37) gave amended $\delta$ values of 2.12 ($\tau$ 7.88) and 6.25 ($\tau$ 3.75). The concentration of sulfuric acid is critical because in acid more dilute than 70% (36)—cited also as 50% (37)!—the hydrated form ($R_2NCH_2OH$ and its protonated species) predominated, whereas in 96% sulfuric acid, the original diamine was stable in its double-protonated form.

### 2. Enamine Studies

Enamines (9) are ambident nucleophiles that may react with electrophilic reagents either at carbon or nitrogen to give either (or both) (38,

38a) immonium salts (**10**) (these salts may possess azacarbonium ion character) or vinyl quaternary ammonium salts (**11**).

$$\diagdown C=C-N\diagup + E^+ \longrightarrow \begin{cases} E-C-C=\overset{+}{N}\diagup \longleftrightarrow E-C-\underset{+}{C}-N\diagup \\ \hspace{2.5cm} (\mathbf{10}) \\ \diagdown C=C-N^+-E \\ \hspace{1.5cm} (\mathbf{11}) \end{cases}$$

(**9**)

Spectroscopic evidence (39) is available in this regard concerning the protonation of such enamine species, a reaction which has caused some controversy (40,41). Elguero and co-workers found that a freshly prepared solution of 1-$N$-morpholino-1-isobutene (**12**), in the presence of 6$N$ hydrochloric acid, shows the presence of two forms, the N- and C-protonated species **13** and **14**. The methyl protons in **13** are doublets at 78 and 84 Hz ($J \sim 2$ Hz) and for **14** a doublet at 114 Hz ($J = 7$ Hz). The solution

(**13**)   (**14**)   (**15**)

rapidly developed a doublet at 130 Hz ($J = 7$ Hz) identifiable as originating from isobutyraldehyde and/or its hydrate. Below concentrations of 6$N$ HCl it was not possible to detect the C-protonated species in accord with the results of Stamhuis et al. (40). In the case of the $N$-dimethylamino enamine of the same aldehyde, i.e., **15**, it was possible to detect the C-protonated form in solutions of 1.5$N$ HCl (or stronger). In 3$N$ solution, e.g., the immonium salt form **16** shows a doublet at 118 Hz ($J = 7$ Hz) (carbon–methyl) and two poorly resolved doublets at $-22$ and $-28$ Hz

(**16**)   (**17**)

($J \sim 1$ Hz) (nitrogen–methyl). Similarly, the $N$-methylanilino enamine of isopropyl methyl ketone provided some interesting data. In a freshly prepared solution in 12$N$ HCl was observed the following: the $N$-protonated form **17** with signals at 84 and 91 Hz (isopropyl carbon–methyl), 66 Hz (carbon–methyl), and $-12$ Hz (N$^+$–methyl); two geometrical isomers of the C-protonated form **18** and **19** with the assignments indicated,

```
123 Hz  (CH₃)₂CH      CH₃  33 Hz        108 Hz  (CH₃)₂CH      CH₃  56 Hz
        d, J = 7 Hz   C                         d, J = 7 Hz   C
                      ‖                                       ‖
                      N⁺                                      N⁺
        C₆H₅         CH₃  −39 Hz        −49 Hz  CH₃          C₆H₅
             (18)                                  (19)
```

and isopropyl methyl ketone with a doublet at 128 Hz ($J = 7$ Hz) and another partly resolved doublet at 60 Hz. In this 12$N$ HCl the N-protonated form appears within 5 min, whereas the two C-protonated species appear after 5 hr. In 3$N$ solution only the isomer **18** and the ketone are observed.

Stamhuis and Maas (41a) studied the kinetics of hydrolysis of 1-$N$-morpholino-1-isobutene, 1-$N$-piperidino-1-isobutene, and 1-$N$-pyrrolidino-1-isobutene in aqueous buffer solutions at different temperatures. They confirmed that a rapid equilibrium was established in the solution between the enamine and its N-protonated conjugate acid. They suggested that the rate-determining step in the hydrolysis was a slow proton transfer to the $\beta$-carbon atom to produce the immonium-azacarbonium ion, which reacted with solvent very rapidly. The absence of any ultraviolet absorption between 2200 and 2300 Å ($>\text{C}=\overset{+}{\text{N}}<$) (42) was interpreted by these workers as revealing no buildup of this C-protonated form.

More recently (41b) Maas et al. reexamined the kinetics of hydrolysis of the three isobutene enamines just mentioned. They obtained spectroscopic evidence that confirmed Elguero's (39) data and kinetic evidence that led them to reconsider the rate-determining step in enamine hydrolysis. They found that the rate-determining step was dependent on the pH of the solution and the nature of the amine moiety of the enamine. In all three cases conditions exist in which the rate-determining step is protonation of the $\beta$-carbon atom; but over a range of pH, this step can become rapid and subsequent hydration and dehydration steps slow.

In some other reactions of enamines, immonium species have also been involved. Thus bromination of a substituted enamine produced the immonium species **20** with absorption at 1626 cm$^{-1}$ ($>\text{C}=\overset{+}{\text{N}}<$) in the

```
        C—C=N⁺   Br⁻
        |
        Br
          (20)
```

infrared (43,44). It was discovered that the methylation reactions of other enamines (e.g., of the 1-azabicycloalkene type) also involve both N- and

C-methylation, and this conclusion was supported by nmr and infrared data (45). The reaction of enamines with acyl halides has also been observed to give immonium species (46).

### 3. Imine Protonations

The question arises both in this section, and throughout our review, about the precise nature of imine protonations; i.e., in a species such as $R_2N^+CH_2\ X^-$, to what extent is its actual carbonium ($C^+$) rather than just its ionic character revealed. One answer lies in the display of chemical behavior exhibited by such compounds (25,47), e.g., their ability to take part in Diels-Alder reactions (48) and to $p$-substitute dimethylaniline (26).

$$[R_2NCH_2^+] + \diagup\!\!\!\diagdown \longrightarrow \left[ R_2N\cdots \begin{array}{c} CH_2-CH_2 \\ \\ \cdots CH_2 \end{array} \right] \longrightarrow R_2N^+\begin{array}{c} CH_2 \\ | \\ CH_2 \end{array}\begin{array}{c} C \\ \\ C \end{array}$$

However, such evidence is not unequivocal. Detailed spectroscopic evidence has now become available to answer this query. Olah (49) has examined the nmr spectra of a range of protonated aldimines and ketimines in fluorosulfonic acid, fluorosulfonic acid–antimony pentafluoride, and deuteriosulfuric acid–antimony pentafluoride with a view to the detection of the contributions of immonium (**A**) versus truly aminocarbonium (**B**) forms. The data for one of the six compounds he examined are as follows. *N*-Benzylidene methylamine was prepared as a mixture of the two geome-

$$\begin{array}{c} R_1 \\ \diagdown \\ C=N \\ \diagup \\ R_2 \end{array}\!\!Rr + H^+ \longrightarrow \begin{array}{c} R_1 \\ \diagdown \\ C=N^+ \\ \diagup \\ R_2 \end{array}\!\!\begin{array}{c} R_3 \\ \diagdown \\ H \end{array} \longleftrightarrow \begin{array}{c} R_1 \\ \diagdown \\ C^+-N \\ \diagup \\ R_2 \end{array}\!\!\begin{array}{c} R_3 \\ \diagdown \\ H \end{array}$$

(A)              (B)

tric isomers **21** and **22**. The nmr spectrum in sulfur dioxide showed two quadruplets for the methine proton at $-8.38$ ppm for the *anti* form and at $-8.15$ ppm for the *syn* isomer (**22**), and the two methyl doublets at $-3.47$ and $-3.27$ ppm for **21** and **22**, respectively. The coupling constants $J_{HCNCH_3}$ are 1.9 and 1.6 Hz for **21** and **22**, respectively. The *anti* form (neat) shows the methine proton as a quadruplet at $-8.04$ ppm ($J_{H-H}$ 1.6 Hz), the phenyl protons (main signal) at $-7.54$ ppm, and the methyl protons (doublet) at $-3.33$ ppm ($J_{H-H}$ 1.6 Hz).

$$\begin{array}{c} H \\ \diagdown \\ C=N \\ \diagup \\ C_6H_5 \end{array}\!\!CH_3 \qquad\qquad \begin{array}{c} H \\ \diagdown \\ C=N \\ \diagup \\ C_6H_5 \end{array}\!\!\begin{array}{c} \\ \\ CH_3 \end{array} \qquad\qquad \begin{array}{c} H \\ \diagdown \\ C=N^+ \\ \diagup \\ C_6H_5 \end{array}\!\!\begin{array}{c} CH_3 \\ \\ H \end{array}$$

(21)                (22)              (23)
*anti*                *syn*

In the protonated *trans* imine 23 the methine proton is shifted to $-8.8$ ppm, a broad doublet with $J_{H-H}$ 17 Hz, the phenyl proton to $-8.03$ ppm, and the methyl protons to $-3.84$ ($J_{H-H}$ 4.5 Hz). These data indicate the *trans* immonium form (**23**) based on the 17-Hz coupling. The chemical shift and coupling data suggest that the immonium form is predominant over the aminocarbonium ion form, even though the latter should be stabilized by the neighboring phenyl group. Similar conclusions were obtained from the investigation of protonated *N*-benzylidene aniline, *N*-benzylidene *p*-chloroaniline, *N*-α-methylbenzylidene methylamine, benzhydrilidene methylamine, and *N*-isopropylidene methylamine.

## 4. Mass Spectra

Because of the extensive data in the area of mass spectra and the many possibilities of producing azacations, we include merely some arbitrary references (50).

Djerassi's group have examined the mass spectra of various hydrazones (51a) and semicarbazones (51) and observe in the latter instance, e.g., such ions as ($CH_3CH{=}N^+{=}NH \leftrightarrow CH_3CH^+{-}N{=}NH$), $HC{=}N^+{-}NHCONH_2$, etc. In that regard, incidentally, Nakata and Takematsu (51b) have observed that, under mass spectral conditions, certain semicarbazones are converted to azines (whose signals are then recorded).

In another recent development, Thornton (52) reported the cleavage of tetrazines as involving the following type of scission:

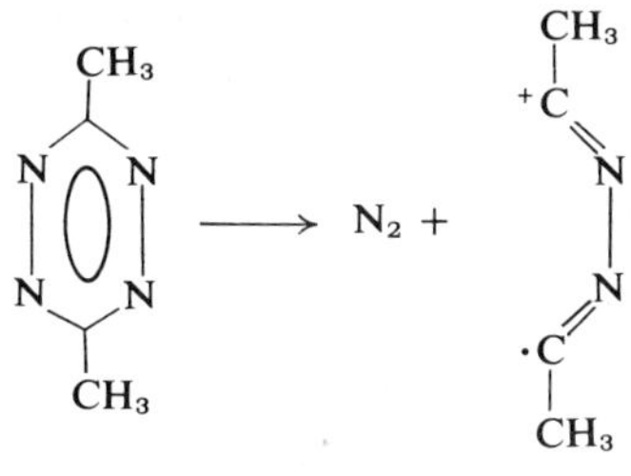

## C. Aminocarbonium Ions as Crystalline Salts

### 1. α-Aminohalides

As we have mentioned, Böhme has extensively examined the use of species such as $R_2NCH_2X$ in generating the azacarbonium ions $R_2NCH_2^+$. He has prepared the parent halides $R_2NCH_2Cl$ by a variety of techniques including, e.g., reacting the more readily accessible α-dialkylaminoethers with inorganic acids or acid halides (53). He has obtained evidence

($-\overset{|}{C}=N-$ stretching at 1681 cm$^{-1}$ in the infrared) that such compounds are in fact the crystalline azacarbonium salts (54)

$$[R_2NCH_2^+ \longleftrightarrow R_2N^+{=}CH_2]\ Cl^-$$

Evidence has also been obtained that some related materials exist as crystalline azacarbonium salts. Thus both Zollinger (55) and Arnold (56) have suggested a polar structure for the amido chlorides (**24**), writing them preferentially with structure **25**. The evidence in favor of the ion-pair

$$R-\underset{\underset{Cl}{|}}{\overset{\overset{Cl}{|}}{C}}-NR_2' \rightleftarrows \left[ R-\underset{+}{\overset{\overset{Cl}{|}}{C}}-NR_2' \longleftrightarrow R-\overset{\overset{Cl}{|}}{C}{=}\overset{+}{N}R_2' \right] Cl^-$$

(**24**)                              (**25**)

formulation (**25**) includes (*a*) infrared $\left(\!\!\diagup\!\!\!\!\diagdown\!\!\!\!\!\! C{=}N-\right)$ absorption in the region of 1667 cm$^{-1}$, (*b*) the insolubility of the materials in nonpolar solvents and their solubility in polar media, (*c*) their chemical behavior with the carbon atom (C$^+$) behaving as a site for nucleophilic attack, (*d*) the fact that solutions of the materials conduct electricity (57). The evidence for saltlike or polar structure is even greater in the case of the related carbamido chlorides (58). Again the compounds reveal the presence of a carbon–nitrogen double bond (infrared absorption at around 1660 cm$^{-1}$), and their relatively high melting points and insolubility in nonpolar solvents all accord with a polar structure (59).

$$R_2N-\underset{\underset{Cl}{|}}{\overset{\overset{Cl}{|}}{C}}-NR_2 \rightleftarrows$$

$$\left[ \underset{\diagup}{\diagdown}\!\underset{+}{N}{-}\overset{\overset{Cl}{|}}{C}{-}N\underset{\diagdown}{\diagup} \longleftrightarrow \underset{\diagup}{\diagdown}N{-}\underset{+}{\overset{\overset{Cl}{|}}{C}}{=}N\underset{\diagdown}{\diagup} \longleftrightarrow \underset{\diagup}{\diagdown}N{=}\overset{\overset{Cl}{|}}{\underset{+}{C}}{-}N\underset{\diagdown}{\diagup} \right] Cl^-$$

Carbodiimides react with hydrochloric acid to form adducts such as ArNHC=NAr and (ArNH)$_2$CCl$_2$ (60), and in the light of the evidence
         |
        Cl

just cited it would appear that these should have ionic character. This point merits reinvestigation. Cyanamide dihydrochloride has been written as the prototype of the type of ions involved NH$_2^+$=C=NH$_2^+$ 2Cl$^-$, on the basis, e.g., of the absence of HCl absorption at 2500–2700 cm$^{-1}$ in the infrared. Interestingly, infrared data suggest that chloroformamidines formed by the addition of acyl halides to carbodiimides thus

$$R-N=C=N-R + R'COCl \longrightarrow R-N=\underset{\underset{COR'}{|}}{\overset{\overset{Cl}{|}}{C}}-N-R$$

are covalent (62). It may be that for ionization to be detected in the solid state with such compounds, both nitrogen atoms have to be reasonably basic.

In early work Reiber and Stewart (30) isolated a variety of immonium salts by the reaction of the corresponding aminonitrile with silver ion, thus

$$R_2C(NR'_2)CN + Ag^+X^- \longrightarrow [R_2C=N^+R'_2]\ X^-$$

They examined the facile hydrolyses of these materials to the corresponding ketone and amine and measured the rapidity of such hydrolyses. Interestingly, they observed some explicit carbon attack by a nucleophile in the following reaction:

$$(CH_3)_2C=N^+(C_2H_5)_2 \xrightarrow{CH_3MgI} (CH_3)_3C-N(C_2H_5)_2$$

In recent studies (62a) reactions of formamidinium salts with organolithium reagents have been described, and a number of new aminocarbonium ion species of this category have been reported (62b,62c). The influence of aminocarbonium ion character on the reactions of 2-chlorobenzimidazolium salts has also been described (62d).

## 2. Immonium Salts

Perhaps the clearest and most unambiguous work on the subject of immonium salts has been the fine work of Leonard and his group. He adapted (63) a previously established technique to prepare crystalline perchlorates of azacarbonium ions (or as he called them "ternary iminium salts"). The general reaction he employed was the simple condensation of a ketone or aldehyde with a secondary amine perchlorate, thus

$$R_2C=O + {}^+NH_2R'_2\ ClO_4^- \longrightarrow [R_2C^+=NR'_2]\ ClO_4^- + H_2O$$

As an example of the ease with which conversion occurs, the mixing of pyrrolidine perchlorate with a slight excess of acetone liberates heat and produces crystalline $N$-isopropylidene pyrrolidinium perchlorate in a matter of seconds. The nmr spectra of some typical salts were analyzed in detail. For example, the spectrum of $N$-isobutylidene pyrrolidinium perchlorate (**26**) consisted of the following signals, expressed in Hertz:

(A) 501 ($J = 9.0$), (B) 249.5, (C) 178, (D) 134, and (E) 79 ($J = 7.0$), the assignments referring to the protons designated. The **A** proton was shown

$$\begin{array}{cc}
\text{(D)} & \text{(B)} \\
\text{CH}_2\text{—CH}_2 \\
| \\
| \\
\text{CH}_2\text{—CH}_2
\end{array}
\underset{/}{\overset{\backslash}{\text{N}^+}}{=}\underset{}{\text{C}}\text{—}\underset{\backslash}{\overset{/}{\text{C}}}\begin{array}{l}\text{H} \quad \text{CH}_3 \text{ (E)} \\ \text{—H} \quad \text{(C)} \\ \text{CH}_3 \text{ (E)}\end{array}
\qquad
\begin{array}{c}\text{CH}_3 \\ \backslash \\ / \\ \text{CH}_3\end{array}\text{C}=\overset{+}{\text{N}}\begin{array}{c}\text{CH}_3 \\ / \\ \backslash \\ \text{CH}_3\end{array} \quad \text{ClO}_4^-$$

(26)       (27)

to be coupled to **C** ($J_{\text{A-C}}$ 9.0 Hz) and to **B** ($J_{\text{A-B}}$ 2.0 Hz). The magnitude of $J$ for spin-spin coupling through three single bonds and the $\text{\textbackslash}\text{C}{=}\overset{+}{\text{N}}/$ bond is in the range previously observed for long-range coupling in systems of the type $-\underset{\text{H}}{\overset{|}{\text{C}}}{=}\underset{\text{H}}{\overset{|}{\text{C}}}{-}\text{C}\text{\textbackslash}$. The proton **A** was also possibly coupled to **D**, since decoupling from **D** by irradiation at 134 Hz sharpened the signal of **A**. The **B** protons were shown to be coupled to **A** and **D**. The coupling of the **C** proton to the **E** protons of the methyl group was clearly observed when irradiation at 178 Hz caused the signal for the **E** protons to collapse from a doublet ($J_{\text{C-E}}$ 7.0 Hz) to a singlet.

Another compound for which spin-spin decoupling revealed long-range splitting was *N*-isopropylidenedimethylaminium perchlorate (27). The nmr spectrum consisted of two signals split into a quintuplet or septuplet. When the low-field signal was followed and the high-field protons were decoupled, or vice versa, the remaining signal collapsed to a sharp singlet. The coupling constant $J = 1.0$ Hz observed was, therefore, assignable to long-range spin-spin coupling over four single bonds and one $\text{\textbackslash}\text{C}{=}\overset{+}{\text{N}}/$ bond. Long-range coupling was, in fact, observed in the nmr spectra of all but one of the iminium salts Leonard examined. Similar azacarbonium salts to those prepared by Leonard in the manner just described may be made by the alkylation of aldimines or ketimines (64) or by direct combination of an aldehyde or ketone with secondary amine complex salts, e.g., hexahalostannates (65–68), halobismuthates (65–67), haloantimonates (65–68), hexahaloplatinates (66,67), or iodide–silver iodide complexes (69).

These "ternary iminium salts," incidentally, react easily with diazomethane to form aziridinium salts (70), themselves capable of showing the characteristics of β-aminocarbonium salts, thus

$$\text{\textbackslash}\text{C}{=}\text{N}^+\text{\textbackslash} + \text{CH}_2\text{N}_2 \longrightarrow \left[ \underset{\text{CH}_2}{\text{C}\text{—}\text{N}^+_\text{\textbackslash}} \longleftrightarrow \underset{\text{CH}_2}{\text{\textbackslash}\text{C}^+ \quad \text{N}\text{\textbackslash}} \right]$$

In recent studies a new dialkylideneammonium cation that is analogous to an allene has been reported (70a). The synthesis and reactions of dichloromethylenedimethylammonium chloride ("phosgene immonium chloride") have also been described (70b), as well as electrochemical reductions of a range of immonium salts (70c).

### 3. Enamine Salts

It is evident that the same type of immonium compound can be made in at least three different ways, thus

1. $\displaystyle\mathop{C}^{\diagdown}_{\diagup}=N-R + HX \longrightarrow \mathop{C^+}^{\diagdown}_{\diagup}-\mathop{N}^{\,|}_{H}-R\ X^-$

2. $\displaystyle\mathop{C}^{\diagdown}_{\diagup}=O + NHR_2 \cdot HX \longrightarrow \mathop{C}^{\diagdown}_{\diagup}=N^+R_2\ X^-$

3. $\displaystyle\mathop{C}^{\diagdown}_{\diagup}=C-NR_2 + HX \longrightarrow \mathop{C}^{\diagdown}_{\diagup H}-C-NR_2\ X^-$

In this chapter we have covered these routes separately. We discussed the first under the heading imine protonations. The second has been covered in the previous section, and we now turn to the third—enamine salt formation. For convenience we have also divided this single topic into studies under equilibrium conditions, discussed earlier, and the actual isolation and study of crystalline enamine salts, the present topic.

Leonard (71) examined the infrared spectra of a series of cyclic and acyclic $\alpha,\beta$- and $\beta,\gamma$-unsaturated tertiary amines, and their salts (largely perchlorate). He observed that the absorption maximum in the double-bond stretching region does not shift appreciably in going from a $\beta,\gamma$-unsaturated amine to its salt. On the other hand, a decided shift (ca. 20–50 cm$^{-1}$) toward higher frequency (from ca. 1650 to 1680 cm$^{-1}$) was observed in going from an $\alpha,\beta$-unsaturated amine, or enamine, to its salt, corresponding to the structural transformation

$$\mathop{C}^{\diagdown}_{\diagup}=C-N \longrightarrow \mathop{C}^{\diagdown}_{\diagup H}-C=N^+-$$

Similarly, Opitz and co-workers (42) used both ultraviolet and infrared spectra to establish that the hexachlorostannates and hexachloroantimonates of a variety of enamines possessed the immonium (C-protonated) structure. Again strong $\mathop{C}^{\diagdown}=\overset{+}{N}-$ stretching (between 1660 and 1690 cm$^{-1}$) was recorded in the infrared region with the ultraviolet maxima between 2200 and 2300 Å.

With some polyhalogenated enamines such as **28**, stable salts with mineral acids have been reported (72). Thus, although the infrared

$$\underset{Cl}{\overset{Cl}{>}}C=C\underset{N(C_2H_5)_2}{\overset{Cl}{<}}$$
(28)

$$\text{phthalimido-NCH}_2-\underset{\overset{+}{N}(C_2H_5)_2}{\overset{\|}{C}}-Ar\ ClO_4^-$$
(29)

spectrum of compound **28** showed absorption at 1612 cm$^{-1}$, its hydrochloride, hydrobromide, and perchlorate salts displayed intense absorption at 1669 cm$^{-1}$. This shift of 57 cm$^{-1}$ is again consistent with the data obtained by Leonard (71,73). A similar result was the protonation of a phthalimido vinyl compound to form the crystalline colorless perchlorate **29**, which shows no —NH absorption in the infrared but —C=N$^+$— at 1650–1660 cm$^{-1}$ (74). Very recent work in this area includes reactions of enediamines with halogens, which lead to di-imimonium species (74a); mercuration of eneamines, leading to C-mercurated iminium salts, demercuration of which provides a useful route to tertiary amines, (74b); and halogenation of tertiary eneamines, leading to $\beta$-halogen immonium salts (74c).

### 4. Vilsmeier-Haack Complexes

The reactions of amides with acyl halides, e.g., phosphorus oxychloride, to produce formylating species may be called chloroformylation (75) or the Vilsmeier-Haack reaction (76,77). The reaction represents an electrophilic substitution and the amide-halide adduct can have azacarbonium ion character. The overall sequence may be represented by a typical example

$$\text{ArH} + \text{HCONR}_2 \cdot \text{POCl}_3 \longrightarrow [\text{ArCHNR}_2]^+ \text{OPOCl}_2^- \xrightarrow{H_2O} \text{ArCHO}$$

The adduct most commonly employed as a formylating species is that derived from POCl$_3$ and DMF, and the arguments concerning its structure can be extended to other related complexes. We can envisage the formation of an ionic electrophilic entity from DMF and POCl$_3$ as follows:

1. $\underset{CH_3}{\overset{CH_3}{>}}N-\underset{}{\overset{H}{C}}=O + ClPCl_2 \longrightarrow \left[(CH_3)_2N-\overset{H}{C^+}-O-\underset{Cl}{\overset{O}{\overset{\|}{P}}}-Cl\right] Cl^-$
(A)

2. A $\longrightarrow \left[(CH_3)_2N-\overset{H}{C^+}-Cl\right]$ OPOCl$_2^-$
(B)

There has been considerable discussion of whether **A** or **B** represents the structure of the complex. Lorenz (78) suggested structure **B**, but

Arnold (79), Zollinger (80), and Brederick (81), among others, favored structure **A**. Thus Zollinger, e.g., reported imide absorption at 1664 cm$^{-1}$ [an absorption substituent sensitive, although in fact it does not serve to distinguish between **A** and **B**] and carbon–oxygen–phosphorus bond absorptions at 1160 and 1039 cm$^{-1}$. On the basis of conductivity measurements, Cramer (82) suggested a comparable structure for the DMF adduct of phenyl dichlorophosphate ($C_6H_5OPOCl_2$). With organic acyl halides similar complexes have been detected as, e.g., the hygroscopic crystalline adduct **30** formed between benzoyl chloride and formamide (81).

$$\left[ \begin{array}{c} OCOC_6H_5 \\ H-C \\ \overset{+}{N}H_2 \end{array} \longleftrightarrow \begin{array}{c} OCOC_6H_5 \\ H-C^+ \\ NH_2 \end{array} \longleftrightarrow \begin{array}{c} O^+COC_6H_5 \\ H-C \\ NH_2 \end{array} \right] Cl^-$$

(30)

Martin and Martin in a recent fundamental paper (83) reexamined the structural evidence for the DMF-POCl$_3$ and DMF-COCl$_2$ adducts, using nmr data. On the basis of chemical shifts, absence of hydrogen–carbon–oxygen–phosphorus coupling and some chemical evidence they conclude that Lorenz's suggestion was in fact correct and that both adducts possessed structures of type **C**, X being OPOCl$_2^-$ and Cl$^-$, respectively.

$$[(CH_3)_2N-CH^+-Cl \longrightarrow (CH_3)_2N^+=CH-Cl] \; X^-$$
(C)

They examined three factors: the influence of solvent and other environmental factors on the nmr spectra they obtained, the nmr spectra of the hygroscopic adducts, and the nmr spectra of adducts of the formylating complexes and some vinyl ethers. Some of their data are as follows:

$$\begin{array}{c} (2.78-2.79 \text{ ppm}) \\ CH_3 \quad O \\ \diagdown \quad \parallel \\ N-C-H \quad (8.02 \text{ ppm}) \\ \diagup \\ CH_3 \\ (2.960-2.965 \text{ ppm}) \end{array}$$

$$\left[ \begin{array}{c} CH_3 \\ \diagdown \\ N-CH_1{}^+-Cl \\ \diagup \\ CH_3 \end{array} \right] Cl^- \qquad \left[ \begin{array}{c} CH_3 \\ \diagdown \\ N-CH_1{}^+-Cl \\ \diagup \\ CH_3 \end{array} \right] OPOCl_2{}^-$$
(D) (B)

Although as expected (84) the N-methyl groups in dimethylformamide showed CH doublets at different fields because of restricted rotation about the carbon–nitrogen bond, this was not observed with the formylating adducts; **D** displayed a CH$_1$ signal at 11.10 ppm, and CH$_3$ protons at 4.06 ppm, whereas **B** showed CH$_1$ at 10.10 and CH$_3$ protons at 3.95 ppm, respectively. The Martins pointed out that if there were a CH–O–P bond, one would expect CH doublet formation ($J \simeq 17$ Hz) (85).

The secondary adducts of these formylating species possessed more complex nmr spectra, as anticipated. Thus with ethyl vinyl ether the following adduct was isolated with DMF–COCl$_2$:

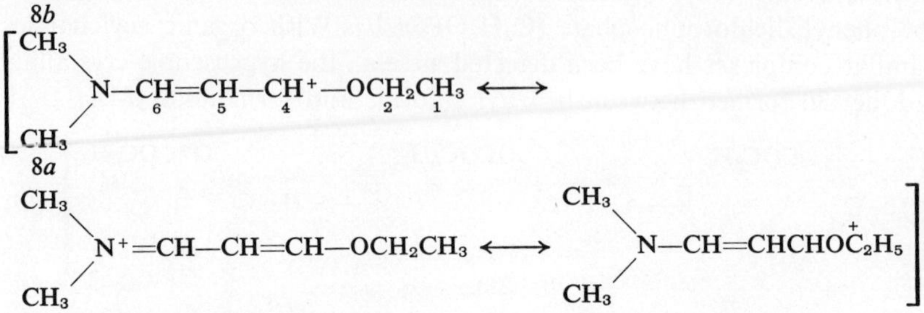

$$\begin{bmatrix} \underset{8b}{\text{CH}_3} \\ \phantom{xx}\diagdown \\ \phantom{xxxx}\text{N}-\underset{6}{\text{CH}}=\underset{5}{\text{CH}}-\underset{4}{\text{CH}^+}-\underset{2}{\text{OCH}_2}\underset{1}{\text{CH}_3} \longleftrightarrow \\ \phantom{xx}\diagup \\ \underset{8a}{\text{CH}_3} \end{bmatrix}$$

$$\text{CH}_3\diagdown\phantom{xxx}\text{N}^+=\text{CH}-\text{CH}=\text{CH}-\text{OCH}_2\text{CH}_3 \longleftrightarrow \text{CH}_3\diagdown\phantom{xxx}\text{N}-\text{CH}=\text{CHCHO}\overset{+}{\text{C}}_2\text{H}_5 \Bigg]$$
$$\text{CH}_3\diagup \phantom{xxxxxxxxxxxxxxxxxxxxxxxxxxxxxx} \text{CH}_3\diagup$$

In this salt, and the related material produced from ethyl vinyl ether and DMF–POCl$_3$, the nitrogen–methyl protons were again doublets, most of the other proton signals being multiplets. These, using the Martins' numbering system, were of the following values: $\delta_1$, 1.44; $\delta_2$, 4.44; $\delta_4$, 8.49; $\delta_5$, 6.28; $\delta_6$, 9.29; $\delta_{8a}$, 3.56; $\delta_{8b}$, 3.73 ppm, with $J_{4-5}$ 11.3 Hz and $J_{5-6}$ 11 Hz. The mechanism for the formation of such adducts may be written as follows:

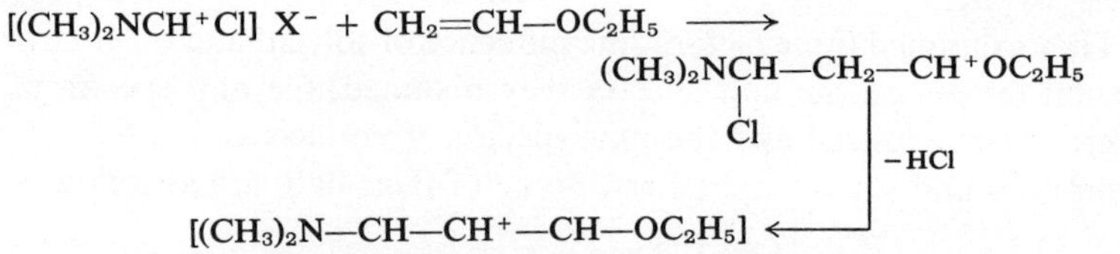

$$[(\text{CH}_3)_2\text{NCH}^+\text{Cl}]\ \text{X}^- + \text{CH}_2=\text{CH}-\text{OC}_2\text{H}_5 \longrightarrow$$
$$(\text{CH}_3)_2\text{NCH}-\text{CH}_2-\text{CH}^+\text{OC}_2\text{H}_5$$
$$\phantom{xxxxxxxxxxxxxxxxxxxxxxxx}|$$
$$\phantom{xxxxxxxxxxxxxxxxxxxxxx}\text{Cl}$$
$$\phantom{xxxxxxxxxxxxxxxxxxxxxxxxxxxx}\Big|-\text{HCl}$$
$$[(\text{CH}_3)_2\text{N}-\text{CH}-\text{CH}^+-\text{CH}-\text{OC}_2\text{H}_5] \longleftarrow$$

The detailed and careful evidence of the Martins does seem to resolve the nature of both the primary formylating adducts and the azacarbonium ions produced from these by electrophilic attack on olefinic substrates.

In a private communication (86), Zollinger stated:

Our conclusions on the constitution of the azacarbonium ion from dimethylformamide and phosphorus oxychloride were based on a C–O–P band in the i.r. and the difference in reactivity of this carbonium ion versus the respective carbonium ion made from dimethylformamide and phosgene.

Martin's arguments are based mainly on nmr and I think that they are more conclusive than our results. They do not, however, explain the difference in the reactivity. My present conclusion therefore is that Martin's proposal for the constitution of the phosphorus-containing carbonium ion is correct but that the reactivity is influenced by the anion, in other words that it is not the bare carbonium ion but an ion pair which enters the Vilsmeier reaction.

Some additional evidence is available concerning the intermediates produced during the reaction also. Thus Smith (87), in studying the formylation of indole succeeded in identifying the cation **31** as an intermediate, basing his identification on ultraviolet spectra and product isolation.

$$\left[\underset{N}{\underset{|}{\bigodot}}\overset{CH-NR_2}{\underset{\oplus}{|}}\right] \qquad \left[(CH_3)_2N-CH=N-\overset{Cl}{\underset{CN}{\underset{|}{C}}}=C-CH=N^+(CH_3)_2\right] Cl^-$$

(31) (32)

When the Vilsmeier reagent (**D**) was treated with malononitrile, an interesting substitution and addition reaction took place to give the immonium salt **32** (88). The nmr spectrum of this salt in trifluoroacetic acid showed four signals for the $N$-methyl species $\tau$ 6.52 (3-H), 6.45 (3-H), 6.37 (3-H), and 6.30 (3-H) and the two methine protons as singlets at $\tau$ 1.55 (1-H) and 1.30 (1-H).

As examples of recent Vilsmeier-Haack processes in which synthetic utility rather than structural features are involved, we may mention the reactions of such complexes with olefins (89) and with hydrazones (90).

Recently $N,N$-dimethylthioformamide has been found to be a versatile reagent in the Vilsmeier reaction (90a). The synthetic utility of Vilsmeier-Haack type complexes of vinyl amide systems has also been described (90b). Treatment of substituted formamides with triethyloxonium tetrafluoroborate has recently been published (90c). Substituted amino methylium tetrafluoroborate salts are formed, and these, when treated with water, cleave to the amine and ethyl formate.

## 5. Polyamino Compounds

The argument regarding when an azacarbonium ion is a carbonium ion rather than an ammonium ion has been brought up earlier in discussing the protonation of enamines and imines. The question is really one of degree. It we look at the sequence

$$\underset{(33)}{-\overset{|}{\underset{|}{C}}\overset{\delta+}{---}\overset{\delta-}{N}\diagdown} \qquad \underset{(34)}{\diagdown\overset{\delta+}{C}=\!=\!=\overset{\delta-}{N}-} \qquad \underset{(35)}{\diagdown\overset{+}{C}=\underset{H}{N}\diagdown} \longleftrightarrow \diagdown\overset{+}{C}-\underset{H}{N}\diagdown$$

it seems a logical progression in increasing electrophilicity of the carbon atom. Even then, however, as Olah's work confirmed, the extent to which formal positive charge is actually localized to any degree on the carbon atom may be quite small. Clearly, with other attached groups capable of charge delocalization, the extent of carbonium ion character is again greatly reduced. Thus the following systems, although formally capable of azacarbonium character, are not discussed in this regard here because of the probable low level to which this character is developed:

$$\underset{(36)}{HC\overset{N<}{\underset{N<}{\oplus}}} \quad \underset{(37)}{HC\overset{NH-}{\underset{OR}{\oplus}}} \quad \underset{(38)}{-NH=C\overset{NH-}{\underset{NH-}{\oplus}}} \quad \underset{(39)}{RS=C\overset{NH-}{\underset{NH-}{\oplus}}}$$

All we offer in this connection are some relevant references. For certain formamidine salts (36), Wellman and Harris (91) have recently reported some nmr spectra and have used these data to examine the *cis–trans* isomerism displayed by the system. The alkylated formamides, which have been extensively used in synthesis by Bredereck and his group (92), also possess extensive charge delocalization as do, to an even greater extent, the triply substituted guanidine (93) and isothiouronium salts (94).

Among the unusual azacations of the guanidinium type is the azide stabilized system recently reported (95), the so-called triazidomethylium salt $(N_3)_3C^+$ $SbCl_6^-$. This was prepared from the reaction

$$3SbCl_4N_3 + CCl_4 \longrightarrow [(N_3)_3C^+] \, SbCl_6^-$$

as a pale yellow shock-sensitive salt, m.p. 145°, whose infrared spectrum indicates that the cation has $C_{3v}$ symmetry with a resonance stabilization similar to guanidinium cations.

For recent work, which illustrates the synthetic utility and the prolific involvement of such extensively delocalized azacarbonium species, see (95a–95f).

## II. AMIDOCARBONIUM IONS [RCONHCH$_2^+$]

By comparison with the efficiency with which an amino group can stabilize adjacent positive charge, we would expect an amido function, being much less basic, to be markedly less efficient. We would therefore predict that amidocarbonium ions would be more difficult to produce than the amino species but that they would be more electrophilic once obtained. Nothing approaching the same degree of quantitative study has been given to these species as has been given to the amino cations discussed earlier, but there has been extensive qualitative study.

Such amido cations (RCONCH$_2^+$) might be produced from three
$\qquad\qquad\qquad\qquad\qquad\quad\;\;\;|$
$\qquad\qquad\qquad\qquad\qquad\quad\;\;R'$
general sources, all of the type RCONR'CH$_2$X, where X is OH (or a derivative thereof, e.g., OR, OCOR), where X is amino NR$_2$ or a related group (NHCOR, NR$_3^+$), or where X is halogen. The production of such azacations from amidomethylol amides is by far the most important route. The reactions of such cations with aromatic compounds has been called

the Tscherniac (96), or more usually, the Einhorn reaction (97). Such cations can also substitute certain reactive aliphatic compounds, both aliphatic and aromatic substitution being conveniently labeled amidoalkylation (98). The topic has been well reviewed (98), and we merely consider those reactions wherein amidocarbonium ions are most likely. For simplicity we also restrict our discussion to the aromatic substitution reactions.

Such a substitution reaction can be written as involving the following steps:

1. $RCONHCH_2OH + H^+ \rightleftharpoonsRCO\overset{+}{N}H_2CH_2OH$
2. $RCONHCH_2OH + H^+ \rightleftharpoons RCONHCH_2OH_2^+$
3. $RCONHCH_2OH_2^+ \overset{K}{\rightleftharpoons}$
   $[RCONH^+=CH_2 \longleftrightarrow RCONH—CH_2^+] + H_2O$
4. $ArX + [RCONHCH_2^+] \overset{k}{\longrightarrow} RCONHCH_2ArX + H^+$

Using amidomethylolamides as reagents in concentrated sulfuric acid as reaction medium, then, the evidence available suggests that such azacations are in fact produced and that reactivity in the aromatic substitution process is determined by the reactivities of such preformed carbonium ions (98b). Thus a general qualitative reactivity sequence with regard to the effect of variation in the amido function on reactivity is as follows:

phthalimido-$NCH_2OH > CCl_3CONHCH_2OH > ClCH_2CONHCH_2OH$
$> C_6H_5CONHCH_2OH$

This order is the reverse of that expected for the stabilities of the resulting carbonium ions (some $\sigma^*$ values being $CCl_3$, $+2.65$; $CH_2Cl$, $+1.05$; and $C_6H_5$, $+0.60$) and suggests that in such concentrated acidic media the effects of substituents on the equilibrium constants ($K$) is more than compensated for by the increasing electrophilicity of the substituting species (i.e., by the increases in $k$).

With regard to the aromatic nucleophilic component, the order of reactivity observed is that operative in electrophilic aromatic substitution (99), with the electrophile being regarded as highly reactive; benzene, e.g., undergoes facile reaction to give the mono- and all three disubstituted amidomethyl products. For the scope of the synthetic use to which the reaction has been put, the reviews (98) should be consulted. An additional attractive synthetic feature is that the acylamido group originally introduced can be easily hydrolyzed, the overall conversion being in effect aminomethylation, thus

$ArH \longrightarrow ArCH_2NHCOR \longrightarrow ArCH_2NH_2$

As we have mentioned, the reactions of α-acylaminohalides can also involve carbonium ions. Thus Cherbuliez and Feer (100) observed that compounds such as $RCONHCH_2Cl$ are as reactive as acid chlorides and that they react readily, e.g., in Friedel-Crafts reactions (101). Böhme (102) extended the reactions of these acylaminohalides to include substitution reactions at acidic –CH sites under basic conditions. Certain of these reactive halides can be generated *in situ* (103), so that

$$ArCH=NR + R'COCl \longrightarrow ArCH(Cl)-N(R)COR'$$
$$\longrightarrow [ArCH=N^+(R)COR']\ Cl^-$$

The same conclusions about the relation between structural features and mode of reactivity can be made in these cases as we have already stated for the amidomethylol compounds. Thus the more electron-withdrawing R is in $RCONHCH_2Cl$, the less likely it is that the halide will react by an $S_N1$ mechanism. Contrast the behavior, e.g., even of the related phthalimido and maleimido species **40** and **41**. Compound **40** undergoes ready solvolysis in water or alcohol (104), but the maleimide halide is only "moderately reactive" and more prone to $S_N2$ than $S_N1$ reactions (105).

(40)       (41)

As noted previously, amidomethylamines $RCONHCH_2NR'_2$ (106,107) can be similarly used to generate the corresponding azacations.

A type of ion intermediate between amino- and amidocarbonium ions has been reported recently by Gross (108). He found that either of the following routes produced the crystalline salt **42**:

$$(R_2N)_2CHCONR_2 + SOCl_2 \longrightarrow$$
$$[R_2N^+=CHCONR_2]\ Cl^-$$
$$(42)$$
$$\begin{array}{c}RO\\ \diagdown\\ \phantom{R_2N}CHCONR_2\\ \diagup\\ R_2N\end{array} \xrightarrow{SOCl_2 \text{ or } AcCl}$$

This material showed infrared absorption at 1660–1670 cm$^{-1}$ due to $>C=\overset{+}{N}<$.

## III. NITRILIUM IONS $\left[RC\overset{+}{=}NR'\right]$

The nitrilium ion species may best be discussed in terms of their methods of synthesis. There may be some overlap because of this whenever the

## AZACARBONIUM IONS

particular species can be made from a variety of materials. Furthermore, although solid salts have been isolated and studied in many of the modes of generation of nitrilium ions, these are not covered separately.

### A. From Nitriles

The reactions from nitriles can be summarized as the attack of various electrophilic species on nitriles. Some of these syntheses (109) have the additional advantage that the nitrilium products can be isolated and their properties studied to substantiate their cationic character. The electrophilic species used to generate the nitrilium salts include electrophilic carbon, protonic acids, and Lewis acids.

#### 1. With Electrophilic Carbon Moieties

**a. By Direct Alkylation with Oxonium Salts (110).** The synthesis of nitrilium ions with electrophilic carbon moieties by direct alkylation with oxonium salts (110) may be accomplished thus

$$CH_3CN + [(C_2H_5)_3O^+] \ BF_4^- \longrightarrow [CH_3C^+=NC_2H_5] \ BF_4^- + (C_2H_5)_2O$$

In the example cited, the $N$-ethyl acetonitrilium fluoroborate was obtained as colorless needles in 89% yield.

A formally similar procedure involves alkylation, using metallic catalysts, e.g., antimony(V) and Fe(III) chlorides, to produce the electrophilic reagent (110), e.g.,

$$RCN + MCl_n + R'Cl \longrightarrow [RC=NR']^+ \ MCl_{n+1}^-$$

The structure of the alkyl halide obviously exerts a profound influence on the velocity of the reaction, the order of reactivity being the expected one: tertiary halides $> 2° > 1°$. As a measure of the ease of reaction $N$-isopropylbenzonitrilium hexachloroantimonate was formed from benzonitrile and isopropyl chloride in 83% yield (after a 24-hr reaction at room temperature) and crystallized with m.p. 138° (decomp.) (110). Sometimes the cationic species produced underwent intramolecular cyclization (111), e.g.,

Similarly synthesized, and usually in higher yields, were the corresponding N-acyl compounds, e.g.,

$$C_6H_5CN + ZnCl_2 + C_6H_5COCl \longrightarrow [C_6H_5C=NCOC_6H_5]^+ \, ZnCl_3^-$$

With excess acyl chloride, compounds involving cations of type
$$\begin{bmatrix} R-C=NCOR \\ | \\ N=CR \end{bmatrix}^+$$
were obtained. With an activated nitrile Lewis acid, catalysis was not required; thus dialkyl cyanamides reacted (112) with benzoyl chloride to form crystalline 3,5-diazapyrylium salts (**43**) and acylnitrilium ions (**44**).

$$R_2N-C\equiv N + C_6H_5COCl \longrightarrow$$

(structure of **43**: a six-membered ring with N, N, O, bearing $C_6H_5$ and $NR_2$ substituents, with $R_2N$ attached)

(**43**)

$$+ \, [R_2N-C\equiv N^+COC_6H_5] \cdot Cl^-$$

(**44**)

With *p*-nitrobenzoyl chloride, the crystalline covalent adduct **45** is formed, and phosgene yields the salt **46**.

$$\underset{(45)}{p\text{-}NO_2C_6H_4\underset{\overset{\|}{O}}{C}-N=\underset{\overset{|}{Cl}}{C}-NR_2} \qquad \underset{(46)}{R_2N-\underset{\overset{|}{Cl}}{C}=N-\underset{\overset{|}{Cl}}{C}=N-\underset{\overset{|}{Cl}}{C}X=\overset{+}{N}R_2 \, Cl^-}$$

**b. The Ritter Reaction.** The phenomenon occurring when olefins (or alcohols) react with nitriles in the presence of strong acids to form amides has been called the Ritter reaction (113). The reaction most likely involves preformation of a carbonium ion which then attacks the nitrile, forming a nitrilium salt or a derivative thereof, which can then be hydrolyzed. The separate steps are as follows:

1. $RCH=CHY + H^+X^- \longrightarrow R-CH^+-CH_2Y$

2. $R-CH^+-CH_2Y + R'CN \longrightarrow \begin{bmatrix} R-CH-CH_2Y \\ | \\ N=\overset{+}{C}-R' \end{bmatrix} X^-$

3. $\begin{bmatrix} RCH-CH_2Y \\ | \\ N=\overset{+}{C}-R' \end{bmatrix} X^- + H_2O \longrightarrow R'CONHCHCH_2-Y \\ \phantom{R'CONHCHCH_2-Y\,} | \\ \phantom{R'CONHCHCH_2-Y\,} R$

The reaction has received considerable synthetic exploitation and has been the subject of some mechanism studies also. For the mechanistic role of the carbonium ion component, the papers of Christol and co-workers (114) provide a useful guide, but perhaps the most valuable material, from the point of view of key references to other work and also because of its clean-cut studies on the mechanism of such Ritter reactions, are the papers of Glikmans and co-workers (115). These workers examined the response of the reaction of isobutene and acrylonitrile to environmental variation with anhydrous acetic acid; added strong acids comprised the main solvent medium. They found that the rate of reaction = $k$[isobutene][acrylonitrile]$H_0^{0.89}$ and concluded that, under their conditions, the rate-determining step was attack of a $t$-butyl carbonium ion pair on the nitrile. They isolated the salt **47** in good yield as the product of such an attack.

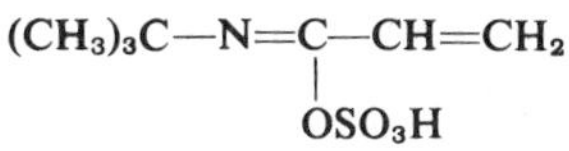

$$(CH_3)_3C-N=C-CH=CH_2 \qquad \left[(CH_3)_3C-N\overset{+}{=}C-CH=CH_2\right] OSO_3H^-$$
$$\qquad\qquad\quad |$$
$$\qquad\qquad OSO_3H$$
$$\qquad\qquad\;\, (47) \qquad\qquad\qquad\qquad\qquad (48)$$

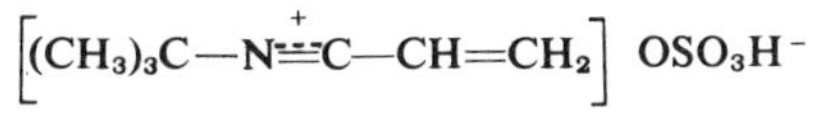

This white salt, m.p. 186°, showed a strong band in the ultraviolet at 2520 Å and in the infrared —C=N— stretching at 1665 cm$^{-1}$, among other bands. Most of the data they report could also be accommodated by the nitrilium salt formulation **48**, which they did not explicitly consider.

There have been a number of interesting variations on the Ritter reaction. Thus formaldehyde has been used as the carbonium ion source, as in the following scheme:

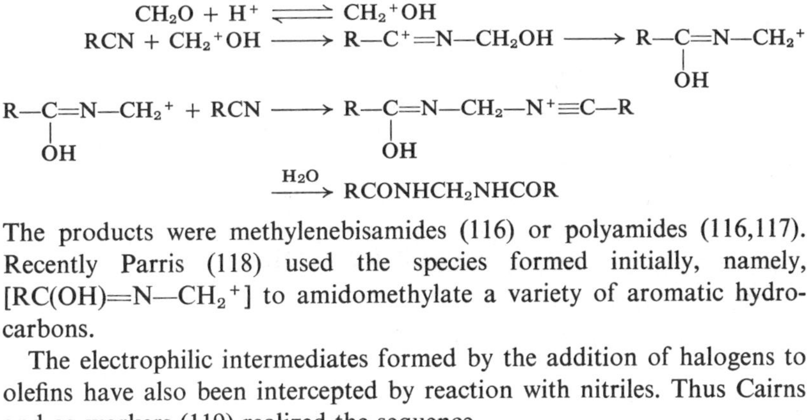

$$CH_2O + H^+ \rightleftharpoons CH_2{}^+OH$$
$$RCN + CH_2{}^+OH \longrightarrow R-C^+=N-CH_2OH \longrightarrow R-C=N-CH_2{}^+$$
$$\qquad\qquad\qquad\qquad\qquad\qquad\qquad\qquad\qquad\qquad\qquad\;\; |$$
$$\qquad\qquad\qquad\qquad\qquad\qquad\qquad\qquad\qquad\qquad\quad OH$$

$$R-C=N-CH_2{}^+ + RCN \longrightarrow R-C=N-CH_2-N^+\equiv C-R$$
$$|\qquad\qquad\qquad\qquad\qquad\qquad\qquad\;\; |$$
$$OH \qquad\qquad\qquad\qquad\qquad\qquad\quad OH$$
$$\qquad\qquad\qquad\qquad \xrightarrow{H_2O} RCONHCH_2NHCOR$$

The products were methylenebisamides (116) or polyamides (116,117). Recently Parris (118) used the species formed initially, namely, [RC(OH)=N—CH$_2{}^+$] to amidomethylate a variety of aromatic hydrocarbons.

The electrophilic intermediates formed by the addition of halogens to olefins have also been intercepted by reaction with nitriles. Thus Cairns and co-workers (119) realized the sequence

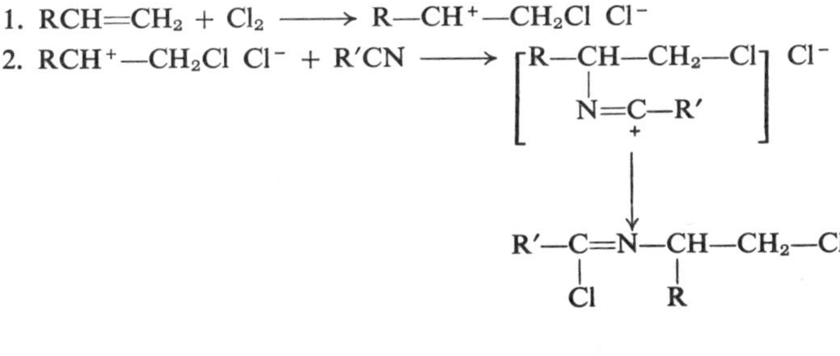

1. $RCH=CH_2 + Cl_2 \longrightarrow R-CH^+-CH_2Cl\;Cl^-$
2. $RCH^+-CH_2Cl\;Cl^- + R'CN \longrightarrow \begin{bmatrix} R-CH-CH_2-Cl \\ | \\ N=C-R' \\ {}_+ \end{bmatrix} Cl^-$

$$\qquad\qquad\qquad\qquad\qquad\qquad\qquad\quad\downarrow$$
$$\qquad\qquad\qquad\qquad\qquad R'-C=N-CH-CH_2-Cl$$
$$\qquad\qquad\qquad\qquad\qquad\qquad |\qquad\quad |$$
$$\qquad\qquad\qquad\qquad\qquad\quad Cl\qquad R$$

This process has been improved by Hassner (120), whose group demonstrated that the process can constitute a stereospecific addition to olefins, exemplified by

$$\text{C}_6\text{H}_{10} + X_2 \longrightarrow \text{[cyclohexane-}X^+ X^-\text{]} \xrightarrow[\text{AgClO}_4]{\text{RCN}} \left[\begin{array}{c}\text{cyclohexane with } X \\ \text{and } N\!\!\equiv\!\!\overset{+}{C}R\end{array}\right] ClO_4^-$$

The nitrilium species was then trapped in a variety of ways, e.g.,

$$\left[\begin{array}{c}\text{cyclohexane with } X \\ N\!\!\equiv\!\!\overset{+}{C}R\end{array}\right] ClO_4^- \xrightarrow{\begin{array}{c}H_2O \\ \\ N_3^-\end{array}} \begin{array}{c}\text{cyclohexane with } X, \text{NHCOR} \\ \\ \text{cyclohexane with } X, N\text{—}C\text{—}R \text{ (tetrazole)}\end{array}$$

A rather similar ring opening, possibly involving nitrilium species, is the following conversion of aziridinium to imidazolinium salts reported by Leonard (121)

$$\overset{R}{\underset{R'}{\text{N}^+}} + R''CN \longrightarrow \left[\begin{array}{c}-N{<}\overset{R}{\underset{R'}{\;}} \\ N\!\!=\!\!\overset{+}{C}\!\text{—}R''\end{array}\right] \longrightarrow \overset{R}{\underset{N}{\text{N}^+}}\overset{R'}{\underset{\;}{\;}}C\!\text{—}R''$$

For recent reviews and discussions of the preparation and reactions of small charged azaring heterocycles, whose reactions clearly involve azacarbonium ions, see Leonard et al. (121a–121c).

**c. By Arylation.** The synthesis of nitrilium ions by arylation using diazonium fluoroborates (110), is as follows:

$$RCN + ArN_2^+ BF_4^- \longrightarrow [RC^+\!\!\equiv\!\!N\text{—}Ar]\, BF_4^-$$

As would be expected, this reaction is sensitive to environmental changes, and by-products may include aryl fluorides as well as quinazolines.

## 2. By Protonation

The adducts of hydrogen halides and various nitriles have been long known and extensively studied (123). The adducts may be of several kinds, and the stoichiometry of HX addition and the structures of the various materials produced has been the subject of some controversies. Hantzsch

(124), e.g., regarded the adducts with mineral acids as the crystalline nitrilium salts [R—C≡NH$^+$]X$^-$ and [R—C≡NH$^+$] HX$_2^-$, but these conclusions have been challenged (125). Some careful work on the adducts derived from acetonitrile has been reported recently by Janz and Danyluk (126). These authors isolated as crystalline materials the compounds CH$_3$CN·2HCl (m.p. 32°), CH$_3$CN·2HBr (m.p. 83–85°), and CH$_3$CN·2HI (m.p. 90–100°). These compounds decomposed on exposure to the atmosphere. The infrared spectrum of CH$_3$CN·2HCl, determined in acetonitrile as solvent, showed strong absorption between 2400 and 2700 cm$^{-1}$ (where the solvent absorbs as well); other bands most consistent with the material were those of the nitrilium salt [CH$_3$C≡NH$^+$] HCl$_2^-$. In the adduct with HBr the infrared spectrum showed greatly diminished C≡N stretching at 2275 cm$^{-1}$, and C=N— stretching appeared at 1630 cm$^{-1}$. Janz and Danyluk concluded that their data for this compound best fitted the imidic halide structure CH$_3$CBr=NH·HBr. Conductivity data for these materials were also obtained and provided information on the progression of the species RCN·2HX from an outer charge-transfer complex to an inner or saltlike form.

Such cations as RCNH$^+$ have electrophilic activity and have been regarded as intermediates in both the Gattermann (127,127a) and Houben-Hoesch (123) reactions. The order of reactivity in the latter reaction is the sequence HC≡NH$^+$ > RCNH$^+$ > C$_6$H$_5$CNH$^+$ (123). Deno (128) has recently examined the nmr spectra of some such cations by observing the nmr spectra of the appropriately substituted nitrile in concentrated sulfuric acid solutions. The reaction involves the following equilibrium:

$$RCN + H^+ \rightleftharpoons [R-C\equiv NH^+ \longleftrightarrow R-C^+=NH]$$

Table III summarizes some of the data Deno obtained. It is evident that the signals for the several species involved were sufficiently different to allow ready identification.

Deno's work was characterized by two additional features. First, he was able to confirm the p$K_a$ data for some nitriles, reported earlier by Hantzsch

TABLE III
Carbon–Hydrogen Nmr Signals, ppm downfield from TMS

| R in RCN | RCN in 70% H$_2$SO$_4$ | RCNH$^+$ in 13% SO$_3$ | RCONH$_3^+$ in 96% H$_2$SO$_4$ | 65% SO$_3$ |
|---|---|---|---|---|
| CH$_3$ | 2.10 | 2.69 | 2.51 | 2.61 |
| CH$_3$CH$_2$ | 2.42 ($\alpha$H) | 3.00 | 2.76 | 2.78 |
| ClCH$_2$ | 4.28 | 4.48 | 4.57 | 4.62 |

(129) but disputed by others (130). When the strength of sulfuric acid required to half-protonate the substrate was used as an index of basicity, the nmr data established the following half-protonation media: $CH_3CH_2CN$ (98% $H_2SO_4$), $CH_3CN$ and $C_6H_5CN$ (100% $H_2SO_4$), and $ClCH_2CN$ (30% oleum). Despite this basicity sequence, the protonated species hydrated in the reverse order (the latter reaction overcompensating for the basicity differences), as Table IV reveals.

**TABLE IV**
Half-times, min at 35°, for Formation of Protonated Amide from Nitrile and/or Protonated Nitrile

| Nitrile | Sulfuric acid in water, % | | |
|---|---|---|---|
| | 70 | 90 | 96 |
| $CH_3CN$ | 125 | 45 | 25 |
| $CH_3CH_2CN$ | 120 | 45 | 30 |
| $C_6H_5CN$ | — | 40 | 35 |
| $ClCH_2CN$ | 20 | 15 | 5 |

In recent fundamental work, protonation of nitriles in $FSO_3H$–$SbF_5$–$SO_2$ solution has been investigated, and the nitrilium ions formed were studied using $H^1$, $C^{13}$, and $N^{15}$ nmr spectroscopy (130a). The synthetic utility of such species in the synthesis of substituted oxazoles has also been described (130b).

### 3. By Reaction with Metallic Cations

In discussing reaction with metallic cations, we are concerned with the feasibility of the reaction

$$RCN + M^+ \longrightarrow R-C^+=N-M$$

As already mentioned, such a reaction performed in the presence of mineral acid constitutes a route to the electrophilic species involved in the Houben-Hoesch reaction (123), where the sequence concerned is

$$RCN + HX \rightleftharpoons RCX=NH \xrightarrow{MX_n} [R-C=NH]^+ \; MX_{n+1}^-$$

Some of the intermediary nitrilium salts have in fact been isolated (131).

Without the presence of mineral acid, complexes between nitriles and metals are still formed, nitriles being well-established ligands. The position of the $C{\equiv}N$ stretching frequency in organic molecules is located at

2220–2250 cm$^{-1}$ (132), and contributions of both forms —C≡N ↔ —C$^+$=N$^-$ are involved. Additional compound formation between such organic nitriles and inorganic halides results in an increase (of between 40 and 100 cm$^{-1}$) in the —C≡N stretching frequency and a marked increase in absorption by the CN group (133). In these compounds the nitrile has enhanced electrophilic character (134) without actually being in the nitrilium state. Thus the infrared spectra of nitrile ligands with a variety of transition and other metals show that the CN bond order is not two (as in a nitrilium species) but if anything has become "more" triple-bonded in character (135). This means, in effect, that the contribution indicated in **A** is replaced by the species **B** in which nitrilium character is

$$R-C\equiv N \longleftrightarrow R-C^+=N^- \qquad R-C\equiv N^+-BCl_3^-$$
$$\text{(A)} \qquad\qquad\qquad \text{(B)}$$

inhibited. Recently a number of new complexes in which nitrilium ion character may be present have been described. These include complexes of acetonitrile with boron trihalides (135a), complexes of dimethylcyanamide and acetonitrile with Ni$^{2+}$ and Lewis acids (135b), and chelation of methylamine to hexakis (methylisocyanide) iron II (135c).

### B. Involving CN Bond-Order Change from 2 to 3

In Section III-A we examined some possible routes to nitrilium ions involving the process

$$RCN + A^+ \longrightarrow [R-C^+=N-A \longleftrightarrow R-C\equiv N^+-A]$$

i.e., a change in which development of nitrilium character is accompanied by a reduction in the CN bond order. We now consider some reactions in which nitrilium species, if produced, would result in an increase in CN bond order.

#### *1. From Imidic Halides*

Nitrilium ions have been produced from imidic halide substrates via salt formation and under solvolysis conditions.

Imidic halides react readily (110,136) with metallic halides to form nitrilium salts, thus

$$\underset{\underset{Cl}{|}}{R-C}=N-R' + SbCl_5 \longrightarrow [RCNR']^+ SbCl_6^-$$

A typical example of this process is the reaction of *N*-methylbenzimidic chloride with antimony pentachloride, both in methylene chloride solution, with the formation of *N*-methylbenzonitrilium hexachloroantimonate, m.p. 219–220° (decomp.) in 75% yield. Halides such as aluminum chloride and titanium tetrachloride have also been used. Gordon and Turrell (137) prepared $[C_6H_5-C\equiv NC_6H_5]^+$ $SbCl_6^-$ as a salt, m.p. 230–233°, whose infrared spectrum (in Nujol and perfluorokerosene) had an intense $-C\equiv N^+-$ stretching at 2300 cm$^{-1}$, and $[CH_3-C\equiv N^+CH_3]$ $SbCl_6^-$, m.p. 178–181°. They analyzed the infrared spectrum of the latter compound in detail and noted intense absorption due to $-C\equiv N^+-$ stretch at 2416 cm$^{-1}$.

The solvolysis of diarylimidic halides (**49**) has recently been shown (138) to involve nitrilium ions also. Working with aqueous acetone as solvent, Ugi's group examined the kinetics of solvolysis of some 33 imidic halides with wide structural variations. They examined salt effects, the influence of added nucleophiles and, to a limited extent (seven compounds), the effect of variation in the substituents X and Y in the materials (**49**) on the

$$p\text{-}XC_6H_4\underset{\underset{\text{(49)}}{|}}{\overset{Cl}{C}}=N-C_6H_4Y(p) + H_2O \longrightarrow [XC_6H_4C\equiv N^+-C_6H_4Y]\ Cl^-$$

$$\downarrow$$

$$XC_6H_4CONH-C_6H_4Y$$

rates of solvolysis. When we fitted Ugi's data to a Hammett ($\rho\sigma$) plot, we obtained a $\rho$ of $-1.2$ ($\pm 0.1$) for the reaction response to variation in X and a $\rho$ of $-0.7$ ($\pm 0.3$) to variation of Y. The negative sign of $\rho$ in both cases supports Ugi's postulate of an ionization mechanism. Very recently kinetic evidence has been reported for the involvement of a nitrilium ion in the isomerization of N-aryl-1-aziridinecarboximidoyl chlorides to carbodimides (138a).

### 2. From Oximes; As Intermediates in the Beckmann Rearrangement

The essential steps in the Beckmann rearrangement (139,140) are (a) conversion of an oxime to a rearranging species, (b) rearrangement, and (c) product formation. Since a variety of reagents, substrates, and reaction conditions have been used to produce rearrangement, no single mechanism can be assigned to cover all these variations. However, many of the Beckmann rearrangement reactions follow the scheme

1. $R-\underset{\underset{N-OA}{\|}}{C}-R' \longrightarrow [R-N=C-R']^+\ OA^-$

2. $[R-N=C-R']^+\ OA^- \longrightarrow \left[\underset{\underset{OA}{|}}{RN=C-R'}\right] \longrightarrow \underset{\underset{A}{|}}{R}-\underset{\underset{O}{\|}}{N}-C-R'$

The nitrilium ion pairs formed initially can be intercepted by added extraneous nucleophiles, e.g., by alcohols to form imido esters (141) and by azide ion to form tetrazoles (142). Both Huisgen (143) and Heldt (144) have suggested that the transition state for the formation of such nitrilium species involves a bridged nonclassical cation. A variety of evidence, including (a) the response of the rearrangement of tosyl oximes to solvent variation (145), (b) the way in which the kinetics of rearrangement of the picryl ethers of oximes of benzocycloalkenones varied with ring size (146), and (c) the influence of substituent effects on the rates of Beckmann transformations (146,147), throws additional light on the nature of this transition state.

From some Beckmann reactions, crystalline nitrilium salts have been isolated. Thus benzophenone oxime and its methyl ether react with excess boron trifluoride or antimony pentachloride, with aryl migration, to form a nitrilium salt as follows (148):

$$(C_6H_5)_2C=NOX + BF_3 \longrightarrow [C_6H_5C\equiv NC_6H_5]^+ \; BF_4^-$$

A 6-hr heating of the oxime with boron trifluoride-etherate yields the diphenylnitrilium salt in 85% yield. The same reaction takes place when benzanilide is used as starting material.

An alternative mode of reaction of oximes, called the Beckmann fission (or fragmentation) reaction (149,149a), can also be visualized as leading to aminocarbonium ions. A generalized example is

$$\underset{NR_2'}{\underset{|}{R-\underset{|}{C}}}-\overset{\overset{NOA}{\|}}{C}-R'' \longrightarrow R_2-C=N^+R_2' + NCR'' + OA^-$$

The reaction can be considered as an elimination process (149,149a) and may share the same transition state as the normal Beckmann process (150) (i.e., it may also possess nitrilium character). In recent work the Beckmann rearrangement of substituted acetophenone oximes in concentrated sulphuric acid has been shown to involve oxime-o-sulphonic acid intermediates from which nitrilium ions are formed by fission of $HSO_4^-$ (150a,150b,150c). The deuterium isotope effect for C—H bond insertion of iminium ions derived from oximes has also been measured (150d). Recently the nitrilium intermediate involved in the Beckmann rearrangement of 2-arylcyclohexanone oxime tosylates was trapped as a pyridinium cation (150e). Synthetic studies of the Beckmann rearrangement with ortho-substituted benzophenone oximes (150f) and α-difluoramino-fluorimines (150g) have also been reported.

Other related rearrangements that may involve nitrilium intermediate- are the Schmidt rearrangement (151) and the rearrangement of N-chloro-ketimines as catalyzed, e.g., by antimony pentachloride (152),

$$R_2C=N-Cl + SbCl_5 \longrightarrow [RC≡NR]^+ \; SbCl_6^-$$

Grob et al. (153) have recently isolated such salts in reactions of this type. The materials they isolated showed the characteristic intense infrared absorption at 2310 cm$^{-1}$ due to [—C≡N$^+$—].

Even more recently, Crawford (154) isolated a similar salt in the reaction

$$\underset{SCH_3}{\underset{|}{C_6H_4}}\overset{CH_3}{\underset{}{C}}=NONa \xrightarrow[Et_2O]{TsCl} CH_3-C≡N^+-\underset{SCH_3}{C_6H_4} \; TsO^-$$

(50)

This material (50), which underwent spontaneous exothermic decomposition at 20°, displayed a very intense characteristic (—C≡N$^+$—) band at 2330 cm$^{-1}$ in its infrared spectrum.

## IV. AMINONITRILIUM IONS [RC≡N—NHR′]$^+$

In the aminonitrilium ion species, the carbonium ion character may be delocalized in three ways

$$\underset{(a)}{R-C^+=N-NH-R'} \longleftrightarrow \underset{(b)}{R^+=C=N-NH-R'} \longleftrightarrow \underset{(c)}{R-C≡N^+-NH-R'}$$

and each represents a way in which the nitrilium ions [RC≡N$^+$R′] can also achieve stabilization. In the present instance, however (i.e., with the insertion of an additional nitrogen atom), canonical form C may be expected to be considerably enhanced because of the adjoined amino function. This amino function can permit another mode of stabilization also, thus

$$\underset{(a)}{R-C^+=N-NH-R'} \rightleftharpoons \underset{(d)}{R-CH=N^+=N-R'} \longleftrightarrow \underset{(e)}{R-CH=N-N=R'^+}$$

That is, following a tautomeric shift, cationic character can be delocalized into R′.

Since no aminonitrilium cations appear to have been obtained as crystalline salts, the evidence for their existence is essentially kinetic. The substrates for their production under these transient conditions have been largely substituted hydrazone species.

## A. From Haloazines

### 1. Ketazines

The action of chlorine at $-60°$ on ketazines leads to dichloroazo-compounds (155), thus

$$R_2C=N-N=CR_2 + Cl_2 \longrightarrow R_2C-N=N-CR_2$$
$$\phantom{R_2C=N-N=CR_2 + Cl_2 \longrightarrow} \underset{Cl}{|} \phantom{N=N} \underset{Cl}{|}$$

Benzing recorded (156) some nucleophilic replacements with such halides, using nucleophiles such as acetate, mercaptides, thiocyanate, cyanide, and azide ions, and reported (157) then on the kinetics and mechanism of their decomposition in aqueous acetone.

Working with 2,2′-dichloro-2,2′-azopropane (**51**) as prototype, he established that the reaction involved was

$$(CH_3)_2\underset{\underset{Cl}{|}}{C}-N=N-\underset{\underset{Cl}{|}}{C}-(CH_3)_2 + 2H_2O \longrightarrow$$
$$(\mathbf{51}) \qquad (CH_3)_2C=O + (CH_3)_2CHOH + N_2 + 2HCl$$

He followed the rate of this decomposition by measuring the rate of nitrogen evolution. This rate was depressed by added chloride ion and accelerated by the presence of other electrolytes. The rates were not liable to acid catalysis, in fact the presence of HCl slowed down the rate of nitrogen evolution.

He regarded (rightly in our opinion) the rate-determining step in the complex series of reactions involved in the fragmentation process as the dissociation to carbonium ions, thus

1. $R_2\underset{\underset{Cl}{|}}{C}-N=N-\underset{\underset{Cl}{|}}{C}R_2 \rightleftharpoons$

$$\left[R_2C^+-N=N-\underset{\underset{Cl}{|}}{C}R_2 \longleftrightarrow R_2C=N^+=N-\underset{\underset{Cl}{|}}{C}R_2\right] Cl^-$$
$$(\mathbf{52})$$

2. $\mathbf{52} + 2H_2O \longrightarrow (CH_3)_2\underset{\underset{OH}{|}}{C}-N=N-\underset{\underset{Cl}{|}}{C}(CH_3)_2$
$$(\mathbf{53})$$

3. $\mathbf{53} \longrightarrow \left[(CH_3)_2\underset{\underset{OH}{|}}{C}-N=N^+=C(CH_3)_2\right] Cl^-, \text{etc.}$

In some elegant recent work McBride (157a) has examined the stereospecific aspects both of ketazine halogenation and of substitution of the products. His work suggests the involvement of various kinds of aminocarbonium ions in both processes.

## 2. Aldazines

With aldazines, halogenation yields hydrazidic halides (158) thus

$$(RCH=N)_2 + Cl_2 \longrightarrow R-\underset{\underset{Cl}{|}}{C}=N-N=CHR \longrightarrow R-\underset{\underset{Cl}{|}}{C}=N-N=\underset{\underset{Cl}{|}}{C}-R$$

Theoretically, azacarbonium ions may be involved in two reactions of these compounds—first in the actual halogenation step, as follows:

$$RCH=N-N=CHR + Cl_2 \longrightarrow \left[ R-\underset{\underset{Cl}{\diagdown\diagup}}{CH\text{---}N}-N=CHR \right] Cl^-$$

and second in the solvolytic reactions of the hydrazidic halides obtained. We have substantial evidence to offer concerning both kinds of reactions with other substrates, but we also have some preliminary data (159) concerning these two sources of azacarbonium ions with aldazine substrates.

We first examined the halogenation step, using a bromination technique for convenience in the rate measurements. Six symmetrically p-disubstituted benzalazines served as substrates; the solvent was 70% aqueous acetic acid containing 0.10M potassium bromide, and the reaction was followed using the amperometric technique we have described elsewhere (160). The reaction was more complex than the corresponding hydrazone reaction (160) in several respects. First, two equivalents of halogen were taken up, rather than a single equivalent as in the latter reaction. Second, the products were not as stable as in the hydrazone case, fragmentation of the initially formed materials being suggested by the brisk evolution of nitrogen that accompanied the azine bromination step. Finally, the influence of substituents on the rate of azine bromine attack was not that expected for an electrophilic substitution reaction inasmuch as it had a positive $\rho$ ($+1.64$) rather than the expected negative one (160). Thus our evidence is insufficient to permit us to comment on the possible intervention of azacarbonium ions in the halogenation stage, and we are giving this reaction additional study.

We have examined (159a) the mechanisms of solvolysis of both the monochloroaldazines (54) to the hydrazides (55) and of the corresponding dichloroazines $(ArCCl=N)_2$ (55A) to oxadiazoles. In 60% aqueous dioxane we studied pH rate profiles for the conversion of compounds (54) to (55); in the region (pH 4–6) where the rates are independent of pH we found strong evidence—common-ion effects, Winstein-Grunwald $m = 0.93$, $\rho$ (for variation of X in 54) $= -1.70$—for azacarbonium ion formation.

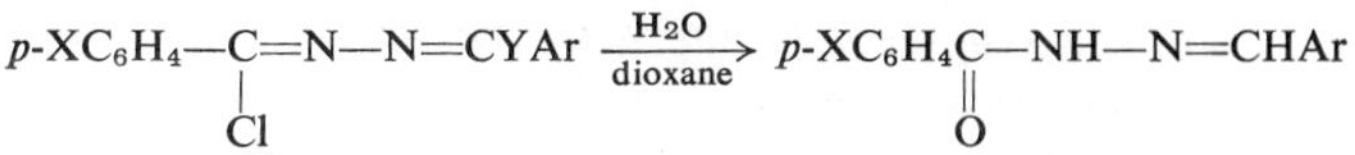

$p$-XC$_6$H$_4$—C(Cl)=N—N=CYAr $\xrightarrow{\text{H}_2\text{O}}_{\text{dioxane}}$ $p$-XC$_6$H$_4$C(=O)—NH—N=CHAr

(54) Y = H  (55)
(55A) Y = Cl

In the same pH region compounds (55A) are remarkably unreactive. In base, however, compounds (55A) are reactive and form oxadiazoles quantitatively. While the rate data are generally complex, at moderate [HO$^-$] ($<5 \times 10^{-2}$M) substituent variations in X cancel out in the various steps involved and cause little overall effect in rate.

Some interesting synthetic work on such haloazines has also been recently recorded (159b).

### B. From Hydrazidic Halides

#### 1. N-Arylhydrazidic Halides: ArC(Br)=NNHAr′ (159c)

We have examined both the kinetics of formation (160) and the kinetics of solvolysis (161) of the N-arylhydrazidic halide species. The synthesis whose rates we examined was the reaction

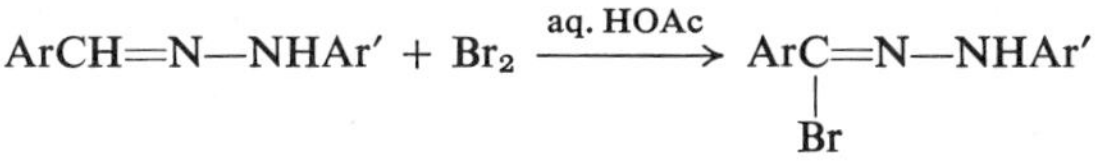

$$\text{ArCH=N—NHAr}' + \text{Br}_2 \xrightarrow{\text{aq. HOAc}} \text{ArC(Br)=N—NHAr}'$$

The kinetics were followed amperometrically, in 70% aqueous acetic acid, and the medium also contained an excess of bromide ion. The effect of substituent variation in Ar on the rate of reaction was quite small ($\rho = -0.62$), whereas, surprisingly, the effect of substituent variation in Ar′ on the rate was much larger ($\rho = -2.2$). One explanation for our data, under our reaction conditions, features the initial formation of an azabromonium ion, stabilized as follows:

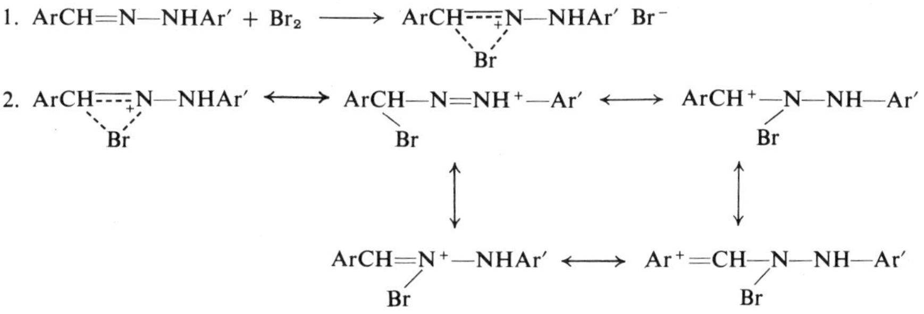

1. ArCH=N—NHAr′ + Br$_2$ ⟶ ArCH(Br)⋯N—NHAr′ Br$^-$

2. ArCH(Br)⋯N—NHAr′ ⟷ ArCH(Br)—N=NH$^+$—Ar′ ⟷ ArCH$^+$(Br)—N—NH—Ar′

   ↕  ↕

   ArCH=N$^+$(Br)—NHAr′ ⟷ Ar$^+$=CH—N(Br)—NH—Ar′

The effective delocalization of charge away from Ar and onto Ar′ is reflected in the appropriate value of $\rho$. The $\Delta S^{\ddagger}$ values we obtained are also consistent with such a picture and with the values obtained for the analogous bromination of styrene (162).

The solvolysis data involved an examination of the rates of hydrolysis in 80% aqueous dioxane, of $N$-arylhydrazidic halides; the reaction was

$$p\text{-}XC_6H_4\overset{\underset{\displaystyle |}{Br}}{C}=N-NHC_6H_4NO_2(p) + H_2O \longrightarrow$$

$$p\text{-}XC_6H_4-\overset{\underset{\displaystyle \|}{O}}{C}-NHNH-C_6H_4NO_2(p) + HBr$$

The rate data can again be fitted to a Hammett plot, with $\rho = -0.93$ ($r = 0.984$, $s = 0.036$) at 50° and $\rho = -0.92$ ($r = 0.987$, $s = 0.04$) at 75°. During the rate runs, the instantaneous values of the rate constants drifted downward slowly (because of common ion rate depression). Again we visualize the rate-determining step as ionization, thus

$$p\text{-}XC_6H_4-\overset{\underset{\displaystyle |}{Br}}{C}=N-NHAr \longrightarrow [p\text{-}XC_6H_4-C\equiv N^+-NHAr]\ Br^-$$

The intermediary carbonium ion is again stabilized as follows:

$$p\text{-}X-C_6H_4-C^+=N-NH-Ar' \longleftrightarrow p\text{-}X^+=C_6H_4=C=N-NH-Ar'$$
$$\longleftrightarrow p\text{-}X-C_6H_4-C\equiv N^+-NH-Ar'$$

In more recent work (163) we have examined some additional features of this solvolysis. We investigated in particular the reaction

$$C_6H_5CBr=N-NHC_6H_4X(p) \xrightarrow{H_2O} C_6H_5CONHNHC_6H_4X(p) + HBr$$

That is, we evaluated the influence of variation in the $N$-aryl moiety on the reaction. Surprisingly, the effect was slight, with $\rho = -0.63$. The observation of a primary salt effect, a large common ion effect, and an $m$ value for this reaction of 0.88 (for a series of binary dioxane–water mixtures) all support the intervention again of azacarbonium ions. We have also examined (163a) the role of *ortho* groups (in the C-phenyl ring) on these solvolyses. We find, in general, that a variety of *ortho*-substituents produce rate accelerations, irrespective of the nature of the substituent (ortho-nitro producing the greatest effect). We ascribe these results to field effects.

## 2. N-Tetrazolylhydrazidic Halides

When solvolyzed in aqueous alcohol, the $N$-tetrazolylhydrazidic halides (164,165) underwent a fragmentation reaction resulting in the formation of triazolyl azides (166). The mechanism of this reaction was initially visualized as involving an anchimerically assisted expulsion of bromide ion by

$$p\text{-XC}_6\text{H}_4\text{—CH=NNH—C}\begin{array}{c}\text{N—N}\\ \phantom{x}\\ \text{HN—N}\end{array} \longrightarrow p\text{-XC}_6\text{H}_4\text{—}\overset{\text{Br}}{\underset{|}{\text{C}}}\text{=NNH—C}\begin{array}{c}\text{N—N}\\ \phantom{x}\\ \text{HN—N}\end{array}$$

$$\downarrow$$

$$p\text{-XC}_6\text{H}_4\text{—C=N}\diagdown\text{NH}\diagup\text{N=C}\diagdown\text{N}_3$$

the neighboring tetrazole group, with simultaneous fission of the tetrazole ring. However, kinetic studies (166) gave a Hammett $\rho$ value of $-1.8$ as the substituent response for the process, thus indicating considerable carbonium ion character in the rate-determining step. Delocalization of the positive charge toward the tetrazole ring again accounts for the relatively low $\rho$ value. We now think that the cyclization reaction entails a primary

$$\left[ p\text{-XC}_6\text{H}_4\text{—C}^+\text{=NNH—C}\begin{array}{c}\text{N—N}\\ \phantom{x}\\ \text{HN—N}\end{array} \longleftrightarrow \quad p\text{-XC}_6\text{H}_4\text{—C}\equiv\text{N}^+\text{—NH—C}\begin{array}{c}\text{N—N}\\ \phantom{x}\\ \text{HN—N}\end{array}\right] \text{Br}^-$$

ionization (partial or complete) of the carbon–bromine bond with subsequent attack of the cationic center by available nucleophiles, the tetrazole ring being the most efficient of these. Competition for the cationic center was effectively demonstrated (166,167) by the addition of selected nucleophiles. When, e.g., a highly nucleophilic environment such as aniline was used for the reaction, the tetrazole ring involvement was eliminated. We have recently published our extended work on these systems (167a). Additional evidence in favor of azocarbonium ion formation was provided by salt effects, with $\rho$ for bromide leaving being $-1.8$ as mentioned, but $\rho = -0.85$ for chloride as leaving group.

Insertion of a methyl group at the 1-position of the tetrazole ring increased the rate of the reaction (with a $\rho$ value of $-2.2$) and had the added effect of preventing tetrazole ring fragmentation; the products were a new ring system and triazolotetrazoles (**56**) (167), but the overall reaction still involved an azacarbonium ion mechanism.

$$p\text{-}XC_6H_5-C\underset{\underset{N=N}{\overset{|}{N-C}}}{\overset{N-N}{\diagdown}}N-CH_3$$

**(56)**

Insertion of benzyl groups at either the 1- or 2-tetrazolyl ring positions changes the bromination kinetics to being syn-anti isomerisations (167b), while the hydrazidic bromides formed still display the kind of reactivity just described for the methyl systems.

We have extended our studies of heterocyclic hydrazidic halides to include 1,2,4-triazoles (167c), imidazolidenes (167d), and 1,3,4-oxadiazoles (167e).

### 3. Nitrilimines

Rather similar species to those described previously can be obtained by the action of bases on hydrazidic halides, thus

$$\underset{\underset{Cl}{|}}{Ar-C}=N-NH-Ar' \xrightarrow[C_6H_6]{Et_3N} [Ar-C^+=N-N^--Ar']$$

These reactive intermediates have been called nitrilimines, and they may be considered the conjugate bases of aminonitrilium ions. They have been elegantly exploited by Huisgen and his group in 1,3-dipolar reactions (168). Several examples of synthetic procedures developed are

$$C_6H_5-C^+=N-N^--C_6H_5 \xrightarrow{CH_6C_5N} C_6H_5-C\underset{N=C-C_6H_5}{\overset{N-N-C_6H_5}{\diagdown}} \quad \text{Ref. 169}$$

$$2\cdot C_6H_5-C^+=N-N^--C_6H_5 \longrightarrow C_6H_5-C\underset{N=N}{\overset{N-N}{\diagdown}}C-C_6H_5 \quad \text{Ref. 170}$$

$$C_6H_5-C^+=N-N^--C_6H_5 \xrightarrow[\text{CH}_2=\overset{CH_3}{\text{C}}\text{COOCH}_3]{} C_6H_5-C\underset{\underset{COOCH_3}{\overset{|}{CH_2-C-CH_3}}}{\overset{N-N-C_6H_5}{\diagdown}} \quad \text{Ref. 171}$$

Using pH profiles, we have examined (169) both the rates of azocarbonium formation and of 1,3-dipolar ion formation in the base-catalyzed hydrolysis of such hydrazidic halides. The major reaction pathway followed depends upon the pH of the medium. At high [H$^+$] the pH independent reaction was identified as unimolecular dissociation of the hydrazidic halide to form an azocarbonium ion. At higher pH, the base-catalyzed process involved loss of bromide ion from the hydrazidic bromide anion to give a 1,3-dipolar ion. Electron-donating substituents attached either to the hydrazidic C-aryl ring or to the hydrazone nitrogen atom aid azocarbonium ion formation. Opposing effects operate in 1,3-dipolar ion formation, which is favored by electron-withdrawing substituents attached to the nitrogen atom and by electron-donating substituents attached to the C-aryl function.

## REFERENCES

1. H. Hellmann and G. Opitz, *α-Aminoalkylierung*, Verlag Chemie, Weinheim, Germany, 1960.
2. For another review of this reaction, see F. F. Blicke, *Organic Reactions*, Vol. I, Wiley, New York, 1942, p. 303 et seq.; see also B. Reichert, *Die Mannich Reaktion*, Springer-Verlag, Berlin, 1959.
3. K. Bodendorf and G. Koralewski, *Arch. Pharm.*, **271**, 101 (1933).
4. S. V. Lieberman and E. C. Wagner, *J. Org. Chem.*, **14**, 1001 (1949).
5. E. R. Alexander and E. J. Underhill, *J. Am. Chem. Soc.*, **71**, 4014 (1949).
6. T. F. Cummings and J. R. Shelton, *J. Org. Chem.*, **25**, 419 (1960).
7. (a) J. H. Burckhalter, J. N. Wells, and W. J. Mayer, *Tetrahedron Lett.*, **1964** (21), 1353. See also J. H. Burckhalter and R. I. Leib, *J. Org. Chem.*, **26**, 4078 (1961).
   (b) Cf. J. E. Fernandez, *Tetrahedron Lett.*, **1964** (39), 2889.
8. J. E. Fernandez, J. S. Fowler, and S. J. Glaros, *J. Org. Chem.*, **30**, 2787 (1965); also J. E. Fernandez and J. S. Fowler, *J. Org. Chem.*, **29**, 402 (1964), and J. E. Fernandez, *Tetrahedron Lett.*, **1964** (39), 2889.
9. L. L. Schaleger and F. A. Long, *Advances in Physical Organic Chemistry*, Vol. I, V. Gold, Ed., Academic Press, New York, 1963, p. 1 et seq.
10. M. Seefelder, *Chem. Ber.*, **99**, 2678 (1966).
10a. V. M. Belikov, Y. N. Belokon, M. M. Dolgaya, and N. S. Martinkova, *Tetrahedron*, **26**, 1199 (1970).
10b. M. Masui, K. Fugita, and H. Ohmori, *Chem. Commun.*, **1970**, 182.
10c. B. B. Thompson, *J. Pharm. Sci.*, **57**, 715 (1968).
10d. W. L. Nobles and L. D. Potty, *J. Pharm. Sci.*, **57**, 1097 (1968).
11. For reviews of this topic see C. A. Grob, *Theoretical Organic Chemistry*, Butterworths, London, 1959, p. 114 et seq.; *Bull. Soc. Chim. France*, **1960**, 1360; *Gazz. Chim. Ital.*, **92**, 902 (1962).
11a. C. A. Grob, *Angew. Chem. Int. Ed.*, **8**, 535 (1969).
12. H. K. Hall, Jr., and C. H. Lueck, *J. Org. Chem.*, **28**, 2818 (1963); cf. also P. S. Traylor and F. H. Westheimer, *J. Am. Chem. Soc.*, **87**, 553 (1965).
13. M. Halmann, A. Lapidot, and D. Samuel, *J. Chem. Soc.*, **1962**, 1944.
14. F. Klages and G. Lukasczyk, *Chem. Ber.*, **96**, 2066 (1963).
14a. K. Yates and J. C. Riordan, *Can. J. Chem.*, **43**, 2328 (1965).
15. D. Buckley, S. Dunstan, and H. B. Henbest, *J. Chem. Soc.*, **1957**, 4880.

16. D. Buckley, S. Dunstan, and H. B. Henbest, *J. Chem. Soc.*, **1957**, 4901.
17. M. Polonovskii and M. Polonovskii, *Compt. Rend.*, **184**, 331 (1927); *Bull. Soc. Chim. France*, **39**, 1147 (1926).
18. M. Polonovskii and M. Polonovskii, *Bull. Soc. Chim. France*, **41**, 1190 (1927); cf. also R. Huisgen, W. Heydkamp, and F. Bayerlein, *Chem. Ber.*, **93**, 363 (1960).
19. R. Huisgen and W. Kolbeck, *Tetrahedron Lett.*, **1965** (12), 783; cf. also V. J. Traynelis and R. F. Martellow, *J. Am. Chem. Soc.*, **80**, **1958**, 6950.
20. Cf. H. E. De La Mare, *J. Org. Chem.*, **25**, 2114 (1960), footnote 12.
21. N. J. Leonard and D. F. Morrow, *J. Am. Chem. Soc.*, **80**, 371 (1958).
22. P. A. S. Smith and R. N. Loeppky, *J. Am. Chem. Soc.*, **89**, 1147 (1967).
23. R. Damico and C. D. Broaddus, *J. Org. Chem.*, **31**, 1607 (1966).
24. D. H. Rosenblatt et al., *J. Am. Chem. Soc.*, **89**, 1158, 1163 (1967); cf. also D. M. Gardner, R. Helitzer, and D. H. Rosenblatt, *J. Org. Chem.*, **32**, 1115 (1967).
24a. M. Masui, H. Sayao, and Y. Tsuda, *J. Chem. Soc. (B)*, **1968**, 973.
24b. C. P. Andrieux and J. M. Saveant, *Bull. Soc. Chim. France*, **1969**, 1254.
24c. W. E. Thun and W. R. McBride, *J. Org. Chem.*, **34**, 2997 (1969).
25. For a recent review outlining some synthetic reactions, see H. Gross and E. Hoft, *Angew. Chem. Int. Ed.*, **6**, 3335 (1967).
26. H. Böhme and K. Hartke, *Chem. Ber.*, **93**, 1310 (1960).
27. H. Böhme, H. J. Bohn, E. Kohler, and J. Rochr, *Justus Liebigs Ann. Chem.*, **664**, 130 (1963).
28. G. M. and R. Robinson, *J. Chem. Soc.*, **123**, 532 (1923).
29. Cf. A. Schonberg, E. Singer, and W. Knofel, *Chem. Ber.*, **99**, 3813 (1966).
30. See, e.g., H. G. Reiber and T. D. Stewart, *J. Am. Chem. Soc.*, **62**, 3026 (1940); H. Larramona, *Compt. Rend.*, **241**, 319 (1955).
31. J. Canceill and J. Jacques, *Bull. Soc. Chim. France*, **1965**, 903; cf. also J. Canceill, M. J. Brienne, and J. Jacques, *ibid.*, **1966**, 3612.
32. T. D. Stewart and W. E. Bradley, *J. Am. Chem. Soc.*, **54**, 4172 (1932).
33. T. D. Stewart and W. E. Bradley, *J. Am. Chem. Soc.*, **54**, 4183 (1932).
34. P. Le Henaff, *Bull. Soc. Chim. France*, **1965**, 3113.
35. J. E. Fernandez and R. Sutor, *J. Org. Chem.*, **32**, 477 (1967).
36. P. S. Skell and J. de Luis, unpublished work, cited by N. C. Deno, *Chem. Eng. News*, October 5 (96), 1964.
37. P. S. Skell and J. de Luis, in Reference 36, referred to by N. C. Deno, *Progress in Physical Organic Chemistry*, Vol. II, S. Cohen, A. Streitwieser, Jr., and R. W Taft, Eds., Interscience, New York, 1964, p. 183.
38. See J. Szmuszkovicz, *Advan. Org. Chem.*, **4**, 3 (1963).
38a. D. Cantacuzene and M. Tordeux, *Tetrahedron Lett.*, **1971**, 4807.
39. J. Elguero, R. Jacquier, and G. Tarrago, *Tetrahedron Lett.*, **1965** (51), 4719; *Tetrahedron Lett.*, **1966** (10), 1112.
40. E. J. Stamhuis, W. Maas, and H. Wynberg, *J. Org. Chem.*, **30**, 2160 (1965).
41. (a) E. J. Stamhuis and W. Maas, *J. Org. Chem.*, **30**, 2156 (1965); (b) W. Maas, M. J. Janssen, E. J. Stamhuis, and H. Wynberg, *J. Org. Chem.*, **32**, 1111 (1967).
42. G. Opitz, H. Hellmann, and H. W. Schubert, *Justus Liebigs Ann. Chem.*, **623**, 117 (1959).
43. R. L. Peterson, J. L. Johnson, R. P. Holysz, and A. C. Oth, *J. Am. Chem. Soc.*, **79**, 1115 (1957).
44. L. Paul, E. Schuster, and G. Hilgetag, *Chem. Ber.*, **100**, 1087 (1967).
45. M. G. Reinecke and L. R. Kray, *J. Org. Chem.*, **30**, 3671 (1965).
46. G. H. Alt and A. J. Speziale, *J. Org. Chem.*, **31**, 1340, 2073 (1966).

47. See, e.g., N. J. Leonard, K. Jann, J. V. Paukstelis, and C. K. Steinhardt, *J. Org. Chem.*, **28**, 1499 (1963); N. J. Leonard, J. V. Paukstelis, and L. E. Brody, *J. Org. Chem.*, **29**, 3383 (1964); V. F. Bystrar et al., *Dokl. Akad. Nauk SSSR*, **148**, 839 (1963); *Chem. Abstr.*, **59**, 6336e (1963).
48. H. Böhme, K. Hartke, and A. Muller, *Chem. Ber.*, **96**, 607 (1963); cf. also H. Böhme and H. Orth, *Chem. Ber.*, **99**, 2842 (1966).
49. G. A. Olah and P. Kreienbuhl, *J. Am. Chem. Soc.*, 1967, in press.
50. (a) See, e.g., J. H. Beynon, *Mass Spectrometry and its Application to Organic Chemistry*, Elsevier, New York, 1960, p. 387 et seq.; H. Budzikiewicz, C. Djerassi, and D. H. Williams, *Interpretation of Mass Spectra of Organic Compounds*, Holden-Day Inc., San Francisco, 1964; *Structure Elucidation of Natural Products by Mass Spectrometry*, Vol. I, *Alkaloids*, Holden-Day Inc., San Francisco, 1964. (b) C. Djerassi et al., *J. Org. Chem.*, **31**, 3666 (1966).
51. (a) C. Djerassi et al., *Chem. Ber.*, **99**, 2284 (1966). (b) H. Nakata and A. Tatematsu, *Chem. Commun.*, **1967**, 208.
52. S. J. Weininger and E. R. Thornton, *J. Am. Chem. Soc.*, **89**, 2050 (1967).
53. H. Böhme, L. Koch, and E. Kohler, *Chem. Ber.*, **95**, 1849 (1962).
54. Cf. also H. Böhme, *Angew. Chem. Int. Ed.*, **4**, 603 (1965).
55. H. H. Bosshard and H. Zollinger, *Helv. Chim. Acta*, **42**, 1659 (1959).
56. Z. Arnold, *Chem. Listy*, **52**, 2013 (1958); *Collect. Czech. Chem. Commun.*, **24**, 4048 (1959).
57. F. Klages and W. Grill, *Justus Liebigs Ann. Chem.*, **594**, 21 (1955).
58. For a review of the chemistry of these materials, see H. Bredereck, R. Gompper, H. G. v. Schuh, and G. Theilig, *Angew. Chem.*, **71**, 753 (1959); H. Eilingsfeld, M. Seefelder, and H. Weidinger, *Angew. Chem.*, **72**, 836 (1960); *Angew. Chem, Int. Ed.*, sample issue, pp. 45–55 (1961); see also H. Eilingsfeld, M. Seefelder, and H. Weidinger, *Chem. Ber.*, **96**, 2899 (1963), and related papers.
59. Cf. H. Eilingsfeld, G. Neubauer, M. Seefelder, and H. Weidinger, *Chem. Ber* **97**, 1232 (1964).
60. For pertinent references, see F. Kurzer and K. Douraghi-Zadeh, *Chem. Rev.*, **67**, 107 (1967).
61. B. I. Sukhornkov and A. I. Finkelstein, *Opt. i Spectrosk.*, **7**, 393 (1959).
62. K. Hartke and E. Polou, *Chem. Ber.*, **99**, 3155, 3163 (1966).
62a. C. F. Hobbs and H. Weingarten, *J. Org. Chem.*, **36**, 2881 (1971).
62b. H. Bohme, G. Auterhoff, and W. Hover, *Chem. Ber.*, **104**, 3350 (1971).
62c. W. Walter and K. P. Ruess, *Chem. Ber.*, **102**, 2640 (1969).
62d. P. Seconi, P. Vivarelli, and A. Ricci, *J. Chem. Soc. (B)*, **1970**, 254.
63. N. J. Leonard and J. V. Paukstelis, *J. Org. Chem.*, **28**, 3021 (1963).
64. For references, see J. Goerdeler, *Methoden der Organischen Chemie*, Vol. XI/2 (Houben-Weyl), Georg Thiele Verlag, Stuttgart, Germany, 1958, pp. 616–618; see also C. R. Hauser and D. Lednicer, *J. Org. Chem.*, **24**, 46 (1959).
65. M. Lamchen, W. Pugh, and A. M. Stephen, *J. Chem. Soc.*, **1954**, 4418.
66. W. Pugh, *J. Chem. Soc.*, **1954**, 2423.
67. M. Lamchen, W. Pugh, and A. M. Stephen, *J. Chem. Soc.*, **1954**, 2429.
68. G. Opitz and W. Menz, *Justus Liebigs Ann. Chem.*, **652**, 139 (1962); see also Reference 42 and G. Opitz, H. Hellmann, and H. W. Schubert, *Justus Liebigs Ann. Chem.*, **623**, 117 (1959).
69. R. Kuhn and H. Schretzmann, *Chem. Ber.*, **90**, 557 (1959).
70. See, e.g., N. J. Leonard and K. Jann, *J. Am. Chem. Soc.*, **84**, 4806 (1962).
70a. B. Samuel and K. Wade, *Chem. Commun.*, **1968**, 1081.

70b. H. G. Viehe and Z. Janousek, *Angew. Chem. Int. Ed.*, **10**, 573, 574, and 575 (1971).
70c. C. P. Andrieux and J. M. Saveant, *Bull. Soc. Chim. France*, **1968**, 4671.
71. N. J. Leonard and V. W. Gash, *J. Am. Chem. Soc.*, **76**, 2781 (1954); see also N. J. Leonard, A. S. Hay, R. W. Fulmer, and V. W. Gash, *ibid.*, **77**, 439 (1955), and N. J. Leonard, C. K. Steinhardt, and C. Lee, *J. Org. Chem.*, **27**, 4027 (1962).
72. A. J. Speziale and R. C. Freeman, *J. Am. Chem. Soc.*, **82**, 909 (1960).
73. N. J. Leonard and F. P. Hauck, Jr., *J. Am. Chem. Soc.*, **79**, 5279 (1957).
74. L. Paul, E. Schuster, and G. Hilgetag, *Chem. Ber.*, **100**, 1087 (1967).
74a. L. Duhamel, P. Duhamel, and G. Ple, *Tetrahedron Lett.*, **1972**, 85.
74b. R. D. Bach and D. K. Mitra, *Chem. Commun.*, **1971**, 1433.
74c. H. Ahlbrecht and M. T. Reiner, *Tetrahedron Lett.*, **1971**, 4901.
75. K. E. Schultz, J. Reisch, and U. Stoess, *Angew. Chem. Int. Ed.*, **4**, 1081 (1965),
76. A. Vilsmeier and A. Haack, *Chem. Ber.*, **60**, 119 (1927).
77. For reviews, see M. R. de Maheas, *Bull. Soc. Chim. France*, **1962**, 1989; G. A. Olah and S. J. Kuhn, in *Friedel-Crafts and Related Reactions*, Vol. III, Pt. 2, G. Olah, Ed., Interscience, New York, 1964, pp. 1211 et seq.
78. H. Lorenz and R. Wizinger, *Helv. Chim. Acta*, **28**, 600 (1948).
79. Z. Arnold and F. Sorm, *Collect. Czech. Chem. Commun.*, **23**, 452 (1958); Z. Arnold, *ibid.*, **24**, 4048 (1959); Z. Arnold and J. Zemlicka, *ibid.*, **24**, 2378 (1959).
80. H. H. Bosshard and H. Zollinger, *Helv. Chim. Acta*, **42**, 1653 (1959); cf. also H. H. Bosshard, E. Jenny, and H. Zollinger, *ibid.*, **44**, 1203 (1961).
81. H. Bredereck, R. Gompper, H. Rempfer, K. Klemm, and H. Keck, *Chem. Ber.*, **92**, 329 (1959); see also H. Bredereck, R. Gompper, K. Klemm, and H. Rempfer, *ibid.*, **92**, 837 (1959); C. Jutz, *ibid.*, **91**, 850 (1958).
82. F. Cramer, S. Rittner, W. Reinhard, and P. Desai, *Chem. Ber.*, **99**, 2252 (1966).
83. G. Martin and M. Martin, *Bull. Soc. Chim. France*, **1963**, 1637; see also H. Normant and G. Martin, *ibid.*, **1963**, 1646.
84. V. J. Kowaleswski and D. G. Kowaleswski, *J. Chem. Phys.*, **32**, 1272 (1960).
85. G. Mavel, *J. Chim. Phys.*, **1962**, 683, 762; G. Mavel and G. Martin, *Compt. Rend.*, **1962**, 254, 260; *ibid.*, **1961**, 253, 2523.
86. Private communication to F. L. Scott, July 3, 1967.
87. G. F. Smith, *J. Chem. Soc.*, **1954**, 3842; cf. also W. C. Anthony, *J. Org. Chem.*, **25**, 2049 (1960).
88. C. Jutz and W. Muller, *Angew. Chem. Int. Ed.*, **5**, 1042 (1966).
89. C. Jutz and W. Muller, *Chem. Ber.*, **100**, 1536 (1967).
90. W. Hoyle, *J. Chem. Soc. (C)*, **1967**, 690.
90a. J. G. Dingwall, D. H. Reid, and K. Wade, *J. Chem. Soc. (C)*, **1969**, 913.
90b. H. Bredereck, F. Effenberger, D. Zeyfang, and K. A. Hirsch, *Chem. Ber.*, **101**, 4036 (1968); H. Bredereck, G. Simchen, and R. Wahl, *Chem. Ber.*, **101**, 4048 (1968).
90c. H. G. Nordmann and F. Krohnke, *Angew. Chem. Int. Ed.*, **8**, 984 (1969).
91. K. M. Wellman and D. L. Harris, *Chem. Commun.*, **1967**, 256.
92. H. Bredereck, F. Effenberger, and E. Henseleit, *Chem. Ber.*, **98**, 2754, 2887 (1965).
93. R. Mecke and W. Kutzelnigg, *Spectrochim. Acta*, **16**, 1225 (1960).
94. C. G. Overberger and H. A. Friedman, *J. Org. Chem.*, **30**, 1926 (1965).
95. U. Muller and K. Dehnike, *Angew. Chem. Int. Ed.*, **5**, 841 (1966).
95a. R. W. Hoffmann, *Angew. Chem. Int. Ed.*, **7**, 754 (1968).

95b. N. Wiberg, *Angew. Chem. Int. Ed.*, **7**, 766 (1968).
95c. T. Nakai and M. Okawara, *Chem. Commun.*, **1971**, 907.
95d. G. A. Olah, D. L. Brydon, and R. D. Porter, *J. Org. Chem.*, **35**, 313, 317, 328 (1970).
95e. R. F. Smith, D. S. Johnson, C. L. Hyde, T. C. Rosenthal, and A. C. Bates, *J. Org. Chem.*, **36**, 1155 (1971).
95f. W. Kutzelnigg and R. Mecke, *Spectrochim. Acta*, **17**, 530 (1961).
96. J. Tscherniac, German Patents, 134,979 and 134,980, 1902.
97. Cf. A. Einhorn, *Justus Liebigs Ann. Chem.*, **361**, 113 (1908).
98. For reviews, see (a) H. Hellmann, *Newer Methods of Preparative Organic Chemistry*, Vol. II, W. Foerst, Ed., Academic Press, New York, 1963, pp. 277 et seq.; (b) H. E. Zaugg and W. B. Martin, *Organic Reactions*, Vol. XIV, Wiley, New York, 1965, pp. 52 et seq.; see also H. E. Zaug, A. M. Kotre, and J. E. Fraser, *J. Org. Chem.*, **34**, 11 (1969).
99. See, e.g., R. O. C. Norman and R. Taylor, *Electrophilic Substitution in Benzenoid Compounds*, Elsevier, Amsterdam, 1965.
100. E. Cherbuliez and E. Feer, *Helv. Chim. Acta*, **5**, 678 (1922).
101. E. Cherbuliez and G. Sulzer, *Helv. Chim. Acta*, **8**, 567 (1925).
102. H. Böhme, R. Broese, and F. Eiden, *Chem. Ber.*, **92**, 1258 (1959); see also H. Böhme, A. Dick, and G. Driesen, *ibid.*, **94**, 1879 (1961); H. Böhme, R. Broese. A. Dick, F. Eiden, and D. Schunemann, *Chem. Ber.*, **92**, 1599 (1959).
103. H. Böhme and K. Hartke, *Chem. Ber.*, **96**, 595, 600 (1963); H. Böhme et al., *ibid.*, **98**, 1463 (1965).
104. J. Sakellarios, *J. Am. Chem. Soc.*, **70**, 2822 (1948).
105. P. O. Tawney, R. H. Snyder, R. P. Conger, K. A. Leibbrand, C. H. Stiteler, and A. R. Williams, *J. Org. Chem.*, **26**, 15 (1961).
106. H. Hellmann, I. Loschmann, and F. Lingers, *Chem. Ber.*, **87**, 1690 (1954).
107. R. O. Atkinson, *J. Chem. Soc.*, **1954**, 1329.
108. H. Gross, J. Gloede, and J. Freiberg, *Justus Liebigs Ann. Chem.*, **702**, 68 (1967).
109. For a review of some of these synthetic procedures, see J. Goerdeler, Reference 64, p. 618.
110. H. Meerwein, P. Laasch, R. Mersch, and J. Spille, *Chem. Ber.*, **89**, 209 (1956); for a comprehensive list of Meerwein's papers in this area, see *Chem. Ber.*, **131**, 100 (1967).
111. M. Lora-Tamayo, G. G. Munoz, and R. Madronero, *Chem. Ind. (London)*, **1959**, 65; *Chem. Ber.*, **93**, 289 (1960); *Bull. Soc. Chim. France*, **1958**, 1331, 1334; *Chem. Ber.*, **97**, 2230, 2234 (1964).
112. K. Bredereck and R. Richter, *Chem. Ber.*, **99**, 2454, 2461 (1966).
113. See, e.g., J. J. Ritter and P. P. Minieri, *J. Am. Chem. Soc.*, **70**, 4045 (1948); J. J. Ritter and J. Kalish, *ibid.*, **70**, 4048 (1948); J. J. Ritter, *ibid.*, **70**, 4253 (1948); F. R. Benson and J. J. Ritter, *ibid.*, **71**, 4130 (1949); R. M. Lusskin and J. J. Ritter, *ibid.*, **72**, 5577 (1950); H. Plant and J. J. Ritter, *ibid.*, **73**, 4076 (1951); J. J. Ritter, U.S. Patent, 2,573,673 (October 30, 1951); E. J. Tillmanns and J. J. Ritter, *J. Org. Chem.*, **22**, 839 (1957), among other papers.
114. See, e.g., H. Christol, A. Laurent, and G. Salladie, *Bull. Soc. Chim. France*, **1963**, 877; H. Christol, A. Laurent, and M. Mousseron, *ibid.*, **1961**, 2319; R. Jacquier and H. Christol, *ibid.*, **1957**, 596, and other papers in the series.
115. G. Glikmans, B. Torck, M. Hellin, and F. Coussemant, *Bull. Soc. Chim. France*, **1966**, 1376; *ibid.*, **1966**, 1383.

116. E. E. Magat, B. F. Faris, J. E. Reith, and L. F. Salisbury, *J. Am. Chem. Soc.*, **73**, 1028 (1951); E. E. Magat and L. F. Salisbury, *ibid.*, **73**, 1035 (1951).
117. D. T. Mowry and E. L. Ringwald, *J. Am. Chem. Soc.*, **72**, 4439 (1950).
118. C. L. Parris and R. M. Christenson, *J. Org. Chem.*, **25**, 1888 (1960).
119. T. L. Cairns, P. J. Graham, P. L. Barrick, and R. S. Schreiber, *J. Org. Chem.*, **17**, 751 (1952).
120. A. Hassner, L. A. Levy, and R. Gault, *Tetrahedron Lett.*, **1966** (27), 3119.
121. N. J. Leonard and L. E. Brady, *J. Org. Chem.*, **30**, 817 (1965).
121a. D. R. Crist and N. J. Leonard, *Angew. Chem. Int. Ed.*, **8**, 962 (1969).
121b. N. J. Leonard, D. B. Dixon, and T. R. Keenan, *J. Org. Chem.*, **35**, 3488 (1970).
121c. T. R. Keenan and N. J. Leonard, *J. Am. Chem. Soc.*, **93**, 6567 (1971).
122. H. Meerwein, P. Laasch, R. Mersch, and J. Nentwig, *Chem. Ber.*, **89**, 224 (1956).
123. For a review of some of their chemistry, and also additional material relevant to this section, see W. Ruske, in *Friedel-Crafts and Related Reactions*, Vol. III, Pt. 1, G. A. Olah, Ed., Interscience, New York, 1964, pp. 383 et seq.; for the material relevant to the present point, see pp. 396 et seq.
124. A. Hantzsch and W. Gerdel, *Chem. Ber.*, **64**, 667 (1931).
125. See, e.g., L. E. Hinkel and G. J. Treharne, *J. Chem. Soc.*, **1945**, 866.
126. G. J. Janz and S. S. Danyluk, *J. Am. Chem. Soc.*, **81**, 3848, 3850, 3854 (1959).
127. See W. E. Truce, *Organic Reactions*, Vol. IX, pp. 39, 40 (1957); G. A. Olah and S. J. Kuhn, in *Friedel-Crafts and Related Reactions* (Ref. 77), pp. 1202 et seq.
127a. E. Allenstein, A. Schmidt, and V. Beyl, *Chem. Ber.*, **99**, 431 (1966).
128. N. C. Deno, R. W. Gaugler, and M. J. Wisotsky, *J. Org. Chem.*, **31**, 1967 (1966).
129. A. Hantzsch, *J. Physik. Chim.*, **65**, 41 (1909).
130. L. P. Hammett and A. J. Deyrup, *J. Am. Chem. Soc.*, **54**, 4239 (1932); H. Le Maire and H. J. Lucas, *ibid.*, **73**, 5198 (1951).
130a. G. A. Olah and T. E. Kiovsky, *J. Am. Chem. Soc.*, **90**, 4666 (1968).
130b. G. Kille and J. P. Fleury, *Bull. Soc. Chim. France*, **1968**, 4631.
131. F. Klages, R. Ruhnau, and W. Hauser, *Justus Liebigs Ann. Chem.*, **626**, 60 (1959).
132. M. F. Amr El-Sayed and R. K. Sheline, *J. Inorg. Nucl. Chem.*, **6**, 187 (1958).
133. H. G. Coerver and C. Curran, *J. Am. Chem. Soc.*, **80**, 3522 (1958).
134. Compare, e.g., P. Oxley, F. W. Short, and M. Partridge, *J. Chem. Soc.*, **1947**, 1110; N. C. Stephenson, *J. Inorg. Nucl. Chem.*, **24**, 801 (1962).
135. W. Gerrard, M. F. Lappert, H. Pyszora, and J. W. Wallis, *J. Chem. Soc.*, **1960**, 2182.
135a. B. Swanson, D. F. Shriver, and J. A. Ibers, *Inorg. Chem.*, **8**, 2182 (1969).
135b. H. F. Henneike and R. S. Drago, *Inorg. Chem.*, **7**, 1908 (1968).
135c. J. Miller, A. L. Balch, and J. H. Enemark, *J. Am. Chem. Soc.*, **93**, 4613 (1971).
136. F. Klages and W. Grill, *Justus Liebigs Ann. Chem.*, **594**, 21 (1955).
137. L. E. Gordon and G. C. Turrell, *J. Org. Chem.*, **24**, 269 (1959).
138. I. Ugi, F. Beck, and U. Fetzer, *Chem. Ber.*, **95**, 126 (1962).
138a. D. A. Tomalia, T. J. Giacobbe, and W. A. Sprenger, *J. Org. Chem.*, **36**, 2142 (1971).
139. For an excellent critical review, see P. A. S. Smith, in *Molecular Rearrangements*, Vol. I, P. de Mayo, Ed., Interscience, New York, 1963, pp. 483 et seq.
140. See also W. Z. Heldt and L. G. Donaruma, *Organic Reactions*, Vol. 11, Chap. 1 (1964).
141. P. Oxley and W. E. Short, *J. Chem. Soc.*, **1948**, 1514.
142. R. L. Burke and R. M. Herbst, *J. Org. Chem.*, **20**, 726 (1955).

143. R. Huisgen, J. Witte, H. Walz, and W. Jira, *Justus Liebigs Ann. Chem.*, **604**, 191 (1957).
144. W. Z. Heldt, *J. Am. Chem. Soc.*, **80**, 5972 (1958).
145. W. Z. Heldt, *J. Org. Chem.*, **26**, 1695 (1961).
146. R. Huisgen, J. Witte, and I. Ugi, *Chem. Ber.*, **90**, 1844 (1957).
147. See, e.g., P. J. McNulty and D. E. Pearson, *J. Am. Chem. Soc.*, **81**, 612 (1959).
148. H. Meerwein, *Angew. Chem.*, **67**, 374 (1955).
149. Cf., e.g., C. A. Grob, *Bull. Soc. Chim. France*, 1960, 1360; A. F. Ferris, *J. Org. Chem.*, **25**, 12 (1960); L. Bauer and R. E. Hewitson, *J. Org. Chem.*, **27**, 3982 (1962); C. A. Grob and P. W. Schiess, *Angew. Chem. Int. Ed.*, **6**, 1 (1967); A. Hassner and E. Nash, *Tetrahedron Lett.*, **1965** (9), 525.
149a. K. G. Artz and G. L. Grob, *Helv. Chim. Acta*, **51**, 807 (1968); W. Eisele, C. A. Grob, E. Renk, and H. von Tschammer, *Helv. Chim. Acta*, **51**, 816 (1968).
150. Reference 139, pp. 503, 504.
150a. B. J. Gregory, R. B. Moodie, and K. Schofield, *Chem. Commun.*, **1968**, 1380.
150b. B. J. Gregory, R. B. Moodie, and K. Schofield, *Chem. Commun.*, **1969**, 645.
150c. Y. Yukawa and T. Ando, *Chem. Commun.*, **1971**, 1601.
150d. P. T. Lansbury and P. C. Briggs, *Chem. Commun.*, **1969**, 1152.
150e. A. C. Huitric and S. D. Nelson, *J. Org. Chem.*, **34**, 1231 (1969).
150f. H. Watanabe, C. L. Mao, and C. R. Hauser, *J. Org. Chem.*, **34**, 1786 (1969).
150g. T. E. Stevens, *J. Org. Chem.*, **34**, 2451 (1969).
151. Reference 139, p. 509.
152. W. Theilacker and H. Mohl, *Justus Liebigs Ann. Chem.*, **563**, 99 (1949).
153. C. A. Grob, H. P. Fischer, W. Raudenbusch, and J. Zergenyi, *Helv. Chim. Acta*, **47**, 1003 (1964).
154. R. J. Crawford and C. Woo, *J. Org. Chem.*, **31**, 1655 (1966).
155. S. Goldschmidt and B. Acksteiner, *Justus Liebigs Ann. Chem.*, **618**, 173 (1958).
156. E. Benzing, *Justus Liebigs Ann. Chem.*, **631**, 1 (1960).
157. E. Benzing, *Justus Liebigs Ann. Chem.*, **631**, 10 (1960).
157a. D. S. Malament and J. M. McBride, *J. Am. Chem. Soc.*, **92**, 4586, 4593 (1970).
158. R. Stolle, *J. Prakt. Chem.*, **85**, 386 (1912).
159. Unpublished work, Miss A. Fleming, M.Sc. thesis, University College, Cork, 1967.
159a. P. A. Cashell, A. F. Hegarty, and F. L. Scott, *Tetrahedron Lett.*, **1971**, (50), 4767.
159b. W. T. Flowers, D. R. Taylor, A. E. Tipping, and C. N. Wright, *J. Chem. Soc. (C)*, **1971**, 1986, 3097.
159c. For a recent review of hydrazidic halide chemistry see R. N. Butler and F. L. Scott, *Chem. Ind.*, **1970**, 1216.
160. Cf. A. F. Hegarty and F. L. Scott, *J. Chem. Soc. (B)*, **1966**, 672, 1031; *Chem. Commun.*, **1967**, 521; F. L. Scott, F. A. Groeger, and A. F. Hegarty, *J. Chem. Soc. (B)*, **1971**, 1141.
161. F. L. Scott and J. B. Aylward, *Tetrahedron Lett.*, **1965** (13), 841; J. B. Aylward and F. L. Scott, *J. Chem. Soc. (B)*, **1969**, 1080.
162. K. Yates and W. V. Wright, *Tetrahedron Lett.*, **1965** (24), 1927.
163. F. L. Scott, M. Cashman, and A. F. Hegarty, *J. Chem. Soc. (B)*, **1971**, 1607.
163a. A. F. Hegarty, M. Cashman, J. B. Aylward, and F. L. Scott, *J. Chem. Soc. (B)*, **1971**, 1879.
164. F. L. Scott and M. N. Holland, *Proc. Chem. Soc.*, **1962**, 106.

165. See also F. L. Scott and D. A. Cronin, *Chem. Ind. (London)*, **1964**, 1757.
166. F. L. Scott and D. A. Cronin, *Tetrahedron Lett.*, **1963** (11), 715.
167. R. N. Butler and F. L. Scott, *J. Chem. Soc. (C)*, **1967**, 239.
167a. F. L. Scott, D. A. Cronin, and J. K. O'Halloran, *J. Chem. Soc. (C)*, **1971**, 2769.
167b. J. C. Tobin, A. F. Hegarty, and F. L. Scott, *J. Chem. Soc. (B)*, **1971**, 2198.
167c. T. A. F. O'Mahony, P. Quain, A. F. Hegarty, and F. L. Scott, to be published.
167d. F. L. Scott and J. K. O'Halloran, *Tetrahedron Lett.*, **1970** (47), 4083.
167e. R. N. Butler, T. M. Lambe, and F. L. Scott, *J. Chem. Soc., Perkin I*, **1972**, 269; *Tetrahedron Lett.*, **1971**, 1729.
168. For some reviews, see R. Huisgen, *Bull. Soc. Chim. France*, **1965**, 3431; *Angew. Chem.*, **75**, 742 (1963); *Proc. Chem. Soc.*, **1961**, 357.
169. R. Huisgen, R. Grashey, M. Seidel, G. Wallbillich, H. Knupfer, and R. Schmidt, *Justus Liebigs Ann. Chem.*, **653**, 105 (1962).
170. R. Huisgen, E. Aufderhaar, and G. Wallbillich, *Chem. Ber.*, **98**, 1476 (1965).
171. R. Huisgen, R. Sustmann, and G. Wallbillich, *Chem. Ber.*, **100**, 1786 (1967).
172. A. F. Hegarty, M. P. Cashman, and F. L. Scott, *J. Chem. Soc., Perkin II*, **1972**, 44.

CHAPTER 31

# Protonated Heteroaliphatic Compounds

GEORGE A. OLAH AND A. M. WHITE
*Department of Chemistry, Case Western Reserve
University, Cleveland, Ohio*

AND

DANIEL H. O'BRIEN
*Department of Chemistry, Texas A & M University,
College Station, Texas*

| | |
|---|---|
| I. Introduction | 1698 |
| II. Experimental Methods for the Study of Protonated Compounds | 1698 |
|    A. Acid–Solvent Systems | 1699 |
|    B. Cryoscopic and Conductometric Methods | 1703 |
|    C. Spectroscopic Methods | 1704 |
|       1. Vibrational Spectroscopy | 1705 |
|       2. Nmr Spectroscopy | 1706 |
| III. Protonated Alcohols | 1710 |
| IV. Protonated Thiols | 1716 |
| V. Protonated Ethers | 1717 |
|    A. Aliphatic Ethers | 1720 |
|    B. Protonated Alkyl Phenyl Ethers | 1725 |
| VI. Protonated Sulfides | 1730 |
| VII. Diprotonated Alkoxy Alcohols | 1733 |
| VIII. Protonated Aldehydes and Ketones | 1734 |
| IX. Protonated Amides and Thioamides | 1748 |
|    A. General | 1748 |
|    B. The Site of Protonation | 1749 |
|       1. Infrared Evidence | 1749 |
|       2. Nmr Evidence | 1750 |
| X. Protonated Carboxylic Acids, Esters, and Anhydrides | 1754 |
| XI. Protonated Dicarboxylic Acids and Anhydrides | 1763 |
| XII. Protonated Carbonic Acid and Derivatives | 1764 |
| XIII. Protonated Carbamic Acid and Derivatives | 1767 |
| XIV. Protonated Imines | 1771 |
| XV. Protonated Enamines | 1771 |
| XVI. Protonated Ketoximes | 1772 |
| XVII. Protonated Nitriles | 1773 |
| XVIII. Protonated Nitro Compounds | 1773 |
| XIX. Protonated Amino Acids | 1774 |
| References | 1776 |

Reprinted with permission of the American Chemical Society from *Chem. Rev.*, **70**, 561 (1970).

## I. INTRODUCTION

Protonated organic compounds play a vital role as intermediates in acid-catalyzed reactions. This review emphasizes information about the structure of these intermediates as a means toward a better understanding of such reactions. In particular, for each class of compound we will discuss the site of protonation, the conformation of the protonated compound, the electronic distribution in these positively charged species, and some of the reactions they undergo. The recent advances in this field (1a–1c) have been a direct result of the development of superacid systems which, because of their high acidity and low nucleophilicity, enable the preparation of stable solutions of many protonated organic compounds. Nuclear magnetic resonance spectroscopy can be used to observe the acidic protons in the ions directly without the complications arising from proton exchange with the solvent, as generally occurs rapidly in weaker acid systems.

In our review we have perhaps overemphasized the results from nmr studies of superacid solutions of protonated organic compounds and as a consequence failed to discuss in sufficient detail many of the significant results obtained in other acid systems. Our reasons for this approach are our own interest in superacid systems and also our desire to present an as up-to-date account of the field as is compatible with brevity and readability.

We have also been somewhat selective about the classes of compounds dealt with and only discuss oxygen, nitrogen, and sulfur protonation of aliphatic compounds. This has stemmed from the fact that our interest in the direct observation of protonated heteroaliphatic molecules arose from their importance to studies of stable carbonium ions. In fact many of these ions can be considered as substituted carbonium ions. Thus protonated acetone, acetic acid, and carbonic acid correspond to dimethylhydroxycarbonium ion, dihydroxymethylcarbonium ion, and trihydroxycarbonium ion, respectively. In addition, reactions of many of these ionic species involve formation of true carbonium ions, as, for example, the dehydration of protonated alcohols to give alkylcarbonium ions. We have omitted discussions of the carbon protonation of hydrocarbons as these stand out as separate topics.

## II. EXPERIMENTAL METHODS FOR THE STUDY OF PROTONATED COMPOUNDS

The experimental methods for the study of protonated compounds in solution can be grouped into two categories. The first are methods which allow detection of these species but which give little information concerning their structure. This group includes, for example, the cryoscopic and conductometric methods, both of which have been valuable in the detection of protonated species but which have given little or no information as to the site of protonation. The second category are those methods which permit structural assignments to be made, and, in particular, allow the site of protonation to be determined. This category includes the spectroscopic methods of which nmr has probably contributed the most to our knowledge of the structure of protonated species. While infrared and Raman spectroscopy could potentially provide such information, particularly since exchange of acidic protons with solvent molecules is not a limiting factor as it is in many nmr experiments, difficulties, both in obtaining and interpreting such spectra, have limited the application of these techniques. Electronic (ultraviolet and visible) spectroscopy has been of considerable benefit to the study of equilibria, particularly between the organic base and its conjugate acid, but again detailed interpretation is difficult and has been restricted to certain classes of conjugated compounds.

It is not the purpose of this review to deal in detail with the physical methods that have been used in the study of protonation, particularly since there are many excellent reviews (2) of these methods already available. Rather it is the aim to examine the more frequently used methods in relation to the type of organic substrate being protonated and, in particular, in relation to the type of acid system being employed. Since the physical method employed frequently dictates to a large extent the acid systems that can be used, it is convenient to commence with a review of the acid systems that have been developed for examination of protonated species in solution.

### A. Acid–Solvent Systems

The choice of an acid system for the observation of protonated species in solution is dependent upon a number of factors. Of prime importance is the selection of an acid system which will meet the demands of the method of observation being employed. Secondly, the protonated species must be reasonably stable in the acid system. Lastly, the acid system must be sufficiently acidic to protonate the organic base in question.

The ability of a given acid to donate a proton to a base is conveniently expressed by an acidity function (3,4). In largely aqueous media, and in dilute solution with respect to the organic substrate, the hydrogen ion concentration, as given by the pH, accurately expresses the protonating ability of the medium. In more acidic solvents, where the activity coefficients (referred to dilute solution in water) of the species present become significantly different from unity, this is no longer true and various methods have been used to set up acidity functions in such media.

For an acidity function $H_x$, and a base having a thermodynamic equilibrium constant, $K_a$, the concentrations of the base $C_b$ and that of its conjugate acid $C_a$ are given by

$$H_x = pK_a + \log \frac{C_b}{C_a} \tag{1}$$

Unfortunately, no one acidity function accurately describes the protonation behavior of all organic bases in a given acid system. For present purposes, however, the acidity of acid media will be described in terms of the Hammett acidity function $H_0$, derived originally from the protonation of a series of primary anilines and some oxygen bases in sulfuric acid, and since extended to a large number of acid media (5,6).

The majority of organic bases whose protonation will be discussed in this review have $pK_a$ values which lie between 0 and $-10$. Table I summarizes the known basicity values of some representative aliphatic weak bases. The values given refer to the value of $H_0$ at half-protonation in

## TABLE I
### Basicities of Some Representative Aliphatic Bases[a]

| Class | Compound | $pK_a$ |
|---|---|---|
| Carbonyl protonation | | |
| Carboxylic acid | Acetic acid | $-6.1$ |
| Carboxylic acid ester | Ethyl acetate | $-6.5$ |
| Ketone | Acetone | $-7.2$ |
| Aldehyde | Alkyl aldehydes | ca. $-8$ |
| Amides | Acetamide | 0 |
| Ether O-protonation | | |
| Ethers | Diethyl ether | $-3.6$ |
| Alcohols | Methanol | $-2.0$ |
| Sulfur protonation | | |
| Mercaptans | Methyl mercaptan | $-6.8$ |
| Sulfides | Dimethyl sulfide | $-5.4$ |
| Nitrogen bases | | |
| Amines | Methylamine | 10.6 |
| Nitriles | Acetonitrile | $-11$ |
| Other | | |
| Nitro | Nitromethane | $-11.9$ |
| Olefin | 1,1-Diphenylethylene | $-5.5$ |
| Phosphines | $n$-Butylphosphine | 0.0 |
| | Dimethylphosphine | 3.9 |
| | Trimethylphosphine | 8.7 |

[a] Values taken from extensive data tabulated by Arnett (9).

sulfuric acid, and thus the $pK_a$ value will only have thermodynamic significance if the base accurately obeys the $H_0$ acidity function. While this is known not to be true in some cases [e.g., olefins (7) and amides (8)], these values do give a guide to the relative basicities of various types of organic bases, both with respect to each other and with respect to the various acid systems to be discussed.

For comparison with Table I the $H_0$ values for some pure protonic acids are listed in Table II. As can be seen, the majority of the classes of compound listed in Table I could be examined as their protonated form in any of the acids listed in Table II provided no further reaction of the protonated species occurred. However, for many observations it is necessary to either increase or decrease the acidity of the medium.

**TABLE II**
$H_0$ **Values of Some Strong Acids (11,14)**

| Acid | $H_0$ |
|---|---|
| $HSO_3F$ | $-12.8$ |
| $HSO_3Cl$ | $-12.8$ |
| $H_2SO_4$ | $-11.0$ |
| HF | $-10.2$ |

According to the solvent–system definition of acids and bases (10), an acid in a particular solvent system is a compound which increases the concentration of the characteristic cation, while a base increases the concentration of the characteristic anion. The characteristic cation or anion is established, in the case of protonic solvents, by the autoprotolysis equilibrium and in the case of nonprotonic solvents by self-ionization which may or may not have any physical reality for the pure solvent. Thus, to decrease the acidity of one of the protonic solvents in Table II it is necessary to add a proton acceptor to the system. An example of this is the sulfuric acid–water system in which a range of acidities between $-11$ and $+7$ (on the $H_0$ scale) may be obtained by suitably adjusting the concentrations of the components. To increase the acidity it is necessary to select a compound which will increase the concentration of the characteristic cation. Thus considering the autoprotolysis equilibrium for HF (eq. 2) it can be seen

$$2HF \rightleftharpoons H_2F^+ + F^- \qquad (2)$$

that a compound which behaves as a fluoride ion acceptor will increase the concentration of the $H_2F^+$ ion and thereby increase the acidity of the medium. A further example of this type of behavior is provided by the oleum system, which, at least in simple terms, can be considered as being more acidic than sulfuric acid owing to the increased concentration of the $H_3SO_4^+$ ion by complexing of the $SO_3$ with the $HSO_4^-$ ion giving $HS_3O_{10}^-$ (11).

$$2H_2SO_4 = H_3SO_4^+ + HSO_4^- \xrightarrow{SO_3} HS_3O_{10}^- \quad (3)$$

The effects of different fluoride ion acceptors have been compared (12) and have been ranked according to the ability of HF solutions of these Lewis acids to dissolve electropositive metals. The strongest Lewis acid investigated was found to be $SbF_5$. Weaker but still to be considered as strong acids in HF were $AsF_5$, $BF_3$, and $PF_5$. A similar conclusion was reached by studies of the extraction of o- and p-xylenes into solutions in HF of various fluoride ion acceptors, which is a measure of the ability of the media to protonate the aromatic (13).

A number of difficulties are encountered in obtaining direct measurements of acidity in these mixed acid systems. One is that it is difficult to find a weak enough base for measurement of the equilibrium between the base and its conjugate acid using conventional spectroscopic techniques. Nitro compounds have been used. It is not at all clearly established, however, that nitro compounds protonate according to the $H_0$ acidity function (9), and thus it is doubtful whether these compounds give a true extension of the $H_0$ scale for acids stronger than sulfuric acid. With these reservations in mind, some evaluations of the acidity function $H_0$ in some mixed acid systems are given in Table III.

Fluorosulfuric acid containing $SbF_5$ is besides $HF-SbF_5$ probably the most acidic solvent system yet found, although its acidity has not been accurately established (11,14). A number of observations in our laboratories suggest that the acidity increases up to at least 1 mol of added $SbF_5$. (We estimate $H_0$ for $1:1M$ $FSO_3H-SbF_5$ as $-17.5$ to $-18$.) The remarkably high acidity of $1:1M$ $FSO_3H-SbF_5$ has resulted in its trivial naming as

**TABLE III**
$H_0$ **Values of Some Mixed Acid Systems** (15)

| Acid system | $H_0$ |
|---|---|
| $H_2SO_4 + 1.0M\ SO_3$ | $-12.2$ |
| $H_2O + 1.0M\ BF_3$ | $-11.4$ |
| $HF + 0.02M\ NbF_5$ | $-12.5$ |
| $HF + 0.36M\ NbF_5$ | $-13.5$ |
| $HF + 0.36M\ SbF_5$ | $-14.3$ |
| $HF + 3.00M\ SbF_5$ | $-15.3$ |

"magic acid®." It should be noted that fluorosulfonic acid is synonymous with fluorosulfuric acid. Both names are currently used, although the latter appears to be becoming more acceptable.

Certain added solvents have little or no effect on the concentration of either the characteristic cation or anion of the solvent and can thus be used as inert diluents. Thus addition of $SO_2$ to $FSO_3H$ does not affect the conductivity as would be expected if it were acting as a base, and it is thus believed not to be protonated under these conditions (16). Similarly, the Raman and spectra of $SO_2$ in HF are virtually unchanged from those of neat $SO_2$—again evidence that $SO_2$ is not behaving as a base (15). This has important consequences in the recording of low-temperature nmr spectra in mixed acids of high acidity. By addition of $SO_2$ or $SO_2ClF$ it is possible to circumvent the broadening of peaks due the high viscosity of the media at low temperature without appreciably diminishing the acidity of the system, although a slight decrease has been observed (17).

It is thus possible, in a particular acid system, to increase or decrease the acidity at will, or to dilute the acid without markedly effecting the overall acidity.

With these points in mind, the most important of the physical methods that have been used to study protonation will be discussed.

## B. Cryoscopic and Conductometric Methods

Historically, the study of protonation in strong acid systems was first investigated by the cryoscopic method in sulfuric acid (18,19). Sulfuric acid, due to its convenient freezing point (10.371°) (20) and its large cryoscopic constant [6.12°/(mole)(kg)] (21), is a very suitable solvent for cryoscopic measurements. Care has to be taken in such determinations either to repress, or suitably correct for, the self-ionization of the solvent. This can be achieved by carrying out the determinations in slightly aqueous acid. There is a danger, however, in that certain solutes, such as nitro compounds (22), show nonideal behavior in slightly aqueous sulfuric acid and, in addition, solutes which are capable of dehydration will give misleading freezing point depressions in the presence of small amounts of water. An example of this latter behavior is acetic anhydride, which in 99.8% sulfuric acid gives a twofold depression of freezing point but in the absence of water gives a fourfold depression (23). This is interpreted as due to dehydration of the acid by the anhydride, the ionization scheme being

$$(CH_3CO)_2O + H_3O^+ + H_2SO_4 = 2CH_3CO_2H_2^+ + HSO_4^- \qquad (4)$$

The results of cryoscopic determinations give the $\nu$ factor, which is defined as the number of kinetically separate dissolved particles that are produced by the addition of the solution of one molecule of the solute (24).

The $\nu$ factor can be obtained for any solute in 100% sulfuric acid from expression 5 which takes into account the self-ionization of the solvent (18). $\theta$ is the freezing point depression measured from the freezing point of

$$\nu = \frac{\theta}{6.12m}(1 + 0.002\theta - 0.098ms) - \frac{md}{m} \qquad (5)$$

the hypothetically undissociated sulfuric acid (10.625°), $m$ is the molal concentration of the solute, $md$ is the total molal concentration of ions and molecules arising from self-dissociation, and $s$ is the number of moles of solvent used up in the ionization of 1 mol of the solute. Values of $md$ for different electrolytes have been tabulated (25).

Clearly the $\nu$ factor can be ambiguous as to the precise nature of the ionization step, and furthermore gives no information concerning the structure of the species formed. Some of this ambiguity can be removed by measurements of freezing point depressions in various cryoscopic mixtures of sulfuric acid (26–28), and in this way certain complex modes of ionization have been distinguished. Additional data can be provided by conductivity measurements in sulfuric acid (29,30). It has been found that 99% of the current in solutions of bases in sulfuric acid is carried by the hydrogen sulfate ions; similarly in solutions of acids in sulfuric acids, the current is carried by the $H_3SO_4^+$ ions. Thus by measurements of the conductivity of solutions in sulfuric acid the value of $\gamma$ or the number of $HSO_4^-$ or $H_3SO_4^+$ ions produced per molecule of solute can be determined.

## C. Spectroscopic Methods

Of the various spectroscopic methods available for study of protonation of weak organic bases, ir and nmr spectroscopy have received the most attention and give the most information concerning the structure of the ions under examination. While much of the early work on protonation was conducted using uv and visible spectroscopic techniques, the accent in this area has been on the use of changes in the electronic spectra on protonation to determine relative concentrations of protonated and unprotonated substrate and thus the basicity of the compound as well as the acidity of the medium. Electronic spectra are very valuable in studies of the protonation of aromatic and heteroaromatic systems and can, on detailed interpretation, give information on the charge distribution of the protonated molecule. The majority of protonated aliphatic compounds, however, do not show absorption above 200 m$\mu$, and as a result only a few studies have been reported.

## 1. Vibrational Spectroscopy

The ir and Raman spectral techniques have provided much detailed information on the position of protonation in a number of heteroorganic compounds. These techniques are of particular value in the protonation of carbonyl compounds, a class of compound that has received considerable attention because of the intermediacy of the protonated species in many acid-catalyzed reactions.

The application of ir spectroscopy to the study of protonation is not without its pitfalls, difficulties arising both experimentally and in the assignment of bands in the spectrum. Considerable controversy, for example, was generated in the literature (31) concerning the site of protonation of amides due to an apparent increase in the carbonyl stretching frequency on protonation. Protonation on the carbonyl oxygen would be expected to lead to a decrease in this frequency, and the observed increase was attributed to nitrogen protonation. The resolution of this question, chiefly by the application of nmr spectroscopy, shows that carbonyl protonation occurs and will be discussed later. In a number of cases the ir evidence as to the site of protonation is less ambiguous. Thus the carbonyl stretching frequency of acetic acid appears at 1715 cm$^{-1}$ in water and gradually diminishes in intensity with increase in the sulfuric acid concentration (32,33). At the same time a new band appears at lower frequency (1600 cm$^{-1}$) which is Raman inactive and is assigned to an antisymmetric carbon–oxygen stretching band in the carbonyl protonated species. In the carboxylate anion, which has the same symmetry, an antisymmetric carbon–oxygen stretch appears at 1584 cm$^{-1}$.

The ir spectra of protonated heteroorganic bases all have a broad, strong band in the region 2000–3500 cm$^{-1}$ due to the X–H stretching

**TABLE IV**

X–H and X–D Deformation Frequencies in Protonated Heteroaliphatic Compounds (34)

| Class | X | XH, cm$^{-1}$ | XD, cm$^{-1}$ |
|---|---|---|---|
| Carbonyls | O | 1257 | 983 |
| Thiocarbonyls | S | 968 | 730 |
| N-Heterocycles | N | 1258 | 944 |
| Tertiary amines | N | 1402 | 1062 |
| Secondary amines | N | 1585 | 1178 |
| Primary amines | N | 1580 | 1182 |
| Amine oxides | O | 1463 | 1084 |
| Ethers | O | 1072 | 820 |

mode (34). The position of this band depends on the hydrogen bonding with the anion and on the nature of the anion and thus gives little information on the nature of the atom protonated. The X–H bending mode is more characteristic of the type of atom protonated, and this has been correlated with the position of protonation in a number of organic bases (34). A summary of X–H deformation frequencies in a number of classes of protonated molecules is given in Table IV.

The experimental difficulties, and their resolution, associated with spectroscopic observations in a number of highly acidic media have been discussed in detail (35,36).

## 2. Nmr Spectroscopy

Nmr spectroscopy offers a unique possibility for examining in detail the structure of protonated molecules in solution. Proton magnetic resonance has been used extensively in the elucidation of structure and, as well as structural information, indications of the charge density at various sites in the ion can be obtained. In addition nmr can be used, in a number of cases, to determine rates of inter- and intramolecular processes, including determinations of rotational barriers in protonated molecules. Many comprehensive treatments are available on the general application and theory of nmr spectroscopy, and it will be the purpose of this section only to elaborate points which are of particular pertinence to studies of protonation behavior.

Of especial importance is the limiting factor associated with the exchange of acidic protons in the protonated molecule with the solvent. For observation of separate resonances in the nmr spectrum, the lifetime of the acidic proton on the site of protonation must be at least $10^{-1}$–$10^{-2}$ sec. The lifetime of the proton in a solvent of given acidity will be dependent on the basicity of the site protonated and also on the temperature. For a given site (37,38), the rate of exchange of the acidic proton with the solvent will, in general, decrease with increasing acidity of the medium. The most favorable case for observing an acidic proton by nmr will thus be at low temperature and in a medium high acidity. It is for this reason that superacid systems and the use of low-temperature nmr spectroscopy has led to the direct observation of a considerable number of protonated weak bases, under conditions where the exchange rate with the solvent is low enough to observe the acidic proton directly. The use of low temperatures, both to decrease the rate of exchange and also to prevent further reaction of the ion, leads to experimental difficulties due to the viscosity of the acid system. In this regard the availability of diluents such as $SO_2$ and $SO_2ClF$ which apparently have only a small effect on the acidity, enables temperatures as low as $-120°$ to be attained in media such as 1:1$M$ FSO$_3$H–SbF$_5$ ("magic acid®").

For the FSO$_3$H–SbF$_5$, FSO$_3$H–SbF$_5$–SO$_2$, and FSO$_3$H–SbF$_5$–SO$_2$ClF acid system, normal glass sample tubes may be used. In acid systems containing HF, however, glass is unsuitable due to reaction to give silicon tetrafluoride. Sample tubes for use with HF can be made of polyethylene, polypropylene, Teflon, or Kel-F. Quartz tubes may also be used, but sealing should be avoided due to slow production of SiF$_4$ gas, which may result in explosions (39).

Another practical consideration in recording nmr spectra in strong acid systems is the choice of a suitable reference standard. Tetramethylsilane, the usual reference standard used in nmr work, is decomposed by sulfuric acid and other strong acids and is therefore not suitable as an internal standard (40). Capillaries containing TMS as an external standard have proved highly suitable and can also be used with nmr spectrometers which utilize the internal lock method of field-frequency stabilization. It is most convenient to quote chemical shift values in δ (ppm from TMS), because of the fact that the highly acidic protons encountered in these systems would have negative chemical shifts on the $\tau$ scale. To convert from external TMS to internal TMS as standard, relationship 6 is used (41)

$$\delta_{TMS}^{int} = \delta_{TMS}^{ext} + 0.21 \quad (6)$$

for 1:1$M$ FSO$_3$H–SbF$_5$–SO$_2$ at $-60°$. δ values are negative for protons absorbing to low field of TMS, although according to current convention the sign is omitted.

Several internal standards can also be used. The H$_3$O$^+$ peak in a number of strong acid systems is invariant with concentration and temperature and appears at δ 10.05 (from internal TMS) (41). The tetramethylammonium ion (as the SbF$_6^-$ or BF$_4^-$ salt) is stable under strong acid conditions and absorbs at δ 4.05–4.15 (from internal TMS). Methylene chloride can be used as a standard but will react with some superacids, particularly in the presence of SO$_2$, to give protonated formaldehyde and a variety of other intermediates (42).

Sample preparation for nmr measurements of protonated species requires special attention due to the high concentration (5–10%) of substrate normally employed. The heat of protonation resulting from dissolving such large concentrations of precursors in the acid necessitates slow addition and cooling of both compound and acid. It is also inadvisable to allow protonated and unprotonated material to come into contact as this can result in complications such as polymerization. A suggested method for preparing samples in FSO$_3$H–SbF$_5$–SO$_2$ (or SO$_2$ClF) solution is to add a solution of the compound in SO$_2$ (or SO$_2$ClF) slowly to a FSO$_3$H–

SbF$_5$–SO$_2$ (SO$_2$ClF) solution, both solutions being cooled to $-78°$ in a dry ice–acetone cooling bath. Rapid mixing, such as is obtained by means of a vortex-type stirrer, is also recommended for obtaining "clean" nmr spectra of the protonated organic substrate. An excess of superacid will protonate any moisture in the system to form H$_3$O$^+$, and it is, therefore, not always necessary to take extra precautions to work in an inert, dry atmosphere. Whenever such conditions are needed, however, usual vacuum line and drybox techniques should be applied.

Finally, it is sometimes the case that the solvent acid peak obscures the region of the nmr spectrum of interest. This problem can be overcome by variation of the acid system, by a change in temperature or by variation in the concentration of the diluent used.

It is pertinent, in this introductory section, to review briefly some of the salient features associated with the nmr spectra of ions and in particular the features associated with the acidic protons, observed at low temperature, in superacid media. Generally, these will be protons attached to heteroatoms and will show certain distinct features. The chemical shift of protons attached to oxygen reflects the charge on the proton and the double-bond character of the oxygen. Typical shifts for such protons are presented in Table V for various classes of protonated molecules. Protons attached to sulfur occur at higher field than those on oxygen; thus in protonated thiols the SH proton is in the range $\delta$ 5.9–6.6 and in thiocarboxylic acids the range is $\delta$ 6.5–7.5. NH protons are often broadened due to the presence of the nitrogen quadrupole. The shift appears to follow the bond order of the nitrogen, protonated nitriles ($\delta$ 10.5–11.5), protonated imines ($\delta$ 9.5–10.0), and protonated amines ($\delta$ 8.0–9.0).

Hindered rotation about the carbon–heteroatom bond often leads to the observation of *cis–trans* isomerism in the protonated molecule. A close parallel has been found between vicinal coupling constants in such ions and the corresponding coupling constants in olefins, *trans* HCXH coupling constants being larger than *cis* HCXH coupling constants. This parallelism has been used in assignment of the spectrum to particular isomers in many instances, and a number of examples of this approach will be given later. This approach has been recently criticized (43) as a result of theoretical calculations on the stability of the various isomers of protonated aldehydes and carboxylic acids. There is, however, now a considerable amount of evidence which supports the approach (for a more detailed discussion see Ref. 44), and there now seems to be little doubt as to its validity. The analogy with uncharged olefinic systems is less valid for four-bond allylic coupling constants, and, while in the olefinic systems *cis* allylic coupling constants are invariably larger than the *trans* coupling constants, the reverse has been found to be true in several ions (45).

## TABLE V

| Class | Ion | Shift (δ) from ext TMS | Range of chemical shifts (δ) |
|---|---|---|---|
| Aldehydes | $\mathrm{CH_3\!-\!\overset{+}{C}(H)\!=\!OH}$ | 15.1 | 15–17 |
| Ketones | $\mathrm{(CH_3)_2\overset{+}{C}\!=\!OH}$ | 14.9 | 13–15 |
| Carboxylic acids | $\mathrm{CH_3\!-\!C(OH_B)\!=\!\overset{+}{O}H_A}$ | 13.0 ($H_B$)<br>12.3 ($H_A$) | 12–13.5 |
| Carbonates | protonated carbonate (bridged $H\!-\!O\!\cdots\!C\!\cdots\!O\!-\!H$, $=OH$) | 10.5 | 9.5–11.5 |
| Alcohols | $\mathrm{CH_3OH_2^+}$ | 9.4 | 9.0–10.0 |
| Ethers | $\mathrm{CH_3\overset{+}{O}(H)CH_3}$ | 9.0 | 8.0–9.0 |

Carbon magnetic resonance has much potential application to the observation of protonated species due to the much larger chemical shifts which are encountered with carbon as compared to proton spectra (46). In addition, a close correlation of the chemical shift with charge density has been found, and $^{13}$C–H coupling constants can give information on the hybridization of the carbon atom concerned. $^{13}$C resonance has, however, achieved only limited application at present due to experimental difficulties associated with low natural abundance and low sensitivity of $^{13}$C. Protonated species have to be observed in relatively dilute solution, and thus, without the use of $^{13}$C enrichment of the substrate, highly sensitive techniques coupled with time-averaging computer techniques for observation of spectra have to be employed. Undoubtedly this area will receive much more attention in the future, with more widespread use of the application of Fourier transform methods to natural abundance $^{13}$C nmr spectroscopy. Internuclear double resonance (indor) (47) has been used for the observation of $^{13}$C spectra of simple protonated organic molecules (45,48,49). This method entails observation of the proton spectrum while simultaneously irradiating the $^{13}$C region. In cases where there is a coupling of the $^{13}$C nucleus with a proton nucleus, changes in the proton spectrum will result when the irradiating frequency coincides with transition frequencies of the $^{13}$C nucleus. By scanning the $^{13}$C region and simultaneously monitoring a $^{13}$C satellite peak in the proton spectrum, the $^{13}$C spectrum can be obtained, together with the chemical shift of the $^{13}$C nucleus. This method has the advantages implicit in observing direct proton resonance but has the disadvantages that a carbon–hydrogen coupling must be present and that difficulties arise in other than simple molecules.

For protonation on oxygen and nitrogen, $^{17}$O and $^{14}$N or $^{15}$N resonance could potentially provide much information, but this field has to date received only little attention (44), although $^{14}$N shifts in protonated amides have been reported (50).

A more extensive discussion of $^{13}$C, $^{17}$O, $^{14}$N, $^{15}$N as well as $^{1}$H resonance spectra can be found in Ref. 44.

## III. PROTONATED ALCOHOLS

In sulfuric acid it has been shown by cryoscopic measurements that methyl and ethyl alcohol give stable solutions of the hydrogen sulfates (51). Many other alcohols show similar initial behavior, but the solutions are not stable at room temperature. Alcohols form acid salts (eq. 7) which

$$\text{ROH} + 2\text{H}_2\text{SO}_4 \longrightarrow \text{RHSO}_4 + \text{H}_3\text{O}^+ + \text{HSO}_4^- \tag{7}$$

in some cases can be isolated, and in solution it has been presumed that oxygen protonated species exist (52). The first direct nmr evidence for the existence of protonated alcohols in strong acid solutions was found in 1961. The nmr spectrum of ethanol in $BF_3$–HF solution at $-70°$ gave a well-resolved triplet at about δ 9.90 for the protons on oxygen coupled to the methylene protons (38). In $HSO_3F$ this fine structure is not observed, even at $-95°$, due to fast exchange (53).

The nmr spectra of a series of aliphatic alcohols have been investigated in the stronger acid system, $HSO_3F$–$SbF_5$ using sulfur dioxide as diluent (54,55). Methyl, ethyl, n-propyl, isopropyl, n-butyl, isobutyl, sec-butyl, n-amyl, neopentyl, n-hexyl, and neohexyl alcohol all give well-resolved nmr spectra at $-60°$, under these conditions.

$$\text{ROH} \xrightarrow[-60°]{\text{FSO}_3\text{H}-\text{SbF}_5-\text{SO}_2} \text{ROH}_2^+ \tag{8}$$

The strength of this acid system is reflected by the fact that even at 25° solutions of primary alcohols in $HSO_3F$–$SbF_5$ show fine structure for the proton on oxygen (Fig. 1). This indicates that at this relatively high temperature the exchange rate is still slow on the nmr time scale. The chemical shift data for a number of protonated alcohols are presented in Table VI, and the spectrum of protonated methanol is illustrated in Figure 1. The $OH_2^+$ protons of the normal alcohols appear at a lower field ($\delta$ 9.3–9.5) than the isomeric secondary alcohols ($\delta$ 9.1). This is due to a different charge distribution as confirmed by the $C_1$ protons appearing at higher field with the normal alcohols ($\delta$ 5.0–4.7) than the $C_1$ methine proton of the secondary alcohols ($\delta$ 5.4 and 5.5).

As would be expected, the proton on oxygen of protonated alcohols ($\delta$ 9.1 to 9.5) is slightly more deshielded than the proton on oxygen of protonated ethers ($\delta$ 7.8–9.0). The hydrogens on the carbon adjacent to oxygen show a significant downfield shift of about 1.5 ppm when compared to the protonated alcohol.

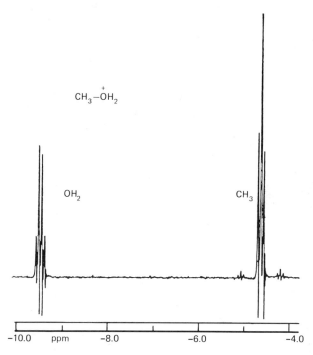

Fig. 1. Pmr spectrum of protonated methyl alcohol in $FSO_3H$–$SbF_5$–$SO_2$ solution at −60° [according to G. A. Olah, J. Sommer, and E. Namanworth, *J. Am. Chem. Soc.*, **89**, 3576 (1967)].

## TABLE VI
### Shifts and Coupling Contents of Protonated Alcohols at −60° in $HSO_3F$–$SbF_5$–$SO_2$ [a]

| Alcohol | $OH_2$ | $H_1$ | $H_2$ | $H_3$ | $H_4$ | $H_5$ | $J_{HCOH}$ |
|---|---|---|---|---|---|---|---|
| $\overset{1}{C}H_3OH$ | 9.4 (q) [b] | 4.7 (t) | | | | | 3.6 |
| $\overset{2}{C}H_3\overset{1}{C}H_2OH$ | 9.3 (t) | 4.9 (m) | 1.9 (t) | | | | 3.6 |
| $\overset{3}{C}H_3\overset{2}{C}H_2\overset{1}{C}H_2OH$ | 9.4 (t) | 4.7 (m) | 1.8 (m) | 0.8 (t) | | | 3.5 |
| $\overset{2}{C}H_3\overset{1}{C}HCH_3$<br>   $\ \ |$<br>   OH | 9.1 (d) | 5.5 (m) | 1.7 (d) | | | | 3.0 |
| $\overset{4}{C}H_3\overset{3}{C}H_2\overset{2}{C}H_2\overset{1}{C}H_2OH$ | 9.4 (t) | 5.0 (m) | 2.0 (m) | 1.4 (m) | 1.1 (t) | | 3.5 |
| $\overset{3}{C}H_3\diagdown$<br>$\qquad\quad\overset{2}{C}H\overset{1}{C}H_2OH$<br>$CH_3\diagup$ | 9.4 (t) | 4.7 (m) | 2.3 (m) | 1.1 (d) | | | 3.6 |
| $\overset{4}{C}H_3\overset{3}{C}H_2\overset{1}{C}HOH\overset{2}{C}H_3$ | 9.1 (d) | 5.4 (m) | 1.6 (d) | 1.9 (m) | 0.9 (t) | | 3.0 |
| $\overset{5}{C}H_3\overset{4}{C}H_2\overset{3}{C}H_2\overset{2}{C}H_2\overset{1}{C}H_2OH$ | 9.3 (t) | 4.9 (m) | 2.3 (m) → 1.1 ($H_2$–$H_4$) | | | 0.9 | 3.4 |
| $\overset{4}{C}H_3\diagdown$<br>$\qquad\quad\overset{3}{C}H\overset{2}{C}H_2\overset{1}{C}H_2OH$<br>$CH_3\diagup$ | 9.4 (t) | 4.3 (m) | 1.8 (m) ($H_2$–$H_3$) | | 0.9 (d) | | 3.6 |

| Compound | | | |
|---|---|---|---|
| $(\overset{2}{C}H_3)_3\overset{1}{C}CH_2OH$ | 9.5 (t) | 4.5 (t) | 1.0 (s) | 3.7 |
| $\overset{6}{C}H_3\overset{5}{C}H_2\overset{4}{C}H_2\overset{3}{C}H_2\overset{2}{C}H_2\overset{1}{C}H_2OH$ | 9.2 (t) | 4.8 (m) | 2.8 (m) to 0.8 ($H_2$–$H_6$) | 3.5 |
| $(\overset{3}{C}H_3)_2\overset{2}{C}\overset{1}{C}H_2OH$ | 9.3 (t) | 4.9 (m) | 1.9 (t) | 0.8 (s) | 3.5 |

[a] From TMS external capillary.
[b] Multiplicity: s, singlet; d, doublet; t, triplet; q, quartet; and m, multiplet.

Aliphatic glycols in $FSO_3H$–$SbF_5$–$SO_2$ solution give diprotonated species at low temperatures (56). In diprotonated diols the protons on oxygen are found at lower fields than in protonated alcohols reflecting the presence of two positive charges. This is especially true for ethylene glycol ($\delta$ 11.2) where the positive charges are adjacent. As the separation of the positive charges becomes greater with increasing chain length, the chemical shift of the protons on oxygen of protonated diols approaches that of protonated alcohols.

The reactivity of protonated alcohols (55) and protonated diols (56) in strong acids has been studied by nmr spectroscopy. Protonated methyl alcohol shows surprising stability in $HSO_3F$–$SbF_5$ and can be heated to 50° without undergoing significant decomposition. Protonated ethyl alcohol is somewhat less stable and begins to decompose at about 30°. The cleavage of protonated n-propyl alcohol has been followed in the temperature range of 5–25°, giving a mixture of trimethyl carbonium ion and isopropyldimethyl carbonium ion (eq. 9). Higher protonated alcohols

$$CH_3CH_2CH_2\overset{+}{O}H_2 \xrightarrow[HSO_3F-SbF_5]{5-25°} (CH_3)_3C^+ + (CH_3)_2CH\overset{+}{C}(CH_3)_2 + H_3O^+ \qquad (9)$$

cleave to stable tertiary carbonium ions. For protonated primary and secondary alcohols, the initially formed primary and secondary carbonium ions rapidly rearrange to the more stable tertiary carbonium ions under the conditions of the reaction. For example, protonated n-butyl alcohol cleaves to n-butyl cation which rapidly rearranges to trimethylcarbonium ion (eq. 10).

$$CH_3CH_2CH_2CH_2\overset{+}{O}H_2 \xrightarrow{k_1} [CH_3CH_2CH_2CH_2^+] \xrightarrow{k_2} (CH_3)_3C^+ \qquad k_2 \gg k_1 \qquad (10)$$

The cleavage to carbonium ions, shown to be first order, is enhanced by branching of the chain: protonated 1-pentanol is stable up to 0°, isopentyl alcohol is stable up to $-30°$, and neopentyl alcohol cleaves at $-50°$. The stability of the protonated primary alcohols also decreases as the chain length is increased. This is shown by comparison of the rate constants of the cleavage at 15° (Table VII).

When the $FSO_3H$–$SbF_5$–$SO_2$ solutions of diprotonated glycols are allowed to warm up, pinacolone rearrangements, formation of allylic carbonium ions, and cyclization reactions of diprotonated glycols can be directly observed by nmr spectroscopy. Diprotonated ethylene glycol rearranges to protonated acetaldehyde in about 24 hr at room temperature. Protonated 1,2-propanediol undergoes a pinacolone rearrangement to

## TABLE VII
### Rate of Cleavage of Protonated Normal Aliphatic Alcohols at 15°

| Protonated alcohol | Rate constant × $10^{-3}$ min$^{-1}$ |
|---|---|
| n-Methyl | Stable to +50° |
| n-Ethyl | Stable to +30° |
| n-Propyl | 20.5 |
| n-Butyl | 48.1 |
| n-Pentyl | 68.4 |
| n-Hexyl | 91.4 |

protonated propionaldehyde probably through the initial cleavage of water from the secondary position (eq.11). Diprotonated 2,3-butanediol

$$CH_3CH-CH_2 \underset{+OH_2 \; +OH_2}{|\quad\quad|} \xrightarrow{-H_2O} CH_3-\overset{+}{C}H-CH_2-O\overset{H}{\underset{H}{\overset{+}{\diagup}}} \longrightarrow CH_3CH_2C\overset{\overset{+}{OH}}{\underset{H}{\diagdown}} \quad (11)$$

rearranges to protonated methyl ethyl ketone either through a direct hydride shift (eq. 12) or through a bridged intermediate (eq. 13).

$$CH_3-\underset{H \; +OH_2}{\overset{+OH_2}{\underset{|\quad|}{C}}-CH-CH_3} \longrightarrow CH_3-\overset{H \; H}{\underset{H}{\overset{\diagdown\overset{+}{O}\diagup}{C}}}-\overset{+}{C}HCH_3 \longrightarrow CH_3-\overset{+O}{\overset{\diagup H}{\underset{\|}{C}}}-CH_2CH_3 \quad (12)$$

$$CH_3-\underset{H \; H}{\overset{\overset{+}{O}H_2 \; \overset{+}{O}H_2}{\underset{|\quad|}{C---C}}-CH_3} \longrightarrow CH_3-\overset{\overset{H}{\underset{+}{O}}}{\underset{H \; H}{C----C}}-CH_3 \longrightarrow$$

$$\underset{\overset{\|}{CH_3-C-CH_2-CH_3}}{^+OH} \quad (13)$$

Diprotonated 2,4-pentanediol loses water and rearranges to form a stable allylic carbonium ion (eq. 14). Diprotonated 2,5-hexanediol, above

$$CH_3CHCH_2CHCH_3 \underset{+\quad\;\;\;\;+}{\underset{OH_2\quad OH_2}{|\quad\quad\;\;|}} \xrightarrow{-H_2O} [CH_3CH\overset{H}{\overset{\diagup|}{\underset{+}{CH}}}-\overset{}{\underset{+}{C}HCH_3}] \xrightarrow{-H^+}$$

$$CH_3CH=CH-\underset{+}{C}HCH_3 \longleftrightarrow CH_3\underset{+}{C}H-CH=CHCH_3 \quad (14)$$

about $-30°$, rearranges to a mixture of protonated *cis-* and *trans-*$\delta,\delta'$-dimethyltetrahydrofurans (eq. 15). This would seem to indicate that there

$$\underset{\underset{+}{OH_2}}{CH_3CHCH_2}-\underset{\underset{+}{OH_2}}{CH_2CHCH_3} \xrightarrow{-H_2O} \underset{\underset{+}{OH_2}}{CH_3CHCH_2}-\underset{+}{CH_2CHCH_3} \underset{+H^+}{\overset{-H^+}{\rightleftarrows}}$$

$$\underset{\underset{H}{O}}{CH_3CHCH_2}-\underset{+}{CH_2CHCH_3} \longrightarrow \underset{H_3C}{\overset{H_2C-CH_2}{\underset{\underset{H}{O_+}}{HC\quad CH}}}CH_3 \quad (15)$$

are either significant amounts of the monoprotonated form present or that the carbonium ion formed can easily lose a proton before ring formation takes place.

## IV. PROTONATED THIOLS

Aliphatic thiols are completely protonated in $FSO_3H-SbF_5$ diluted with $SO_2$ at $-60°$ (57) (see Fig. 2).

$$RSH \xrightarrow[-60°]{FSO_3H-SbF_5-SO_2} RSH_2{}^+ SbF_5-FSO_3{}^- \quad (16)$$

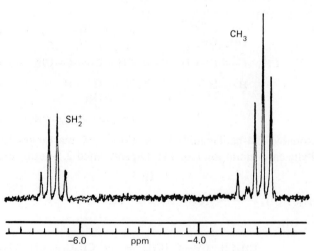

Fig. 2. Pmr spectrum of protonated methyl mercaptan in $HSO_3F-SbF_5-SO_2$ solution at 60°.

The resonance for the proton on sulfur is at considerably higher field (δ 5.9–6.6) than the corresponding proton on oxygen in protonated alcohols (δ 9.1–9.5) reflecting the larger size of the sulfur atom compared to oxygen. Table VIII summarizes the nmr data for a series of protonated aliphatic thiols. The protonated thiols are considerably more stable than protonated alcohols. Protonated *t*-butyl thiol shows no appreciable decomposition at $-60°$ in $HSO_3F$–$SbF_5$–$SO_2$, while tertiary alcohols could not be observed under the same conditions and even secondary alcohols decompose at a significant rate. Protonated thiols decompose at higher temperatures to give protonated hydrogen sulfide (singlet, δ 6.60) and stable carbonium ions. For example, protonated *t*-butyl thiol slowly cleaves to trimethylcarbonium ion and protonated hydrogen sulfide (eq. 17) when the temperature is increased to $-30°$ ($t_{1/2} \sim 15$ min). Protonated

$$(CH_3)_3CSH_2^+ \xrightarrow{-30°} (CH_3)_3C^+ + H_3S^+ \quad (17)$$

*t*-amyl thiol also cleaves at this temperature to the dimethylethylcarbonium ion (eq. 18).

$$\underset{\underset{CH_3}{|}}{\overset{\overset{CH_3}{|}}{CH_3CH_2-C-SH_2^+}} \xrightarrow{-30°} \underset{\underset{CH_3}{|}}{\overset{\overset{CH_3}{|}}{CH_3CH_2-C^+}} + H_3S^+ \quad (18)$$

Protonated secondary thiols are stable at even higher temperatures. Protonated isopropyl thiol cleaves slowly at 0° in $FSO_3H$–$SbF_5$ (1:1*M*) solution. No well-identified carbonium ions were found in the nmr spectra due to the instability of the isopropyl cation under these conditions. Protonated *sec*-butyl thiol cleaves to trimethylcarbonium ion at this temperature.

$$\underset{\underset{CH_3}{|}}{CH_3CH_2CHSH_2^+} \xrightarrow{0°} [CH_3CH_2\overset{+}{C}HCH_3] \longrightarrow (CH_3)_3C^+ \quad (19)$$

Protonated primary thiols are stable at much higher temperatures. Protonated *n*-butyl thiol slowly cleaves to trimethylcarbonium ion only at $+25°$.

$$CH_3CH_2CH_2CH_2SH_2^+ \xrightarrow{+25°} [CH_3CH_2CH_2CH_2^+] \longrightarrow (CH_3)_3C^+ \quad (20)$$

## V. PROTONATED ETHERS

The proton acceptor properties of both aliphatic and aromatic ethers have been studied extensively by a wide variety of techniques. The results of these investigations have been reviewed (51,58,59). It is well known that

## TABLE VIII

### Nmr Chemical Shifts and Coupling Constants of Protonated Thiols at $-60°$ in $HSO_3F-SbF_5-SO_2$

| Thiol | $H_1$ | $H_2$ | $H_3$ | $H_4$ | $SH_2^+$ | $J_{H-H^+}$, Hz |
|---|---|---|---|---|---|---|
| HSH | | | | | 6.60 (1) | — |
| $\overset{1}{C}H_3SH$ | 2.95 (3)[b] | | | | 6.45 (4) | 8.0 |
| $\overset{2}{C}H_3\overset{1}{C}H_2SH$ | 3.37 (6) | 1.48 (3) | | | 6.22 (cm) | 8.0 |
| $\overset{3}{C}H_3\overset{2}{C}H_2\overset{1}{C}H_2SH$ | 3.40 (cm) | 1.98 (cm) | 1.00 (3) | | 6.37 (cm) | 8.0 |
| $\overset{2}{C}H_3\!\!\diagdown\!\!\overset{1}{C}HSH$ / $CH_3\!\!\diagup$ | 3.98 (cm) | 1.73 (2) | | | 5.93 (2) | 7.5 |
| $\overset{4}{C}H_3\overset{3}{C}H_2\overset{2}{C}H_2\overset{1}{C}H_2SH$ | 3.47 (6) | 1.70 (cm) | 1.70 (cm) | 0.90 (3) | 6.45 (3) | 7.6 |
| $\overset{3}{C}H_3\!\!\diagdown\!\!\overset{1}{C}HCH_2SH$ / $CH_3\!\!\diagup$ | 3.32 (cm) | 2.25 (cm) | 1.03 (cm) | | 6.40 (3) | 8.0 |
| $\overset{3}{C}H_3\overset{2b}{C}H_2\overset{1}{C}HSH$ / $\underset{2a}{C}H_3$ | 4.08 (6) | a. 1.82 (2) b. 2.10 (cm) | 1.15 (3) | | 6.35 (2) | 7.0 |

| | | | |
|---|---|---|---|
| (CH₃)₃CSH | — | 1.75 (1) | 6.42 (1) |
| CH₃<br>│<br>³CH₃²ᵇCH₂CSH<br>│<br>CH₃<br>²ᵃ | — | a. 1.75 (1)<br>b. 1.97 (*cm*) 1.07 (3) | 6.32 (1) |

[a] From external capillary of TMS.
[b] Multiplicity of peaks shown in parentheses; *cm* = complex multiplet.

ethers form low-melting solid complexes with acids in which the proton is attached to the basic site, the oxygen atom (59). In solution, cryoscopic (60), ir (61–65), conductance (66,67), and solubility measurements (68) indicate that a 1:1 complex is formed between the ether and the acid. Recent investigations in strong acid systems have confirmed these earlier findings.

$$R\text{—}O\text{—}R + H\text{—}Y \rightleftharpoons R\text{—}\overset{+}{\underset{|}{O}}\text{—}R + Y^- \tag{21}$$

$$\phantom{R\text{—}O\text{—}R + H\text{—}Y \rightleftharpoons R\text{—}}\overset{H}{\phantom{O}}$$

## A. Aliphatic Ethers

Until recently the basicity of saturated aliphatic ethers had not been compared quantitatively to the protonation equilibria of other organic bases. The difficulty was caused primarily by the inability of saturated aliphatic ethers to give significant spectral changes in the uv and thereby act as Hammett indicators and be amenable to calculation of base strength on the familiar Hammett pH–$H_0$ scale (3,69). This difficulty has been overcome in a series of recent investigations by a combination of the usual Hammett acidity calculations with solvent extraction and analysis by gas chromatography (70–75). The method involves the distribution of the ether between a nonpolar, inert organic solvent and an aqueous phase containing a variable amount of sulfuric acid. The distribution of the ether between the phases was measured from low sulfuric acid concentration where the ether is essentially unprotonated to high acid concentration where the ether is essentially completely protonated. The distribution constants were determined by means of gas chromatographic analysis of the inert phase. The reliability of the method has been tested by comparison with results obtained using the $H_0$–indicator method (70) and by comparing the order of basicities with the order determined from the effect of various aliphatic ethers on the O–D stretching frequency of deuteriomethanol (71). Some representative $pK_a$'s for the conjugate acids of several aliphatic ethers are presented in Table IX. Earlier investigations had concluded that the basicity of aliphatic ethers would be primarily controlled by steric effects and that an increase in the size of the alkyl groups would result in base weakening (76,77). However, inspection of Table I shows that no single factor controls the order of basicity, but rather it appears to be a delicate balance between steric, inductive, and solvation effects (74). In the series of methyl ethers studied, an increase in the size of the other alkyl group causes a trend toward greater basicity. In this series the steric effect is relatively constant and the base-strengthening inductive effect predominates. As the size of both alkyl groups is increased, steric factors become more important. The line between inductive and steric control of

## TABLE IX
**Dissociation Constants of Conjugate Acids of Some Aliphatic Ethers**

| Ether | | | |
|---|---|---|---|
| $R_1$ | $R_2$ | $pK_a$ | Ref. |
| $CH_3$ | $CH_3$ | −3.83 | 71,74 |
| $CH_3$ | $CH_3CH_2$ | −3.82 | 74 |
| $CH_3$ | $CH(CH_3)_2$ | −3.47 | 74 |
| $CH_3$ | $CH_3CH_2CH_2CH_2$ | −3.50 | 74 |
| $CH_3$ | $C(CH_3)_3$ | −2.89 | 74 |
| $CH_3CH_2$ | $CH_3CH_2$ | −3.59 | 71,74 |
| $CH_3CH_2$ | $CH_3CH_2CH_2CH_2$ | −4.12 | 74 |
| $CH_3CH_2$ | $C(CH_3)_3$ | −2.84 | 74 |
| tetrahydrofuran | | −2.08 | 71,74 |
| tetrahydropyran | | −2.79 | 71,74 |
| oxepane | | −2.02 | 75 |
| $CH_3$ | $C_6H_5$ | −6.54 | 70–73 |

basicity is crossed when the groups become as big as about *n*-propyl and decreasing basicity is observed. This decrease in basicity has been ascribed to steric hindrance due to "F-strain" and to solvation of the oxonium ion (74). The much greater basicity of cyclic aliphatic ethers has been explained as primarily caused by substantial relief of internal strain upon protonation (75).

The basicity value for diethyl ether as determined by solvent extraction (71,74) has been challenged (78–80). Values of the $pK_a$ were −5.7 (78,79) and −6.2 (80) based on the titration of diethyl ether in glacial acetic acid using perchloric acid with Sudan III as indicator (78,79) and on the inflection of a plot of the change of the chemical shift difference between the methyl triplet and the methylene quartet versus $H_0$ (80). However, examination of the experimental data shows that the choice of the inflection point is difficult and the agreement with the other data may be fortuitous.

Further, such a low value for the basicity of diethyl ether would make it a base comparable in strength to aromatic ethers such as anisole, which seems highly unlikely.

Recent investigation of the nmr and ir spectra of oxonium salts in methylene chloride has led to the identification of the resonance due to the proton on oxygen. These solutions of hexachloroantimonates were found to form both "free" complexes in which the proton is bonded to only one oxygen (1) and complexes in which the proton is shared between two ether

$$(C_2H_5)_2\overset{+}{O}H\ SbCl_6^- \qquad (C_2H_5)_2\overset{+}{O}H\cdots O(C_2H_5)_2 SbCl_6^-$$

$$\text{(1)} \qquad\qquad\qquad\qquad \text{(2)}$$

$$\delta\ 9.34 \qquad\qquad\qquad\qquad \delta\ 3.58$$

molecules (2) (81). The ir spectrum was studied and used to confirm the existence of the "free" complex and the bridged species. The proton on oxygen resonances were singlets, indicating that the proton is rapidly exchanging under the conditions of the experiment. Investigations using the extremely strong acid system, $HSO_3F$–$SbF_5$–$SO_2$ at low temperatures leads to comparable values for the chemical shift of the proton on oxygen for a variety of aliphatic ethers of $\delta$ 7.88–9.03 (Table X). Because of the stronger acid system and the low temperature, however, the exchange rate is slowed sufficiently so that the expected splitting of the proton on oxygen by the adjacent hydrogens is observed (82) (Fig. 3).

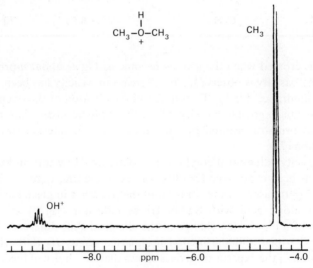

Fig. 3. Pmr spectrum of protonated dimethyl ether in $FSO_3H$–$SbF_5$–$SO_2$ solution.

## TABLE X

Nmr Chemical Shifts and Coupling Constants of Protonated Ethers at −60° in $HSO_3F$–$SbF_5$–$SO_2$, ppm [a]

| Ether | $H_1$ | $H_2$ | $H_3$ | $H_4$ | $OH^+$ | $J_{H-OH^+}$, Hz |
|---|---|---|---|---|---|---|
| $(\overset{1}{C}H_3)_2O$ | 4.49 (2) | | | | 9.03 (7) | 3.4 |
| $(\overset{2}{C}H_3\overset{1}{C}H_2)_2O$ | 4.73 (8) | 1.53 (3) | | | 8.61 (5) | 3.6 |
| $(\overset{3}{C}H_3\overset{2}{C}H_2\overset{1}{C}H_2)_2O$ | 4.63 (6) | 1.90 (6) | 0.92 (3) | | 8.60 (5) | 3.8 |
| $\overset{2}{C}H_3\!\!-\!\!\overset{1}{C}H\!\!-\!\!\overset{}{}\,O$ <br> $\phantom{xx}CH_3$ (cyclic, with both $CH_3$) | 5.18 (m) | 1.55 (2) | | | 7.88 (3) | 4.1 |
| $(\overset{4}{C}H_3\overset{3}{C}H_2\overset{2}{C}H_2\overset{1}{C}H_2)_2O$ | 4.52 (6) | ~1.60 (m) | ~1.60 (m) | ~0.90 (m) | 8.56 (5) | 3.8 |
| $\overset{4}{C}H_3\overset{3}{C}H_2\overset{2}{C}H_2\overset{1a}{C}H_2O\overset{1b}{C}H_3$ | a. 4.73 (6) <br> b. 4.37 (2) | ~1.66 (m) | ~1.66 (m) | 0.90 (3) | 8.86 (6) | 3.7 |
| $\overset{3}{C}H_3\overset{2a}{C}H_2\overset{1a}{C}HO\overset{1b}{C}H_3$ <br> $\phantom{xxxx}|$ <br> $\phantom{xxx}\underset{2b}{CH_2}$ | a. 4.96 (m) <br> b. 4.30 (2) | a. 1.97 (m) <br> b. 1.61 (2) | 0.97 (3) | | 8.47 (5) | 3.8 |

[a] From external capillary of TMS; figures in parentheses represent multiplicity of peaks; $m$ = multiplet.

The cleavage of protonated ethers in strong acid systems has not been studied extensively. Kinetic investigation of the cleavage of ethers in 99.6% sulfuric acid using cryoscopic methods showed that cleavage takes place by unimolecular fission of the conjugate acid of the ether to form the most stable carbonium ion and an alcohol. The carbonium ion and alcohol formed rapidly unite with hydrogen sulfate ion. The overall rate in sulfuric acid, however, appears to be dependent upon the concentration of sulfur trioxide. To reconcile these observations, the mechanism of equations 22–24 has been proposed (83).

$$R\text{—}O\text{—}R' + H_2SO_4 \rightleftharpoons R\overset{H}{\underset{+}{\text{—}O\text{—}}}R' + HSO_4^- \qquad (22)$$

$$R\overset{H}{\underset{+}{\text{—}O\text{—}}}R' + SO_3 \rightleftharpoons R\overset{+}{\underset{SO_3H}{\text{—}O\text{—}}}R' \qquad (23)$$

$$R\overset{+}{\underset{SO_3H}{\text{—}O\text{—}}}R' \xrightarrow{RDS} R^+ + R'HSO_4 \qquad (24)$$
$$\phantom{R\overset{+}{\underset{SO_3H}{\text{—}O\text{—}}}R'} \xrightarrow{HSO_4^-} RHSO_4$$

Inductive and resonance effects seem to be more important in the alkyl chain of the potential carbonium ion than in the other alkyl group.

In a solution of $HSO_3F\text{–}SbF_5$, $n$-butyl methyl ether does not show any significant change, either cleavage or rapid exchange, as indicated by the nmr spectrum, up to $+40°$. At this temperature, $n$-butyl methyl ether cleaves and a sharp singlet appears at $\delta$ 4.0. This can be attributed to the

TABLE XI

Kinetics of Cleavage of $CH_3CH_2\overset{CH_3\ \ H}{\underset{+}{\diagdown\ \diagup}}CHOCH_3$

| Temperature, °C | $k$, sec$^{-1} \times 10^4$ |
|---|---|
| −29.0 | 3.64 |
|  | 3.68 |
|  | 5.11 |
|  | Av. 4.14 ± 0.64 |
| −17.5 | 10.9 |
|  | 8.22 |
|  | 7.57 |
|  | 9.65 |
|  | Av. 9.08 ± 1.2 |
| $E_a = 8.5 \pm 3$ kcal | |

rearrangement of the *n*-butyl cation, formed in the cleavage, to trimethylcarbonium ion (eq. 25).

$$CH_3\overset{+}{\underset{H}{O}}CH_2CH_2CH_2CH_3 \longrightarrow CH_3\overset{+}{O}H_2 + [CH_3CH_2CH_2CH_2]^+$$

$$\longrightarrow (CH_3)_3C^+ \quad (25)$$

Ethers in which one of the groups is secondary begin to show appreciable cleavage at $-30°$. Protonated *sec*-butyl methyl ether cleaves cleanly at $-30°$ to protonated methanol and trimethylcarbonium ion. Ethers in which one of the alkyl groups is tertiary cleave rapidly even at $-70°$.

It was found possible to measure the kinetics of cleavage of protonated *sec*-butyl methyl ether by following the disappearance of the methoxy doublet in the nmr spectrum with simultaneous formation of protonated methanol and trimethylcarbonium ion. Kinetic data are summarized in Table XI. The cleavage shows pseudo-first-order kinetics. Presumably the rate-determining step is the formation of methylethylcarbonium ion followed by rapid rearrangement to the more stable trimethylcarbonium ion ($k_1 \ll k_2$) (eq. 26).

$$CH_3CH_2\underset{\underset{CH_3}{|}}{\overset{\overset{H}{|}}{C}}\overset{+}{O}CH_3 \xrightarrow{k_1} CH_3\overset{+}{O}H_2 + CH_2CH_2\overset{+}{C}HCH_3 \xrightarrow{k_2} (CH_3)_3C^+ \quad (26)$$

### B. Protonated Alkyl Phenyl Ethers

Recently, a lively controversy has developed concerning the site of protonation of phenols and aromatic ethers. Three possible protonated species can be envisioned: a $\pi$ complex of the proton with the aromatic ring (3); a $\sigma$ complex of the proton with one of the ring carbons (4); and

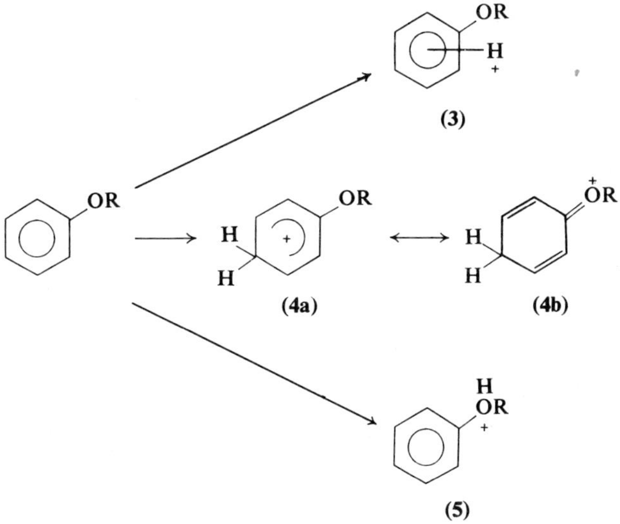

(3)　(27)

(4a) ↔ (4b)　(28)

(5)　(29)

## TABLE XII
### Protonation of Phenols and Alkyl Phenyl Ethers: Oxygen versus Carbon Protonation

| Compound | Acid system[a] | Temperature, °C | Experimental method | Position of protonation | Ref. |
|---|---|---|---|---|---|
| Phenol (OH) | 1<br>2 | 0<br>−64 | Uv<br>Nmr, uv | O<br>C | 72<br>84 |
| m-Cresol (OH, CH₃) | 2 | −67 | Nmr, uv | C | 84 |
| Anisole (OCH₃) | 1<br>2<br>3<br>3<br>4 | 0<br>−64<br>−10<br>−40 to −80<br>— | Uv<br>Nmr, uv<br>Nmr<br>Nmr<br>Nmr | O<br>C<br>C<br>O, C<br>C | 72<br>84–86<br>86–88<br>84 |
| Diphenyl ether (OC₆H₅) | 2 | −85 | Nmr | C | 84 |
| 4-Methylanisole (OCH₃, CH₃) | 3 | −100 | Nmr | O | 86 |
| 2,6-Dimethylanisole (CH₃, OCH₃, CH₃) | 3 | −100 | Nmr | O | 86 |

| Structure | | | | | |
|---|---|---|---|---|---|
| 2-OCH₃, 1,4-CH₃ benzene | 4 | −20 | Nmr | C | 86 |
| 1-OCH₃, 3,5-CH₃ benzene | 4 | −20 | Nmr | C | 86 |
|  | 4 | −78 | Nmr | C | 87 |
| benzoxepine | 1 | 0 | Uv | O | 72 |
| 4-OCH₃ benzyl (HOCH₂CH₂-) | 5 | −60 | Nmr | O[b] | 88 |
| 1,3-(OCH₃)₂ benzene | 4 | −70 | Nmr | C | 86 |
| 1,3,5-(OH)₃ benzene | 6 | 25 | Nmr, uv | C | 84, 89–91 |
|  | 2 | −80 | Nmr | C | 84 |

*(continued)*

## TABLE XII (Continued)

| Compound | Acid system[a] | Temperature, °C | Experimental method | Position of protonation | Ref. |
|---|---|---|---|---|---|
| HO–⟨⟩–OH, OCH₃ | 2 | 25 | Uv | C | 89 |
| RO–⟨⟩–OH, OR | 6 | 25 | Nmr, uv | C | 91 |
| R = CH₃, CH₂CH₃ RO–⟨⟩–OR, OR | 6<br>4<br>6 | 25<br>−80<br>25 | Nmr, uv<br>Nmr<br>H³ exchange | C<br>C<br>C | 89–91<br>86<br>92 |
| R = CH₃, CH₃CH₂ OH–⟨⟩–COOH, OH; HO–⟨⟩–OH | 2<br>6 | −63<br>−55 | Nmr<br>Nmr | C<br>C | 84<br>84 |

[a] Acid systems: 1, $H_2SO_4$ (aqueous, 30–90%); 2, $HSO_3F$; 3, $HF + BF_3$; 4, $HF$; 5, $HSO_3F$–$SbF_5$–$SO_2$; 6, $HClO_4$ (aqueous, 40–75%).
[b] Di-O-protonated species observed.

protonation of the oxygen to form **5**. The evidence has pointed toward the existence in solution of either **4** or **5** and has been based primarily on uv and nmr data. Table XII (72,84–92) summarizes the protonation behavior of a number of phenols and alkyl phenyl ethers in strong acid systems. When one considers the wide differences in temperature, acid system, and structure of the substrate, differences in the site of protonation are not unexpected. For example, a number of alkyl phenyl ethers and phenols were found to undergo exclusive oxygen protonation at 0° in 30–90% aqueous sulfuric acid (72). These results were based on the disappearance of the $n-\pi^*$ transition at about 270 m$\mu$. Similar spectral changes have been found when aniline is protonated (93) and in both cases may be attributed to the removal of the nonbonded electrons on the heteroatom from the resonance system upon protonation. However, under apparently comparable conditions (uv, 25°, 100% sulfuric acid; nmr, −64°, $HSO_3F$), phenol and anisole give exclusive carbon protonation (84,85). These conclusions were based on uv evidence and the integrated intensity of the proton resonances in the nmr spectrum. The lack of the peak at about 270 m$\mu$ in the previous work (72) and the lack of the expected band at about 350 m$\mu$ for anisole and phenol similar to that found for protonated phloroglucinol were attributed to the substantial contribution of the quinoid structure of the carbon-protonated species (**4b**) (84). This apparently contradictory behavior of phenol and alkyl phenyl ethers upon protonation was clarified recently when it was pointed out that the conjugate acid of anisole might be expected to change from primarily the O-protonated species to the C-protonated species in going from 60% sulfuric acid to 100% sulfuric acid (94). The oxygen protonated form would be expected to be more stabilized by hydrogen bonding to water than the carbon protonated form. Thus the decrease in the amount of water upon increasing the sulfuric acid concentration would be expected to favor carbon protonation.

The subtle interplay of several factors seems to determine whether oxygen or carbon protonation is observed. Examination of Table XII for general trends shows that oxygen protonation is observed for unsubstituted phenol and unsubstituted alkyl phenyl ethers usually at relatively low temperatures. Substitution of electron-releasing groups on the aromatic ring causes carbon protonation to predominate. The strength of the acid system and the temperature-dependent rate of exchange of the proton on oxygen compared to the rate of exchange on carbon is important. This is especially true for nmr investigations because differences in these exchange rates will determine which type of protonation will be *observed* experimentally. It has been stated that, generally, oxygen-protonated cations exchange their protons much more readily than carbon-protonated cations (86).

Observation of both oxygen-protonated and carbon-protonated species in the same solution bears out these subtle differences (86). The lack of oxygen protonation in the ring-substituted phenols and alkyl phenyl ethers can be explained by substantial inductive increase of carbon basicity while the oxygen basicity remains relatively constant. For example, anisole displays both oxygen protonation and carbon protonation in HF or $HF-BF_3$ from 0 to $-80°$ but 2,5- and 3,5-dimethylanisole show only carbon protonation down to $-80°$ (86). For protonated anisole, the strong temperature dependence of the mode of protonation is shown by the change of the ratio of C-protonated to O-protonated species from 1.5 at $-80°$ to over 50 at $0°$ (86).

## VI. PROTONATED SULFIDES

Protonated aliphatic sulfides have been studied at low temperatures by nmr spectroscopy in strong acid systems (57). They show well-resolved

$$RSR \xrightarrow[-60°]{FSO_3-SbF_5-SO_2} R-\underset{+}{\overset{H}{S}}-R \quad SbF_5FSO_3^- \qquad (30)$$

nmr spectra, the proton on sulfur being observed at about 6 ppm (Table XIII and for a representative spectrum, that of protonated diethyl sulfide, see Fig. 4).

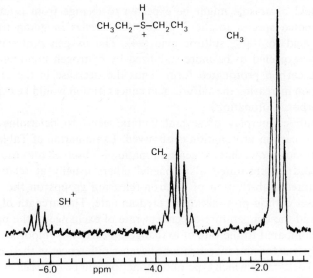

Fig. 4. Pmr spectrum of protonated diethyl sulfide in $FSO_3H-SbF_5-SO_2$ solution at $-60°$.

## TABLE XIII
## Nmr Chemical Shifts and Coupling Constants of Protonated Sulfides at $-60°$ in $HSO_3F$–$SbF_5$–$SO_2$

| Sulfide | $H_1$ | $H_2$ | $H_3$ | $H_4$ | $SH^+$ | $J_{H-SH^+}$, Hz |
|---|---|---|---|---|---|---|
| $\overset{1}{(CH_3)_2}S$ | 3.08 (2)[b] | | | | 6.52 (7) | 8.0 |
| $(\overset{2}{C}H_3\overset{1}{C}H_2)_2S$ | 3.57 (cm) | 1.67 | | | 6.23 (5) | 8.0 |
| $(\overset{3}{C}H_3\overset{2}{C}H_2\overset{1}{C}H_2)_2S$ | 3.33 (cm) | 2.00 (cm) | 1.07 (3) | | 6.18 (5) | 8.1 |
| $\begin{array}{c}\overset{2a}{C}H_3\\ \phantom{x}\\ \overset{2b}{C}H_3\end{array}\!\!\!\!\rangle CH\!\!\rangle_{\!2}S$ | 3.98 (6) | a. 1.62 (2)<br>b. 1.57 (2) | | | 5.80 (3) | 7.5 |
| $\overset{2}{C}H_3\!\!\rangle\overset{1a}{C}H\overset{1b}{S}CH_3\!\!\rangle CH_3$ | a. 3.89 (cm)<br>b. 2.90 (2) | 1.63 (2) | | | 6.07 (cm) | 8.0 |
| $(\overset{4}{C}H_3\overset{3}{C}H_2\overset{2}{C}H_2\overset{1}{C}H_2)_2S$ | 3.33 (cm) | ~1.70 (cm) | ~1.70 (cm) | 1.00 (cm) | 6.13 (5) | 8.0 |
| $(\overset{3}{C}H_3\overset{2a}{C}H_2\overset{1}{C}H)_2S$<br>$\phantom{xxx}\underset{2b}{CH_3}$ | 3.70 (cm) | a. 2.00 (cm)<br>b. 1.71 (cm) | 1.10 (3) | | 5.73 (cm) | 7.7 |

*(continued)*

TABLE XIII (Continued)

| Sulfide | $\delta$, ppm[a] | | | | $J_{H-SH^+}$, Hz |
|---|---|---|---|---|---|
| | $H_1$ | $H_2$ | $H_3$ | $H_4$ | $SH^+$ | |

| Sulfide | $H_1$ | $H_2$ | $SH^+$ | $J_{H-SH^+}$, Hz |
|---|---|---|---|---|
| $\overset{2}{C}H_3$<br>\|<br>$(CH_3\overset{1}{C}-)_2S$<br>\|<br>$CH_3$ | | 1.83 (1) | 5.83 (1) | — |
| $\overset{2}{C}H_3$<br>\|<br>$(CH_3\overset{1}{C}-)SCH_3$<br>\|<br>$CH_3$ | 2.87 (2) | 1.67 (1) | 6.00 (4) | 8.0 |
| $\overset{2b}{C}H_3\;\;\overset{2a}{C}H_3$<br>\|<br>$CH_3\overset{1}{C}SCH$<br>\|    \|<br>$CH_3\;\;\;\;CH_3$ | 4.05 (cm) | a. 1.62 (2)<br>b. 1.73 (1) | 6.25 (2) | 7.0 |

[a] From external capillary of TMS.
[b] Multiplicity of peaks shown in parentheses; cm = complex multiplet.

Recently, the nmr spectrum of protonated thiane-3,3,5,5-$d_4$ has been studied in $FSO_3H$–$SO_2$ in order to determine the conformational position of the proton on sulfur in this six-membered ring and to study the ring inversion process (95). The proton on sulfur resides exclusively in the axial position.

The protonated sulfides are more stable to cleavage than the corresponding protonated ethers and also more stable than the protonated thiols. Protonated methyl $t$-butyl ether is completely cleaved to trimethylcarbonium ion and protonated methanol even at $-70°$. Protonated methyl $t$-butyl sulfide is completely stable at $-60°$. When the temperature is increased to $-15°$, protonated methyl $t$-butyl sulfide very slowly cleaves to trimethylcarbonium ion and protonated methyl thiol.

$$(CH_3)_3C-\overset{H}{\underset{+}{S}}-CH_3 \xrightarrow{-15°} (CH_3)_3C^+ + CH_3SH_2^+ \qquad (31)$$

Protonated di-$t$-butyl sulfide shows very little cleavage at $-60°$. At $35°$ it cleaves slowly ($t_{1/2} \sim 1$ hr) to trimethylcarbonium ion and protonated hydrogen sulfide, the latter giving a peak at $\delta$ 6.60.

$$(CH_3)_3C-\overset{H}{\underset{+}{S}}-C(CH_3)_3 \xrightarrow{-35°} (CH_3)_3C^+ + (CH_3)_3CSH_2^+$$
$$\downarrow$$
$$(CH_3)_3C^+ + H_3S^+$$

Protonated secondary sulfides show extraordinary stability toward the strongly acidic medium. Protonated isopropyl sulfide shows no appreciable cleavage up to $+70°$ in a solution of $FSO_3H$–$SbF_5$ (1:1).

## VII. DIPROTONATED ALKOXY ALCOHOLS

Aliphatic alkoxy alcohols have been studied in $FSO_3H$–$SbF_5$–$SO_2$ solution by nmr and were found to be diprotonated (96). The spectra show the alkoxy proton at about 11 ppm and the hydroxyl protons at about 10 ppm, slightly deshielded from the corresponding protons in the protonated ethers and alcohols, respectively. Cleavage of these dications via three different pathways was reported, and while the mechanism of these reactions has not been studied, the mode of cleavage observed apparently reflects the stability of the leaving groups and their ability to undergo rapid conversion to a stable ion. These three cleavage pathways are illustrated by the examples in equations 32–34. Ions in brackets are presumed to be intermediates although they were not observed.

$$CH_3-\overset{+}{\underset{H}{O}}-CH_2-CH_2-\overset{+}{O}H_2 \xrightarrow[FSO_3H-SbF_5]{70°} \left[ CH_3-\overset{+}{\underset{H}{O}}-CH_2-\overset{+}{C}H_2 \right] + H_3O^+ \quad (32)$$

$$\downarrow$$

$$\underset{+}{O}=C\underset{CH_3}{\overset{CH_3}{\diagdown}}\overset{H}{\diagup}$$

$$CH_3-\overset{+}{\underset{H}{O}}-CH\underset{CH_2\overset{+}{O}H_2}{\overset{CH_3}{\diagup}} \xrightarrow{-10°} [CH_3-\overset{+}{C}H-CH_2-\overset{+}{O}H_2] + CH_3OH_2^+$$

$$\downarrow$$

$$CH_3CH_2-C\underset{H}{\overset{\overset{+}{O}H}{\diagup}} \quad (33)$$

$$CH_3(CH_2)_3-\overset{+}{\underset{H}{O}}-CH_2CH_2\overset{+}{O}H_2 \xrightarrow{-10°} [CH_3(CH_2)_3]^+ + H_2\overset{+}{O}CH_2CH_2\overset{+}{O}H_2 \quad (34)$$

$$\downarrow$$

$$(CH_3)_3C^+$$

## VIII. PROTONATED ALDEHYDES AND KETONES

Extensive kinetic and spectral evidence indicates that in acidic media nucleophilic attack at the carbonyl carbon takes place with prior protonation at the carbonyl oxygen (97). In relatively weak acid systems (dilute, aqueous sulfuric acid to <90% sulfuric acid) uv data have been used to study the equilibrium between acetone and its conjugate acid (98). These data show a shift of the $n-\pi^*$ transition (275 mμ) to shorter wavelength with increasing acid concentration. This shift was attributed to protonation. However, at higher acid concentrations, the self-condensation of acetone to mesityl oxide and phorone makes the interpretation of the ultraviolet data more difficult. The surprisingly high basicity of acetone ($pK_{BH^+}$ = 1.58) (98) reported in this investigation led to the reinvestigation of the protonation of a number of aliphatic ketones in sulfuric acid (99). This work demonstrates that the shift of the band at about 275 mμ at low acid concentrations is due to a medium effect rather than protonation. Increase in the acid strength above about 65% sulfuric acid results in the complete disappearance of this peak because of protonation of the oxygen. The $pK_{BH^+}$ values based on these spectrophotometric results (99) lead to more reasonable values for acetone and other aliphatic ketones of about $-7.0$.

The basicity of variously substituted benzaldehydes and acetophenones in sulfuric acid (100,101) and perchloric acid (102) have been determined from uv data. These basicity values have been correlated with $\sigma^+$ (101), $\sigma_m$, and variations of the carbonyl stretching frequency (101,103).

In comparing the uv and ir data to the more recent application of nmr to the protonation of carbonyls in strong acid systems, it can be seen that the uv and ir data can be interpreted in terms of bond order and hybridization of the carbonyl carbon–oxygen bond, but little structural information results. On the other hand, nmr yields quite definite structural information but care must be taken in relating chemical shift and coupling data directly to bond-order and hybridization considerations, especially since other factors such as the effect of protonation on the magnetic anisotropy of the carbonyl bond and the conformation about the α-carbon–carbonyl carbon must be taken into account.

Combined with the earlier uv and ir studies, nmr leads to a more detailed knowledge of the structure of protonated carbonyls in strong acid solution and demonstrates for the first time the existence of isomers of the protonated species.

A summary of the nmr data for a representative group of protonated aldehydes and ketones is presented in Table XIV (42,53,104–108). When one allows for the wide variety of acid systems and solvents used and the differences in concentration and temperature, the close correspondence of the chemical shifts for the proton on oxygen for a particular carbonyl is quite remarkable. These values generally agree with 1 ppm. Examination of data from only one source, where conditions of acid system, temperature, and concentration are usually more constant, shows that the agreement is much better than this.

The extent of the contribution of the resonance forms of protonated carbonyls (**6a, b**) has been discussed on the basis of the nmr data (42,53, 104–110). These arguments have been based on the large deshielding of the proton on oxygen, the size of the coupling between the proton on oxygen and the aldehydic hydrogen for protonated aldehydes, the existence of

$$R-C\overset{\displaystyle O-H}{\underset{\displaystyle R'}{\phantom{=}}} \longleftrightarrow R-C\overset{\displaystyle O-H}{\underset{\displaystyle R'}{\phantom{=}}} \qquad (35)$$

(6a) (6b)

isomers for protonated carbonyls, and the appearance of allylic-like coupling between the proton on oxygen and the hydrogens on the α carbon. These data strongly indicate that there is substantial double-bond character in the carbonyl carbon–oxygen bond for protonated carbonyls. These data

## TABLE XIV
### Characteristic Chemical Shifts and Coupling Constants of Protonated Aldehydes and Ketones[a]

Aldehyde: $R-\underset{2}{C}H=\underset{1}{\overset{+}{O}H}$

| R | $H_1$ | $J_{1,2}$ | $H_2$ | Acid system[b] | Temperature, °C | Ref. |
|---|---|---|---|---|---|---|
| H | 16.73 | 8.7 (syn)<br>21.1 (anti) | −10.18 | 1 | −60 | 42 |
| CH$_3$<br>(H$_1$, H$_2$ syn) | 15.11<br>16.01<br>15.47<br>14.78 | 9.0<br>8.1<br>8.5<br>8.5 | 10.10<br>10.00<br>10.02<br>9.77 | 1<br>2<br>3<br>1 | −60<br>−82<br>−20<br>−52 | 42<br>104<br>105<br>106 |
| CH$_3$<br>(H$_1$, H$_2$ anti) | 15.47<br>15.74<br>15.14 | 19.5<br>19<br>18.5 | c<br>c<br>c | 1<br>3<br>1 | −60<br>−20<br>−52 | 42<br>105<br>106 |
| CH$_3$CH$_2$<br>(H$_1$, H$_2$ syn) | 15.03<br>16.04<br>15.12 | 8.8<br>8.1<br>9 | −10.15<br>9.97<br>— | 1<br>2<br>1 | −60<br>−80<br>−65 | 42<br>104<br>106 |
| (CH$_3$)$_2$CH | 15.22 | 8.8 | 10.25 | 1 | −60 | 42 |
| (CH$_3$)$_3$C | 15.30 | 9.0 | 10.17 | 1 | −60 | 42 |

**TABLE XIV** (*Continued*)

Ketone: $\mathrm{R-C(R')=\overset{+}{O}H}$
         $\phantom{\mathrm{R-C(R')=O}}1$

| R | R' | H$_1$ *syn* to R | H$_1$ *syn* to R' | Acid system[b] | Temperature, °C | Ref. |
|---|---|---|---|---|---|---|
| CH$_3$ | CH$_3$ | 14.93 | — | 1 | −60 | 107 |
|  |  | 14.24 | — | 1 | −59 | 106 |
|  |  | 14.68 | — | 3 | −20 | 105 |
|  |  | 14.45 | — | 4 | +25 | 53 |
| CH$_3$ | CH$_3$CH$_2$ | 14.26 | 13.91 | 1 | −60 | 107 |
|  |  | 14.03 | 13.65 | 1 | −60 | 106 |
|  |  | 14.67 | 14.40 | 3 | −20 | 105 |
| CH$_3$ | CH$_3$CH$_2$CH$_2$ | 14.30 | 14.16 | 1 | −60 | 107 |
| CH$_3$ | CH(CH$_3$)$_2$ | 14.41 | 13.95 | 1 | −60 | 107 |
|  |  | 14.69 | 14.34 | 3 | −20 | 105 |
| CH$_3$ | C(CH$_3$)$_3$ | 14.23 | — | 1 | −60 | 107 |
|  |  | 14.72 | — | 3 | −20 | 105 |
| CH$_3$ | CH$_3$CH$_2$ | 14.06 | — | 1 | −60 | 107 |
|  |  | 14.43 | — | 3 | −20 | 105 |
| CH(CH$_3$)$_2$ | CH(CH$_3$)$_2$ | 13.86 | — | 1 | −60 | 107 |
| C(CH$_3$)$_3$ | C(CH$_3$)$_3$ | 13.51 | — | 1 | −60 | 107 |
| CH$_3$ | C$_6$H$_5$ | 13.03 | — | 1 | −52 | 106 |
|  |  | 13.46 | — | 5 | −82 | 53 |

[a] Chemical shifts in ppm relative to TMS; coupling constants in Hz.
[b] Acid systems: 1, HSO$_3$F–SbF$_5$–SO$_2$; 2, HF–BF$_3$; 3, HF–SbF$_5$; 4, HSO$_3$F–SbF$_5$; 5, HSO$_3$F.
[c] Exact chemical shift not observed; probably obscured by predominating *syn* isomer.

have been treated in only a qualitative manner, however, and until a more quantitative relationship can be developed, it should be repeated that care should be taken in relating chemical shift data directly to hybridization and bond order considerations.

Evidence for the charge distribution in aldehydes and ketones has also been obtained using $^{13}$C spectroscopy. A correlation between calculated $\pi$-electron density (using a Hückel molecular orbital approach) and $^{13}$C chemical shift has been reported and is shown in Figure 5 (48). The $^{13}$C

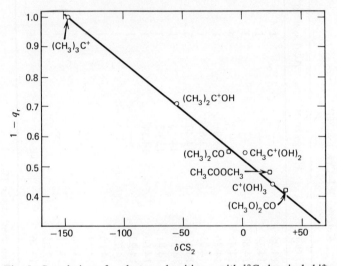

Fig. 5. Correlation of $\pi$-electron densities $q_r$ with $^{13}$C chemical shifts.

shifts in protonated formaldehyde, protonated acetaldehyde, and protonated acetone show that the effect of successively replacing the hydrogens in protonated formaldehyde by methyl groups is to cause an upfield shift of 15 ppm (49). This is a slightly greater effect than in alkanes and indicates that the charge on the $sp^2$ carbon in these three ions is very similar and that the methyl group is not exerting any special stabilizing effect on the ion—any inductive or hyperconjugative effects being similar to those in alkanes. The $^{13}$C resonances of the methyl groups in protonated acetaldehyde and protonated acetone are almost identical with those in their uncharged precursors (49). This observation, coupled with information from the proton spectra of these ions, indicates that the positive charge in these ions is distributed between the $sp^2$ carbon, the methyl and aldehydic protons, and the hydroxyl group, the methyl carbon having a charge close to that in the uncharged precursor. A similar charge distribution has recently been calculated theoretically (43). These calculated charge densities are shown below for acetaldehyde and protonated acetaldehyde together with the assumed bond lengths and angles in the isomers depicted (Fig. 6).

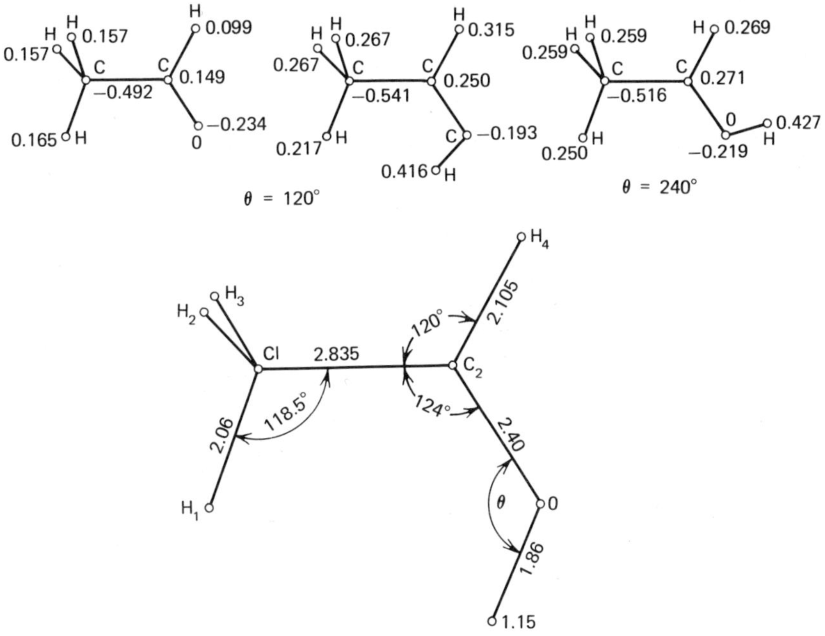

Fig. 6. Gross atomic charges for acetaldehyde and protonated acetaldehyde.

In connection with the evaluation of the charge distribution in protonated aldehydes and ketones the nmr spectra of protonated benzaldehyde and protonated acetophenone are quite revealing. It has been found that whereas in protonated acetophenone the two *ortho* protons in the phenyl ring are equivalent at temperatures as low as $-90°$ in $FSO_3H-SbF_5-SO_2$ solution, protonated benzaldehyde under the same conditions shows nonequivalence of both *ortho* and *meta* protons. Collapse of the two *ortho* multiplets due to the two protons becoming equivalent occurs at $-30°$. Barring accidental equivalence in the case of protonated acetophenone, this result requires that the barrier to rotation about the carbon–phenyl bond be substantially higher in the case of protonated benzaldehyde.

While this could possibly be due to the bulk of the methyl group as compared to hydrogen, it can also be taken to indicate a greater contribution of resonance structures placing the positive charge on the phenyl ring in the case of protonated benzaldehyde, and thus that the methyl group in protonated acetophenone is exerting a stabilizing effect on the ion. This can be shown in a different manner from the $^{19}$F chemical shifts in *meta*- and *para*-substituted fluorobenzenes in which the substituent can be protonated. These shifts can be related to the electronic effects of the substituents, and substituent constants (both resonance and inductive) have been obtained for a wide variety of substituents (111). This approach has been applied in superacid solution in which the substituent is protonated, and some of these results are given in Table XV (41). In the case of

### TABLE XV
### $^{19}$F Nmr Shifts for *m*- and *p*-Fluorophenyl Derivatives Attached to a Positively Charged Center

| Substituent | $\phi_m$ | $\phi_p$ | $\Delta\phi_p - \phi_m$ | $\sigma_I$ | $\sigma_R°$ | $(\sigma_I + \sigma_R°)$ |
|---|---|---|---|---|---|---|
| $-\overset{+}{C}(CH_3)_2$ | 6.35 | 52.39 | 46.0 | 0.98 | 1.56 | 2.54 |
| $-\overset{+}{C}(H)=O$ | 11.03 | 45.13 | 34.1 | 1.64 | 1.16 | 2.80 |
| $-\overset{+}{C}(H)(OH)$ | 6.71 | 40.61 | 33.9 | 1.03 | 1.15 | 2.18 |
| $-\overset{+}{N}(OH)=O$ | 8.60 | 39.60 | 31.0 | 1.30 | 1.05 | 2.35 |
| $-\overset{+}{C}(CH_3)(OH)$ | 4.63 | 32.43 | 27.8 | 0.74 | 0.94 | 1.68 |
| $-\overset{+}{C}(OH)(OH)$ | 4.90 | 24.92 | 20.0 | 0.77 | 0.67 | 1.44 |
| $-\overset{+}{N}\equiv N$ | 10.19 | 32.00 | 11.8 | 1.52 | 0.40 | 1.92 |
| $-\overset{+}{N}H_3$ | 5.01 | 3.75 | −1.3 | 0.79 | −0.04 | 0.75 |

the protonated *m*- and *p*-fluorobenzaldehyde and -acetophenone, it can be seen that the substitution of methyl for hydrogen reduces significantly both the inductive ($\sigma_I$) and resonance ($\sigma_R°$) interaction of the cationic center with the phenyl ring showing that methyl is behaving as an electron donor with respect to hydrogen.

Protonated formaldehyde is the simplest heteroatom containing organic species that has been studied and can be generated by passing formaldehyde from the pyrolysis of paraformaldehyde over a stirred solution of $FSO_3H–SbF_5$ in $SO_2$ at $-76°$ or by reaction of methylene chloride with $FSO_3H–SbF_5$ diluted with $SO_2$ (containing some water) at $-100°$ (112).

$$H_2CCl_2 \xrightarrow{FSO_3H-SbF_5-SO_2} [\overset{+}{C}H_2-Cl] \underset{H^+}{\overset{H_2O}{\rightleftharpoons}} [H_2\overset{+}{O}CH_2-Cl] \rightleftharpoons$$

$$\rightleftharpoons CH_2=\overset{+}{O}H + HCl \qquad (36)$$

$$CH_2O \xrightarrow{FSO_3H-SbF_5-SO_2} CH_2=\overset{+}{O}H$$

The proton and $^{13}C$ spectrum of the $^{13}C$-enriched ion have been studied in some detail (45), and the signs and magnitudes of all the coupling constants have been obtained and are given in Table XVI.

The assignment of $H_1$ and $H_2$ given is consistent both with the magnitudes of the vicinal coupling constants and those of the two direct carbon-hydrogen coupling constants. The large, positive geminal proton-proton coupling constant compares with a value of $+42$ in formaldehyde itself, and the decrease of this value on protonation has been discussed in terms of the molecular orbital approach to the calculation of coupling constants (45). The *cis* and *trans* vicinal coupling constants of 4.0 and 21.7 Hz compare with 11.4 and 19.1 Hz found in ethylene.

### TABLE XVI
### $^1H$ and $^{13}C$ Nmr Data for Protonated Formaldehyde

| | | δ values | | Coupling constant, Hz | |
|---|---|---|---|---|---|
| $H_1$     $H_3$ | | $\delta(H_1)$ | 9.820 | $J_{12}$ | $+21.7$ |
| $^{13}C=O$ | | $\delta(H_2)$ | 9.940 | $J_{13}$ | $+9.0$ |
| $H_2$ | | $\delta(H_3)$ | 16.70 | $J_{23}$ | $+19.0$ |
| $(FSO_3H-SbF_5-SO_2, -60°)$ | | $\delta(^{13}C)$ | $-29.2$ | $J_1(^{13}C)$ | $+198.4$ |
| | | | | $J_2(^{13}C)$ | $+209.8$ |
| | | | | $J_3(^{13}C)$ | $-8.7$ |

The nmr spectra of protonated acetaldehyde and methyl ethyl ketone have been reported by a number of investigators (42,104–107). These spectra are representative of the nmr evidence which can lead to conclusions concerning the structure of protonated carbonyls and are reproduced in Figures 7 and 8. The complex resonances between δ 14 and 17 can be assigned to the proton on oxygen. For protonated aldehydes, where the integrated intensity of the proton on oxygen and the carbonyl hydrogen is 1:1, proof that these low-field resonances are caused by the proton on oxygen and not the carbonyl hydrogen was obtained from the spectrum of protonated acetaldehyde-$d_4$.

The multiplicity of the low-field resonances leaves little doubt that more than one isomer exists for protonated acetaldehyde and for protonated ketones in which the alkyl groups are not the same. For protonated acetaldehyde (Fig. 7) a doublet appears centered at about 15.47 ppm coupled to the carbonyl hydrogen ($J_{HC=\overset{+}{O}H}$ = 8.5 Hz) for the isomer in greater abundance. Slightly downfield from this doublet is another doublet showing fine structure with a larger coupling to the carbonyl hydrogen ($J_{HC=\overset{+}{O}H}$ = 19.5 Hz). These proton on oxygen resonances have been

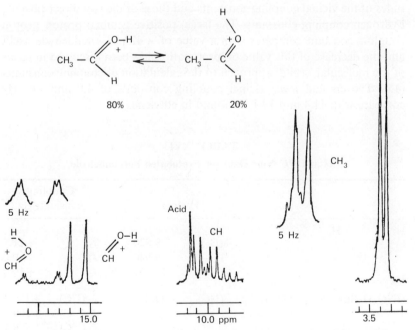

Fig. 7. Nmr spectrum of protonated acetaldehyde ($HSO_3F$–$SbF_5$–$SO_2$; $-60°$).

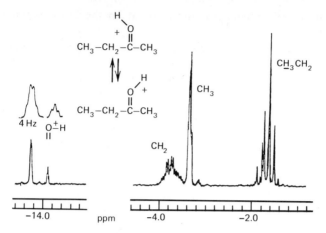

Fig. 8. Nmr spectrum of protonated methyl ethyl ketone (HSO$_3$F-SbF$_3$-SO$_2$; −60°).

assigned to the *syn* (or *cis*) **7** and *anti* (or *trans*) **8** arrangement of the

proton on oxygen to the carbonyl hydrogen. The magnitudes of these coupling constants are comparable with analogous *cis* and *trans* coupling constants found for uncharged, isoelectronic alkenes (113).

Nonempirical molecular orbital calculations have been carried out (43) to determine the relative stabilities of the two isomers of protonated acetaldehyde. Configuration **7** (R = CH$_3$) was found to be less stable than **8** (R = CH$_3$) by 1.4 kcal/mole in reverse order to that observed on the basis of the assignments given above. For protonated formic acid the most stable system was found to be **9**, followed by **10**, which is 1.5 kcal/mole

higher in energy. The other possible structure, **11**, was found to be another 6.0 kcal/mole higher in energy. These latter two structures are again in the reverse order of stabilities to those found on the basis of the assigned structures.

Three explanations for these discrepancies have been presented and discussed: first, that the calculations are too inaccurate; second, that the assignments made experimentally are incorrect; and third, that the difference lies in the fact that the calculations are for an isolated ion while experimental observations have been made in solution. No firm conclusion has been reached as to which of these explanations is the correct one though the body of experimental evidence obtained seems to have such a high degree of consistency as to rule out the possibility of the assignments made experimentally being incorrect.

For ketones in which the alkyl groups are different ($R_1 \neq R_2$), two proton on oxygen resonances appear (12, 13). Protonated methyl ethyl

$$R_1-C\overset{\overset{+}{O-H}}{\underset{R_2}{\diagdown}} \qquad R_1-C\overset{\overset{H}{\diagdown}\overset{+}{O}}{\underset{R_2}{\diagdown}}$$

(12) (13)

ketone (Fig. 8) shows the resonance for the proton on oxygen of the isomer with the proton on oxygen *syn* to the methyl group at approximately 14.30 ppm and for the isomer with the proton on oxygen *syn* to the ethyl group at 13.99 ppm. The explanation given (114) for the appearance to two OH resonances in the case of protonated unsymmetrical benzophenones is probably incorrect—an equilibrium between 12 and 13 is more consistent with the spectra obtained. It will be noted that for the isomer in greater abundance (13, $R_1 = CH_3$; $R_2 = CH_2CH_3$) the proton on oxygen appears at lower field than for the isomer in which the proton on oxygen is *syn* to the larger ethyl group (12, $R_1 = CH_3$; $R_2 = CH_2CH_3$). For the protonated ketones studied, the proton on oxygen of the isomer in greatest abundance was found consistently at lower field. For acetaldehyde, the only aldehyde for which isomerism was observed when protonated, the opposite was true. The chemical shift of the proton on oxygen of protonated carbonyls perhaps reflects changes in the carbon–oxygen bond anisotropy upon protonation and any steric deshielding experienced by the proton on oxygen caused by the *syn* alkyl group.

Closer examination of some of the proton on oxygen resonances leads to evidence which further supports the isomer assignment. The proton on oxygen resonance for the *anti* isomer of protonated acetaldehyde (8, R = $CH_3$, and Fig. 7) centered at 15.47 ppm is further split into a doublet of quartets. For protonated methyl ethyl ketone, the proton on oxygen resonance *syn* to the methyl is a quartet and the proton on oxygen *syn* to the ethyl is a triplet (Fig. 8). These couplings are small, on the order of 0.8–1.2 Hz. Similar couplings were found for the other protonated ketones (107) and have been assigned to long-range, allylic-like coupling of the proton on oxygen with the hydrogens on the α-carbon. Similar coupling for the *syn* isomer of protonated acetaldehyde in which the proton on oxygen is *trans* to the methyl (7, R = $CH_3$, and Fig. 7) was too small to be resolved and is probably less than 0.3 Hz (42). In isoelectronic olefinic systems *cis* allylic coupling is usually larger than *trans* allylic coupling (115). Thus, observation of allylic-like couplings in protonated carbonyls further justifies the isomer assignment and the double-bond character of the carbon–oxygen bond. The appearance of these small allylic couplings in other protonated organic compounds such as esters, acids, and amides has been found to be of value in determining the structure of the protonated species; however, cases have been found in which the magnitude of the *syn* and *anti* allylic coupling constants are reversed and caution should be exercised (45) in drawing conclusions from these four-bond couplings. The exceptions usually are ions in which the methyl group is attached directly to a heteroatom as in methoxycarbonium ion, $CH_3OCH_2^+$.

The isomer in greater abundance, as determined by integration of the proton on oxygen resonances, is the isomer which would have been predicted from simple steric considerations, with the proton on oxygen *syn* to the smaller group. For protonated aldehydes, the increase in steric requirements in going from methyl (acetaldehyde) to ethyl (propionaldehyde) is apparently great enough not to allow observation of the isomer with the proton on oxygen *syn* to the ethyl. For protonated ketones this trend is also observed. Only one isomer is observed for protonated methyl *t*-butyl ketone in which the proton on oxygen is *syn* to the methyl group (107). Qualitative isomer distributions for some protonated carbonyls are presented in Table XVII.

Attempts have been made to measure the energy of activation for the *syn–anti* conversion, and it has been estimated that the barrier to rotation for the proton on oxygen about the carbonyl carbon–oxygen bond has a minimum value of about 17 kcal/mole (105). Attempts to obtain precise values have not yielded conclusive results. As the temperature is raised to about $+35°$, the proton on oxygen resonances begin to broaden. However, before a sharp singlet is formed, indicating free rotation about the carbonyl carbon–oxygen bond, intermolecular proton exchange with the acid system also becomes rapid (104,105).

It is possible that the lowest barrier to interconversion of isomers in protonated acetaldehyde is not rotation about the carbon–oxygen bond but motion of the hydroxyl proton within the plane. Nonempirical

## TABLE XVII
### Isomer Distribution for Protonated Aldehydes and Ketones,

$$R_1\text{—}C(R_2)\text{=}\overset{+}{O}H$$

| $R_1$ | $R_2$ | syn to $R_1$, % | syn to $R_2$, % | Temperature, °C | Ref. |
|---|---|---|---|---|---|
| $CH_3$ | H | ~20 | ~80 | −20 | 105 |
|  |  | ~18 | ~82 | −56 | 106 |
|  |  | 20 | 80 | −60 | 42 |
| $c\text{-}C_3H_5$ | H | ~25 | ~75 | −50 | 106 |
| $CH_3CH_2$ | H | ~0 | ~100 | −60, −65 | 42,106 |
| $CH_3CH_2$ | $CH_3$ | ~33 | ~67 | −20 | 105 |
|  |  | ~20 | ~80 | −60 | 106 |
|  |  | 19.4 | 80.6 | −60 | 107 |
| $(CH_3)_2CH$ | $CH_3$ | ~16 | ~84 | −20 | 105 |
|  |  | 15.4 | 84.6 | −60 | 107 |
| $(CH_3)_3C$ | $CH_3$ | ~6 | ~94 | −20 | 105 |
|  |  | ~0 | ~100 | −60 | 107 |

molecular orbital calculations (43) have been done which suggest the latter alternative. In protonated aldehydes a rotation about the carbon–oxygen bond is rather difficult and has an energy barrier of 25–30 kcal/mole. Isomerization in the plane, however, requires only 17–18 kcal/mole. In protonated formic acid the carbon–oxygen bond is not as strong as in aldehydes, and a rotation around the carbon–oxygen bond requires only 15 kcal/mole, while the motion within the plane still requires 17 kcal/mole. The experimental barrier in the case of protonated formic acid has been found to be 15.3 kcal in HF–SbF$_5$ (116). The carbon–oxygen bond-order dependence of the calculated barriers to rotation and motion of the hydroxyl proton within the plane is illustrated in Figure 9.

Analysis of the nmr spectra of protonated carbonyls leads to evidence for changes in the conformation about the α-carbon–carbonyl carbon bond. The coupling between the α-hydrogens in *syn* protonated acetaldehyde (5) and the carbonyl hydrogen [$J_{CH_3\text{-}CH}$ = 3.4 Hz (42,104–106)] and the lack of this coupling in the other aldehydes indicate subtle changes in the preferred conformation about this bond. For protonated aldehydes other than acetaldehyde where only the *syn* isomer is observed, a measurable increase in the *trans* allylic coupling is found (42). This indicates a change in the preferred conformation as alkyl substitution is made at the α-carbon. The conformation about the α-carbon–carbonyl carbon bond for protonated carbonyls has been discussed (104,105), the conclusions

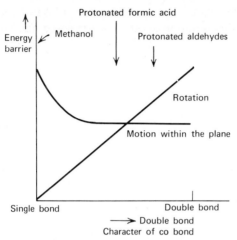

Fig. 9. Schematic comparison of energy barriers for rotation and motion within the plane [according to P. Ros, *J. Chem. Phys.*, **49**, 4902 (1969)].

being based largely on analogy with the conformations found for unprotonated carbonyls (117). However, until these small variations in coupling constants are examined in more detail in relation to structural variations, the conformation about this bond for protonated carbonyls will be in doubt.

Little work has been done concerning the reactivity of protonated carbonyls in strong acid systems. The cleavage of the carbonyl carbon–α-carbon bond is difficult, and the protonated species is apparently relatively stable in excess strong acid even to high temperatures. It has been found for aldehydes that if the α-alkyl group is tertiary, rapid rearrangement takes place. Protonated pivaldehyde undergoes rapid rearrangement even at $-70°$ to protonated methyl isopropyl ketone probably through a tertiary carbonium ion (eq. 37) (42). Similar reactions in dilute

$$\underset{\underset{CH_3}{|}}{\overset{\overset{CH_3}{|}}{H_3C-C-C}}\overset{OH}{\underset{H}{\overset{+}{\diagup}}} \longrightarrow \left[ \underset{H}{\overset{H_3C\ \ OH}{H_3C-\underset{+}{C}-C-CH_3}} \right] \longrightarrow \underset{CH_3}{\overset{\overset{CH_3}{|}}{H_3C-CH-C}}\overset{OH}{\underset{CH_3}{\overset{+}{\diagup}}}$$

(37)

aqueous acid systems have been known for some time (118,119). Undoubtedly rearrangements of even secondary and primary alkyl groups of protonated aldehydes making use of strong acid systems at higher temperatures, and longer reaction times are possible.

Protonated cyclohex-2-enone is converted (120) to 3-methylcyclopent-2-enone in HF–SbF$_5$ solution at 50°. The proposed mechanism is shown in equation 38.

$$\text{cyclohex-2-enone} + H^+ \rightleftharpoons \text{protonated} \rightarrow \text{methylcyclopentenyl cation} \rightleftharpoons \text{3-methylcyclopent-2-enone} \quad (38)$$

## IX. PROTONATED AMIDES AND THIOAMIDES

### A. General

Because of the fundamental importance of the peptide link in biologica systems, the nature of the species obtained from amides in acid systems has been extensively studied. It has been shown that there is substantial double-bond character in the carbon–nitrogen bond in amides (121–123). The barrier to free rotation about this bond has been estimated at 12 ± 3 kcal/mole for $N,N$-diethylacetamide from nmr spectroscopy (122). This indicates that there is appreciable contribution from the resonance form **14b** in amides, and from this point of view it is reasonable to assume that

$$\underset{(14a)}{R-C(=O)NH_2} \longleftrightarrow \underset{(14b)}{R-C(-O^-)=\overset{+}{N}H_2}$$

proton addition will occur at the oxygen. Further, proton addition at oxygen does not destroy this resonance interaction with the free pair of electrons on nitrogen (**15, 16**) and satisfactorily explains restricted rotation in protonated amides as well as for amides themselves. If protonation

$$\underset{(15)}{R-C(=\overset{+}{O}H)NH_2} \longleftrightarrow \underset{(16)}{R-C(-OH)=\overset{+}{N}H_2}$$

$$R-C(=O)NH_2 \xrightarrow{H^+} (15) \quad \text{or} \quad \xrightarrow{H^+} \underset{(17)}{R-C(=O)\overset{+}{N}H_3}$$

occurs at nitrogen, free rotation would be expected about the carbon–nitrogen bond. Similar arguments have been used for the preferred site of protonation in thioamides, ureas, and thioureas. Cryoscopic measurements have shown that amides (124,125) and thioamides (126) are monoprotonated in strong acid systems such as sulfuric acid.

Oxygen protonation of amides and ureas and sulfur protonation of thioamides and thioureas has been challenged by several investigators during the last decade (127–130). The principal evidence favoring O-protonation has been based on nmr spectroscopy while the evidence for N protonation was provided by ir spectroscopy and some basicity measurements of substituted aromatic amides using uv measurements (103,130). These conflicting arguments have been summarized through 1961 in an excellent review (31), and it is concluded that the weight of the experimental evidence favors O-protonation. Since this review appeared, improved experimental techniques have substantiated this view. In this section we will not attempt to review the older data in detail and will only stress those earlier investigations which will give perspective concerning this controversy.

## B. The Site of Protonation

In retrospect, the argument (sometimes somewhat heated) concerning the preferred site of protonation of amides and thioamides resulted from a complete dependence on one experimental method and the failure of that method to give conclusive evidence concerning the site of protonation. One is reminded of the similar controversy concerning the site of protonation of phenols and alkyl phenyl ethers, and, like the protonation of phenols and alkyl phenyl ethers, there now appears to be evidence for the simultaneous existence of both oxygen and nitrogen protonated amides in strong acid systems.

### 1. Infrared Evidence

The principal evidence in favor of N-protonation is provided by ir spectroscopy. For the acid salts of acetamide (127), urea (127,129), and thiourea (127), the infrared shows the presence of additional vibrations in the N–H region, ascribed to the presence of $NH_3^+$, the apparent displacement of the carbonyl stretching frequency to higher values, and the absence of a definite oxygen–hydrogen (or sulfur–hydrogen) vibration. For example, acetamide hydrochloride shows a vibration at about 1715 $cm^{-1}$ compared to the band at 1675 $cm^{-1}$ in the spectra of acetamide. This shift was attributed to the increase in the double-bond character of the carbonyl carbon–oxygen bond upon protonation of the nitrogen atom

(127). Comparable shifts have been found for other protonated amides. Such evidence for N protonation is difficult to contradict if it can indeed be proven that the assigned bands are due to carbonyl stretching.

Considerable doubt has been cast on these assignments and the interpretation that follows them by recent ir studies. A comparison of the spectra of dicyclohexylurea and $^{18}$O-labeled dicyclohexylurea showed that the expected isotopic shift to lower frequency caused the band at 1628 cm$^{-1}$ to shift to 1611 cm$^{-1}$ (131). This evidence indicates that the band at about 1680 cm$^{-1}$, usually associated with carbonyl stretching in amides and ureas, may be misassigned. The ir spectra of N-acyltrialkylammonium halides (18) further indicates the incorrect assignment of the carbonyl

$$R-C\overset{\displaystyle O}{\underset{\overset{+}{N}R_3}{\diagup}}$$

(18)

stretching frequency in the earlier work. Structurally these salts would be similar to protonated amides if protonation occurred at the nitrogen. These compounds show a strong band at about 1810 cm$^{-1}$ assigned to the carbonyl stretching vibration (132) and no band in the vicinity of 1700 cm$^{-1}$. The ir spectra of amide complexes with boron halides indicates that coordination occurs at oxygen rather than nitrogen (133).

In the earlier work (127,129) it was assumed that thiourea and thioacetamide also protonate on nitrogen. This conclusion was based on the presence of a band at 2350 cm$^{-1}$ in both acetamide hydrochloride and thioacetamide hydrochloride, assigned to the $NH_3{}^+$ group. These conclusions have been reexamined (134–137). The uv spectra of thioacetamide shows the disappearance of the $n-\pi^*$ transition in concentrated sulfuric acid solution indicating protonation of the sulfur (136). The N–H band assignments which were thought to be characteristic of the $NH_3{}^+$ are reinterpreted in light of this new evidence (136).

## 2. Nmr Evidence

The nmr evidence concerning the site of protonation of amides, ureas, thioamides, and thioureas has been consistently interpreted in terms of O protonation. While the interpretation of the ir data in favor of N protonation resulted from incorrect band assignment, the early nmr work suffered from the inability to directly observe the proton on either oxygen or nitrogen. This was caused by the weakness of the acid system used resulting in rapid exchange rates and the quadrupole broadening caused by the nitrogen atom.

It was recognized early that for the protonation of amides four limiting cases were possible and would be significant in the interpretation of the nmr spectra (125).

1. N-protonation with slow proton exchange. This would lead to free rotation about the carbon–nitrogen bond. For protonated $N,N$-dimethylamides, the nitrogen methyl groups would be expected to be a doublet caused by the splitting by the proton on nitrogen.
2. N-protonation with rapid exchange. This would lead to a singlet for two methyls on nitrogen if we can assume free rotation about the carbon–nitrogen bond.
3. O-protonation with slow exchange. This should show a resonance for the proton on oxygen. It should be borne in mind that the bond order of the carbon–nitrogen bond could lead to magnetically nonequivalent alkyl groups on nitrogen, and correct interpretation may depend upon determining whether the multiplicity is due to spin–spin splitting (case 1) or magnetic nonequivalence (case 3).
4. O-protonation with fast exchange would yield no definite OH resonance.

In all the cases listed, significant shifts of the resonances compared to the unprotonated amides is to be expected because of the presence of the positive charge.

The nmr of $N$-methylacetamide in dilute aqueous acid solutions led to the conclusion that both the N protonated and O protonated species are present (125,138). Comparison of the exchange rates indicates that the amount of N-protonated species is small compared to the O-protonated species.

The chemical shifts of protonated amides in strong acid systems are summarized in Table XVIII (125,139,140). The initial investigations of the nmr of amides in strong acids can best be described by case 4. This work was done in 100% sulfuric acid at 29°. Compared to the acid systems discussed in this review, such conditions of temperature and acid strength would be expected to lead to fast exchange. For example, the spectra of $N,N$-dimethylformamide in 100% sulfuric acid resemble that of the unprotonated amide (125). The chemical shift difference between the $N$-methyl groups is about the same (0.14 ppm for the amide and 0.12 for the protonated amide) and is field dependent, indicating that the methyls are magnetically nonequivalent. The methyl groups are further split into doublets by coupling to the formyl hydrogen. The magnitude of these couplings (1.0 and 1.5 Hz) and the absence of this coupling in protonated $N,N$-dimethylformamide-1-$d$ indicates that this coupling is not caused by

## TABLE XVIII
Pmr Chemical Shifts for Some Protonated Amides and Thioamides[a]

| Amide (thio) | Acid system[b] | Temperature, °C | =$\overset{+}{O}$H (=$\overset{+}{S}$H) | NH | NCH$_3$ | | Ref. |
|---|---|---|---|---|---|---|---|
| HCONH$_2$ | 1 | −84.5 | 10.72 | 8.28 | — | | 139 |
| HCONHCH$_3$ | 2 | 29 | — | — | 4.11 | 4.03 | 125 |
| | 1 | −85 | 10.49 | 8.58 | 3.33 | | 139,140 |
| HCON(CH$_3$)$_2$ | 1 | 29 | — | — | 4.18 | 4.11 | 125 |
| | 2 | −80 | 9.98 | — | 3.53 | 3.43 | 139 |
| CH$_3$CONH$_2$ | 2 | −80 | 10.40 | 8.30 | — | | 139 |
| CH$_3$CONHCH$_3$ | 1 | 29 | — | — | 3.70 | 2.71 | 125 |
| | 2 | −85 | 10.15 | 8.27 | 3.25 | | 140 |
| CH$_3$CON(CH$_3$)$_2$ | 1 | 29 | — | — | 4.13 | | 125 |
| | 2 | −79 | 9.80 | — | 3.45 | | 139 |
| C$_6$H$_5$CONH$_2$ | 2 | −86 | 10.34 | 8.72 | — | — | 140 |
| CH$_3$CSNH$_2$ | 2 | −81 | 5.72 | 9.25 | — | — | 140 |

[a] Chemical shifts in ppm relative to TMS.
[b] Acid systems: 1, 100% H$_2$SO$_4$; 2, 100% HSO$_3$F.

a proton on nitrogen. It is interesting to compare these small couplings with the long-range allylic couplings found for protonated aldehydes (**20**)

$$\begin{array}{cc} \text{CH}_3\text{(a)} \diagdown \quad \diagup \text{OH} & \text{CH}_3 \diagdown \quad \diagup \text{H} \\ \text{N}^+ = \text{C} & \text{C} = \overset{+}{\text{O}} \\ \text{CH}_3\text{(b)} \diagup \quad \diagdown \text{H} & \text{H} \diagup \quad \diagdown \\ \text{(19)} & \text{(20)} \\ J_{\text{CH}_3\text{(a)}-\text{CH}} = 1.5 \text{ Hz} & J_{\text{CH}_3-\text{OH}^+} = 1.2 \text{ Hz} \\ J_{\text{CH}_3\text{(b)}-\text{CH}} = 1.0 \text{ Hz} & \end{array}$$

(see p. 1745). The coupling of the formyl hydrogen with the methyl group has been found to be 0.8 and 0.5 Hz for unprotonated $N,N$-dimethylformamide (123). The presence of allylic coupling in protonated aldehydes and ketones has been interpreted as indicating substantial contribution from **6b**. In protonated amides, the increase in this allylic coupling compared to the unprotonated amide indicates a substantial contribution of **16** where the positive charge resides on nitrogen.

Similar nmr evidence has been given for the sulfur protonation of $N,N$-dimethylthiobenzamide (137).

While the earlier work was unable to find the resonance for the proton on oxygen because of fast exchange, the use of stronger acid systems and lower temperatures has confirmed the oxygen protonation of amides and their thio analogs (139,140). Acetamide, $N,N$-dimethylacetamide, formamide, and $N,N$-dimethylformamide were examined in fluorosulfonic acid at temperatures ranging from $+25$ to $-92°$. For protonated acetamide, at $-80°$, three resonances appear at $\delta$ 10.72, 8.30, and 2.67 and are assigned to the proton on oxygen, the protons on nitrogen, and the methyl protons. Integration of the areas of these peaks yields a ratio of 1.07:2.10:3.00. The other protonated amides give similar results, consistent only with O protonation. The proton on oxygen is consistently at about 10. It is of interest to compare this value with the value for the proton on oxygen for protonated carbonyls and protonated esters. For protonated carbonyls, the proton on oxygen appears at $\delta$ 14–16, and for protonated esters, where there is considerable delocalization of the positive charge toward the alcoholic oxygen, the proton on oxygen appears at about 12. The appearance of the proton on oxygen in protonated amides at an even more shielded position of about $\delta$ 10 further strengthens the argument that there is considerable contribution from resonance form **16** where the positive charge resides on nitrogen.

The spectra of protonated thioacetamide and thioacetanilide in fluorosulfonic acid at low temperatures have been observed (140). The proton on sulfur appears at $\delta$ 5.72. This is slightly shielded when compared to the proton on sulfur resonance for protonated thiols and sulfides (57) and, like the protonated amides, indicates delocalization as in **22** toward the nitrogen atom (10). Protonation of ureas is discussed further in a later section.

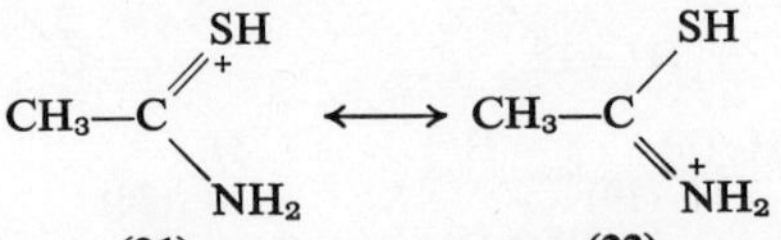

$$CH_3-C\overset{\overset{+}{S}H}{\underset{NH_2}{\diagdown}} \longleftrightarrow CH_3-C\overset{SH}{\underset{\overset{+}{N}H_2}{\diagdown}}$$

(21)                (22)

The one reported exception to O protonation of amides (141) is 2,2-dimethyl-6-oxoquinuclidene (**23**). In this example structural features enhance the basicity of the nitrogen atom compared with that of a normal amide and prevent resonance stabilization of the O-protonated form.

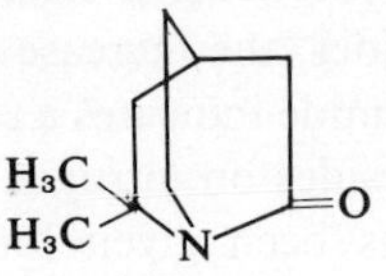

(23)

It has also been found possible to obtain N-protonated amides by reaction of oxocarbonium ions (acylium ions) with amines, in sulfur dioxide solution at low temperature (eq. 39).

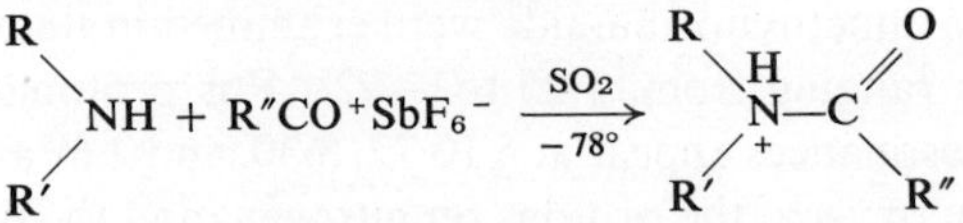

$$\underset{R'}{\overset{R}{\diagdown}}NH + R''CO^+SbF_6^- \xrightarrow[-78°]{SO_2} \underset{R'}{\overset{R}{\diagdown}}\overset{H}{\underset{+}{N}}-C\overset{O}{\underset{R''}{\diagup}} \qquad (39)$$

At higher temperatures rearrangement to the O-protonated amide occurs, this rearrangement being irreversible. The result shows that whereas the O-protonated amide is thermodynamically the most stable ion, the energy barrier for the interconversion from N to O protonation is high enough for direct observation of the former at low temperatures (142).

## X. PROTONATED CARBOXYLIC ACIDS, ESTERS, AND ANHYDRIDES

In sulfuric acid, aliphatic carboxylic acids behave as simple bases being half-protonated in approximately 70% aqueous acid. Uv, ir, Raman, and nmr spectroscopy (32,33,53,143) show that protonation on the carbonyl group occurs. Formic acid is decomposed in sulfuric acid. However, it can be observed in its protonated form, and is stable as such, in fluorosulfonic acid–antimony pentafluoride solution (144) and in hydrogen fluoride–boron trifluoride solution (145). The nmr spectrum of formic acid, at low temperatures, in these acid systems shows the presence of two conformers, **24** (R = H) predominating over **25** (R = H) by a factor of about 2:1. A third possible conformer, **26** (R = H), is not observed.

Evidence for these structures derives from the fact that the coupling between the methine and hydroxyl protons is larger (15 Hz) for a *trans*

(24)  (25)  (26)

relationship than for a *cis* relationship (3.5 Hz). Thus in isomer **24** (R = H) the methine proton is a doublet of doublets (15 and 3.5 Hz), while in isomer **25** (R = H) the methine proton is a triplet (3.5 Hz). The hydroxyl protons occur at lower field than in protonated alcohols and ethers but are more shielded than those in protonated ketones and aldehydes (see Table XIX*a*).

In protonated acetic acid under the same conditions (106) the isomer **25** (R = $CH_3$) is present to the extent of only about 5%, and in all other carboxylic acids studied only isomer **24** is found. The configurational equilibrium energy barriers in protonated formic and acetic acid have been determined (116) and are given in Table XIX*b*. Some disagreement exists in the literature as to the predominant isomer present in protonated pivalic acid. In $HF-BF_3$ solution at $-75°$, only a single OH absorption has been reported (146). This was attributed to the large bulk of the tertiary butyl group causing the hydroxyl protons to adopt an all-*trans* configuration (**26**) (R = $C(CH_3)_3$). Since no other carboxylic acid has been found to exist in this conformation and since in $FSO_3H-SbF_5$ solution, at $-60°$, two OH absorptions are found (144), there is no reason to believe that protonated pivalic acid exists in other than conformer **24** (R = $C(CH_3)_3$) and that the result obtained in $HF-BF_3$ is a result of the rotation about the C=$OH^+$ bonds not being "frozen out" on the nmr time scale. This explanation has been given to account for the observation of a single OH resonance in protonated benzoic acid (53,147) in all strong acid systems in which it has been studied. Delocalization of charge into the phenyl ring would be expected to diminish the C=OH bond order with a consequent decrease in the barrier to rotation. A similar effect is found in some protonated unsaturated carboxylic acids in which charge delocalization can also occur.

On the basis of the coupling constants in protonated formic acid (144,145) and also the $J^{13}{}_{COH}$ coupling constants in protonated formic,

### TABLE XIX[a]
### Pmr Chemical Shifts, ppm[a] of Protonated Acids in 1:1M $FSO_3H$–$SbF_5$

| | Protonated acid, $J$, Hz[b] | | | |
|---|---|---|---|---|
| | OH | $CH_3$ | $CH_2$ | CH |
| $HCO_2H$ (58) | 13.40, $d$ (3.5) | | | 9.70, $t$ (3.5) |
| (57) | {13.57, $d$ (3.5) | | | 9.30, $d$ of $d$ |
| | 13.02, $d$ (15.0) | | | (15.0 and 3.5) |
| $CH_3CO_2H$ | {13.03 | 3.18 | | |
| | 12.33 | | | |
| $CH_3CH_2CO_2H$ | {12.73 | 1.95, $t$ (7.0) | 3.67, $q$ (7.0) | |
| | 12.43 | | | |
| $CH_3(CH_2)_2CO_2H$ | {12.75 | 1.30, $t$ (7.0) | 2.15, $sx$ (7.0) | |
| | 12.42 | | 3.32, $t$ (7.0) | |
| $CH_3(CH_2)_3CO_2H$ | {12.98 | 1.5–2.7, $m$ | 3.58, $t$ (6.5) | |
| | 12.65 | | | |
| $(CH_3)_2CHCO_2H$ | {12.67 | 1.75, $d$ (7.0) | | 3.52, $sp$ (7.0) |
| | 12.40 | | | |
| $(CH_3)_2CHCH_2CO_2H$ | {13.13 | 1.67, $d$ (6.0) | 3.45, $d$ (6.0) | 2.83, $m$ |
| | 12.75 | | | |
| $(CH_3)_3CCO_2H$ | {12.60 | 2.11 | | |
| | 12.42 | | | |

[a] Referred to external TMS.
[b] Multiplicity: $d$, doublet; $t$, triplet; $q$, quartet; $m$, multiplet; $p$, pentuplet; $sx$, sextet; $sp$, septet.

## TABLE XIX*b*
### Activation Parameters of Configurational Equilibration (116)

| Protonated acid [a] | $E_a$, kcal/mole | Log $A$, sec$^{-1}$ | Solvent |
|---|---|---|---|
| Acetic acid | 13.5 ± 0.6 | 14.4 ± 0.7 | HF–BF$_3$[b] |
| Acetic acid | 12.6 ± 0.6 | 13.2 ± 0.6 | HFSO$_3$–SbF$_5$–SO$_2$[c] |
| Acetic acid | 11.2 ± 0.2 | 12.0 ± 0.2 | HF–SbF$_5$[d] |
| Formic acid | 15.3 ± 0.9 | 15.7 ± 0.9 | HF–SbF$_5$[d] |

[a] 1$M$ solution.
[b] BF$_3$ pressure 1 atm at $-80°$.
[c] Molar ratio 7:8:9, respectively.
[d] Excess SbF$_5$ 0.5 mole/liter.

acetic, and propionic acids (147), the "inner" proton (H$_A$ in structure **24**) has been assigned to the highest field resonance. This assignment is in agreement with the model for the anisotropy of the carbonyl group recently proposed (148) on the assumption that similar effects are valid for the protonated group.

No measurable four-bond coupling has been detected between the alkyl group protons and the hydroxyl protons in saturated carboxylic acids although four-bond couplings have been observed in two of the protonated unsaturated carboxylic acids that have been studied in FSO$_3$H–SbF$_5$–SO$_2$ solution (149). The highest field OH resonance (H$_A$ in **27**) is coupled by 1.5 Hz to H$_B$ in acrylic and crotonic acid. The magnitude of this four-bond

(27)

coupling reflects the favorable planar $W$-coupling path between these two protons (115) and is a further example of the fact that coupling through protonated carbonyl oxygen is similar in magnitude and spatial requirements to $sp^2$ carbon atoms. A similar effect is found in protonated thioformic acid in which, in 4:1$M$ FSO$_3$H–SbF$_5$, three isomers, **28–30** (R = H), have been detected (150). In isomer **28**, a 3.5-Hz coupling between H$_A$ and H$_B$ is found. This coupling is not present in either of the other two isomers.

The predominant isomer in protonated thioformic acid is **29** (60%), **28** and **30** being present to the extent of 30 and 10%, respectively. In protonated thioacetic acid the isomer ratio is 3:1:1 for **29**, **28**, and **30** (R = H), and in protonated thiopropionic acid a 7:1:2 isomer ratio is observed. The

(28)   (29)   (30)

decrease in the amount of isomer **28** observed with increasing the size of the R group is consistent with the observations in the protonated carboxylic acid series, although in the latter the decrease is much more marked. A plausible explanation for this difference is that, in the isomer, interaction between the lone pairs on the two heteroatoms is a controlling factor and that sulfur–oxygen interaction is less than that between oxygen–oxygen. The actual ratios observed reflect the steric requirements of the group R, the protons on oxygen and sulfur, and the lone pairs of these atoms, and it is evident from the only small free-energy difference between the isomers that the difference between these steric requirements is small. Variations in the solvent (e.g., $FSO_3H$–$SbF_5$ and HF–$BF_3$) has little effect on the relative isomer abundances indicating that solvation of the ions is not influencing the stability of one or more of these isomers.

The protonation of esters in both sulfuric (151) acid and mixed acid systems (53,152) follows the same pattern as the carboxylic acids. Protonation is on the carbonyl oxygen, and in $FSO_3H$–$SbF_5$–$SO_2$ solution at low temperature different rotational isomers can be identified in some cases (152). Thus methyl formate, in $FSO_3H$–$SbF_5$ solution, has an nmr spectrum showing the presence of two isomers. These isomers can be assigned (45) to **31** and **32**, **31** being the predominant isomer by a ratio of about 9:1. In protonated ethyl formate the *syn–syn* isomer, **33**, has also been detected (152).

(31)   (32)   (33)
90%    10%

In sulfuric acid the protonation is complicated by cleavage of the esters at ambient temperatures (153,154), and in the mixed acid systems cleavage (152) can be observed by raising the temperature from the low temperatures at which protonation is usually observed.

The cleavage of esters under highly acidic conditions occurs via unimolecular cleavage of the protonated ester, involving either alkyl–oxygen or acyl–oxygen scission (154). Evidence for this has come from labeling experiments, from kinetic studies, and more recently by nmr spectroscopy.

In sulfuric acid the methyl and ethyl esters of both benzoic and acetic acid hydrolyze to the corresponding acids via acyl oxygen fission in the rate-determining step (the $A_{Ac}1$ mechanism). For the isopropyl and $t$-butyl esters alkyl–oxygen fission occurs (the $A_{Al}1$ mechanism) (154,155).

$$R'CO\overset{+}{O}HR \longrightarrow R'\overset{+}{C}O + ROH \quad A_{Ac}1 \quad (40)$$
$$\phantom{R'CO\overset{+}{O}HR \longrightarrow} \underset{H_2O}{\longrightarrow} R'CO_2H$$

$$R'CO\overset{+}{O}HR \longrightarrow R'CO_2H + R^+ \quad A_{Al}1 \quad (41)$$
$$\phantom{R'CO\overset{+}{O}HR \longrightarrow R'CO_2H + } \underset{H_2O}{\longrightarrow} ROH$$

At higher acidities (e.g., in oleum or in other mixed acid systems) it is possible to cleave carboxylic acids in a manner analogous to the rate-determining step of the $A_{Ac}1$ mechanism, the resulting species being an oxocarbonium ion (acylium ion) and protonated water (eq. 42).

$$RCOOH_2^+ \longrightarrow R\overset{+}{C}O + H_3O^+ \quad (42)$$

The latter reaction has been studied both under equilibrium conditions (in oleum) (156) and under kinetic conditions [in $FSO_3H$–$SbF_5$ (144) and in $HF$–$BF_3$ (157)]. Thus acetic acid is half-ionized to the methyl oxocarbonium ion in 15% oleum at 35° as determined by observation of the changes in the position of the methyl resonance with increasing acidity (156). In both $HF$–$BF_3$ (157) and $FSO_3H$–$SbF_5$ (144) protonated acetic acid alone can be observed at low temperature and then, by raising the temperature to between $-30$ and $0°$, the kinetics of the cleavage reaction can be followed by nmr, the methyl signals of the protonated acid and the methyl oxocarbonium ion being separated by about 1 ppm. In $FSO_3H$–$SbF_5$ the protonated acid is completely converted to the oxocarbonium ion while in the $HF$–$BF_3$ acid system the reaction goes to between 40 and 70% completion, indicating the greater acidity of the former acid system over the latter and suggesting that the $HF$–$BF_3$ system is of comparable acidity to 15% oleum. While the rate constants for cleavage of acetic acid to the methyl oxocarbonium ion in $FSO_3H$–$SbF_5$ and in $HF$–$BF_3$ agree well with each other for a series of carboxylic acids, the rate constant in oleum has been estimated to be faster by a factor of $10^4$–$10^5$. It has been suggested that this difference is due to a bimolecular mechanism (eq. 43) operating in the case of the oleum system (157).

A significant steric factor has been found to operate in the cleavage of carboxylic acids in sulfuric acid–oleum systems. Thus 2,4,6-trimethylbenzoic acid is completely ionized to the corresponding oxocarbonium ion in 100% sulfuric acid, while it is necessary to use oleum to ionize benzoic

$$\left[ \begin{array}{c} R-C \begin{array}{c} O-H--O \\ \diagdown \\ O------OH \end{array} \begin{array}{c} O \\ \diagup \\ S \\ \diagdown \\ OH \end{array} \\ H \end{array} \right]^+ \longrightarrow RCO^+ + HSO_4^- + H_3O^+ \quad (43)$$

acid (158). It has been shown that it is the *ortho* methyl groups which cause the added stability of the oxocarbonium ion with respect to hydrolysis back to the carboxylic acid.

As has been described earlier, carboxylic acid esters show a more complex behavior than the acids, under strong acid conditions, due to the two possible modes of cleavage. In $FSO_3H$–$SbF_5$ and in $HF$–$BF_3$ solution it is possible to distinguish between the two modes of cleavage by nmr (152,157). Thus at $+30°$ in this acid system, methyl acetate cleaves via acyl–oxygen fission, and the products, the methyl oxocarbonium ion and protonated methanol, are stable under these conditions and can be identified by nmr (eq. 44).

$$H_3C-C\overset{OH}{\underset{OCH_3}{(+}} \longrightarrow CH_3CO^+ + CH_3OH_2^+ \qquad (44)$$

*sec*-Butyl acetate under the same conditions but at $-30°$ cleaves via alkyl–oxygen fission, and the nmr spectrum shows the presence of protonated acetic acid and the *t*-butyl cation, the latter from rearrangement of the first formed methyl ethyl carbonium ion (eq. 45). Since the protonated acid goes to the oxocarbonium ion only very slowly under these conditions, its observation is proof of alkyl–oxygen cleavage.

$$H_3C-C\overset{OH}{\underset{\underset{CH_3}{OCH_2-C_2H_5}}{(+}}$$

$$H_3C-C\overset{OH}{\underset{OH}{(+}} + [CH_3\,CH_2\,\overset{+}{C}H\,CH_3] \qquad (45)$$

$$\downarrow$$

$$(CH_3)_3C^+$$

Esters of tertiary alcohols were found to cleave so rapidly that only the protonated acid and tertiary carbonium ions could be observed on dissolution of the ester in $FSO_3H$–$SbF_5$ at $-80°$.

Ethyl acetate provided an intermediate case. In the same acid system alkyl–oxygen cleavage was observed (eq. 47); however, in weaker acid media such as $4:1M$ $FSO_3H$–$SbF_5$ acyl–oxygen cleavage was observed (152) (eq. 46). The reason for this mechanism dependence on acidity is not fully understood; however, it is probably significant that in the lower acidity conditions the cleavage does not go to completion (159).

Methyl formate is stable at room temperature in $FSO_3H$–$SbF_5$ solution. At higher temperatures in $FSO_3H$–$SbF_5$ protonated methanol and carbon monoxide are formed, protonated carbon monoxide $HCO^+$ being a pos-

$$H_3C-C\overset{+}{\underset{OC_2H_5}{\diagup OH}} \longrightarrow CH_3CO^+ + C_2H_5OH_2^+ \qquad (46)$$

$$H_3C-C\overset{+}{\underset{OC_2H_5}{\diagup OH}} \longrightarrow H_3C-C\overset{+}{\underset{OH}{\diagup OH}} + [CH_3CH_2^+] \qquad (47)$$

$$\downarrow$$
$$(CH_3)_3C^+$$

sible intermediate in this decomposition. In 100% $H_2SO_4$, methyl sulfate and formyl sulfate are the formed products (151). Secondary and tertiary esters of formic acid cleave in $FSO_3H$–$SbF_5$ solution to give protonated formic acid and the carbonium ion product of the corresponding protonated alcohol. Protonated formic acid itself decomposes at elevated temperatures in superacids to give carbon monoxide and the hydronium ion.

Cleavage of protonated S-alkyl thio esters has been studied in $FSO_3H$–$SbF_5$–$SO_2$ solution (150). Primary and secondary S-alkyl thioacetates were found to cleave via acyl–sulfur fission (eq. 49) while alkyl–sulfur fission was found for tertiary S-alkyl thioacetates (eq. 78).

$$R-C\overset{+}{\underset{SR'}{\diagup OH}} \begin{cases} \longrightarrow R-C\overset{+}{\underset{SH}{\diagup OH}} + R'^+ \qquad (48) \\ \qquad (R' = C(CH_3)) \\ \\ \longrightarrow R-C\equiv O^+ + R'OH_2^+ \qquad (49) \\ (R' = CH_3 \text{ and } (CH_3)_2CH) \end{cases}$$

In the case of O-alkyl thio esters, cleavage could only be observed in the case of esters of tertiary alcohols (160), the products in this case being the protonated thio acid and the corresponding carbonium ion (eq. 50).

$$R-C\overset{+}{\underset{SH}{\diagup OC(CH_3)^+}} \longrightarrow R-C\overset{+}{\underset{SH}{\diagup OH}} + C(CH_3)_3^+ \qquad (50)$$

Esters of primary and secondary alcohols proved to be stable to high temperature, reflecting the stability of oxocarbonium ions as compared to their sulfur analogs.

Anhydrides of carboxylic acids behave as esters of tertiary alcohols in $FSO_3H$–$SbF_5$ solution in that cleavage is immediate on dissolution at $-80°$ leading to one molecule of the protonated acid and one of the oxocarbonium ion in equal proportions (17). The same behavior is found in 100% sulfuric acid and in oleum solution, while in slightly aqueous acid, the oxocarbonium ion reacts with water present leading to two molecules of the protonated acid being observed (161).

$$R-\underset{\underset{O}{\|}}{C}-O-\underset{\underset{O}{\|}}{C}-R \longrightarrow RCO^+ + R-C\overset{OH}{\underset{OH}{\diagdown}}{}^+$$

$$\downarrow H_2O$$

$$RCO_2H_2^+$$

Under certain conditions protonated anhydrides can be observed without cleavage (142). Reaction of oxocarbonium ions salts with carboxylic acids in sulfur dioxide solution leads to formation of a monoprotonated anhydride (34). The nmr spectrum indicates that rapid intermolecular exchange of the proton between the two carbonyl groups occurs, this exchange being favored by strong hydrogen bonding (34). Both symmetrical and unsymmetrical protonated anhydrides have been generated in this manner.

$$\begin{array}{c} R \\ \diagdown \\ C=\overset{+}{O} \\ O \diagup \quad \diagdown H \\ \diagdown C=O \\ \diagup \\ R' \end{array}$$

(34)

Ortho esters of carboxylic acid cleave at $-80°$ in $FSO_3H$–$SbF_5$ solution to dialkoxy carbonium ions and protonated alcohols (41).

$$H_3C-\underset{\underset{OCH_3}{|}}{\overset{\overset{OCH_3}{|}}{C}}-OCH_3 \longrightarrow H_3C-\underset{\underset{OCH_3}{|}}{\overset{\overset{H}{\underset{|}{O}}CH_3}{\overset{+}{C}}}-OCH_3$$

$$\downarrow$$

$$CH_3-C\overset{OCH_3}{\underset{OCH_3}{\diagdown}}{}^+ + CH_3OH_2^+ \qquad (51)$$

Protonated carboxylic acids and esters have been isolated as stable complexes, either as their hexachloroantimonates or hexafluoroantimonates (162). No detailed study of the properties of these solid complexes has, however, been reported, although the reaction of these salts with diazomethane to give dimethoxycarbonium ions has been described (eq. 52).

$$R-C\overset{OH}{\underset{OH}{\diagdown}}{}^+ \;\; SbCl_6^- \xrightarrow{CH_2N_2} R-C\overset{OCH_3}{\underset{OCH_3}{\diagdown}}{}^+ + 2N_2 \qquad (52)$$

## XI. PROTONATED DICARBOXYLIC ACIDS AND ANHYDRIDES

The protonation of saturated dicarboxylic acids in sulfuric acid has been investigated by cryoscopy (163). Succinic acid and higher members of the series were found to be only partially ionized as diacid bases. Malonic acid was only monoprotonated under these conditions while oxalic acid underwent decomposition. Adipic acid has been investigated in oleum by nmr (164), and a downfield shift of the methylene protons in 17% oleum was found which was tentatively attributed to diprotonation. In $FSO_3H-SbF_5$, oxalic acid and higher members of the series were found to be diprotonated (17), the nmr spectra showing that the protonated acids has a structure analogous to the monocarboxylic acids.

$$\begin{array}{c} H-O \\ \diagdown \\ {}^+C-(CH_2)_n-C^+ \\ \diagup \quad\quad\quad\quad \diagdown \\ O \quad\quad\quad\quad\quad O \\ | \quad\quad\quad\quad\quad | \\ H \quad\quad\quad\quad\quad H \end{array}$$
(35)

On raising the temperatures of solutions of protonated dicarboxylic acids, cleavage, again analogous to the monocarboxylic acids, was observed (eq. 53).

$$H_2\overset{+}{O}_2C-(CH_2)_n-CO_2\overset{+}{H}_2 \longrightarrow H_2O_2C-(CH_2)_n-\overset{+}{CO}$$
$$\downarrow$$
$$O\overset{+}{C}-(CH_2)_n-\overset{+}{C}O$$
(53)

Malonic acid did not cleave under these conditions while succinic acid cleaved to the extent of ca. 50% showing that an equilibrium is reached between the diprotonated acid and the monooxocarbonium ion. The first ionization of glutaric acid went to completion under these conditions; however, the second ionization to form a dioxocarbonium ion went only the extent of ca. 50%. Other dicarboxylic acids were converted completely to the dioxocarbonium ions (17).

1,2-Dihydroxycyclobutenedione ("squaric acid") was found to be diprotonated (17). The nmr evidence was taken to indicate that there was no significant contribution of the cyclobutenium dication (a potential Hückeloid aromatic structure) to the stability of the molecule (36).

$$\begin{array}{c} HO \quad\quad\quad OH \\ \diagdown \quad\quad\quad \diagup \\ \boxed{2+} \\ \diagup \quad\quad\quad \diagdown \\ HO \quad\quad\quad OH \end{array}$$
(36)

The cyclic anhydrides derived from dicarboxylic acids have been shown, by cryoscopic measurements, to undergo incomplete protonation in sulfuric acid (51). Thus succinic anhydride (165) gives an *i* factor of 1.5, and an *i* factor of 1.8 for 1,8-naphthalic anhydride (166) has been obtained. In $FSO_3H$–$SbF_5$ solution, at $-80°$, cleavage of cyclic anhydrides has been found to be immediate, giving the monooxocarbonium ions (17).

$$\underset{O}{H_2C}\underset{O}{-}\underset{O}{CH_2} \longrightarrow O=\overset{+}{C}-CH_2-CH_2-C\overset{OH}{\underset{OH}{+}} \qquad (54)$$

The ion derived from succinic anhydride gave a temperature-dependent nmr spectrum which was interpreted as being a result of an intramolecular rearrangement through a cyclic intermediate corresponding to the diprotonated anhydride (17) (eq. 55). In contrast to this result, succinimide has

$$H_2\overset{+}{O}_2{-}C(CH_2)_2{-}CO^+ \rightleftharpoons \begin{array}{c} H_2C{-}CH_2 \\ | \quad\quad | \\ C\phantom{-}O\phantom{-}C \\ HO+ \quad\quad +OH \end{array} \rightleftharpoons O={\overset{+}{C}}{-}(CH_2)_2CO_2H_2{}^+ \qquad (55)$$

been found to exist in $FSO_3H$–$SbF_5$ solution as the diprotonated cyclic structure (167), the replacement of the bridging ether oxygen by an NH allowing the molecule to accommodate the dipositive charge better in the cyclic structure (37). Other imides show similar behavior (167).

$$\begin{array}{c} H_2C{-}CH_2 \\ | \quad\quad | \\ C\phantom{-}N\phantom{-}C \\ HO+ \quad H \quad +OH \end{array}$$
(37)

2,2-Dimethyl-β-propiolactone in HF–$BF_3$ at $-80°$ has been reported to give the oxocarbonium ion (eq. 56) (168).

$$\begin{array}{c} H_3C \\ \phantom{H_3}C{-}C{-}H \\ H_3C \phantom{CC} | \phantom{C} H \\ \phantom{H_3C}C{-}O \\ \phantom{H_3CC}O \end{array} \xrightarrow{HF{-}BF_3} H_3C-\underset{CH_2-\overset{+}{O}H_2}{\overset{CH_3}{\underset{|}{\overset{|}{C}}}}-\overset{+}{C}=O \qquad (56)$$

## XII. PROTONATED CARBONIC ACID AND DERIVATIVES

Dialkyl carbonates have been studied in $FSO_3H$–$SbF_5$ solution and have been shown to be protonated on the carbonyl group giving the dialkoxy-hydroxy carbonium ion (38) (4).

$$\begin{array}{c} \text{H} \\ | \\ R-O-\overset{+}{C}-O \\ | \\ O-R \end{array}$$

(38)

Di-*t*-butyl carbonate cleaves immediately at $-80°$, with alkyl–oxygen fission, giving the *t*-butyl cation and protonated carbonic acid. The structure of the latter has been established from the $^{13}$C nmr spectrum of the central carbon atom which shows a 3.5-Hz quartet being coupled to three equivalent hydroxyl protons.

$$(CH_3)_3C-O\diagdown \atop (CH_3)_3C-O\diagup C^+\!\!=\!\!OH \longrightarrow {HO\diagdown \atop HO\diagup} C^+\!\!=\!\!OH + 2(CH_3)_3C^+ \quad (57)$$

Diisopropyl and diethyl carbonate cleave at higher temperature, also via alkyl–oxygen cleavage, with initial formation of protonated alkyl hydrogen carbonates. The latter can also be formed by protonation of their sodium salts.

$$\begin{array}{c}RO\diagdown \\ RO\diagup\end{array}C^+\!\!=\!\!OH \longrightarrow \begin{array}{c}HO\diagdown \\ RO\diagup\end{array}C^+\!\!=\!\!OH \longrightarrow \begin{array}{c}HO\diagdown \\ HO\diagup\end{array}C^+\!\!=\!\!OH \quad (58)$$
$$+\,R^+ \qquad\qquad +\,R^+$$

Dimethyl and diethyl carbonates have also been observed by nmr in $H_2SO_4$ solution, in which they are also protonated (169), although under these conditions fast exchange precludes observation of the hydroxyl proton.

Protonated carbonic acid can also be obtained by dissolving inorganic carbonates and hydrogen carbonates in $FSO_3H$–$SbF_5$ at $-80°$. It is stable in solution to about $0°$, at which temperature it decomposes to the hydronium ion and carbon dioxide (48) (eq. 59).

$$K_2CO_3 \xrightarrow{H^+} H_3CO_3^+$$
$$H_3CO_3^+ \longrightarrow H_3O^+ + CO_2 \quad (59)$$

*Ortho* esters of carbonic acid cleave at $-80°$ in $FSO_3H$–$SbF_5$ solution giving trialkoxy carbonium ions and protonated alcohols (41), analogous to the cleavage of ortho esters of carboxylic acids under the same conditions (eq. 60).

The trimethoxycarbonium ion has been prepared by isolating protonated

$$\underset{\underset{CH_3O}{\overset{CH_3O}{\diagdown}}}{\overset{\overset{H}{\underset{|}{\underset{+}{C}}}}{\diagup}}\!\!\!\underset{OCH_3}{\overset{OCH_3}{\diagup}} \longrightarrow CH_3OH_2{}^+ + CH_3O-\overset{\overset{OCH_3}{\diagup}}{\underset{OCH_3}{\diagdown}}C\!\!\overset{+}{{}} \qquad (60)$$

dimethyl carbonate as a stable complex (with $SbCl_6{}^-$) and treating it with diazomethane (170).

The thio analogs of protonated carbonic acid have been prepared in $FSO_3H$–$SbF_5$–$SO_2$ solution, and their method of preparation and the chemical shifts of the OH and SH protons are given in Table XX (150). The increasing deshielding of both the OH and SH protons as the number of thiol groups in the ion is increased is consistent with the lesser ability of sulfur as compared to oxygen to delocalize the positive charge on the central carbon atom. Preparation of the ions containing both OH and SH groups also lead to formation of both protonated carbonic acid and protonated trithiocarbonic acid and it was proposed that protonated carbon dioxide, carbonyl sulfide, and/or carbon disulfide were intermediates in this interconversion.

**TABLE XX**

| Equation | | δ OH | δ SH |
|---|---|---|---|
| $BaCO_3 \xrightarrow[SO_2,\ -60°]{1:1\ FSO_3H-SbF_5}$ | (protonated carbonic acid structure) | 11.55 | — |
| $O=C\overset{SK}{\underset{OC(CH_3)_3}{\diagdown}}\longrightarrow$ | (protonated monothiocarbonic acid structure) | 11.99 | 6.73 |
| $S=C\overset{SK}{\underset{OC(CH_3)_3}{\diagdown}}\longrightarrow$ | (protonated dithiocarbonic acid structure) | 12.56 | 7.19 |
| $BaCS_3 \longrightarrow$ | (protonated trithiocarbonic acid structure) | — | 7.66 |

Protonated carbonic acid was shown to behave as an electrophile and with amines under suitable conditions gave carbamic acids, which were observed, protonated, in $FSO_3H$–$SbF_5$ solution (48).

$$C(OH)_3^+ + R_2NH \longrightarrow R_2N-\overset{\displaystyle OH}{\underset{\displaystyle OH}{C^+}} + H_3O^+ \qquad (61)$$

## XIII. PROTONATED CARBAMIC ACID AND DERIVATIVES

In oleum solution, alkyl carbamates cleave via alkyl–oxygen fission giving protonated carbamic acids. The latter are unstable under these conditions and lose carbon dioxide giving the corresponding amine (171).

$$R-NH-\overset{\displaystyle O}{\overset{\|}{C}}-OR' \longrightarrow R \cdot NH \cdot CO_2H_2^+ + R'^+$$
$$\longrightarrow RNH_3^+ HSO_4^- + CO_2 \qquad (62)$$

In $FSO_3H$–$SbF_5$ solution, at $-60°$, alkyl carbamates can be observed in their protonated form by nmr, protonation occurring on the carbonyl group (**40**) (172).

(**39**)  (**40**)  (**41**)

Hindered rotation about the C=N bond results in the observation of *cis* and *trans* isomers in ethyl *N*-methylcarbamate (**40, 41**). In addition, the hydroxyl proton in one of the isomers is coupled to the NH proton ($J = 2.8$ Hz) suggesting that in this isomer these protons bear a *W* relation to each other as in protonated thiol acids (150). The isomer in which this coupling is observed would thus correspond to **41**. It is unlikely, but not proven, that in this isomer rotation about the C–OH bond is "frozen out" on the nmr time scale. If "free rotation" occurs, the observed coupling of 2.8 Hz will be the weighted average of the coupling constants in the various rotamers. The second isomer, which shows no detectable four-bond coupling, will thus have structure **39**.

Carbamic acid esters of both primary and secondary alcohols cleave via alkyl–oxygen scission in $FSO_3H$–$SbF_5$ solution, protonated carbamic acids being stable under these conditions. The resultant spectra show the latter to be carboxyl protonated (eq. 63).

$$\underset{R'}{\overset{R}{>}}N-\underset{OR''}{\overset{OH}{C{\lessgtr}+}} \longrightarrow \underset{R'}{\overset{R}{>}}N-\underset{OH}{\overset{OH}{C{\lessgtr}+}} + R''^{+} \quad (63)$$

Mono-$N$-substituted carbamic acids, like the esters, show hindered rotation about the C=N bond and coupling of one of the hydroxyl protons to the NH proton. This coupling is not observed in protonated carbamic acid itself.

In contrast to the usual observation of O protonation of carbamate esters, $N,N$-diisopropylcarbamate esters (methyl and ethyl) have been shown to be N-protonated in $FSO_3H$ and 98% $H_2SO_4$ solution (173). It has, in this case, been found possible to observe (174) rearrangement from O-protonated $N,N$-diisopropylcarbamate esters to the N-protonated ions. Protonation of the methyl and ethyl esters in fluorosulfonic acid alone, or diluted with either sulfur dioxide of sulfuryl chlorofluoride at $-78°$, gives the O-protonated carbamate ester (42). At $-60°$, rearrangement to the N-protonated ester (43) occurs, the rate being highly dependent on the solvent composition. This rearrangement is irreversible; at equilibrium a mixture of N- and O-protonated ions is observed, the mixture composition being 90% N-protonated and 10% O-protonated. This composition was independent of the solvent used. Carbonyl protonation is thus favored kinetically under these conditions, even though this leads to the thermodynamically least stable ion ($\Delta G° = 1.3$ kcal).

This small free energy difference between the O- and N-protonated carbamates suggests to us that caution should be exercised in extrapolating these results to other solvents. The different acidity function behavior of amides (O protonation) and amines (N protonation) has been interpreted in terms of differing hydration requirements of the two types of cations, solvation by water molecules in aqueous sulfuric acid being more important in the case of O-protonated amides (175). It is thus quite possible that in aqueous acid solvation of the O-protonated $N,N$-diisopropylcarbamate ester reverses the thermodynamic stabilities of the O- and N-protonated

$$\underset{O}{\overset{CH_3O}{>}}C-N\underset{CH(CH_3)_2}{\overset{CH(CH_3)_2}{<}} \xrightarrow[-78°]{FSO_3H} \underset{HO}{\overset{CH_3O}{>}}C=\overset{+}{N}\underset{CH(CH_3)_2}{\overset{CH(CH_3)_2}{<}}$$

$$\downarrow -60° \quad (42)$$

$$\underset{HO}{\overset{CH_3O}{>}}\overset{+}{C}=N\underset{CH(CH_3)_2}{\overset{CH(CH_3)_2}{<}} + \underset{O}{\overset{CH_3O}{>}}C-\overset{+}{\underset{}{N}}\underset{CH(CH_3)_2}{\overset{H\ \ CH(CH_3)_2}{<}}$$

(42)          (43)
10%          90%

(64)

species from the stabilities observed in 98% sulfuric acid and in fluorosulfuric acid. The similar acidity function dependence (173) of ethyl N,N-dimethylcarbamate (which is O-protonated in FSO₃H solution) and ethyl N,N-diisopropylcarbamate may thus be a result of a reversal of the thermodynamic stabilities of the O- and N-protonated forms of the latter ester.

In contrast to monoprotonation of carbamic acid being observed in FSO₃H–SbF₅–SO₂ solution, urea and guanidine under the same conditions are diprotonated (176). Nmr parameters for the diprotonated species (**44** and **45**) are summarized below.

$\delta(OH)$ 12.82
$\delta(NH_2)$ 9.38
$\delta(NH_3^+)$ 9.01

$H_3\overset{+}{N}-C\overset{OH}{\underset{NH_2}{\diagup}}{}^+$

(**44**)

$H_3\overset{+}{N}-C\overset{NH_2}{\underset{NH_2}{\diagup}}{}^+$

(**45**)

$\delta(NH_3^+)$ 8.68
$\delta(NH_2)$ 8.07
       7.85

Cryoscopic measurements (on tetraethylurea) in 100% sulfuric acid also show that diprotonation occurs in this acid system (140). Guanidine, in spite of its first $pK_a$ being higher (13.6) than that of urea, adds only 1.3 protons in sulfuric acid (177). This is apparently due to the high stability of the guanidinium ion due to resonance stabilization of the charge. The position of the first protonation of urea and its derivatives has been the subject of some discussion (127); however, ir, uv, and nmr data on crystalline salts of protonated urea (131,137,178,179) have indicated that the first protonation is on the carbonyl group, as would be expected by analogy with the guanidinium ion, and also with protonated carbonic and carbamic acids.

Both 1,1-dimethylurea and tetramethylurea, when diprotonated in FSO₃H–SbF₅–SO₂ solution, show two nmr signals for the imino methyl groups (176). For this nonequivalence to be observed there must be restricted rotation about the C=N bond. A lower limit for the $\Delta G$ for rotation of 20 kcal was found indicating that the predominant contribution to the structure is from the immonium form (**46**). In diprotonated 1,1-di-

(**46**)

(**47**)

methylguanidine (**47**) this barrier to rotation was found to be 15 kcal. This value demonstrates the greater extent of charge delocalization and hence lower carbon–nitrogen bond order in the diprotonated guanidines as compared to diprotonated ureas. In diprotonated guanidine itself this barrier leads to two NH resonances being observed in the nmr spectra in addition to the peak due to the ammonium protons (176).

An interesting rearrangement has been observed for formylurea which is triprotonated in $FSO_3H$–$SbF_5$–$SO_2$ solution (41). On preparation of the ion at $-78°$ the nmr spectrum shows the conformation to be **48a**. At $-50°$ the ion rearranges (half-life about 1 hr) completely to **48b**. This rearrange-

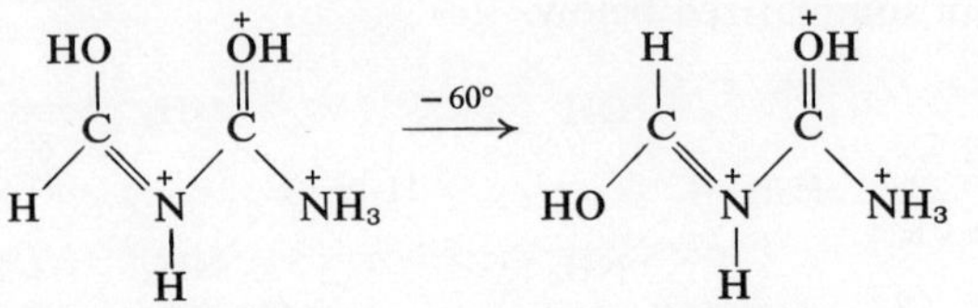

ment is irreversible. Apparently the mechanism of the triprotonation leads the least stable geometrical isomer being formed presumably due to hydrogen bonding in the mono- and diprotonated intermediates.

The first protonation of thiourea is on sulfur (38,137), and there is evidence that it is diprotonated in 100% sulfuric acid and in fluorosulfonic acid (140). In $FSO_3H$–$SbF_5$ both thiourea and selenourea have been shown to be diprotonated.

Biotin, a biologically important compound related to urea, has been shown to be protonated, without further structural changes, in $FSO_3H$–$SbF_3$ (176). It is of interest that the urea base in the molecule in this example is only monoprotonated, although the molecule as a whole is triply protonated. The sites of protonation, all of which can be clearly distinguished by nmr, are the carbonyl group, the sulfide group, and the carboxylic acid function (**49**). The nmr spectrum also shows that protonation of the sulfur occurs *trans* to the valeric acid side chain.

At higher temperatures, the protonated acid group is dehydrated to the oxocarbonium ion, without any other changes in the molecule.

$\delta(OH)$    9.64
$\delta(NH)$    7.04
              7.21
$\delta(SH)$    6.36
$\delta(CO_2H_2)$    12.28
                    12.54

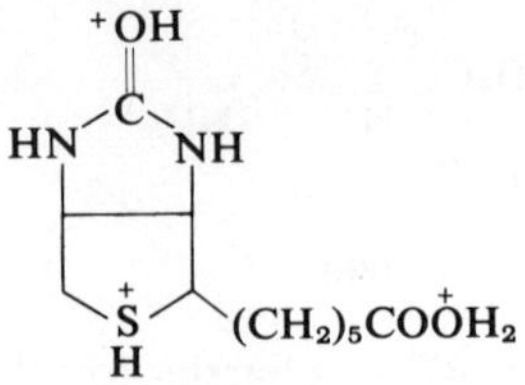

(**49**)

## XIV. PROTONATED IMINES

Imines have been protonated in $FSO_3H$–$SbF_5$, $FSO_3H$, and $D_2SO_4$–$SbF_5$ solution (180). By analogy with ketones and aldehydes, protonated ketimines should show hindered rotation about the C=N bond on the nmr time scale. This has been found to be true, and, for example, in protonated $N$-propylidinemethylimine, up to at least $-20°$ in $FSO_3H$–$SbF_5$ solution, the carbon–methyl groups are found to be nonequivalent (50).

$$\underset{CH_3}{\overset{CH_3}{\diagdown}}C=\overset{+}{N}\underset{H}{\overset{CH_3}{\diagup}}$$

(50)

Although the barriers to rotation in carbonyl compounds and the analogous imines have not been determined, one would predict that because of the greater ability of $NR_2$ over $OR$ to stabilize an adjacent positively charged center (see discussion of protonated ureas and guanidines) that the barrier in imines would be higher for rotation about the C=X bond and that the contribution of the amino carbonium ion resonance structure in protonated imines is minor.

## XV. PROTONATED ENAMINES

The site of protonation in enamines has not been completely resolved (181–185). Enamines are strong bases, their basicity being comparable to amines (182). The protonated molecules have been observed by nmr in dilute hydrochloric acid solution, and it was shown that the site of protonation is apparently acidity dependent (184,185). Thus 1-$N$-morpholine-1-isobutylene in $6N$ HCl solution shows initially two doublets at δ 1.3 and 1.9. The former shows a 2-Hz coupling and is attributed to the N-protonated species (**51a**) and the latter a 7-Hz coupling due to the C-protonated species (**51b**).

(51a)    (51b)

Under these conditions hydrolysis to isobutyraldehyde occurs and results in the appearance of a new doublet in the nmr spectrum on allowing the solution to stand. At acidities less than $6N$ HCl, only the N-protonated form was observed. The $N$-methylanilino enamine of isopropyl methyl ketone in 12N HCl initially shows three species, the N-protonated species **52** and the two geometrical isomers of the C-protonated species **53** and **54**. The former disappears after 5 min at ambient temperatures, while the latter are stable for several hours, hydrolysis to the ketone eventually occurring.

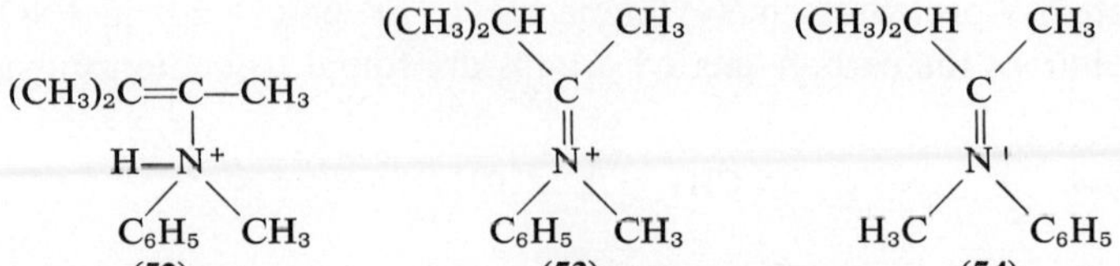

$$(CH_3)_2C=C-CH_3 \quad (52) \qquad (53) \qquad (54)$$

## XVI. PROTONATED KETOXIMES

Nitrogen protonation of acetone and acetophenone oximes has been observed in $FSO_3H$–$SbF_5$–$SO_2$ solution (186). The nmr spectra of protonated acetone oxime (**55**) has two methyl signals showing restricted rotation about the C=N bond as expected. Both are coupled to the NH proton.

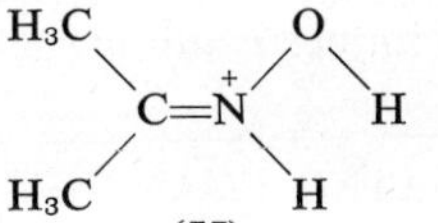

$\delta(CH_3)$ 2.20
           2.26
$\delta(OH)$ 11.1
$\delta(NH)$ 12.0

(55)

On heating a solution of this ion to 100° for 30 min, conversion to *N*-methylacetonitrilium ion (**56**) was observed via a Beckmann rearrangement.

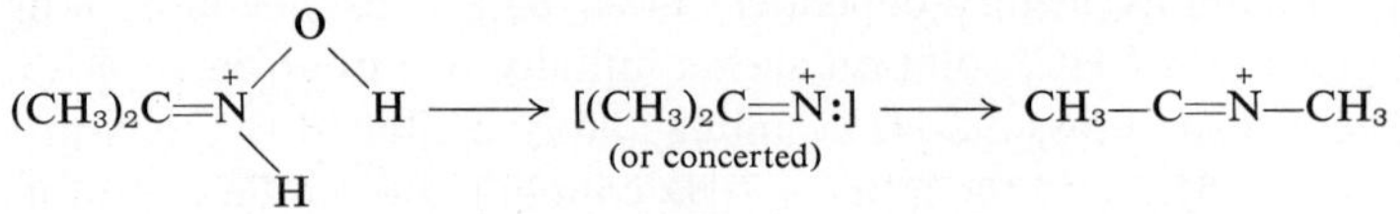

$$(CH_3)_2C=\overset{+}{N}\diagdown_H^{O-H} \longrightarrow [(CH_3)_2C=\overset{+}{N}:] \longrightarrow CH_3-C=\overset{+}{N}-CH_3$$
$$\text{(or concerted)} \qquad\qquad (56)$$

Protonated cyclohexanone oxime (**57**) has been observed on dissolution of nitrosocyclohexane in $FSO_3H$–$SbF_5$–$SO_2$ solution (187).

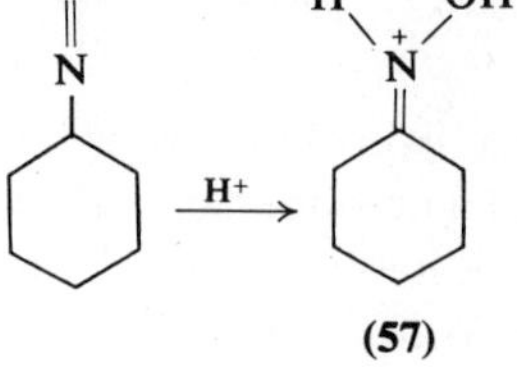

(57)

## XVII. PROTONATED NITRILES

Nmr studies of nitriles in sulfuric acid–oleum solution have shown that they are very weak bases (188). By observing the chemical shift of the alkyl group, half-protonation of acetonitrile was shown to occur in 100% sulfuric acid while 30% oleum was necessary to protonate chloroacetonitrile. In slightly aqueous sulfuric acid and at 35°, the protonated nitriles slowly hydrated with formation of protonated amides.

Hydrogen cyanide, acetonitrile, and other alkyl nitriles have been examined in $FSO_3H$–$SbF_5$ solution at low temperature. Under these conditions the $NH^+$ proton appears as a very broad resonance at ca. 10 ppm. $^{13}C$ chemical shifts and the $J_{^{13}C-H}$ coupling constant in protonated HCN show that the nitrile carbon is still $sp$ hybridized and that the protonated nitrile is therefore a linear species. This is also indicated by the $^{15}N$–H coupling constants in protonated $HC^{15}N$ (186).

Hydrogen cyanide has also been observed in HF solution by nmr. It was found that if 2 moles of HF were added, the resulting ion was protonated difluoromethylamine (189).

$$H\text{—}C\equiv N \xrightarrow{3HF} HF_2C\text{—}\overset{+}{N}H_3HF_2^-$$

## XVIII. PROTONATED NITRO COMPOUNDS

Cryoscopic measurements in sulfuric acid solution have shown that mononitro compounds behave as weak bases (190,191). Nitromethane is 20% ionized and nitrobenzene 40% ionized in 100% sulfuric acid. The low basicity of nitro compounds has been employed in spectroscopic determinations of the acidity of strong acid solutions (see Section II-A). A number of nitro compounds have been protonated in $FSO_3H$–$SbF_5$ in $HF$–$BF_3$ and $HF$–$SbF_5$ solutions (187,192), and under these conditions protonation on oxygen can be clearly demonstrated by nmr, the OH proton appearing at ca. $\delta$ 16. It is of interest that the spectra of protonated nitrobenzenes are temperature dependent (187). Thus at low temperatures, the *ortho* protons in protonated 3,5-dichloro-4-methylnitrobenzene are nonequivalent due to restricted rotation about the carbon–nitrogen bond. At higher temperatures rotation about this bond is fast enough to result in *ortho* protons becoming magnetically equivalent. The barrier to rotation ($\Delta G = 7$ kcal/mole) shows the importance of charge delocalization by the aromatic nucleus in the protonated species. This is also shown by studies of the $^{19}F$ resonance spectra of protonated *m*- and *p*-fluoronitrobenzenes (see p. 1740 and Table XV).

Nitroalkanes have been shown to cleave in $FSO_3H$–$SbF_5$ solution, leading to formation of carbonium (187), nitrosonium, and hydronium ions. Thus 2-fluoro-2-nitropropane gives the fluorodimethyl carbonium ion (eq. 65), and 1-nitro-2-methylpropane gives the *t*-butyl cation (eq. 66).

$$\underset{\underset{CH_3}{|}}{\overset{\overset{CH_3}{|}}{F-C-N}}\overset{O}{\underset{O}{\diagup\!\!\!\!\diagdown}} + \xrightarrow[SO_2,\ -80°]{FSO_3H-SbF_5} F-\underset{\underset{CH_3}{|}}{\overset{\overset{CH_3}{|}}{C^+}} + NO^+ + H_3O^+ \qquad (65)$$

$$\underset{CH_3}{\overset{CH_3}{\diagdown}}CH-CH_2-N\overset{+}{O_2}H \longrightarrow (CH_3)_3C^+ + NO + H_3O \qquad (66)$$

## XIX. PROTONATED AMINO ACIDS

Cryoscopic studies of L-leucine in 100% sulfuric acid indicated partial diprotonation of the base. Subsequently further studies of amino acids in the same media indicated that the extent of diprotonation is increased as the amino group is further removed from the carboxyl group ($v$ 2.2 for L-leucine, 2.7 for D-alanine, and 3.9 for aminocaproic acid) (193).

The first protonation site of amino acids is on the amino group, and the monoprotonated amino acids (in $CF_3COOH$ and $CF_3COOD$) have been studied by proton nmr spectroscopy (at 220 MHz) (194) and by $^{13}C$ spectroscopy (195). In fluorosulfonic acid–antimony pentafluoride solution protonation of both the amino group and the carboxyl function occurs, and in addition other basic sites in the molecule can also be protonated (196). Such polyprotonated amino acids have been studied by nmr spectroscopy. Glycine, e.g., has in $FSO_3H$–$SbF_5$ solution, a spectrum which shows the ion to have structure **58**. Leucine and valine both show similar protonation behavior. In the latter case nonequivalence of the methyl resonances is observed due to asymmetry at the α-carbon (**59**).

$$\overset{+}{H_3N}-CH_2-C\underset{O}{\overset{O}{\diagup\!\!\!\!\diagdown}}+H \qquad \qquad \underset{H_3C}{\overset{}{\diagdown}}\underset{CH_3}{\overset{}{\diagup}} \\ \underset{H}{\phantom{XXXXX}} \qquad \qquad \overset{+}{H_3N}-CH-C\underset{O}{\overset{O}{\diagup\!\!\!\!\diagdown}}+H \\ \phantom{XXXXXXX} \underset{H}{\phantom{XX}}$$

(58)             (59)

Dehydration to give an amino oxocarbonium ion is not observed in the case of the diprotonated α-amino acids. This is expected in view of the stability of diprotonated malonic acid (see Section XI). Increasing the separation of the cationic centers facilitates the dehydration reaction, leading to the formation of γ- and α-aminooxocarbonium ions (eq. 67).

$$H_3N^+CH_2CH_2CH_2CO_2H_2^+ \underset{50\%}{\overset{FSO_3H-SbF_5}{\rightleftarrows}} H_3N^+CH_2CH_2CH_2CO^+ + H_3O^+$$
$$\phantom{H_3N^+CH_2CH_2CH_2CO_2H_2^+}\phantom{xxxxxxxxxxxx}50\%$$

$$H_3N^+(CH_2)_4CO_2H_2^+ \underset{3\%}{\overset{FSO_3H-SbF_5}{\rightleftarrows}} H_3N^+(CH_2)_4CO^+ + H_3O^+ \quad (67)$$
$$\phantom{H_3N^+(CH_2)_4CO_2H_2^+}\phantom{xxxxxxx}97\%$$

Other amino acids studied under these conditions were L-alanine, L-leucine, L-isoleucine, L-cystine, and L-methionine. In the case of L-cystine, no protonation of the disulfide linkage was observed, while in the case of L-methionine an SH proton was observed at δ 6.52.

Diaminocarboxylic and monoaminodicarboxylic acids such as L-lysine, aspartic acid, and glutamic acid are triprotonated, and amino acids with a guanidine group, such as arginine and homoarginine, were tetraprotonated, the guanidine function being itself diprotonated as discussed previously. Asparagine and glutamine are α-amino acids containing amide groups on the side chain, and triprotonation of these was found, the amide group being O-protonated. Investigation of α-amino acids containing phenyl substituents was complicated by the reactivity of the ring; however, under suitable conditions diprotonated phenylalanine, triprotonated L-tyrosine, diprotonated L-3,5-dibromotyrosine, and 3-5-diiodotyrosine could be observed. The heterocyclic amino acids L-proline, L-hydroxyproline, L-histidine, and L-tryptophan were studied. In the latter case protonation at the 3 position of the indole ring was found (60). Protonation generates a second asymmetric center in this molecule, and this gives rise to two doublets being observed for $H_a$.

**(60)**

Some simple peptides were also investigated; glycylglycine, glycylglycylglycine, and glycylglycylglycylglycine were found to be tri-, tetra-, and pentaprotonated, respectively (eq. 68). No cleavage of peptide linkages was observed under these conditions.

$$H_2NCH_2\overset{O}{\overset{\|}{C}}NHCH_2CO_2H \xrightarrow[SO_2, -60°]{FSO_3H-SbF_5} H_3\overset{+}{N}CH_2\overset{\overset{+}{HO}}{\overset{\|}{C}}NHCH_2CO_2\overset{+}{H_2} \quad (68)$$

In this context, anhydrous HF has been found to be a good solvent for proteins, and many can be recovered from solution with essentially full retention of biological activity (for a review, see Ref. 197). This is also true of some enzymes, chlorophyll, vitamin $B_{12}$, and carbohydrates. The solubility of compounds of biological importance in hydrogen fluoride has been reviewed recently (197). It is clear that the solubility of these compounds must be a result of polyprotonation, and this coupled with nmr spectrometers equipped with superconducting magnets may provide a new tool in the elucidation of protein structures.

## REFERENCES

1. (a) N. C. Deno, *Progr. Phys. Org. Chem.*, **2**, 129 (1964); (b) H.-H. Perkampus, *Advan. Phys. Org. Chem.*, **4**, 195 (1966); (c) G. A. Olah and P. v. R. Schleyer, Eds., *Carbonium Ions*, Interscience, New York, Vol. I, 1968; Vol. II, 1970; Vol. III, 1972; Vol. V, in preparation.
2. E. A. Braude and F. C. Nachod, *Determination of Organic Structures by Physical Methods*, Vol. I, Academic Press, New York, 1955; F. C. Nachod and W. D. Phillips, Eds., *ibid.*, Vol. II, 1962.
3. F. A. Long and M. A. Paul, *Chem. Rev.*, **57**, 1 (1957).
4. L. P. Hammett and A. J. Deyrup, *J. Am. Chem. Soc.*, **54**, 2721 (1932).
5. M. J. Jorgensen and D. R. Hartter, *J. Am. Chem. Soc.*, **85**, 878 (1963).
6. N. C. Deno, *Surv. Progr. Chem.*, **2**, 155 (1964).
7. N. C. Deno, P. T. Groves, and G. Saines, *J. Am. Chem. Soc.*, **81**, 5790 (1959).
8. K. Yates, J. B. Stevens, and A. R. Katritzky, *Can. J. Chem.*, **42**, 1957 (1964).
9. E. M. Arnett, *Progr. Phys. Org. Chem.*, **1**, 223 (1963).
10. D. W. Meek, in *The Chemistry of Non-Aqueous Solvents*, Vol. I, J. J. Lagowski, Ed., Academic Press, New York, 1966, p. 6.
11. R. J. Gillespie, in *Friedel-Crafts and Related Reactions*, Vol. I, G. A. Olah, Ed., Interscience, New York, 1963.
12. A. F. Clifford, H. C. Beachall, and W. M. Jack, *J. Inorg. Nucl. Chem.*, **5**, 57 (1957).
13. D. A. McCaulay, W. S. Higley, and A. P. Lien, *J. Am. Chem. Soc.*, **78**, 3009 (1956).
14. R. J. Gillespie, *Acct. Chem. Res.*, **1**, 202 (1968).
15. H. H. Hyman, L. A. Quarterman, M. Kilpatrick, and J. J. Katz, *J. Phys. Chem.*, **65**, 123 (1961).
16. R. J. Gillespie and S. Pez, personal communication; see Reference 14.
17. G. A. Olah and A. M. White, *J. Am. Chem. Soc.*, **89**, 4752 (1967).
18. R. J. Gillespie, in *Studies on Chemical Structure and Reactivity*, J. H. Ridd, Ed., Methuen, London, 1966, p. 173.
19. R. J. Gillespie and E. A. Robinson, in *Carbonium Ions*, Vol. I, G. A. Olah and P. v. R. Schleyer, Eds., Interscience, New York, 1968.
20. S. J. Bass and R. J. Gillespie, *J. Chem. Soc.*, **1960**, 814.
21. R. J. Gillespie, *J. Chem. Soc.*, **1954**, 1851.
22. J. R. Brayford and P. A. H. Wyatt, *J. Chem. Soc.*, **1955**, 3453.
23. J. Leisten, *J. Chem. Soc.*, **1955**, 298.
24. R. J. Gillespie, E. D. Hughes, and C. K. Ingold, *J. Chem. Soc.*, **1950**, 2473.
25. S. J. Bass, R. J. Gillespie, and E. A. Robinson, *J. Chem. Soc.*, **1960**, 821.
26. J. A. Leisten and K. L. Wright, *J. Chem. Soc.*, **1964**, 3173.

27. J. A. Leisten and P. R. Walton, *J. Chem. Soc.*, **1964**, 3180.
28. E. A. Robinson and S. A. A. Quadri, *Can. J. Chem.*, **45**, 2385 (1967).
29. R. J. Gillespie and S. Wasif, *J. Chem. Soc.*, **1953**, 221.
30. W. H. Lee, in *The Chemistry of Non-Aqueous Solvents*, Vol. II, J. J. Lagowski, Ed., Academic Press, New York, 1967, p. 160.
31. A. R. Katritzky and R. A. Y. Jones, *Chem. Ind. (London)*, **1961**, 722.
32. A. Casadevall, G. Couquil, and R. Corriu, *Bull. Soc. Chim. France*, **1964**, 187.
33. S. Hoshino, H. Hosoya, and S. Nagakura, *Can. J. Chem.*, **44**, 1961 (1966).
34. D. Cook, *Can. J. Chem.*, **42**, 2292 (1964).
35. Reference 10, p. 1.
36. T. C. Waddington, Ed., *Non-Aqueous Solvent Systems*, Academic Press, New York, 1965.
37. C. MacLean, J. H. van der Waals, and E. L. Mackor, *Mol. Phys.*, **1**, 247 (1958).
38. C. MacLean and E. L. Mackor, *J. Chem. Phys.*, **34**, 2207 (1961).
39. M. Kilpatrick and J. G. Jones, in Reference 30, p. 151.
40. R. E. Reavill, *J. Chem. Soc.*, **1964**, 519.
41. G. A. Olah and A. M. White, unpublished results.
42. G. A. Olah, D. H. O'Brien, and M. Calin, *J. Am. Chem. Soc.*, **89**, 3582 (1967).
43. P. Ros, *J. Chem. Phys.*, **49**, 4902 (1968).
44. G. A. Olah, J. M. Bollinger, and A. M. White, *Progr. Nucl. Magn. Resonance Spectrosc.*, in press.
45. A. M. White and G. A. Olah, *J. Am. Chem. Soc.*, **91**, 2943 (1969).
46. J. B. Stothers, *Quart. Rev., Chem. Soc.*, **1**, 144 (1965).
47. E. B. Baker, *J. Chem. Phys.*, **37**, 911 (1962).
48. G. A. Olah and A. M. White, *J. Am. Chem. Soc.*, **90**, 1884 (1968).
49. G. A. Olah and A. M. White, *J. Am. Chem. Soc.*, **91**, 5801 (1969).
50. R. E. Richards, *Trans. Faraday Soc.*, **58**, 845 (1962).
51. R. J. Gillespie and J. A. Leisten, *Quart. Rev., Chem. Soc.*, **8**, 40 (1954).
52. Reference 11, p. 689.
53. T. Birchall and R. J. Gillespie, *Can. J. Chem.*, **43**, 1045 (1965).
54. G. A. Olah and E. Namanworth, *J. Am. Chem. Soc.*, **88**, 5327 (1966).
55. G. A. Olah, J. Sommer, and E. Namanworth, *J. Am. Chem. Soc.*, **89**, 3576 (1967).
56. G. A. Olah and J. Sommer, *J. Am. Chem. Soc.*, **90**, 927 (1968).
57. G. A. Olah, D. H. O'Brien, and C. U. Pittman, Jr., *J. Am. Chem. Soc.*, **89**, 2996 (1967).
58. R. L. Burwell, *Chem. Rev.*, **54**, 615 (1954).
59. W. Gerrard and E. D. Macklen, *Chem. Rev.*, **59**, 1105 (1959).
60. L. P. Hammett, *Physical Organic Chemistry*, McGraw-Hill, New York, 1940, p. 47.
61. P. Grange, J. Lascombe, and M. L. Josien, *Spectrochim. Acta*, **16**, 981 (1960).
62. R. M. Adams and J. J. Katz, *J. Mol. Spectrosc.*, **1**, 306 (1957).
63. J. Arnold, J. E. Bertie, and D. J. Millen, *Proc. Chem. Soc.*, **1961**, 121.
64. P. Grange and J. Lascombe, *J. Chim. Phys.*, **60**, 1119 (1963).
65. C. Quivoron and J. Neel, *C. R. Acad. Sci. (Paris)*, **259**, 1845 (1964).
66. V. A. Plotnikov and M. L. Kaplan, *Izv. Akad. Nauk SSSR, Otd. Khim. Nauk*, **1948**, 256; *Chem. Abstr.*, **42**, 7151$f$ (1948).
67. G. Jander and K. Kraffczyk, *Z. Anorg. Allg. Chem.*, **282**, 121 (1955).
68. M. V. Ionin and V. G. Shverina, *Zh. Obshch. Khim.*, **35**, 209 (1965).
69. L. P. Hammett, Reference 60, Chapter IX.

70. E. M. Arnett and C. Y. Wu, *Chem. Ind. (London)*, **1959**, 1488.
71. E. M. Arnett and C. Y. Wu, *J. Am. Chem. Soc.*, **82**, 4999 (1960).
72. E. M. Arnett and C. Y. Wu, *J. Am. Chem. Soc.*, **82**, 5660 (1960).
73. E. M. Arnett, C. Y. Wu, J. N. Anderson, and R. D. Bushick, *J. Am. Chem. Soc.*, **84**, 1674 (1962).
74. E. M. Arnett and C. Y. Wu, *J. Am. Chem. Soc.*, **84**, 1680 (1962).
75. E. M. Arnett and C. Y. Wu, *J. Am. Chem. Soc.*, **84**, 1684 (1962).
76. J. Hine and M. Hine, *J. Am. Chem. Soc.*, **74**, 5266 (1952).
77. P. D. Bartlett and J. D. McCollum, *J. Am. Chem. Soc.*, **78**, 1441 (1956).
78. J. T. Edward, *Chem. Ind. (London)*, **1963**, 489.
79. T. Higuchi, C. H. Barnstein, H. Glassemi, and W. E. Perez, *Anal. Chem.*, **34**, 400 (1962).
80. J. T. Edward, J. B. Leane, and I. C. Wang, *Can. J. Chem.*, **40**, 1521 (1962).
81. F. Klages, J. E. Gordon, and H. A. Jung, *Chem. Ber.*, **98**, 3748 (1965).
82. G. A. Olah and D. H. O'Brien, *J. Am. Chem. Soc.*, **89**, 1725 (1967).
83. D. Jaques and J. A. Leisten, *J. Chem. Soc.*, **1961**, 4963.
84. T. Birchall, A. N. Bourns, R. J. Gillespie, and P. J. Smith, *Can. J. Chem.*, **42**, 1433 (1964).
85. T. Birchall and R. J. Gillespie, *Can. J. Chem.*, **42**, 502 (1964).
86. D. M. Brouwer, E. L. Mackor, and C. Maclean, *Rec. Trav. Chim. Pays-Bas*, **85**, 109 (1966).
87. D. M. Brouwer, E. L. Mackor, and C. Maclean, *Rec. Trav. Chim. Pays-Bas*, **85**, 114 (1966).
88. (a) C. Maclean and E. L. Mackor, *Discuss. Faraday Soc.*, **34**, 196 (1962); (b) G. A. Olah, E. Namanworth, M. B. Comisarow, and B. Ramsey, *J. Am. Chem. Soc.*, **89**, 5259 (1967).
89. W. M. Schubert and R. H. Quacchia, *J. Am. Chem. Soc.*, **84**, 3778 (1962).
90. W. M. Schubert and R. H. Quacchia, *J. Am. Chem. Soc.*, **85**, 1278 (1963).
91. A. J. Kresge, G. W. Barry, K. R. Charles, and Y. Chiang, *J. Am. Chem. Soc.*, **84**, 4343 (1962).
92. A. J. Kresge and Y. Chiang, *Proc. Chem. Soc. London*, **1961**, 81.
93. G. W. Wheland, *Resonance in Organic Chemistry*, Wiley, New York, 1955, p. 289.
94. A. J. Kresge and L. E. Hakka, *J. Am. Chem. Soc.*, **88**, 3868 (1966).
95. J. B. Lambert, R. G. Keske, and D. K. Weary, *J. Am. Chem. Soc.*, **89**, 5921 (1967).
96. G. A. Olah and J. Sommer, *J. Am. Chem. Soc.*, **90**, 4323 (1968).
97. S. Patai, Ed., *The Chemistry of the Carbonyl Group*, Interscience, New York, 1966, Chapter 9.
98. S. Nagakura, A. Minegishi, and K. Stanfield, *J. Am. Chem. Soc.*, **79**, 1033 (1957).
99. H. J. Campbell and J. T. Edward, *Can. J. Chem.*, **38**, 2109 (1960).
100. K. Yates and R. Stewart, *Can. J. Chem.*, **37**, 664 (1959).
101. R. Stewart and K. Yates, *J. Am. Chem. Soc.*, **80**, 6355 (1958).
102. K. Yates and H. Wai, *Can. J. Chem.*, **43**, 2131 (1965).
103. M. Liler, *Spectrochim. Acta*, *Pt. A*, **23**, 139 (1967).
104. H. Hogeveen, *Rec. Trav. Chim. Pays-Bas*, **86**, 696 (1967).
105. D. M. Brouwer, *Rec. Trav. Chim. Pays-Bas*, **86**, 879 (1967).
106. M. Brookhart, G. C. Levy, and S. Winstein, *J. Am. Chem. Soc.*, **89**, 1735 (1967).
107. G. A. Olah, M. Calin, and D. H. O'Brien, *J. Am. Chem. Soc.*, **89**, 3585 (1967).

108. G. A. Olah and M. Calin, *J. Am. Chem. Soc.*, **90**, 938 (1968).
109. D. M. Brouwer, *Chem. Commun.*, **1967**, 515.
110. G. A. Olah and C. U. Pittman, Jr., *J. Am. Chem. Soc.*, **88**, 3310 (1966).
111. R. W. Taft et al., *J. Am. Chem. Soc.*, **85**, 709, 3146 (1963).
112. G. A. Olah, J. M. Bollinger, and J. Brinich, *J. Am. Chem. Soc.*, **90**, 2587 (1968).
113. A. A. Bothner-By and C. Naar-Colin, *J. Am. Chem. Soc.*, **83**, 231 (1961).
114. T. J. Sekuur and P. Kranenburg, *Tetrahedron Lett.*, **1966**, 4793.
115. (a) S. Sternhell, *Rev. Pure Appl. Chem.*, **14**, 15 (1964); (b) *Quart. Rev., Chem. Soc.*, **23**, 236 (1969).
116. H. Hogeveen, *Rec. Trav. Chim. Pays-Bas*, **87**, 1313 (1968).
117. G. J. Karabatsos and N. Hsi, *J. Am. Chem. Soc.*, **87**, 2864 (1965).
118. S. Daniloff and E. Venus-Danilova, *Ber.*, **59**, 377 (1926).
119. H. Hopff, C. D. Nenitzescu, D. A. Isacescu, and I. P. Cantuniari, *Ber.*, **69**, 2244 (1936).
120. H. Hogeveen, *Rec. Trav. Chim. Pays-Bas*, **87**, 1295 (1968).
121. W. D. Phillips, *J. Chem. Phys.*, **23**, 1363 (1955).
122. H. S. Gutowsky and C. H. Holm, *J. Chem. Phys.*, **25**, 1228 (1956).
123. V. J. Kowalewski and D. G. deKowalewski, *J. Chem. Phys.*, **32**, 1272 (1960).
124. J. L. O'Brien and C. Niemann, *J. Am. Chem. Soc.*, **79**, 1386 (1957).
125. G. Fraenkel and C. Franconi, *J. Am. Chem. Soc.*, **82**, 4478 (1960).
126. J. Sandstrom, *Acta Chem. Scand.*, **16** (2), 1616 (1962).
127. E. Spinner, *Spectrochim. Acta*, **15**, 95 (1959).
128. E. Spinner, *J. Phys. Chem.*, **64**, 275 (1960).
129. M. Davies and L. Hopkins, *Trans. Faraday Soc.*, **53**, 1563 (1957).
130. J. T. Edward, H. S. Chang, K. Yates, and R. Stewart, *Can. J. Chem.*, **38**, 1518 (1960).
131. R. Stewart and L. J. Muenster, *Can. J. Chem.*, **39**, 401 (1961).
132. D. Cook, *Can. J. Chem.*, **40**, 2362 (1962).
133. W. Gerrard, M. F. Lappert, H. Pyszora, and J. W. Wallis, *J. Chem. Soc.*, **1960**, 2144.
134. M. J. Janssen, *Rec. Trav. Chim. Pays-Bas*, **79**, 454 (1960).
135. M. J. Janssen, *Rec. Trav. Chim. Pays-Bas*, **79**, 464 (1960).
136. M. J. Janssen, *Spectrochim. Acta*, **17**, 475 (1961).
137. W. Kutzelnigg and R. Mecke, *Spectrochim. Acta*, **17**, 530 (1961).
138. A. Berger, A. Loewenstein, and S. Meiboom, *J. Am. Chem. Soc.*, **81**, 62 (1959).
139. R. J. Gillespie and T. Birchall, *Can. J. Chem.*, **41**, 148 (1963).
140. T. Birchall and R. J. Gillespie, *Can. J. Chem.*, **41**, 2642 (1963).
141. H. Pracejus, *Chem. Ber.*, **92**, 988 (1959).
142. G. A. Olah and P. J. Szilagyi, unpublished results.
143. R. Stewart and K. Yates, *J. Am. Chem. Soc.*, **82**, 4059 (1960).
144. G. A. Olah and A. M. White, *J. Am. Chem. Soc.*, **89**, 3591 (1967).
145. H. Hogeveen, A. F. Bickel, C. W. Hilbers, E. L. Mackor, and C. MacLean, *Rec. Trav. Chim. Pays-Bas*, **86**, 687 (1967); *Chem. Commun.*, **1966**, 898.
146. H. Hogeveen, *Rec. Trav. Chim. Pays-Bas*, **86**, 809 (1967).
147. G. A. Olah and A. M. White, *J. Am. Chem. Soc.*, **89**, 7072 (1967).
148. G. J. Karabatsos, G. C. Sonnischsen, N. Hsi, and D. J. Fenoglio, *J. Am. Chem. Soc.*, **89**, 5067 (1967).
149. G. A. Olah and M. Calin, *J. Am. Chem. Soc.*, **90**, 405 (1968).
150. G. A. Olah, A. Ku, and A. M. White, *J. Org. Chem.*, **34**, 1827 (1969).
151. G. Fraenkel, *J. Chem. Phys.*, **34**, 1466 (1961).

152. G. A. Olah, D. H. O'Brien, and A. M. White, *J. Am. Chem. Soc.*, **89**, 5694 (1967).
153. A. Bradley and M. E. Hill, *J. Am. Chem. Soc.*, **77**, 1575 (1955).
154. C. K. Ingold, *Structure and Mechanism in Organic Chemistry*, Cornell University Press, Ithaca, N.Y., 1953.
155. J. Leisten, *J. Chem. Soc.*, **1956**, 1572.
156. N. C. Deno, C. U. Pittman, and M. J. Wisotsky, *J. Am. Chem. Soc.*, **86**, 4370 (1964).
157. H. Hogeveen, *Rec. Trav. Chim. Pays-Bas*, **86**, 816 (1967).
158. H. P. Treffers and L. P. Hammett, *J. Am. Chem. Soc.*, **59**, 1708 (1937).
159. For a discussion, see G. A. Olah, A. M. White, and D. H. O'Brien, in *Carbonium Ions*, Vol. IV, G. A. Olah and P. v. R. Schleyer, Eds., Interscience, 1973.
160. G. A. Olah and A. Ku, *J. Org. Chem.*, **35**, 331 (1970).
161. E. A. Robinson and S. A. A. Quadin, *Can. J. Chem.*, **45**, 2391 (1967).
162. F. Klages and E. Zarnge, *Ber.*, **92**, 1828 (1959).
163. L. A. Wiles, *J. Chem. Soc.*, **1953**, 996.
164. C. U. Pittman, Ph.D. thesis, The Pennsylvania State University, University Park, 1964.
165. G. Oddo and A. Casalino, *Gazzetta*, **47**, 232 (1917).
166. M. S. Newman and N. C. Deno, *J. Am. Chem. Soc.*, **73**, 3651 (1951).
167. G. A. Olah and R. Schlosberg, *J. Am. Chem. Soc.*, **90**, 6464 (1968).
168. H. Hogeveen, *Rec. Trav. Chim. Pays-Bas*, **87**, 1303 (1968).
169. B. G. Ramsey and R. W. Taft, *J. Am. Chem. Soc.*, **88**, 3058 (1966).
170. F. Klages and H. Mearesch, *Ber.*, **85**, 863 (1952); see also H. Meerwein, in *Methoden der Organischen Chemie* (Houben-Weyl), Vol. VI/3, Georg Thieme, Stuttgart, 1965, p. 325.
171. T. I. Bieber, *J. Am. Chem. Soc.*, **75**, 1409 (1953).
172. G. A. Olah and M. Calin, *J. Am. Chem. Soc.*, **90**, 401 (1968).
173. V. C. Armstrong, D. W. Farlow, and R. B. Moodie, *Chem. Commun.*, **1968**, 1362.
174. G. A. Olah, J. A. Olah, and A. M. White, unpublished results.
175. (a) V. C. Armstrong and R. B. Moodie, *J. Chem. Soc. B*, **1968**, 275; (b) R. B. Homer and R. B. Moodie, *J. Chem. Soc. B*, **1963**, 4377.
176. G. A. Olah and A. M. White, *J. Am. Chem. Soc.*, **41**, 2642 (1963).
177. G. Williams and M. L. Hardy, *J. Chem. Soc.*, **1953**, 2560.
178. C. Holstead, A. H. Lamberton, and P. A. H. Wyatt, *J. Chem. Soc.*, **1953**, 3341.
179. C. R. Redpath and J. A. S. Smith, *Trans. Faraday Soc.*, **58**, 462 (1962).
180. G. A. Olah and P. Kreinbuhl, *J. Am. Chem. Soc.*, **89**, 4756 (1967).
181. E. J. Stamhuis, W. Maas, and H. Wynberg, *J. Org. Chem.*, **30**, 2160 (1965).
182. E. J. Stamhuis and W. Maas, *J. Org. Chem.*, **30**, 2156 (1965).
183. W. Maas, M. J. Janssen, E. J. Stamhuis, and H. Wynberg, *J. Org. Chem.*, **32**, 1111 (1967).
184. J. Elguero, R. Jacquier, and G. Tarrago, *Tetrahedron Lett.*, **1965**, 4719.
185. J. Elguero, R. Jacquier, and G. Tarrago, *Tetrahedron Lett.*, **1966**, 1112.
186. G. A. Olah and T. E. Kiovsky, *J. Am. Chem. Soc.*, **90**, 4666 (1968).
187. G. A. Olah and T. E. Kiovsky, *J. Am. Chem. Soc.*, **90**, 6461 (1968).
188. N. C. Deno, R. W. Gaugler, and M. J. Wisotsky, *J. Org. Chem.*, **31**, 1967 (1966).
189. R. J. Gillespie, personal communication.
190. R. J. Gillespie and C. Solomons, *J. Chem. Soc.*, **1957**, 1796.
191. R. J. Gillespie, *J. Chem. Soc.*, **1950**, 2592.
192. H. Hogeveen, *Rec. Trav. Chim. Pays-Bas*, **86**, 1320 (1967).

193. J. L. O'Brien and C. Niemann, *J. Am. Chem. Soc.*, **73**, 4264 (1951).
194. B. Bak, C. Dambmann, F. Nicolaisen, and E. J. Pederson, *J. Mol. Spectrosc.*, **74**, 78 (1968).
195. W. J. Horsley and H. Sternlicht, *J. Am. Chem. Soc.*, **90**, 3738 (1968).
196. G. A. Olah, D. L. Brydon, and R. D. Porter, *J. Org. Chem.*, **35**, 317 (1970).
197. H. H. Hyman and J. J. Katz, in *Non-Aqueous Solvent Systems*, T. C. Waddington, Ed., Academic Press, New York, 1965, pp. 76–79.

CHAPTER 32

# Bridgehead Carbonium Ions

RAYMOND C. FORT, JR.
*Department of Chemistry, Kent State University, Kent, Ohio*

| | |
|---|---|
| I. Introduction | 1783 |
| II. Theoretical Considerations and Physical Evidence | 1784 |
| III. Solvolytic Reactivity | 1785 |
| IV. Substituent Effects | 1797 |
| V. Deamination | 1799 |
| VI. Exchange Processes | 1802 |
| VII. Bridgehead Ions in Rearrangement Reactions | 1809 |
| VIII. Fragmentation | 1813 |
| IX. Bridgehead Carbonium Ions in Condensed Ring Systems | 1814 |
| X. Miscellaneous | 1821 |
| XI. Summary | 1824 |
| XII. Acknowledgments | 1824 |
| XIII. Supplement | 1824 |
| References | 1829 |

## I. INTRODUCTION

The general subject of bridgehead reactivity has been reviewed in 1966 in depth (1). Only carbonium ions are considered in this chapter, save where comparisons to other reactive intermediates may prove instructive. The factors influencing the stability of bridgehead cations are discussed, along with the utility of bridgehead systems as mechanistic probes. The literature has been covered through December 1968 in depth; some material of later date is included in the Supplement.

The importance of the study of bridgehead chemistry was first pointed out by Bartlett and Knox (2). These workers noted the extreme unreactivity of triptycyl and apocamphyl bridgehead derivatives in nucleophilic substitution. The suggestion followed that the inertness of these compounds was the result of two factors: (*a*) the inability of the substrate to undergo Walden inversion,* and therefore, bimolecular nucleophilic sub-

---

* It is interesting that at the time of Bartlett's original work with bridgehead ions, there were still suggestions (3) that the *only* way in which one anion could replace another in a neutral molecule was via a Walden inversion. According to Ogg (4), "It is suggested that carbonium ions with an open sextet never play an appreciable role in observable organic reactions, and that mechanisms employing such ions must be abandoned."

stitution; and (b) the geometrically derived instability of the carbonium ion which would be involved necessarily in a unimolecular process. [Later investigations (5) have ruled out an abnormal front-side bimolecular substitution.] It is these terms which may be employed still to explain all bridgehead reactivity.

The insight of Bartlett must be applauded, for the strongest evidence for the generally accepted view of carbonium ions as planar, $sp^2$-hybridized entities is a direct descendant of these ideas.

## II. THEORETICAL CONSIDERATIONS AND PHYSICAL EVIDENCE

Empirically, it is easy to predict that a simple alkyl carbonium ion will prefer $sp^2$ hybridization, with the three attached groups lying at 120° angles in a plane. This hybridization allows the electron deficiency to be accommodated "in" an orbital having no $s$ character and places the remaining bonds as far apart as possible, thus minimizing repulsions between them. The isoelectronic trimethyl- and trihalo-derivatives of boron are known to adopt this geometry (6).

A variety of more detailed theoretical calculations confirm this first-order expectation. Treatments employing determinantal wave functions (7), the CNDO method (8), and *ab initio* procedures (9) all give essentially the same result—a *strong* preference for planar, trigonal geometry.

Until recently, direct physical evidence regarding the shape of simple (nonresonance-stabilized) alkyl carbonium ions has been unavailable because of the obvious difficulty in keeping such ions long enough to employ any method of analysis that would supply detailed structural information. The well-known technique developed by Olah and his co-workers (10), by which carbonium ions are stabilized as their $SbF_6^-$ salts in antimony pentafluoride solution has overcome this difficulty.

Infrared and proton magnetic resonance (pmr) examinations of several such salts suggest that the cation is planar. For example, vibrational analyses (10a) of the *t*-butyl and other tertiary cations indicate that their Raman and ir spectra are best interpreted in terms of a planar species. Strong spin couplings across the positive center, and the magnitudes of $^{13}CH$ coupling constants (11,12) (163–169 Hz), also favor the planar formulation.

Although the geometrical preference of carbon cations seems well defined by calculations and observations such as these, we must note that these are, in a sense, *ex post facto* justifications for results deduced from the chemistry of bridged systems. We shall now proceed to a discussion of this chemistry.

## III. SOLVOLYTIC REACTIVITY

An important concern in *any* discussion of $S_N1$ reactivity is the extent to which rates of solvolysis reflect the stability of the carbonium ion formed. Statements such as "triphenylmethyl chloride solvolyzes more rapidly than *t*-butyl chloride because a more stable carbonium ion is formed" are strictly true only under very special conditions. The idea that the more stable a species is, the more stable will be the transition state leading to it, and consequently, the more rapid will be its formation, is a useful one. The basis for this reasoning, as applied to formation of carbonium ions, is simply that insofar as the transition state partakes of the character of a carbonium ion, it will be stabilized by the same factors that stabilize the carbonium ion. Using relative rates to indicate relative stabilities for carbonium ions is really valid only when the transition states for the two reactions fall at the same point along the reaction coordinate; that is, when ionization has proceeded to the same extent in both transition states.

Information about whether this is the case is clearly not available, and all rate comparisons thus suffer a degree of uncertainty. We shall observe the usual (13) custom: for a series of closely related compounds (e.g., bridgehead bromides), rates may be taken to fairly well represent stability.

The literature abounds in qualitative examples of bridgehead carbonium ion reactivity; the variety of reaction conditions conveys some idea of the ease of formation of the carbonium ions. Equations 1–4 illustrate the unreactivity of bicyclo[2.2.1]heptane derivatives, and equations 5–7

(5) [triptycyl-Br] $\xrightarrow[\text{NaOH}]{\text{aq}}$ no reaction   Refs. 16,17

(6) [triptycyl-OH] $\xrightarrow[\text{CH}_3\text{COOH}]{\text{H}_2\text{SO}_4}$ no reaction   Ref. 18

(7) [triptycyl anhydride-Br] $\xrightarrow[\text{NaOH}]{\text{aq}}$ no reaction   Ref. 19

present the unreactivity of triptycyl derivatives. Only a part of the inertness of the latter compounds results from geometry. It has been estimated (18) that the inductive effect of the aromatic rings should destabilize the cation by approximately $10^9$ relative to bicyclo[2.2.2]octyl. The *relative* ease of reaction in this latter system may be seen in equations 8–10.

(8) [bicyclo[2.2.2]octyl-OCH$_3$] $\xrightarrow[\text{SnCl}_4]{\text{CH}_3\text{OH}}$ [bicyclo[2.2.2]octyl-Cl]   Refs. 20,23

(9) [bicyclo[2.2.2]octyl-CH$_2$OH] $\xrightarrow[\text{HCOOH}]{\text{H}_2\text{SO}_4}$ [bicyclo[2.2.2]octyl-COOH]   Ref. 21

(10) [bicyclo[2.2.1]heptyl-Cl] $\xrightarrow[\text{aq HCl}]{\text{ZnCl}_2}$ [product, Cl] 33% + [product, Cl] 67%   Ref. 22

The utility of carbonium ion reactions in the adamantane series has been widely documented (24). A few of the more recent examples are given here (eqs. 11–13); others are referenced (28–35).

[Structure with NO₂] →(Br₂)→ [Structure with Br and NO₂]   Ref. 25   (11)

[Adamantane] →(FSO₂OH / SbF₅)→ [Adamantyl cation⁺]   Ref. 26   (12)

[Adamantane] →(NCl₃ / AlCl₃)→ [Adamantyl-NH₂]   Ref. 27   (13)

Bridgehead positions in larger ring systems are "normal" tertiary centers (eq. 14). Clearly, the larger the rings composing the polycycle, the more

[Bicyclic alcohol HO] →(PCl₃)→ [Bicyclic chloride Cl]   Ref. 36   (14)

easily can a planar (or near-planar) bridgehead be incorporated without strain.

Quantitative data are presented in Table I and summarized in Chart 1 for a more detailed discussion of these ideas.

Of the various factors determining these rates, we examine solvation first. It is recognized (13) that participation of solvent molecules from the rear assists in the ionization of 2° substrates, and of course it is the major pathway for reaction of 1° compounds. Although some recent work (56) suggests a bimolecular contribution to the reactions of trityl derivatives, these are clearly special cases, and solvent participation is generally unimportant in 3° systems. Reactions of bridgehead derivatives provide some of the strongest evidence for this conclusion. Thus, although rear-side solvent participation is precluded *absolutely* by the cage construction of the bi- and tricyclic molecules, solvolyses of bridgehead compounds

## TABLE I
### Bridgehead Solvolytic Reactivity[a]

| System | Leaving group | Solvent | $T$, °C | $k$, sec$^{-1}$ | $\Delta H^{\ddagger}$, kcal | $\Delta S^{\ddagger}$, eu | Ref. |
|---|---|---|---|---|---|---|---|
| t-Butyl | Cl | EtOH | 25 | $9.70 \times 10^{-8}$ | — | — | 37 |
|  | Cl | 80% EtOH | 25 | $9.24 \times 10^{-6}$ | 22.3 | −6.6 | 37 |
|  | Cl | AcOH | 70 | $1.62 \times 10^{-6}$ | — | — | 37 |
|  | Cl | MeOH | 70 | $2.07 \times 10^{-4}$ | — | — | 37 |
| t-Butyl | Br | EtOH | 25 | $4.40 \times 10^{-6}$ | 24.6 | −0.4 | 38 |
|  | Br | 80% EtOH | 25 | $3.58 \times 10^{-4}$ | 21.5 | −2.3 | 38 |
|  | Br | AcOH | 25 | $3.02 \times 10^{-6}$ | 24.4 | −1.9 | 38 |
|  | Br | MeOH | 25 | $3.44 \times 10^{-5}$ | 23.4 | −0.5 | 38 |
| t-Butyl | Br | 70% dioxan | 25 | $4.55 \times 10^{-4}$ | — | — | 38 |
| 1-Bicyclo[1.1.1]pentyl[b] | Cl | 80% EtOH | 25 | $3.03 \times 10^{-5}$ | 12.0 | −4.0 | 39 |
| 1-Bicyclo[2.1.1]hexyl[b] | Br | 40% EtOH | 96.7 | $6.03 \times 10^{-5}$ | 23.8 | −14 | 40 |
| 1-Bicyclo[2.2.1]heptyl | Br[c] | 80% EtOH–NaOH | 25 | $7 \times 10^{-16}$ | 32 | — | 5 |
|  | OTs[d] | AcOH | 25 | $9 \times 10^{-15}$ | — | — | 41 |
|  | OBs | AcOH/NaOAc | 25 | $2.78 \times 10^{-14}$ | 31.7 | −14.3 | 41 |
| 1-Bicyclo[2.2.2]octyl | Br | 70% dioxan | 100 | $7.11 \times 10^{-7}$ | — | — | 42 |
|  | Br | HCO$_2$H/HCO$_2$Na | 79.3 | $1.23 \times 10^{-5}$ | 25.9 | −7.8 | 5 |
|  | Br | AcOH/KOAc | 165 | $2.05 \times 10^{-5}$ | 31.5 | −8.9 | 5 |
|  | Br | MeOH/NaOMe | 145.6 | $1.52 \times 10^{-5}$ | 27.2 | −16.5 | 5 |
|  | Br | 80% EtOH/NaOH | 134.6 | $1.89 \times 10^{-5}$ | 26.4 | −16.1 | 5 |
|  | Br | 80% EtOH | 124.2 | $8.29 \times 10^{-6}$ | 27.3 | −15.6 | 43 |
|  | Br | 40% EtOH–NaOH | 85.6 | $1.71 \times 10^{-5}$ | 23.7 | −14.7 | 5 |

| Compound | X | Solvent | Temp | k | ΔS‡ | Ref |
|---|---|---|---|---|---|---|
| | Br | 40% EtOH | 96.8 | $5.99 \times 10^{-5}$ | $-15$ | 40 |
| | Br | EtOH | 40.0 | $8.70 \times 10^{-10}$ | — | 44 |
| | OBs | AcOH | 75 | $1.13 \times 10^{-4}$ | $-1.7$ | 45 |
| 3.3-Dimethyl-1-bicyclo[2.2.2]octyl | Br | 70% dioxan | 100.4 | $9.87 \times 10^{-7}$ | $-11.0$ | 46 |
| | Br | EtOH/NaOEt | 158.4 | $9.83 \times 10^{-6}$ | — | 46 |
| | Br | EtOH/LiOAc | 158.4 | $8.10 \times 10^{-6}$ | — | 46 |
| | Br | $HCO_2H/HCO_2Na$ | 79.3 | $3.12 \times 10^{-5}$ | $-8.0$ | 5 |
| | Br | AcOH/KOAc | 154.7 | $1.87 \times 10^{-5}$ | $-10.2$ | 5 |
| | Br | MeOH/NaOMe | 145.9 | $3.69 \times 10^{-5}$ | $-19.1$ | 5 |
| | Br | 80% EtOH/NaOH | 146.1 | $9.33 \times 10^{-5}$ | $-15.1$ | 5 |
| | Br | 80% EtOH | 100.0 | $2.70 \times 10^{-6}$ | $-20$ | 47 |
| 1-Bicyclo[3.2.1]octyl | Br | 70% dioxan | 131.7 | $1.80 \times 10^{-6}$ | — | 42 |
| 1-Bicyclo[3.3.1]nonyl | Br | 70% dioxan | 131.7 | $6.95 \times 10^{-7}$ | — | 42 |
| | Cl | 60% EtOH | 25 | $2.02 \times 10^{-6}$ | $-8.2$ | 48 |
| | Br | 80% EtOH | 24.1 | $3.04 \times 10^{-6}$ | $-3.2$ | 49 |
| 5-Methyl-1-tricyclo[3.3.1³,⁷0¹,⁵]-nonane (noradamantane)ᶠ | OTs | AcOH | 100 | $1.72 \times 10^{-7}$ | $-2.6$ | 50 |
| 1-Adamantyl | Cl | 80% EtOH | 25 | $7.59 \times 10^{-9}$ | $-10.2$ | 51 |
| | Br | 80% EtOH | 24.95 | $1.16 \times 10^{-7}$ | — | 52 |
| | Br | 80% EtOH | 25 | $4.38 \times 10^{-7}$ | $-12.0$ | 51 |
| | Br | 60% EtOH | 24.95 | $4.56 \times 10^{-6}$ | — | 52 |
| | Br | 50% EtOH | 35.8 | $1.04 \times 10^{-4}$ | $-7.8$ | 53 |
| | I | 80% EtOH | 25.0 | $8.45 \times 10^{-7}$ | $-8.6$ | 51 |
| | OTs | AcOH | 25.0 | $5.86 \times 10^{-4}$ | — | 51 |

(continued)

**TABLE I** (*Continued*)

| System | Leaving group | Solvent | $T$, °C | $k$, sec$^{-1}$ | $\Delta H^{\ddagger}$, kcal | $\Delta S^{\ddagger}$, eu | Ref. |
|---|---|---|---|---|---|---|---|
| 1-Homoadamantyl[g] | Cl | 80% EtOH | 25.0 | $3.45 \times 10^{-6}$ | — | — | 54 |
|  | Br | 80% EtOH | 25.0 | $1.64 \times 10^{-4}$ | — | — | 54 |
| 1-Congressyl[h] | Br | 80% EtOH | 75.0 | $1.04 \times 10^{-3}$ | — | — | 55 |

[a] Key: MeOH = CH$_3$OH, EtOH = C$_2$H$_5$OH, NaOMe = NaOCH$_3$, NaOEt = NaOC$_2$H$_5$, AcOH = CH$_3$CO$_2$H, NaOAc = NaOCOCH$_3$, LiOAc = LiOCOCH$_3$, KOAc = KOCOCH$_3$, OTs = $p$-CH$_3$C$_6$H$_4$SO$_3^-$, OBs = $p$-BrC$_6$H$_4$SO$_3^-$.
[b] Solvolysis occurs with ring opening; see below.
[c] Rate extrapolated from reaction at 216°.
[d] Estimated assuming tosylate is one-third as fast as brosylate.
[e] Estimated from solvolysis of mixtures.

[f]   [g]   [h]

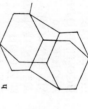

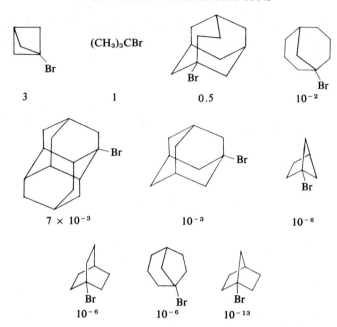

Chart 1. Relative Solvolytic Reactivities of Bromides in 80% Ethanol at 25°.

show the same sensitivity to solvent structure, polarity, and nucleophilicity as do acyclic tertiary compounds (226).

This may be seen, e.g., in the $m$ parameter of the Grunwald-Winstein correlation (37), which measures the sensitivity of the reactivity of a given compound to changes of the ionizing power of the solvent in which it reacts. If the role of solvent were different in the transition states for tertiary acyclic and bridgehead solvolyses, we would expect a significant difference in $m$ values. As Chart 2 reveals, however, such deviations are not observed.

Therefore, we conclude, covalent solvation of *both* sorts of structure is probably negligible in the transition state.

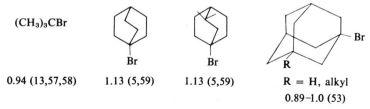

Chart 2. $m$ Values for Bridgehead Bromides in Ethanol–Water.

An interesting confirmation of this reasoning is found in solvolysis of derivatives of the alcohol 1 (60). The shallow dish shape of the molecule should permit the approach of at least one solvent molecule to the rear of the bridgehead, but we find no enhanced reactivity. Indeed, the tosylate 2, and the *p*-nitrobenzoate 3 solvolyze some seven powers of ten less rapidly than their adamantyl counterparts, as we might expect from the strained nature of the ring system (60).

(1) X = OH
(2) X = OTs
(3) X = OCOC$_6$H$_4$NO$_2$-*p*

The appearance potentials for the bridgehead cations, determined in the vapor phase, should allow us to evaluate the intrinsic stability of the ion, free of solvation effects. Chart 3 gives relevant values. That these data

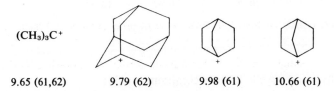

(CH$_3$)$_3$C$^+$

9.65 (61,62)     9.79 (62)     9.98 (61)     10.66 (61)

Chart 3. Appearance Potentials[a] (from Bromides) for Bridgehead Cations.

closely parallel the ionic stabilities as measured by the solvolysis rates (Fig. 1) suggests strongly that whatever the exact effect of solvation, it does not significantly alter the "natural" order of stabilities.

To analyze fully the effect of solvent on the reactivity of these bridged systems, information about ground state solvation would be necessary. The importance of this phenomenon is made clear in the observation (63) that changes in $\Delta H$ of solution for *t*-butyl chloride in ethanol–water mixtures match the changes in $\Delta H$ for solvolysis. No experimental evidence is available to indicate that any portion of the reactivity differences between bridged systems is the result of differences in the activity coefficients of the various molecules. Because of the close similarity of their gross structure—a round hydrocarbon blob with a polar group attached—it is reasonable to expect such variations to be extremely small among any group of, say, bridgehead bromides. We thus conclude that solvation plays a minor role in determining the relative reactivity of bridgehead derivatives.

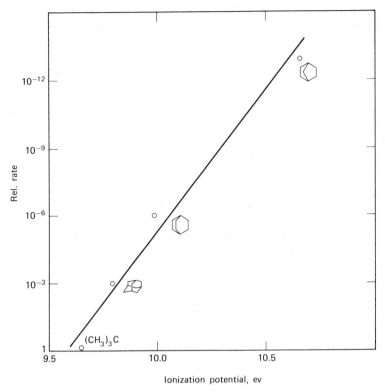

Fig. 1. Correlation of solvolysis rates with appearance potentials.

The possible role of structures like **4** and **5** in bridgehead solvolyses may be noted. These are given credence by observations of $\pi$ routes to bridgehead carbonium ions (discussed later); in a formal sense they represent simply carbon–carbon hyperconjugative contributions and may be likened to a "nonclassical" structure such as **6** or **7** for these ions (5,46). However, this proliferation of dotted lines can be ruled out both by the dictum of William of Occam and by data from substituted adamantanes. In particular, one finds (Table III, p. 1798) that methyl substitution at bridgeheads in adamantane does not enhance solvolysis, as might be expected if charge were delocalized to other atoms of the bridged systems.

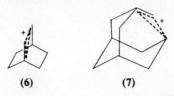

(6)            (7)

Attention may now be directed to the potential energy of the carbonium ion as influenced by geometry. We have argued previously that the preferred configuration of a carbonium ion is that having the substituents disposed in a plane with the maximum possible separation of 120° between them. If the ion is forced to deviate from this optimum geometry, then an increase in its potential energy can be expected; accompanying this increase will be a diminution of stability and a decrease in the rate of any process having a transition state resembling the carbonium ion.

Both Bartlett (2,16,19,64,65) and Doering (59,66,67) arrived at the choice of nonplanarity as the most important single factor in bridgehead carbonium ion chemistry. They also pointed out that it should be possible to estimate the relative reactivities of the various systems by calculating the strain involved in making each of them planar. Crude estimates of the excess energy of the bicyclo[2.2.2]octyl bridgehead cation, e.g., ranged from 6 (59) to 22.5 kcal (2), and a strain of about 12 kcal was calculated for a planar adamantyl cation (51). These estimates were essentially hand computations based on inspection of molecular models. Nonetheless, rate differences between $t$-butyl and polycyclic bridgeheads could be semi-quantitatively matched to increases in angle strain in passing from ground to transition states (1,24,51).

The development of machine computational methods of conformational analysis (68–70) has assisted the more detailed assessment of the role of angle strain. Schleyer and Gleicher (71) have analyzed strain in both the polycyclic hydrocarbons and their derived carbonium ions as the sum of a number of independent terms:

$$E_{\text{total}} = E_{\text{bond length changes}} + E_{\text{angle strain}} + E_{\text{torsional}} + E_{\text{nonbonded interactions}}$$

(15)

The total energy is minimized as a function of the geometry of the total system under consideration. It was found necessary to employ "harder" nonbonded potential functions and a C–C$^+$–C bending force constant four times greater than the usual C–C–C value to obtain internally consistent results.

The calculations yielded values for the strain energy difference between the ground state (the hydrocarbon) and the transition state (the carbonium ion) which correlate well with the observed relative rates of solvolysis (Fig. 2). The average deviation from the straight line of Figure 2 is $10^{\pm 0.8}$,

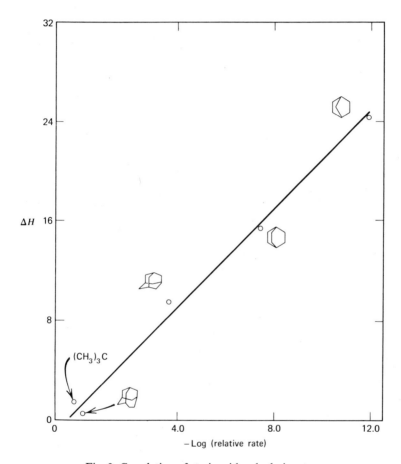

Fig. 2. Correlation of strain with solvolysis rates.

which compares favorably with the $10^{\pm 0.25}$ of the Foote-Schleyer correlation of secondary tosylate solvolysis rates (72,73). (The latter employs far more points.)

In Table II, the changes in each of the strain parameters upon passing from hydrocarbon to ion are presented. Several features deserve comment. The computations again show that the most important effect is the increase

## TABLE II
### Components of Strain Increase,[a] kcal/mole

| System | Bond length strain | Change of angle strain | Nonbonded strain | Torsional strain |
|---|---|---|---|---|
| $t$-Butyl | −0.06 | 0.145 | −0.230 | 3.000 |
| 3-Homoadamantyl | 0.337 | −1.508 | 1.087 | 0.317 |
| 1-Adamantyl | −0.120 | 8.140 | 1.776 | 0.376 |
| 1-Bicyclo[2.2.2]octyl | 0.123 | 8.157 | 4.337 | 0.317 |
| 1-Bicyclo[2.2.1]heptyl | 1.040 | 23.286 | 0.461 | −0.398 |

[a] Strain in carbonium ion–strain in hydrocarbon.

of *angle* strain upon ionization. It is this term which makes the greatest contribution to the overall strain increase. Yet other effects cannot be dismissed. These can be seen by a brief examination of each molecule.

Adamantane has long been considered to be strain free (24). However, recent determinations of the heat of formation of adamantane lead to surprisingly large strain estimates of 6–8 kcal/mole due, primarily, to nonbonded interactions (74). The lessened reactivity relative to $t$-butyl is virtually all the result of angle deformation as the bridgehead flattens.

Homoadamantane is strained both by angle deformations forced by the two-carbon bridge, and by torsional strain in the bridge itself. There is very little change in these interactions upon ionization, and thus the high reactivity of this system results [as previously suggested (1,24)] from the fortuitous near equivalence of two substantial strains.

The bicyclo[2.2.2]octyl bridgehead geometry is calculated to be quite similar to that in adamantane, with the substantial rate difference resulting from a very unfavorable bridgehead–bridgehead (C-1–C-4) nonbonded interaction in the carbonium ion. There is some question, however, about the exact structure of the hydrocarbon and its derivatives. The calculations (71) suggest that the molecule has $D_{3h}$ symmetry—i.e., 0° dihedral angles along the ethano bridges. Although several experimental studies support this conclusion (75–77), several others (78,79) find twisting of 5–10° along these bonds, giving species of $C_3$ or $D_3$ symmetry. It seems unlikely that such twisting would make any substantial difference in the total hydrocarbon strain, since some increase in the C-1–C-4 nonbonded repulsion would be expected (80) to at least partially compensate for the lessened torsional strain.

The structure computed for norbornane is in excellent agreement with experimental determinations (81–85). The molecule is substantially strained, and the increase upon ionization, almost entirely in the angle strain term, is enormous.

On the whole, it would seem that the major features of bridgehead carbonium ion reactivity can be reproduced by considering angle strain and nonbonded repulsions. Other factors, such as solvation, make only small contributions to reactivity differences, probably accounting (51) for no more than a single power of ten out of the $10^{13}$ range of reaction rates.

Several of the systems that appear in Table I and Chart 1 do not fit this simple picture, and these require comment. The bicyclo[1.1.1]pentyl and bicyclo[2.1.1]hexyl systems are far more reactive than is to be expected on the basis of the previous discussion. A large portion of the rate enhancement may be attributed to the relief of strain provided by ring opening. Both compounds give only monocyclic products, equations 16 and 17

$$\text{(16)}$$

$$\text{(17)}$$

(39,40). It has also been suggested that the formation of a cyclobutyl carbonium ion, with charge delocalization, may contribute to the extremely rapid reaction of 1-chlorobicyclo[1.1.1]pentane. Whatever the detailed explanation, it is clear that these are not typical bridgehead systems in carbonium ion processes.

## IV. SUBSTITUENT EFFECTS

It is not the purpose of this chapter to discuss in detail the transmission of electrical effects through bridged molecules—leading references may be found in papers by Stock (86–88), Wilcox (89,90), and Ritchie (91)—but the numerous data now available on substituent effects on bridged cationic reactivity are of some interest. Relevant data are collected in Tables III and IV.

The initial results on the solvolysis of alkyl-substituted bromoadamantanes (92,93) gave the surprising result that methyl groups actually retarded solvolysis. In explanation, it was suggested that depending on the exact hybridization of the carbon to which it was attached, a methyl group might be electron withdrawing relative to hydrogen (1,92). This conception was

## TABLE III[a]
### Solvolysis Rates of Substituted Adamantyl Bromides[b]

| Substituents | $10^5 k$, sec$^{-1}$ | $\dfrac{k_X}{k_H}$ | $\Delta H^{\ddagger}$, kcal | $\Delta S^{\ddagger}$, eu |
|---|---|---|---|---|
| None | 13.8 | 1.00 | 22.9 | −10.8 |
| 3-$t$Bu | 29.0 | 2.10 | 24.5 | −4.7 |
| 3-$i$Pr | 19.3 | 1.40 | 24.5 | −5.5 |
| 3-CH$_3$ | 9.76 | 0.71 | 24.0 | −8.2 |
| 3,5-diCH$_3$ | 6.35 | 0.46 | 24.7 | −9.3 |
| 3,5,7-triCH$_3$ | 4.28 | 0.31 | — | — |
| 3-C$_2$H$_5$ | 13.4 | 0.97 | 23.3 | −9.6 |
| 3-C$_6$H$_5$ | 2.39 | 0.174 | 24.3 | −10.1 |
| COOR | 0.115 | 0.00834 | 25.7 | −12.3 |
| CN | 0.00241 | 0.000175 | 24.7 | −22.5 |
| CO$_2$H | 1.30$^c$ | 0.0165 | — | — |
| Br | 0.240$^c$ | 0.0030 | — | — |

[a] Data from Refs. 45,53,92,93.
[b] In 80% ethanol at 75°.
[c] In 70% dioxan at 100°, Ref. 95.

## TABLE IV[a]
### Solvolysis of 4-Substituted 1-Bicyclo[2.2.2]Octyl Brosylates[b]

| Substituents | $10^5 k$, sec$^{-1}$ | $\dfrac{k_X}{k_H}$ | $\Delta H^{\ddagger}$, kcal | $\Delta S^{\ddagger}$, eu |
|---|---|---|---|---|
| H | 11.3 | 1.00 | 26.2 | −1.7 |
| $t$Bu | 6.24 | 0.55 | 27.1 | −0.3 |
| $i$Pr | 4.75 | 0.43 | 26.8 | −1.6 |
| C$_2$H$_5$ | 4.04 | 0.36 | 26.9 | −1.6 |
| CH$_3$ | 3.38 | 0.30 | 27.4 | −0.6 |
| C$_6$H$_5$ | 0.814 | 0.072 | 27.6 | −2.9 |
| CO$_2$R | 0.0789 | 0.0070 | 27.1 | −8.9 |
| CN | 0.0025 | 0.000221 | 29.1 | −10.1 |

[a] Data from Ref. 45.
[b] In acetic acid at 74°.

later supported by a measurement of the deuterium isotope effect on the dipole moments of some saturated hydrocarbons (94).

The compounds examined by Schleyer and Woodworth (45) were used to test the generality of this behavior. Two facts stand out. First, when substituted into the bicyclo[2.2.2]octyl system, *all* alkyl groups decrease reactivity. This is a significant difference from adamantyl, and it is emphasized if the logs of the relative rates for the two series are plotted against each other. With the point for hydrogen omitted, the data define a straight line of correlation coefficient .999!

Hydrogen is substantially removed from this line. Schleyer and Woodworth interpret this to mean that the difference between hydrogen and the other substituents is not entirely inductive in origin. They argue that replacement of hydrogen by a carbon atom may cause a change in hybridization of the substitution site that would produce a subtle alteration in geometry at the other bridgehead, thus sterically affecting the reaction rate. That is, the alkyl groups *do* have an inductive order of electron release increasing from methyl through *t*-butyl, but the position for hydrogen will be a function of the specific conformational changes produced.

The explanation is an interesting one, but, of course, it requires confirmation. Until some evaluation of the magnitude of such a steric effect can be made, however, it is probably unwise to interpret the rate depressions produced by methyl substitution as *necessarily* requiring that methyl be electron withdrawing relative to hydrogen.

Other substituent effects are unexceptional. It would be useful to make some computations in an attempt to discover the mode of transmission of these polar effects; i.e., inductive versus field. All previous studies having this goal have dealt with processes occurring at a site not part of the ring system, processes that are therefore less sensitive to the influence of the substituents (45). The true cause of the alkyl group effects might be elucidated by such a treatment.

## V. DEAMINATION

The deamination of most bridgehead amines proceeds with considerable facility (2,96–101). Simple reactions are shown in equations 18 and 19. A number of amines react to give rearranged products, several by a Favorskii-like route, as shown in equations 20–23.

$$\text{(structure with C=O and NH}_2\text{)} \xrightarrow[\text{aq AcOH}]{\text{NaNO}_2} \text{(structure with C=O and OH)} + \text{(quinone COOH structure)} \quad \text{Ref. 102} \quad (20)$$

$$\text{(bicyclic NH}_2\text{, O structure)} \xrightarrow[\text{aq AcOH}]{\text{NaNO}_2} \text{(COOH bicyclic structure)} \quad \text{Refs. 103–105} \quad (21)$$

$$\text{(NH}_2\text{, OH structure)} \xrightarrow{\text{HNO}_2} \text{(CHO structure)} \quad \text{Ref. 106} \quad (22)$$

$$\text{(NH}_2\text{ bicyclic)} \xrightarrow{\text{HNO}_2} \text{mixture of olefins (unidentified)} \quad \text{Ref. 40} \quad (23)$$

The ease of deamination of apocamphyl amine (eq. 18) led Bartlett and Knox (2) to wonder whether an $S_Ni$ process (107–109) might be occurring. Although such a mechanism might be invoked to explain the stereochemistry of deamination in some alicyclic systems, it is not really needed to understand the facile reaction of bridgehead amines. The high stability of the nitrogen molecule formed in the decomposition of a diazonium ion reduces the activation energy for this decomposition to the order of 3–5 kcal (110,111), thus allowing the generation of relatively high energy carbonium ions. This interpretation is supported by several studies.

White and co-workers (112–114) have examined the decomposition of $N$-nitrosoamides, which also involves a diazonium ion. They found in one case that an oxygen label was completely scrambled (eq. 24), a result incompatible with $S_Ni$ substitution. In another example (eq. 25) they were able to intercept the bridgehead carbonium ion—one more result arguing against an $S_Ni$ process. This latter capture has also been obtained in the deamination of apocamphylamine in chlorobenzene solution, equation 26 (115).

On the other hand, there is also evidence that the diazonium ion has a longer than normal lifetime when an unstable bridgehead cation would be formed upon decomposition. White et al. (113) attributed the completeness of scrambling in the reaction of equation 24 to this source. In one case,

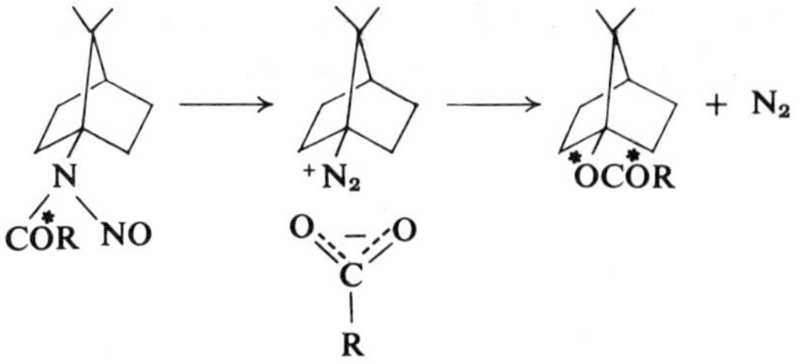


(24)

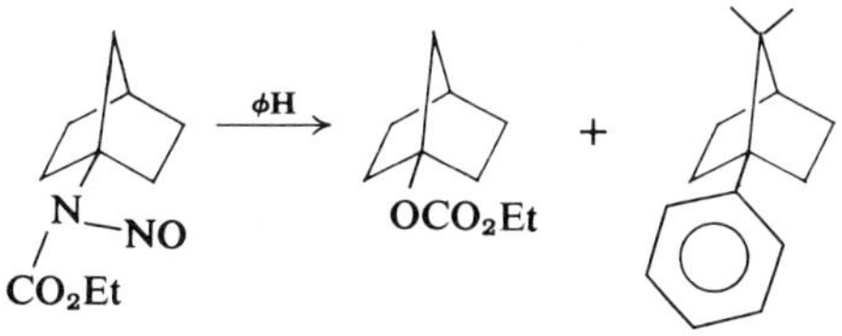


(25)

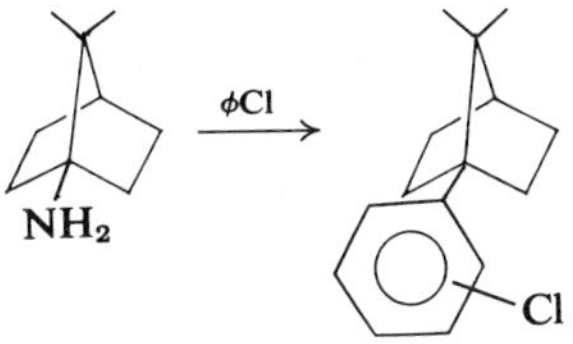


(26)

where $o = 40\%$
$m = 32\%$
$p = 28\%$

it was possible to trap the diazonium ion before decomposition by coupling with 2-naphthol (eq. 27) (116).

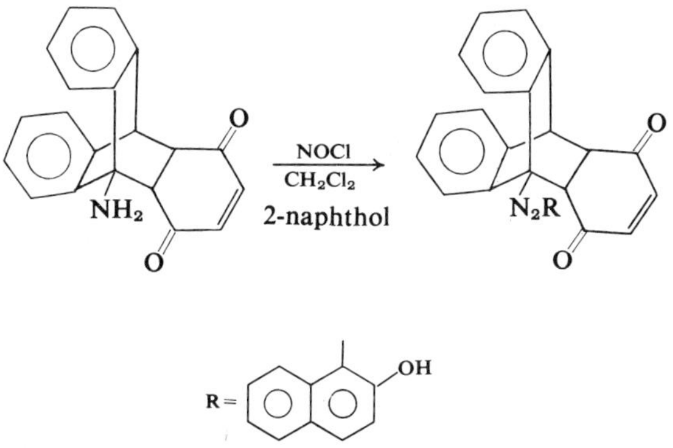


(27)

Finally, one system has been studied (117) in which the bridgehead carbonium ion is so unstable (apparently because of electron withdrawal by chlorine) that deamination proceeds by a *radical* pathway (eq. 28). In this example, it is well to recognize that *electrophilic* attack on nitrobenzene normally gives greater than 90% *meta*-product.

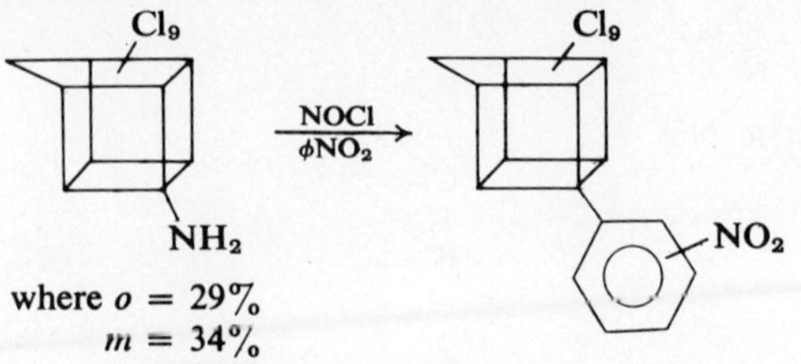

$$\underset{\substack{\text{NH}_2 \\ \text{where } o = 29\% \\ m = 34\% \\ p = 37\%}}{\text{Cl}_9} \xrightarrow[\phi\text{NO}_2]{\text{NOCl}} \underset{\text{NO}_2}{\text{Cl}_9} \qquad (28)$$

## VI. EXCHANGE PROCESSES

We use the term exchange processes to designate a variety of non-solvolytic processes in which one bridgehead substituent (including hydrogen) is replaced by another via an intermediate carbonium ion.

A number of bridgehead cations have been prepared (118–120) in the "magic acid" system of Olah (10) (eq. 29–32). Of particular interest is the

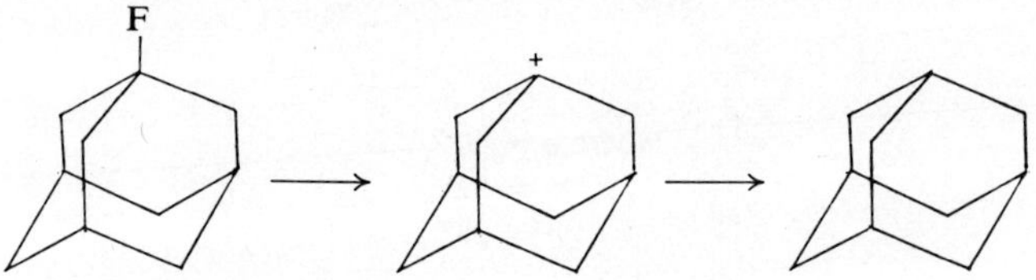

(29)

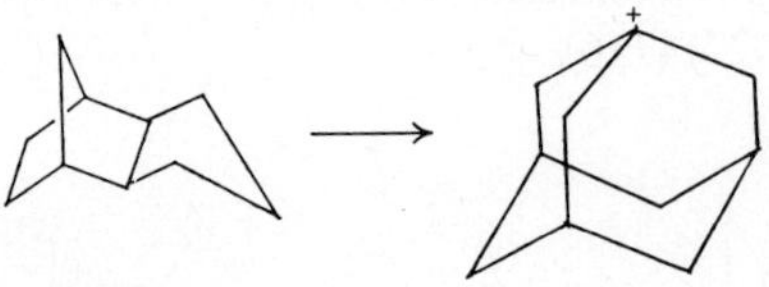

(30)

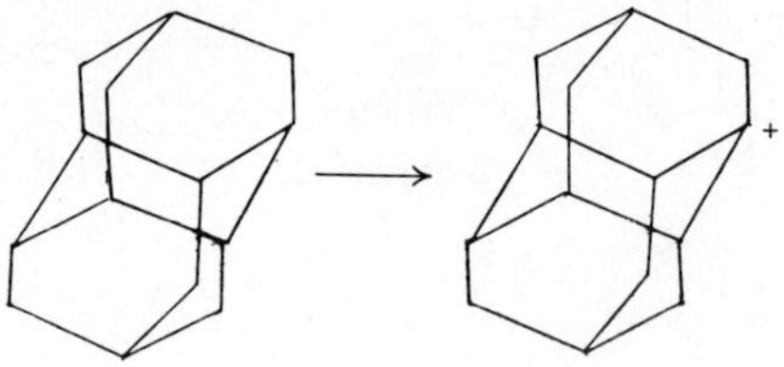

(31)

$$\xrightarrow{-60°} \xrightarrow{-10°} \qquad (32)$$

transformation of equation 30. This rearrangement, when the usual aluminum chloride or aluminum bromide catalyst systems are used, is considered to proceed by repeated formation and quenching of carbonium ions (24,121). In $SbF_5$–$FSO_3H$ solution, such hydride transfers cannot occur (10); the reaction must therefore involve a substantial number of either intermolecular hydride transfers, or intramolecular hydride migrations. A considerable driving force for the formation of the adamantane skeleton is apparent.

Several exchange processes are of considerable utility in introducing functional groups into polycyclics. Aluminum chloride catalyzed hydride–halide exchange (122) is such a reaction. Applied to norbornane (123), it gives only 2-*exo*-chloronorbornane (eq. 33). Adamantyl chlorides can be made in good yields (eq. 34) (53,124).

$$\text{(33)}$$

$$\text{(34)}$$

R = H, $CH_3$

Ionic bromination is a related reaction that probably proceeds by the chain mechanism of equations 35 and also gives excellent yields of bridgehead adamantane derivatives (125). The uncatalyzed reaction results in the introduction of a single bromine; Lewis acids promote more extensive

$$Br_2 \longrightarrow Br^+ + Br^- \qquad (35a)$$
$$Br^+ + RH \longrightarrow BrH + R^+ \qquad (35b)$$
$$R^+ + Br_2 \longrightarrow RBr + Br^+ \qquad (35c)$$

halogenation (eq. 36). Other polycycles that have been functionalized by this route include congressane (eq. 37) (55), noradamantane (eq. 38) (126), bicyclo[3.3.1]nonane (49), and protoadamantane (126a). Considerable selectivity is observed, for the relative proportions of products are determined by the ease of formation of the various carbonium ions.

$$\text{(36)}$$

$$\text{(37)}$$

$$\text{(adamantane)} \xrightarrow{Br_2} \text{(1-bromoadamantane)} \qquad (38)$$

The Koch carboxylation (127) is a reaction in which a carbonium ion is captured by carbon monoxide to give an acylium ion, which is then hydrated to the carboxylic acid (eqs. 39–41). Adamantane (128) and alkyl adamantanes (129) give high yields of bridgehead carboxylic acids (eq. 42).

$$HCO_2H + H_2SO_4 \longrightarrow CO + H_3O^+ + HSO_4^- \qquad (39)$$

$$t\text{-BuOH} + H_2SO_4 \longrightarrow t\text{-Bu}^+ + H_3O^+ + HSO_4^- \qquad (40a)$$

$$t\text{-Bu}^+ + RH \longrightarrow R^+ + t\text{-BuH} \qquad (40b)$$

$$R^+ + CO \longrightarrow RCO^+ \longrightarrow RCO_2H + H^+ \qquad (41)$$

1-Adamantanol or 1-adamantyl bromide may be used as the carbonium ion source, in place of the $t$-butanol. The product may be either that of

$$\xrightarrow{t\text{BuOH, HCOOH}}_{H_2SO_4} \qquad (42)$$

$$R = H, CH_3, C_2H_5$$

kinetic control (most stable carbonium ion) or that of thermodynamic control (most stable product) depending on the molecule being carboxylated and the concentration of the sulfuric acid. Thus carboxylation of isopropyl adamantane or the related alcohol (eq. 43) leads to the more

$$\qquad (43)$$

stable adamantane acid (53,227). On the other hand, carboxylation of the related primary alcohol, hydroxymethyl adamantane, under these same conditions produces the much less stable homoadamantane carboxylic acid

$$\qquad (44)$$

(eq. 44) (24,71). The former reaction may well involve a ring opening and reclosure, as shown in equation 45. This π route to a bridgehead carbonium

$$\text{(45)}$$

ion has been independently observed by several groups. Methylenebicyclo[3.3.1]nonanone may be converted (eq. 46) to substituted adamantanols (130–132), as may the corresponding bis-(methylene) compound

$$\text{(46)}$$

X = OH, Cl, OR

(eq. 47) (133). Formation of other bridgehead ions by this kind of pathway has not been observed, but it would be of considerable interest in itself.

$$\text{(47)}$$

It is possible also to introduce amino functions directly by a variation of the Koch reaction, the Ritter amidation. The sole difference is the use of a nitrile, providing the isoelectronic –CN group to act as a nucleophile in place of carbon monoxide (134). Two specific cases deserve special mention. Adamantane carboxylic acid is converted into formamidoadamantane by the action of sodium cyanide in 96% sulfuric acid (!) (135), illustrating again the reversibility of this kind of reaction (eq. 48). For a

$$\text{(48)}$$

reaction analogous to that noted previously for isopropyladamantane, a DuPont patent (136) claims the preparation of the side-chain reaction product (eq. 49). The reason for the difference in behavior is not clear.

$$\text{HO-Ad} \xrightarrow[\text{H}_2\text{SO}_4]{\text{CH}_3\text{CN}} \text{CH}_3\text{COHN-Ad} \tag{49}$$

Another direct introduction of an amine function involves treatment of a hydrocarbon with a mixture of aluminum chloride and nitrogen trichloride; quantitative yields are reported (eqs. 50 and 51) (137). Isomerization during the reaction is facile.

$$\text{(adamantane)} \longrightarrow \text{1-NH}_2\text{-adamantane} \longleftarrow \text{(endo-tetrahydrodicyclopentadiene)} \tag{50}$$

$$\text{(adamantane)} \longrightarrow \text{2-NH}_2\text{-adamantane} \longleftarrow \text{(perhydroacenaphthylene)} \tag{51}$$

The use of aluminum chloride as a catalyst allows a novel introduction of chlorine and sulfur functions into adamantane (eqs. 52 and 53) (138); as noted so many times, these reactions have not been tried on other polycycles.

$$\text{adamantane} \xrightarrow[\text{CCl}_4]{\text{AlCl}_3} \text{1-Cl-adamantane} + \text{1,3-Cl}_2\text{-adamantane} \tag{52}$$

$$\qquad\qquad\qquad 13\% \qquad\qquad 71\%$$

$$\text{adamantane} \xrightarrow[\text{SOCl}_2]{\text{AlCl}_3} \text{1-SOCl-adamantane} \tag{53}$$

A number of unusual reactions, involving exchange, disproportionation, and overall oxidation have been described and are reported here. The first observation (139) of this sort was that 1-adamantanol is converted by concentrated sulfuric acid to a mixture of adamantane and adamantanone (eq. 54). Further work showed that adamantane also may be converted to

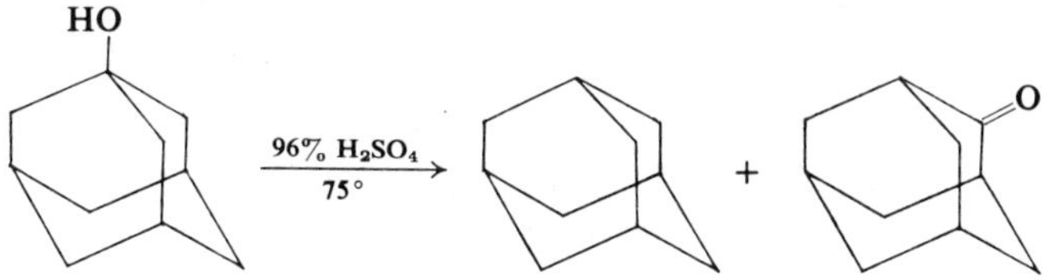

$$\text{1-adamantanol} \xrightarrow[75°]{96\% \text{ H}_2\text{SO}_4} \text{adamantane} + \text{adamantanone} \quad (54)$$

adamantanone by the same reagent. A plausible scheme for this process is presented in Chart 4. In the reaction with adamantane itself, sulfuric acid

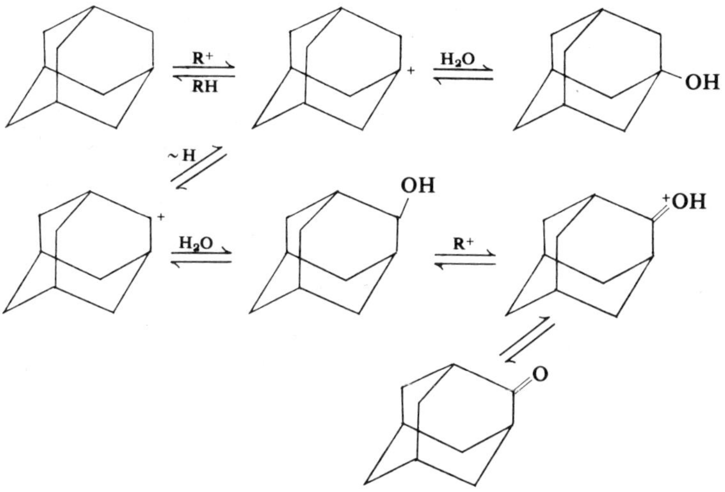

Chart 4.

acts as an oxidizing agent to form the first adamantyl cation; $SO_2$ is evolved copiously. Yields might be enhanced by addition of a carrier species (e.g., an alcohol) to increase carbonium ion concentration. Other variations of this reaction are shown in equations 55–58 (141,142).

In several cases, substantial amounts of rearrangement are observed (eqs. 59–61) (143,144).

All these products can be rationalized by schemes similar to that in Chart 4; however, the driving force for formation of many of the compounds is not apparent. In a 2-substituted adamantane, the substituent suffers a pair of gauche butane-type interactions, and such compounds are therefore less stable than the 1-isomers. Although the foregoing reactions

are carried out under conditions that permit equilibration, 2-substituted compounds predominate. It would appear necessary to suggest that equilibrium has not yet been reached.

## VII. BRIDGEHEAD IONS IN REARRANGEMENT REACTIONS

Despite the strain inherent in most bridgehead carbonium ions, they are formed readily by neopentyl-type rearrangements. The instability of primary carbonium ions and the possibility of some strain *relief* provide the driving force. Some of the ions produced in this fashion are shown in the following equations: 1-norbornyl (40,145,146), 1-bicyclo[2.2.2]octyl (40,46,96,147,148,149), 1-bicyclo[3.2.1]octyl (40,46,148), 1-bicyclo[3.2.2]-nonyl (149–151), 1-adamantyl (120,137,152–155), 3-homoadamantyl (24,71,156,157), and the bicyclo[4.3.1]- and bicyclo[3.3.2]decyl ions (158).

|  |  |  |
|---|---|---|
| ![structure]  CH₂OTs → AcOH → [OAc structure] | Ref. 147 | (66) |
| [OH structure] → HBr → [Br structure] + [Br structure] | Ref. 148 | (67) |
| CHNNHTs → CH₃ONa / N-methylpyrrolidinone → CH₂OH + OH | Ref. 149 | (68) |
| CH₂OTs → aq dioxan → HO-[bicyclic] | Ref. 151 | (69) |
| CH₂OTs (adamantane) → AcOH → OAc (adamantane) | Ref. 152 | (70) |
| CH₂OH / CH₂OH → H⁺ → HO / OH | Ref. 154 | (71) |
| OH → 1. SbF₅/FSO₂OH 2. quench → OH (adamantane) | Ref. 155 | (72) |
| CH₂OH → H⁺ → [cation+] → CH₂OH (adamantane) | Ref. 157 | (73) |

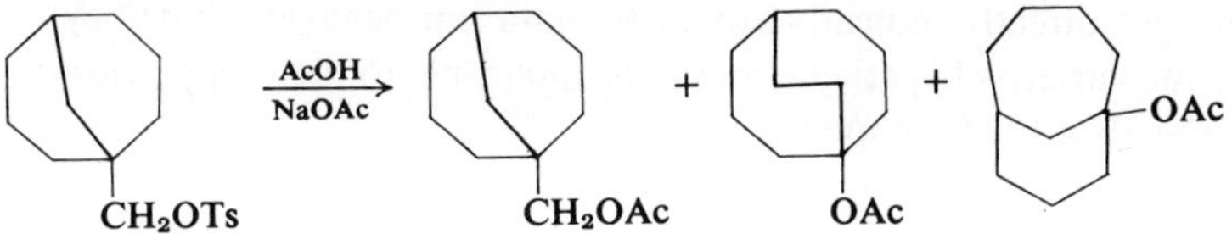

[structure: bicyclic CH₂OTs] →(AcOH, NaOAc) [bicyclic CH₂OAc] + [structure OAc] + [structure-OAc]  Ref. 158  (74)

A quinuclidine derivative also gives this type of rearrangement, but a fragmentation follows (eq. 75) (151).

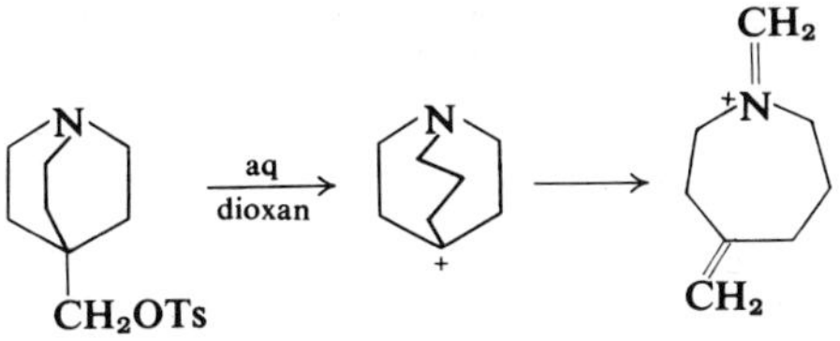

[quinuclidine-CH₂OTs] →(aq dioxan) [intermediate] → [ring-opened product with =CH₂ and +N=CH₂]  Ref. 151  (75)

Rates of solvolysis of these neopentyl-like structures are often greater than that of neopentyl itself (Chart 5); this *may* reflect some degree of

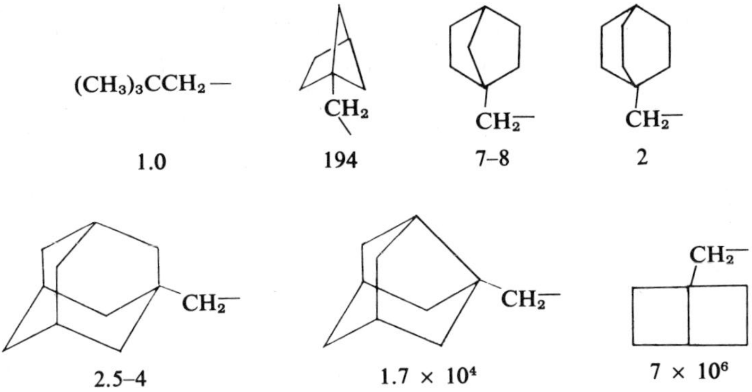

$(CH_3)_3CCH_2-$     1.0

[bicyclic CH₂]       194

[norbornyl CH₂⁻]     7–8

[norbornyl CH₂⁻]     2

[adamantyl CH₂⁻]     2.5–4

[adamantyl CH₂⁻]     $1.7 \times 10^4$

[cubyl CH₂⁻]         $7 \times 10^6$

Chart 5. Relative Solvolysis Rates of Neopentyl Structures Producing Bridgehead Carbonium Ions.

neighboring carbon participation; however, the point is beyond the scope of this chapter. We would simply point out that the reactions proceed easily.

Bridgeheads behave as ordinary alkyl groups in those rearrangements involving migration to an electron-deficient atom, since no substantial amount of positive charge is developed at the migrating carbon. Thus the Hofmann reaction has been successfully applied to amides **8–13** among others, and the Curtius or Schmidt reaction to the acids **14–22**. Other examples of such rearrangements appear in equations 76–79. These reactions are not directly indicative of carbonium ion behavior, but they do illustrate the variety of synthetic methods applicable to the preparation of bridgehead-substituted compounds.

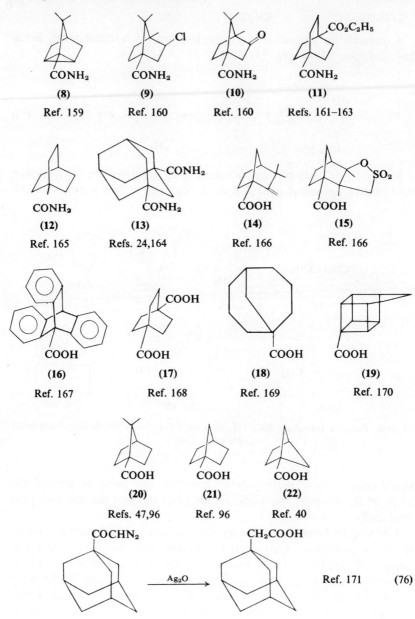

(76)

[Structure] —PCl₅→ [Structure]  Ref. 165  (77)
N–CH₃, HO                NHCOCH₃

[Structure with C=O–Ph] —CF₃CO₃H→ [Structure with OC(=O)Ph]  Ref. 172  (78)

[Structure]–B(OH)₃ —OOH⁻→ [Structure]–OB(OH)₂  Ref. 173  (79)

## VIII. FRAGMENTATION

The fragmentation upon solvolysis of various acyclic compounds containing a γ-nitrogen has been thoroughly characterized; the term "frangomeric effect" has been coined to describe the accompanying rate enhancement (174,175). Several bridgehead derivatives have been employed in demonstrating the stereoelectronic requirements of the process. For example, the quinuclidine **23** (eq. 80) is observed to solvolyze some four

[Structure of quinuclidine with Br → ring-opened iminium with =CH₂ groups]  (80)

(**23**)

powers of ten more rapidly than its carbocyclic analog (176). The acceleration results from participation of the nitrogen lone pair, which is held by the bicyclic system very nearly collinear with the carbon–bromine bond.

1-Bromo-3-aminoadamantanes behave similarly (eq. 81), and investigators observed (177) rate enhancements between 30 and 500—large values relative to the corresponding alkyl bromoadamantanes. Again, the molecular skeleton provides the proper geometry for maximum participa-

[Equation (81): adamantane derivative with Br and NR₂ → bicyclic ketone with =CH₂ + R₂NH, aq EtOH/OH⁻]

tion of the nitrogen lone pair. One nonbridged example will serve to reinforce this point. The *cis*-amino tosylates **24** and **25** solvolyze more rapidly than 3-alkyl cyclohexyl tosylates and yield fragmentation products; the *trans* isomers **26** and **27** react more slowly than the models, and do not fragment (178).

**(24)** TsO, NH₂ cyclohexane
**(25)** TsO, N(CH₃)₂ cyclohexane
**(26)** NH₂, OTs cyclohexane
**(27)** N(CH₃)₂, OTs cyclohexane

Other fragmentations are displayed in equations 82 and 83. The latter reaction may be reversed by addition of bromine to the diene.

[Equation (82): adamantane with NHCOCH₃ and Br, NaOH → bicyclic enone with =CH₂. Refs. 179, 180]

[Equation (83): adamantane with CH₂Br and Br, Zn/DMF → bicyclic diene with two =CH₂ groups. Ref. 181]

## IX. BRIDGEHEAD CARBONIUM IONS IN CONDENSED RING SYSTEMS

Until recently, only qualitative information was available regarding bridgehead carbonium ion reactivity in condensed rings such as decalin and hydrindan. These investigations suggested that the ions were formed as easily as any other 3° carbonium ion. The reactions of equations 84–93 are representative of these studies.

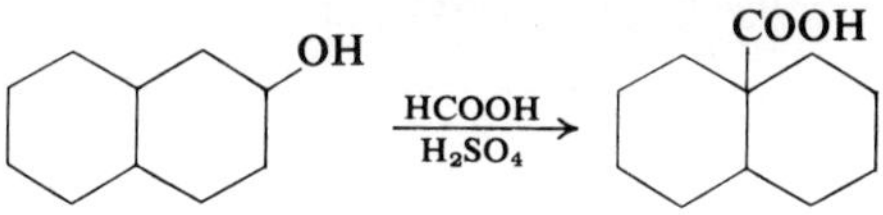

(84) Refs. 182,183

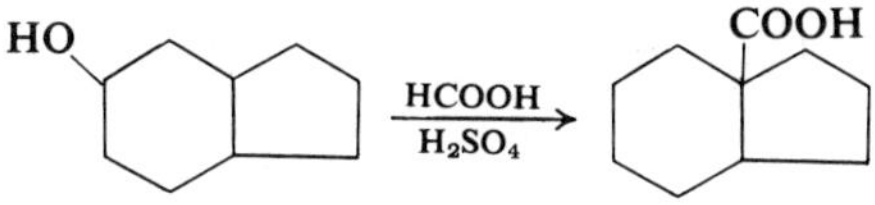

(85) 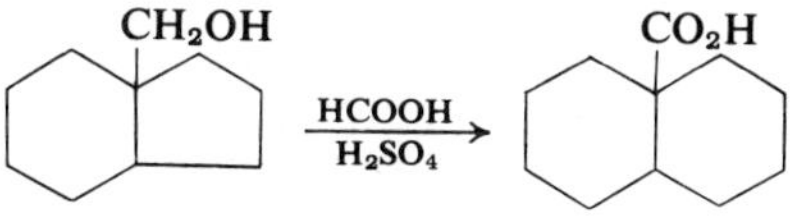 Ref. 184

(86) 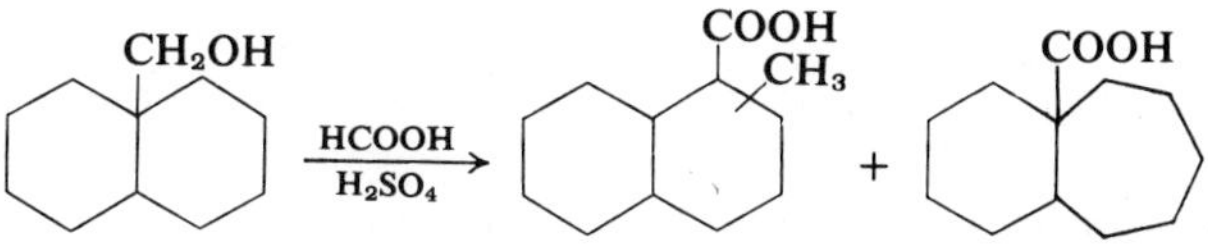 Ref. 184

(87) 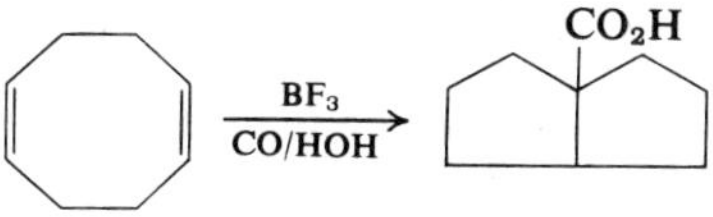 Ref. 184

(88) 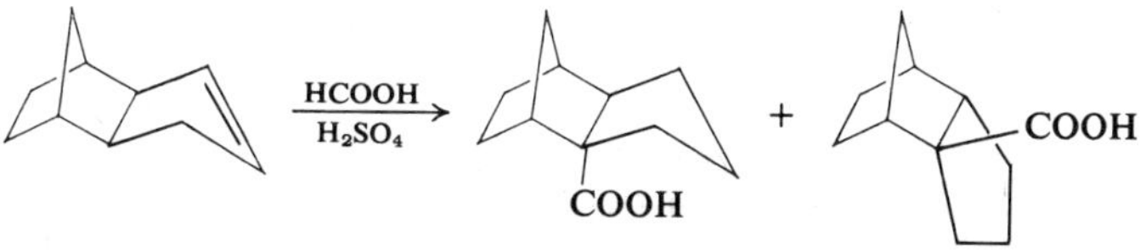 Ref. 185

(89) Ref. 185

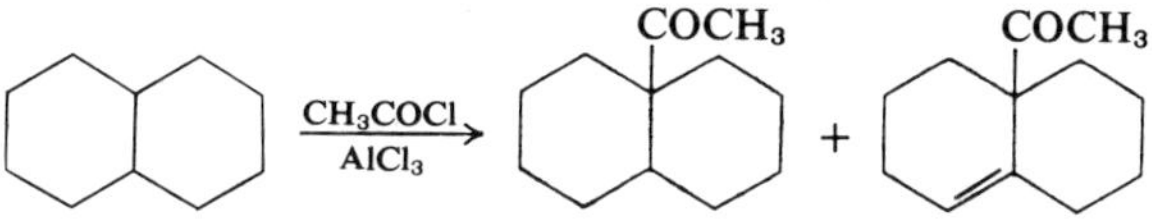

(90) Refs. 187–190

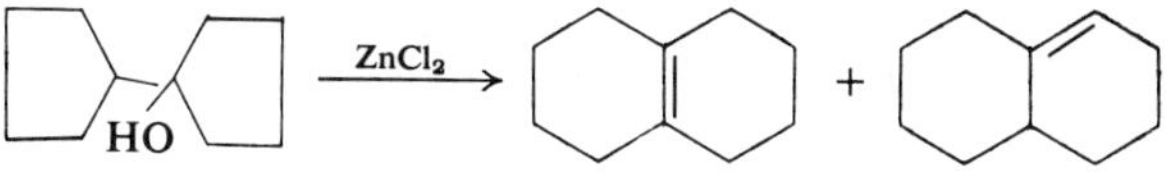

(91) Ref. 191

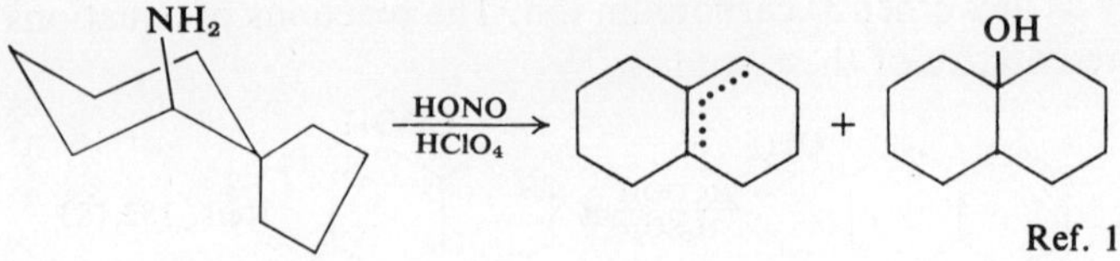

Ref. 192 (92)

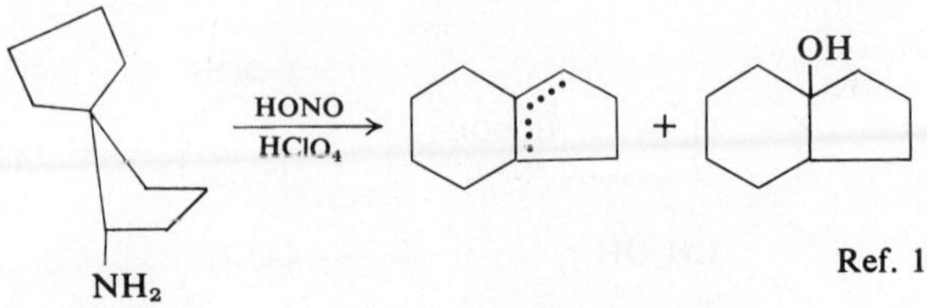

Ref. 193 (93)

Clearly, none of these observations allows any precise conclusions about the stability of the bridgehead ion. Additionally, another feature, the stereochemical course of the reactions, is obscured because virtually all were carried out under conditions that could equilibrate the products. A single rate constant (eq. 94) is of dubious value, since the stereochemistry of the material solvolyzed was not determined (194).

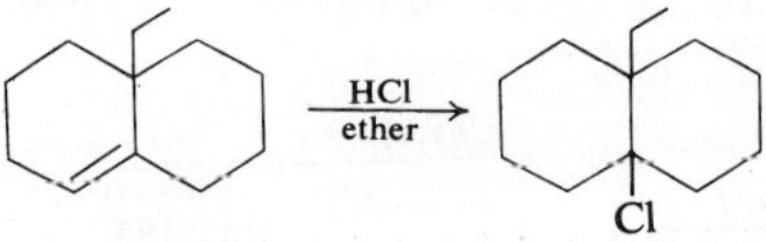

solvolyzes 5 times faster than $t$-butyl (94)

Table V presents rate data from two recent studies, providing an interesting contrast with the behavior of bridged systems (195–197). In Chart 6 these data are compared with the rates of some model systems from Table VI. The condensed ring derivatives are seen to be quite reactive, in contrast to most of the bridged molecules previously discussed. Although this reactivity could be the result of a fortuitous similarity of strain in the covalent ground state and the carbonium ion (24,71), inspection of models or crude strain calculations suggests that a trigonal atom can be introduced at the bridgehead with minimal strain.

The product distributions form the most interesting part of these studies, however. As Tables VII–X reveal, there are sharp differences in product composition as a function of substrate stereochemistry. The carbonium ion intermediates formed from the stereoisomers of a given system therefore cannot be the same. This phenomenon obviously has no counterpart in bridged systems, where the molecular shape is fixed.

The difference in the ions could be simply the location of the gegenion in a tight ion pair. However, the similarity of data for chlorides and $p$-nitrobenzoates in different solvents militates against this explanation, as

## TABLE V
### Solvolysis of Bridgehead Derivatives of Condensed Rings

| System | Leaving group | Solvent | Buffer | $T$, °C | $k$, $\times 10^5$ sec$^{-1}$ | $\Delta H^{\ddagger}$, kcal | $\Delta S^{\ddagger}$, eu |
|---|---|---|---|---|---|---|---|
| cis-9-Decalyl | OPNB[a] | 60% acetone | none | 100 | 2.21 | 27.6 | −11.0 |
| | OPNB | 60% acetone | 2 eq NaOAc | 100 | 2.22 | 29.5 | −1.5 |
| | Cl | 80% EtOH | Et$_3$N | 70.0 | 76.6 | — | — |
| trans-9-Decalyl | OPNB | 60% acetone | none | 100 | 0.690 | 28.1 | −9.0 |
| | OPNB | 60% acetone | 2 eq NaOAc | 100 | 0.705 | 31.5 | 1.9 |
| | Cl | 80% EtOH | Et$_3$N | 70.0 | 56.8 | — | — |
| cis-8-Hydrindanyl | OPNB | 60% acetone | none | 80 | 2.11 | 27.0 | −3.7 |
| | OPNB | 60% acetone | 2 eq NaOAc | 80 | 2.22 | 27.06 | −3.5 |
| trans-8-Hydrindanyl | OPNB | 60% acetone | none | 80 | 7.93 | 26.4 | −2.9 |
| | OPNB | 60% acetone | 2 eq NaOAc | 80 | 8.59 | 27.79 | 1.2 |
| cis-1-Bicyclo[3.3.0]octyl | OPNB | 60% acetone | none | 100 | 0.0086 | 29.9 | −2.1 |
| | OPNB | 60% acetone | 2 eq NaOAc | — | — | — | — |
| trans-1-Bicyclo[3.3.0]octyl | OPNB | 60% acetone | none | 60 | 18.7 | 23.1 | −2.1 |
| | OPNB | 60% acetone | 2 eq NaOAc | — | — | — | — |

[a] OPNB = $p$-O$_2$NC$_6$H$_4$CO$_2^-$.

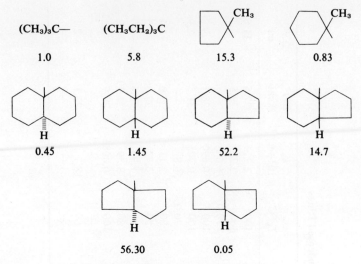

Chart 6. Relative Solvolysis Rates.[a]

[a] For *p*-nitrobenzoates in 60% acetone at 100°.

## TABLE VI
### Solvolysis of Some Model *p*-Nitrobenzoates in 60% Acetone

| *p*-Nitrobenzoate | $T$, °C | $k \times 10^6$, sec$^{-1}$ | $\Delta H^{\ddagger}$, kcal | $\Delta S^{\ddagger}$, eu |
|---|---|---|---|---|
| $(CH_3)_3C$ | 100 | 18.5 | 28.0 | −5.7 |
| $(C_2H_5)_3C$ | 100 | 99.1 | 27.3 | −4.0 |
| cyclopentyl-CH$_3$ | 80 | 23.1 | 25.9 | −6.7 |
| cyclopentyl-nC$_4$H$_9$ | 80 | 42.2 | 25.7 | −6.0 |
| cyclohexyl-CH$_3$ | 100 | 15.3 | 29.5 | −2.0 |

does the formation of both *cis* and *trans* alcohols from a single substrate. It appears more reasonable to attribute the results to the intervention of conformationally distinct carbonium ions.

This conclusion is further reinforced by a direct trapping experiment (197), Table XI, in which triethylsilane is employed as the trapping agent (198). The carbonium ions are captured by hydride transfer.

## TABLE VII[a]
### Solvolysis Products of Decalyl Chlorides[b]

| | Δ9,10 | Δ1,9 | cis-OH | trans-OH |
|---|---|---|---|---|
| cis-Cl | 42 | 51 | 5 | 2 |
| trans-Cl | 78 | 22 | — | — |

[a] Data from Ref. 195.
[b] In 80% ethanol containing 1.2 eq triethylamine.

## TABLE VIII[a]
### Solvolysis Products of Decalyl p-Nitrobenzoates[b]

| | Δ9,10 | Δ1,9 | cis-OH | trans-OH |
|---|---|---|---|---|
| cis-OPNB | 45.0 | 48.0 | 4.0 | 2.0 |
| trans-OPNB | 39.0 | 59.0 | 2.0 | 0.0 |

[a] Data from Ref. 197.
[b] In 60% acetone containing 2 eq sodium acetate.

## TABLE IX[a]
### Solvolysis Products of Hydrindanyl p-Nitrobenzoates[b]

| | Δ | Δ | cis-OH | trans-OH |
|---|---|---|---|---|
| cis-OPNB | 60.7 | 30.4 | 3.5 | 5.4 |
| trans-OPNB | 21.0 | 76.0 | 1.0 | 2.0 |

[a] Data from Ref. 197.
[b] In 60% acetone containing 2 eq sodium acetate.

## TABLE X[a]
### Solvolysis Products of Perhydropentalenyl p-Nitrobenzoates[b]

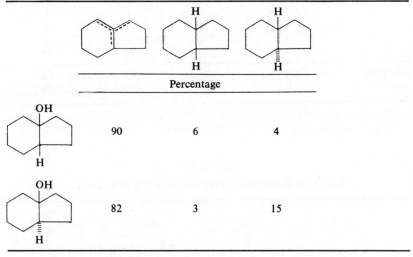

| | | | | |
|---|---|---|---|---|
| cis-OPNB | 88 | 7.0 | 0 | 5.0 |
| trans-OPBN | 14 | 83 | 0 | 3.0 |

[a] Data from Ref. 197.
[b] In 60% acetone containing 2 eq sodium acetate.

## TABLE XI
### Trapping of the 8-Hydrindanyl Carbonium Ion [a]

| | Percentage | | |
|---|---|---|---|
| OH (H axial) | 90 | 6 | 4 |
| OH (H equatorial) | 82 | 3 | 15 |

[a] Reaction with $(CH_3CH_2)_3SiH/CF_3CO_2H/CH_2Cl_2$ at room temperature, method of Ref. 198.

From the data so far available, it is not possible to make any definitive statements about the conformations of the species involved or the barrier to interconversion, although it must be substantial. Examination of models, a less-than-precise experimental method, would indicate that in the decalyl cations, e.g., one of the rings adopts a twist boat conformation. Clearly, most exciting work remains to be done in this new area.

## X. MISCELLANEOUS

The reaction of thionyl halides with alcohols has been shown to proceed by an ion-pair mechanism under some conditions (eq. 95) (199). As we

$$ROH + SOX_2 \longrightarrow ROSOX \longrightarrow [R^+X^-] + SO_2 \qquad (95)$$

might expect, bridgehead alcohols (with the exception of bicyclo[2.2.1]-heptane derivatives) react readily with the reagent; some examples are given in equations 96–100.

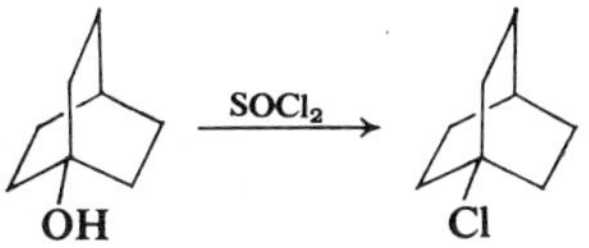

Ref. 150 (96)

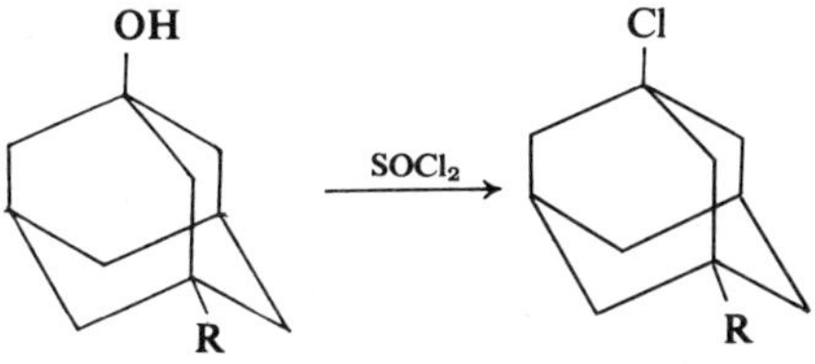

Ref. 24 (97)

R = H, CH₃

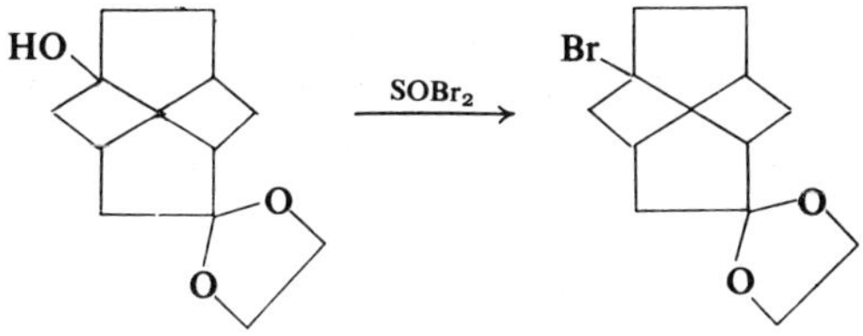

Refs. 200, 20 (98)

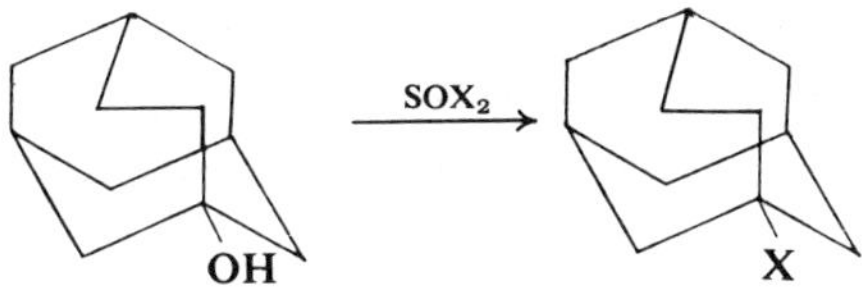

X = Cl, Br

Ref. 202 (99)

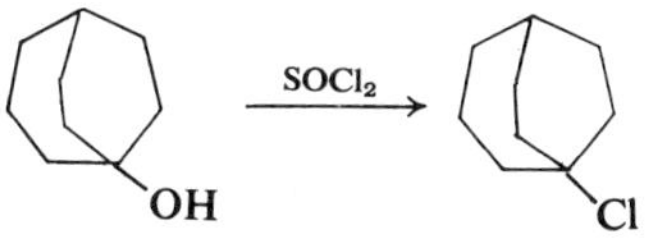

Ref. 150 (100)

Bridgehead-substituted molecules have frequently been used to test for the intermediacy of carbonium ions in organic reactions. Thus it has been suggested (203) that the failure of 1-norbornyl hypochlorite (**28**) to yield

1-norbornyl chloride upon treatment with triphenyl phosphine (eq. 101) implies that **29** is an intermediate. In a nonbridgehead system, a species like **29** would give chloride by either an $S_N1$ or $S_N2$ process; however, both are forbidden to **29**.

$$\underset{\substack{\text{OCl} \\ \textbf{(28)}}}{\diamond} \xrightarrow{\phi_3 P} \underset{\substack{^+OP\phi_3 \\ Cl^- \\ \textbf{(29)}}}{\diamond} \not\longrightarrow \underset{Cl^-}{\diamond^+} + \phi_3 PO \qquad (101)$$

One of the pieces of evidence against the involvement of cations in Raney nickel desulfurization (204) is the observation made by van Tamelen and Grant (205) and presented in eq. 102. Both phenyl apocamphyl sulfide and the derived sulfone are converted readily to hydrocarbon.

$$\underset{S\phi}{\diamond} \xrightarrow{\text{Ni}} \diamond \xleftarrow{\text{Ni}} \underset{SO_2\phi}{\diamond} \qquad (102)$$

In agreement with the supposition that the chromic acid cleavage of secondary benzylic alcohols (eq. 103) involves the carbonium ion $R^+$, the yields of cleavage products are found to decrease as R is changed from *t*-butyl (70%) to 1-adamantyl (30%) to apocamphyl (0%) (206).

$$\text{Ph-CH(OH)-R} \xrightarrow{CrO_3} \text{Ph-CHO} + ROH \qquad (103)$$

It has been suggested (207) that the rapid racemization of sulfonium salt **30** by acetic acid (it racemizes as easily as the *t*-butyl analog) means that C—S bond cleavage to give a carbonium ion is not important in the racemization mechanism.

$$\text{Adamantyl-}\overset{+}{S}(C_2H_5)(CH_3)\ ClO_4^-$$

**(30)**

After examining the half-wave potentials ($E_{1/2}$, generally taken to measure the ease of reduction) for the polarographic reduction of a number of alkyl, cycloalkyl, and bridged bromides, Lambert and co-workers

(208,209) concluded that both $S_N2$-like and $S_N1$-like mechanisms were involved. They suggested that the halide "backed up" to the dropping mercury electrode, with the positive end of the carbon–bromine dipole approaching as closely as possible. If the halide were relatively unhindered, an electron would be transferred in an $S_N2$-like process, producing R· and Br⁻. A more hindered bromide would, at a higher potential, be ionized by the field of the electrode to R⁺ and Br⁻. Further products would then result from reduction of R⁺. In the view of Lambert, the high reduction potentials (in DMF versus the calomel electrode) of bridgehead bromides, which fell in the order of their solvolytic reactivities, supported this mechanism.

Sease and co-workers (210), who examined many of the same compounds (in $CH_3CN$ versus a silver–silver bromide electrode) argue that the process should be represented as a one-electron displacement *on bromine*, yielding R· and Br⁻. The relative inertness of the bridgehead systems is attributed to strain in nonplanar radicals.

The latter view is supported by Sicher (211,212), who has pointed out that the $E_{1/2}$ values for a series of monocyclic bromides bears no relation to either their $S_N1$ reactivity (solvolysis in aqueous dioxan) or to their $S_N2$ reactivity (lithium iodide in acetone). The difficulty in using bridgehead reactivity as a probe in this case is an obvious one: bridgehead free radical stability, once it is corrected for inductive effects, parallels the carbonium ion stability (213). Thus the same behavior should be expected for the systems studied, regardless of the mechanism.

The 1-adamantyl carbonium ion has been prepared by decarbonylation. In one example it was trapped by reaction with benzene (eq. 104); other bicyclic compounds gave no such products (214). In the other (eq. 105)

it was directly observed (215) by pmr in "magic acid" solution. The acylium salt intermediate could also be detected at low temperatures.

The adamantyl cation, regardless of the exact method of its formation, possesses an interesting pmr spectrum (216). The chemical shift of the γ protons, two carbons removed from the positive charge, is greater than that of the β protons, which are adjacent to the charge. This anomalous shift was originally (24,216) explained by a "cage effect," an overlap of the empty $p$ orbital with the rear lobes of the carbon–hydrogen bonds

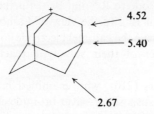

inside the molecular cavity. In view of the lack of supporting evidence for such a phenomenon it also may be attributed to the anisotrophy of the $p$ orbital, which is rigidly held in the planes of the bridgehead carbon–hydrogen bonds, or to "through-bond" transmission.

## XI. SUMMARY

The promise of the original predictions of Bartlett and Knox has clearly been fulfilled in the contribution of studies of bridgehead reactivity to our general knowledge of carbonium ion behavior. In particular, the strongest experimental evidence for our picture of carbonium ions as planar species derives from this source. The extension of the ideas developed to other molecules, and to other reactive intermediates, should provide for much exciting labor.

## XII. ACKNOWLEDGMENTS

The author wishes to express his thanks to the Petroleum Research Fund and the National Science Foundation, which provided support for the work of his group reported herein, and to Kent State University for a Faculty Research Fellowship during the tenure of which a major portion of the manuscript was prepared.

## XIII. SUPPLEMENT

Because of the long interval between the completion of this manuscript and its publication, an attempt is made here to note some of the more recent interesting work. This supplement is not intended to be comprehensive, but it does extend at least partial coverage into 1971.

## Section III

The amination reaction of adamantane (eq. 13) has been reported in detail (217); adamantane precursors lead to aminoadamantane upon treatment with the reagent. Ionic chlorination of adamantane with $AlCl_3/CCl_4$ has been observed (218,219).

Additional solvolytic data are given in Supplemental Table I. The Schleyer-Gleicher approach (71) to relating strain increase upon ionization to solvolysis rate is also successful with these additional systems and several others in Table I that had not been treated previously (220). However, these calculations fail to reproduce the extremely low reactivity of derivatives of alcohol 1.

The source of this inertness is attributed to unfavorable hyperconjugative interactions in the carbonium ion (220,224,225) (which are not considered in the original calculations) as well as to lack of solvent assistance. The conclusion that rear-side solvent participation plays no role in the solvolysis of tertiary substrates has been reinforced by a detailed comparison of the behavior of $t$-butyl chloride and adamantyl bromide; the responses of these two compounds to solvent changes are virtually identical (226).

## Section VI

The fragmentation-recombination mechanism for the interconversion of the two isopropyl adamantane carbonium ions (eq. 45) has been shown to be invalid by the use of $D_2SO_4$ in the carboxylation (227). Deuterium is incorporated into the methyls of the isopropyl group, but not into the methine position, as required by the proposed mechanism. Dilution experiments prove that the ions are interconverted in fact by *inter*molecular hydride shifts.

Likewise, the portion of Chart 4 that indicates a direct migration as interconverting the 2- and 1-adamantyl cations is incorrect. It has been shown that here too the hydride shifts are *inter*molecular (228–230). Indeed, the 1,2 hydride migration in the adamantyl cation appears to be sterically forbidden, for the CH bonds lie at an angle of 60° to the empty $p$ orbital. The barrier to 1,2-hydride shift is greater than 29 kcal/mole (230).

## SUPPLEMENT TO TABLE I

| System | Leaving group | Solvent | $T$, °C | $k$, sec$^{-1}$ | $\Delta H^{\ddagger}$, kcal | $\Delta S^{\ddagger}$, eu | Ref. |
|---|---|---|---|---|---|---|---|
| 1-Bicyclo[3.2.2]nonyl | Cl | 80% EtOH | 70 | $4.64 \times 10^{-6}$ | 23.8 | −14.0 | 220 |
| 1-Bicyclo[3.3.2]decyl | Cl | 80% EtOH | 70 | $7.33 \times 10^{-3}$ | 21.3 | −6.5 | 220 |
| 1-Twistyl | Br | 80% EtOH | 70 | $1.41 \times 10^{-7}$ | 24.6 | −18.4 | 221 |
|  | OTs | AcOH | 70 | $1.02 \times 10^{-4}$ | 23.4 | −9.0 | 221 |
|  | Cl | 80% EtOH | 70 | $2.99 \times 10^{-9}$ | — | — | 221 |
|  | Cl | 50% EtOH | 70 | $1.35 \times 10^{-7}$ | 26.6 | −12.8 | 221 |
| 1-Tricyclo[3.1.0$^{3,7}$]nonyl | OTf$^a$ | AcOH | 70 | $2.96 \times 10^{-2}$ | 22.7 | +0.4 | 222 |
| 3-Tricyclo[3.0.0$^{3,7}$]octyl | Cl | 80% EtOH | 85 | $1.41 \times 10^{-7}$ | — | — | 223 |

$^a$ OTf = OSO$_2$CF$_3$.

## Section VII

Rearrangement of the tricyclo[5.2.1.0⁴,¹⁰]decyl alcohol **31** to 1-adamantanol has been observed (231).

![structure 31 converting to 1-adamantanol via 1. FSO₃H 2. H₂O] (106)

(31)

## Section VIII

Grob has reexamined the behavior of the bromide **23** and the corresponding tosylate (232). The ratio $k_{OTs}/k_{Br}$ (80% EtOH) is very nearly the same as the ratio for the carbocyclic analog, and it is concluded therefore that the large rate enhancement observed in the solvolysis of the quinuclidine derivatives does *not* result from direct participation of nitrogen. However, this interpretation has been criticized since steric effects, rather than electronic ones, may be responsible for the high tosylate/bromide ratios in tertiary systems (220).

## Section IX

The behavior of bridgehead carbonium ions in the decalyl, hydrindanyl, and perhydropentalenyl systems has been described in more detail (233, 234). In one study, further trapping results and the lack of isomerization of *p*-nitrobenzoates during solvolysis are interpreted as favoring the intervention of conformationally isomeric carbonium ions in the reactions of these *p*-nitrobenzoates. The observation that product distribution in the deoxidation of the *cis*- and *trans*-9-decalols depends on initial stereochemistry (235) also tends to support this view. On the other hand (234), it is argued that the kinetics of solvolysis of the 8-hydrindanyl chlorides requires that the carbonium ions be identical and that the differences in product composition result from differing locations of the gegenion in an intimate ion pair.

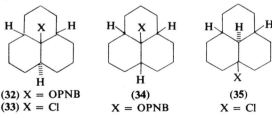

(32) X = OPNB  
(33) X = Cl

(34)  
X = OPNB

(35)  
X = Cl

## TABLE XII
### Solvolysis Rates of Some Bridgehead Condensed Rings

| Compound | Solvent | $T$, °C | $k$, sec$^{-1}$ | $\Delta H^{\ddagger}$, kcal | $\Delta S^{\ddagger}$, eu | Ref. |
|---|---|---|---|---|---|---|
| 32 | 80% Me$_2$CO | 100 | $1.63 \times 10^{-6}$ | 30.7 | −3.1 | 236 |
| 33 | 80% EtOH | 70 | $7.63 \times 10^{-4}$ | 20.2 | −14.2 | 220 |
| 34 | 80% Me$_2$CO | 25 | $1.51 \times 10^{-4}$ | 22.9 | +0.7 | 236 |
| 35 | 80% EtOH | 70 | $2.52 \times 10^{-5}$ | 23.0 | −8.8 | 220 |

The solvolysis of compounds **32** and **33** and **34** and **35** has been reported (220,236). The parameters are given in Table XII. The rates of reaction of **33** and **35**, in contrast to the derivatives of alcohol **1** noted previously, can be rationalized by strain analysis (220). The relatively high rate of **34** compared with **32** is probably the result of relief of strain.

## Section X

It appears that the decomposition of bridgehead chloroformates involves bridgehead carbonium ions (237,238). Thus, e.g., decomposition of 1-adamantyl chloroformate in nitrobenzene solvent gives substantial amounts of *m*-(1-adamantyl)-nitrobenzene (237).

## REFERENCES

1. R. C. Fort, Jr., and P. v. R. Schleyer, *Adv. Alicyclic Chem.*, **1**, 283 (1966).
2. P. D. Bartlett and L. H. Knox, *J. Am. Chem. Soc.*, **61**, 3184 (1939).
3. (a) A. R. Olson, *J. Chem. Phys.*, **1**, 418 (1933); (b) A. R. Olson and H. H. Voge, *J. Am. Chem. Soc.*, **56**, 1690 (1934).
4. R. A. Ogg, *J. Am. Chem. Soc.*, **61**, 1946 (1939).
5. M. Finkelstein, Ph.D. dissertation, Yale University, 1955.
6. L. E. Sutton et al., Eds., *Tables of Interatomic Distances and Configuration in Molecules and Ions*, Spec. Publ. Nos. 11 and 18, The Chemical Society, London, 1958 and 1965.
7. J. W. Linnett and A. J. Poe, *Trans. Faraday Soc.*, **47**, 1033 (1951).
8. K. B. Wiberg, *J. Am. Chem. Soc.*, **90**, 59 (1968); H. Kollmar and H. O. Smith, *Theoret. Chim. Acta*, **20**, 65 (1971).
9. J. E. Williams, Jr., R. Sustmann, L. C. Allen, and P. v. R. Schleyer, *J. Am. Chem. Soc.*, **91**, 1037 (1969); R. Sustmann, J. E. Williams, M. J. S. Dewar, L. C. Allen, and P. v. R. Schleyer, *J. Am. Chem. Soc.*, **91**, 5350 (1969); W. A. Lathan, W. J. Hehre, and J. A. Pople, *J. Am. Chem. Soc.*, **93**, 808 (1971); W. A. Lathan, W. J. Hehre, L. A. Curtiss, and J. A. Pople, *J. Am. Chem. Soc.*, **93**, 6377 (1971).
10. G. A. Olah, E. B. Baker, J. C. Evans, W. S. Tolgyesi, J. S. McIntyre, and I. J. Bastien, *J. Am. Chem. Soc.*, **86**, 1360 (1964), et seq.
10a. G. A. Olah, J. R. Demember, A. Commeyras, and J. L. Bribes, *J. Am. Chem. Soc.*, **93**, 459 (1971).
11. G. A. Olah and M. B. Comisarow, *J. Am. Chem. Soc.*, **88**, 1818 (1966).
12. V. A. Koptyug, I. S. Isaev, and A. I. Rezvukhin, *Tetrahedron Lett.*, **1967**, 823.
13. A. Streitwieser, Jr., *Solvolytic Displacement Reactions*, McGraw-Hill, New York, 1963, Chapter 1.
14. E. A. Prill, *J. Am. Chem. Soc.*, **69**, 62 (1947).
15. J. G. Traynham and J. S. Dehn, *J. Org. Chem.*, **23**, 1545 (1958).
16. P. D. Bartlett and E. S. Lewis, *J. Am. Chem. Soc.*, **72**, 1005 (1950).
17. P. D. Bartlett, S. G. Cohen, J. D. Cotman, Jr., N. Kornblum, J. R. Landry, and E. S. Lewis, *J. Am. Chem. Soc.*, **72**, 1003 (1950).
18. P. D. Bartlett and F. D. Greene, *J. Am. Chem. Soc.*, **76**, 1088 (1954).
19. P. D. Bartlett and S. G. Cohen, *J. Am. Chem. Soc.*, **62**, 1183 (1940).

20. Z. Suzuki and K. I. Morita, *J. Org. Chem.*, **32**, 31 (1967).
21. R. C. Fort and R. E. Franklin, unpublished observation; T. McAllister, Z. Dolešek, F. P. Lossing, R. Gleiter, and P. v. R. Schleyer, *J. Am. Chem. Soc.*, **89**, 5982 (1967).
22. R. R. Sauers and D. H. Ahlstrom, *J. Org. Chem.*, **32**, 2233 (1967).
23. J. Kopecky and J. Smejkal, *Tetrahedron Lett.*, **1967**, 1931, 3889.
24. R. C. Fort, Jr., and P. v. R. Schleyer, *Chem. Rev.*, **64**, 277 (1964); R. C. Bingham and P. v. R. Schleyer, *Fort. Chem. Forschung*, **18**, 1 (1971).
25. H. Stetter, J. Weber, and C. Wulff, *Ber.*, **97**, 3488 (1964).
26. G. A. Olah and J. Lukas, *J. Am. Chem. Soc.*, **90**, 933 (1968).
27. P. Kovacic and P. D. Roskos, *Tetrahedron Lett.*, **1968**, 5833.
28. W. Haaf, *Ber.*, **97**, 3234 (1964); F. N. Shepanov, V. F. Backlan, and S. S. Guts, *Sintez. Prirodn. Soedin, ikh Analogov i Fragmentov, Akad. Nauk SSSR, Otd. Obsch. i Tekhn, Khim.*, **1965**, 95; cf. *Chem. Abstr.*, **65**, 627b (1966).
29. F. N. Stepanov, E. I. Dikolenko, and G. I. Danilenko, *Zh. Organ. Khim.*, **2**, 640 (1966); cf. *Chem. Abstr.*, **65**, 8782h (1966).
30. P. Boldt and W. Thielecke, *Angew. Chem.*, **78**, 1058 (1966).
31. H. Stetter and M. Krause, *Tetrahedron Lett.*, **1967**, 1841.
32. T. Sasaki, S. Eguchi, and T. Toru, *Bull. Chem. Soc. Japan*, **41**, 236 (1968); *ibid.*, **41**, 238 (1968).
33. T. Sasaki, S. Eguchi, and T. Toru, *Chem. Commun.*, **1968**, 780.
34. M. R. Peterson, Jr., and G. H. Wahl, Jr., *Chem. Commun.*, **1968**, 1552.
35. T. Sasaki, S. Eguchi, and T. Toru, *Tetrahedron Lett.*, **1968**, 4135.
36. O. Wallach and W. Walker, *Ann.*, **271**, 285 (1892).
37. E. Grunwald and S. Winstein, *J. Am. Chem. Soc.*, **70**, 846 (1948).
38. A. H. Fainberg and S. Winstein, *J. Am. Chem. Soc.*, **79**, 1602 (1957).
39. K. B. Wiberg and V. Z. Williams, Jr., *J. Am. Chem. Soc.*, **89**, 3373 (1967).
40. K. B. Wiberg and B. R. Lowry, *J. Am. Chem. Soc.*, **85**, 3188 (1963).
41. C. J. Norton, Ph.D. dissertation, Harvard University, 1955.
42. A. B. Sayigh, Ph.D. dissertation, Columbia University, 1954.
43. P. Brenneisen, C. A. Grob, R. A. Jackson, and M. Ohta, *Helv. Chim. Acta*, **48**, 146 (1965).
44. C. A. Grob, *Gazz. Chim. Ital.*, **92**, 902 (1962).
45. P. v. R. Schleyer and C. W. Woodworth, *J. Am. Chem. Soc.*, **90**, 6528 (1968).
46. M. Sprecher, Ph.D. dissertation, Columbia University, 1953.
47. M. Levitz, Ph.D. dissertation, Columbia University, 1951.
48. W. G. Dauben and C. D. Pouter, *J. Org. Chem.*, **33**, 1237 (1968).
49. P. v. R. Schleyer, P. R. Isele, and R. C. Bingham, *J. Org. Chem.*, **33**, 1239 (1968).
50. P. v. R. Schleyer and E. Wiskott, *Tetrahedron Lett.*, **1967**, 2845.
51. P. v. R. Schleyer and R. D. Nicholas, *J. Am. Chem. Soc.*, **83**, 2700 (1961).
52. H. Stetter, J. Mayer, M. Schwarz, and C. Wulff, *Ber.*, **93**, 226 (1960).
53. R. C. Fort, Jr., Ph.D. dissertation, Princeton University, 1964.
54. H. Stetter and P. Goebel, *Ber.*, **96**, 550 (1963).
55. R. C. Fort, Jr., unpublished results; T. M. Gund, M. Nomura, V. Z. Williams Jr., P. v. R. Schleyer, and C. Hoogzand, *Tetrahedron Lett.*, **1970**, 4875; T. M. Gund, P. v. R. Schleyer, and C. Hoogzand, *Tetrahedron Lett.*, **1971**, 1583.
56. U. Miotti and A. Fava, *J. Am. Chem. Soc.*, **88**, 4274 (1966).
57. S. Winstein, E. Grunwald, and H. W. Jones, *J. Am. Chem. Soc.*, **73**, 2700 (1951).
58. S. Winstein, A. H. Fainberg, and E. Grunwald, *J. Am. Chem. Soc.*, **79**, 4146 (1957).

59. W. v. E. Doering, M. Levitz, A. Sayigh, M. Sprecher, and W. P. Whelan, *J. Am. Chem. Soc.*, **75**, 1008 (1953).
60. See Supplement, Refs. 220 and 224.
61. J. L. Franklin and F. H. Field, *J. Chem. Phys.*, **21**, 550 (1953).
62. Z. Dolejsek and F. P. Lossing, National Research Council of Canada; private communication from Dr. Dolejsek. See ref. 21.
63. E. M. Arnett, P. McC. Duggleby, and J. J. Burke, *J. Am. Chem. Soc.*, **85**, 1350 (1963).
64. P. D. Bartlett, *Bull. Soc. Chim. France*, **1951**, C100.
65. P. D. Bartlett, in *Organic Chemistry—An Advanced Treatise*, Vol. III, H. Gilman, Ed., Wiley, New York, 1953, p. 58.
66. W. v. E. Doering and E. F. Schoenewaldt, *J. Am. Chem. Soc.*, **73**, 2333 (1951).
67. W. v. E. Doering, paper presented at the 123rd National Meeting, American Chemical Society, Los Angeles, 1953, *Abstr.*, p. 35M.
68. J. B. Hendrickson, *J. Am. Chem. Soc.*, **83**, 4537 (1961); *J. Am. Chem. Soc.*, **86**, 4854 (1964).
69. K. B. Wiberg, *J. Am. Chem. Soc.*, **87**, 1070 (1965).
70. N. L. Allinger, M. A. Miller, L. W. Chow, R. A. Ford, and J. C. Graham, *J. Am. Chem. Soc.*, **87**, 3430 (1965).
71. G. J. Gleicher and P. v. R. Schleyer, *J. Am. Chem. Soc.*, **89**, 582 (1967).
72. C. S. Foote, *J. Am. Chem. Soc.*, **86**, 1853 (1964).
73. P. v. R. Schleyer, *J. Am. Chem. Soc.*, **86**, 1854, 1856 (1964).
74. P. v. R. Schleyer, J. E. Williams, and K. R. Blanchard, *J. Am. Chem. Soc.*, **92**, 2377 (1970); R. H. Boyd, S. N. Sanwal, S. Shary-Tehrany, and P. McNally, *J. Phys. Chem.*, **75**, 1264 (1971).
75. J. J. Macfarlane and I. G. Ross, *J. Chem. Soc.*, **1960**, 4169.
76. A. H. Nethercot and A. Javan, *J. Chem. Phys.*, **21**, 363 (1953).
77. O. Ermer and J. D. Dunitz, *Chem. Commun.*, **1968**, 567.
78. P. Bruesch and H. H. Gunthard, *Spectrochim. Acta*, **22**, 877 (1966).
79. A. F. Cameron, G. Ferguson, and D. G. Morris, *Chem. Commun.*, **1968**, 316; *J. Chem. Soc. (B)*, **1968**, 1249; O. Ermer and J. D. Dunitz, *Helv. Chim. Acta*, **52**, 1861 (1969).
80. N. L. Allinger, *Tetrahedron*, **22**, 1367 (1966).
81. G. Dallinga and L. H. Toneman, *Recueil*, **87**, 795 (1968); C. Altona and M. Sundaralingam, *J. Am. Chem. Soc.*, **92**, 1995 (1970).
82. G. Ferguson, C. J. Fritchie, J. M. Robertson, and G. A. Sim, *J. Chem. Soc.*, **1961**, 1976.
83. D. A. Brueckner, T. A. Hamor, J. M. Robertson, and G. A. Sim, *J. Chem. Soc.*, **1962**, 799.
84. A. F. Cesur and D. F. Grant, *Acta Crystallogr.*, **18**, 55 (1965).
85. A. C. MacDonald and J. Trotter, *Acta Crystallogr.*, **18**, 243 (1965); *Acta Crystallogr.*, **19**, 456 (1965).
86. H. D. Holtz and L. M. Stock, *J. Am. Chem. Soc.*, **86**, 5183, 5188 (1964).
87. F. W. Baker, R. C. Parish, and L. M. Stock, *J. Am. Chem. Soc.*, **89**, 5677 (1967).
88. G. L. Anderson and L. M. Stock, *J. Am. Chem. Soc.*, **90**, 212 (1968).
89. C. F. Wilcox, Jr., and J. F. McIntyre, *J. Org. Chem.*, **30**, 777 (1965).
90. C. F. Wilcox, Jr., and C. Leung, *J. Org. Chem.*, **33**, 877 (1968).
91. C. D. Ritchie and G. H. Megerle, *J. Am. Chem. Soc.*, **89**, 1452 (1967).
92. R. C. Fort, Jr., and P. v. R. Schleyer, *J. Am. Chem. Soc.*, **86**, 4194 (1964).
93. C. A. Grob, W. Schwarz, and H. P. Fischer, *Helv. Chim. Acta*, **47**, 1385 (1964).

94. V. W. Laurie and J. S. Muenter, *J. Am. Chem. Soc.*, **88**, 2883 (1966).
95. H. Stetter and J. Mayer, *Ber.*, **95**, 667 (1962).
96. W. P. Whelan, Ph.D. dissertation, Columbia University, 1952.
97. M. Wilhelm and D. Y. Curtin, *Helv. Chim. Acta*, **40**, 2129 (1957).
98. H. Hart and R. A. Martin, *J. Am. Chem. Soc.*, **82**, 6362 (1960).
99. P. Lipp and C. Padberg, *Ber.*, **54**, 1316 (1921).
100. T. W. Campbell, V. E. McCoy, J. C. Kauer, and V. S. Foldi, *J. Org. Chem.*, **26**, 1422 (1961).
101. W. Theilacker and K. H. Beyer, *Ber.*, **94**, 2968 (1961).
102. D. E. Applequist and J. P. Kliemann, *J. Org. Chem.*, **26**, 2178 (1961).
103. A. C. Cope and M. E. Synerholm, *J. Am. Chem. Soc.*, **72**, 5228 (1950).
104. A. C. Cope and E. S. Graham, *J. Am. Chem. Soc.*, **73**, 4702 (1951).
105. A. C. Cope, E. S. Graham, and D. J. Marshall, *J. Am. Chem. Soc.*, **76**, 6159 (1954).
106. K. Ebisu, L. B. Batty, J. M. Higaki, and H. O. Larson, *J. Am. Chem. Soc.*, **87**, 1399 (1965).
107. J. H. Ridd, *Quart. Rev. (London)*, **15**, 418 (1961).
108. A. Streitwieser, Jr., and C. E. Coverdale, *J. Am. Chem. Soc.*, **81**, 4275 (1959).
109. D. L. Boutle and C. A. Bunton, *J. Chem. Soc.*, **1961**, 761.
110. A. Streitwieser, Jr., and W. D. Schaeffer, *J. Am. Chem. Soc.*, **79**, 2888 (1957).
111. A. Streitwieser, Jr., *J. Org. Chem.*, **22**, 861 (1957).
112. E. H. White and C. A. Aufdermarsh, *J. Am. Chem. Soc.*, **83**, 1179 (1961).
113. E. H. White, M. J. Billig, and J. M. Bakke, paper presented at the 149th National Meeting of the American Chemical Society, Detroit, 1965; *Abstr.*, p. 37P.
114. E. H. White, H. P. Tiwari, and M. J. Todd, *J. Am. Chem. Soc.*, **90**, 4734 (1968).
115. P. Beak and R. J. Trancik, *J. Am. Chem. Soc.*, **90**, 2714 (1968); P. Beak, R. J. Trancik, J. B. Mooberry, and P. Y. Johnson, *J. Am. Chem. Soc.*, **88**, 4288 (1966).
116. D. Y. Curtin, B. H. Klanderman, and D. F. Tavares, *J. Org. Chem.*, **27**, 2709 (1962).
117. K. V. Scherer, Jr., G. A. Ungefug, and R. S. Lunt, III, *J. Am. Chem. Soc.*, **88**, 2859, 2860 (1966).
118. P. v. R. Schleyer, R. C. Fort, Jr., W. E. Watts, M. B. Comisarow, and G. A. Olah, *J. Am. Chem. Soc.*, **86**, 4195 (1964).
119. P. v. R. Schleyer, W. E. Watts, R. C. Fort, Jr., M. B. Comisarow, and G. A. Olah, *J. Am. Chem. Soc.*, **86**, 5679 (1964).
120. G. A. Olah and J. Lukas, *J. Am. Chem. Soc.*, **90**, 933 (1968).
121. L. Schmerling, *The Chemistry of Petroleum Hydrocarbons*, II, Reinhold, New York, 1955, Chapter 31.
122. P. D. Bartlett, F. E. Condon, and A. Schneider, *J. Am. Chem. Soc.*, **66**, 1531 (1944).
123. L. Schmerling, *J. Am. Chem. Soc.*, **68**, 195 (1946).
124. K. Gerzon, E. V. Krumkalns, R. L. Brindle, F. J. Marshall, and M. A. Root, *J. Med. Chem.*, **6**, 760 (1963).
125. H. Stetter and C. Wulff, *Ber.*, **93**, 1366 (1960); T. M. Gorrie and P. v. R. Schleyer, *Org. Prep. Procedures, Int.*, **3**, 159 (1971).
126. P. v. R. Schleyer, private communication, 1968.
126a. A. Karim, M. A. McKervey, E. M. Engler, and P. v. R. Schleyer, *Tetrahedron Lett.*, 3987 (1971).
127. H. Koch and W. Haaf, *Ann.*, **618**, 251 (1958).

128. H. Koch and W. Haff, *Angew. Chem.*, **72**, 628 (1960).
129. H. Koch and J. Franken, *Ber.*, **96**, 213 (1963).
130. H. Stetter, J. Gartner, and P. Tacke, *Angew. Chem. Int. Ed.*, **4**, 153 (1965).
131. H. Stetter, J. Gartner, and P. Tacke, *Ber.*, **98**, 3888 (1965).
132. H. Stetter, J. Gartner, and P. Tacke, *Ber.*, **99**, 1435 (1966).
133. M. Eakin, J. Martin, and W. Parker, *Chem. Commun.*, **1965**, 206.
134. W. Haaf, *Ber.*, **97**, 3234 (1964).
135. W. Haaf, *Ber.*, **96**, 3359 (1963).
136. Netherlands patent no. 6,408,505, to E. I. DuPont, Wilmington, Delaware; cf. *Chem. Abstr.*, **63**, 516b (1965).
137. P. Kovacic and P. D. Roskos, *Tetrahedron Lett.*, **1968**, 5833.
138. H. Stetter, M. Krause, and W. D. Last, *Angew. Chem.*, **80**, 970 (1968).
139. H. W. Geluk and J. L. M. A. Schlatmann, *Chem. Commun.*, **1967**, 426.
140. H. W. Geluk and J. L. M. A. Schlatmann, *Tetrahedron*, **24**, 5361 (1968).
141. H. W. Geluk and J. L. M. A. Schlatmann, *Tetrahedron*, **24**, 5369 (1968).
142. H. W. Geluk and J. L. M. A. Schlatmann, *Recueil*, **88**, 13 (1969).
143. Netherlands patent no. 6,511,851, to Sun Oil Co.; cf. *Chem. Abstr.*, **65**, 3768b (1966).
144. M. A. McKervey, J. R. Alford, J. F. McGarrity, and E. J. F. Rea, *Tetrahedron Lett.*, **1968**, 5165.
145. W. G. Dauben, J. L. Chitwood, and K. V. Scherer, Jr., *J. Am. Chem. Soc.*, **90**, 1014 (1968).
146. W. G. Dauben and J. L. Chitwood, *J. Am. Chem. Soc.*, **90**, 3835 (1968).
147. R. L. Bixler and C. Niemann, *J. Org. Chem.*, **23**, 742 (1958).
148. R. R. Sauers and D. H. Ahlstrom, *J. Org. Chem.*, **32**, 2233 (1967).
149. J. W. Wilt, C. A. Schneider, H. F. Dabek, Jr., J. J. Kraemer, and W. J. Wagner, *J. Org. Chem.*, **31**, 1543 (1966).
150. C. A. Grob, M. Ohta, R. Renk, and A. Weiss, *Helv. Chim. Acta*, **41**, 1191 (1958).
151. C. A. Grob, R. M. Hoegerle, and M. Ohta, *Helv. Chim. Acta*, **45**, 1823 (1962).
152. P. v. R. Schleyer and E. Wiskott, *Tetrahedron Lett.*, **1967**, 2845.
153. B. R. Vogt and J. R. E. Hoover, *Tetrahedron Lett.*, **1967**, 2841.
154. B. R. Vogt, S. R. Suter, and J. R. E. Hoover, *Tetrahedron Lett.*, **1968**, 1609.
155. H. W. Whitlock, Jr., and M. W. Siefkin, *J. Am. Chem. Soc.*, **90**, 4929 (1968).
156. J. E. Nordlander, S. P. Jindal, P. v. R. Schleyer, R. C. Fort, Jr., J. J. Harper, and R. D. Nicholas, *J. Am. Chem. Soc.*, **88**, 4475 (1966); S. H. Liggero, R. Sustmann, and P. v. R. Schleyer, *J. Am. Chem. Soc.*, **91**, 4571 (1969).
157. B. R. Vogt, *Tetrahedron Lett.*, **1968**, 1575.
158. S. H. Graham and D. A. Jonas, *J. Chem. Soc. (C)*, **1969**, 188.
159. P. Lipp and C. Padberg, *Ber.*, **54**, 1316 (1921).
160. J. Houben and E. Pfankuch, *Ann.*, **489**, 193 (1931).
161. J. D. Roberts and W. T. Moreland, Jr., *J. Am. Chem. Soc.*, **75**, 2267 (1953).
162. F. W. Baker, H. D. Holtz, and L. M. Stock, *J. Org. Chem.*, **28**, 514 (1963).
163. F. W. Baker and L. M. Stock, *J. Org. Chem.*, **32**, 3344 (1967).
164. V. Prelog and R. Seiwerth, *Ber.*, **74**, 1769 (1941).
165. H. P. Fischer and C. A. Grob, *Helv. Chim. Acta*, **47**, 564 (1964).
166. Y. Asahina and H. Kawahata, *Ber.*, **72**, 1540 (1939).
167. H. H. Lau and H. Hart, *J. Am. Chem. Soc.*, **81**, 4897 (1959).
168. J. C. Kauer, R. E. Benson, and G. W. Parshall, *J. Org. Chem.*, **30**, 1431 (1965).
169. J. R. Wiseman, *J. Am. Chem. Soc.*, **89**, 5966 (1967).
170. G. L. Dunn, V. J. DiPasquo, and J. R. E. Hoover, *Tetrahedron Lett.*, **1966**, 3737.

171. H. Stetter, M. Schwarz, and A. Hirschorn, *Ber.*, **92**, 1629 (1959).
172. M. F. Hawthorne, W. D. Emmons, and K. S. McCallum, *J. Am. Chem. Soc.*, **80**, 6393 (1958).
173. H. Minato, J. C. Ware, and T. G. Traylor, *J. Am. Chem. Soc.*, **85**, 3024 (1963).
174. C. A. Grob, *Gazz. Chim. Ital.*, **92**, 902 (1962).
175. C. A. Grob and P. W. Scheiss, *Angew. Chem.*, **79**, 1 (1967).
176. P. Brenneisen, C. A. Grob, R. A. Jackson, and M. Ohta, *Helv. Chim. Acta*, **48**, 146 (1965).
177. C. A. Grob and W. Schwartz, *Helv. Chim. Acta*, **47**, 1870 (1964).
178. U. Burckhardt, C. A. Grob, and H. R. Kiefer, *Helv. Chim. Acta*, **50**, 231 (1967).
179. H. Stetter and P. Tacke, *Angew. Chem.*, **74**, 354 (1962).
180. H. Stetter and P. Tacke, *Ber.*, **96**, 694 (1963).
181. F. N. Stepanov and W. D. Suchowerchow, *Angew. Chem.*, **79**, 860 (1967).
182. R. E. Pincock, E. Grigat, and P. D. Bartlett, *J. Am. Chem. Soc.*, **81**, 6332 (1959).
183. H. Christol and G. Solladie, *Bull. Soc. Chim. France*, **1966**, 1307.
184. H. Koch and W. Haaf, *Ber.*, **94**, 1252 (1961).
185. K. E. Moller, *Angew. Chem.*, **75**, 1122 (1963).
186. H. Koch and W. Haaf, *Ann.*, **638**, 111 (1960).
187. G. Baddeley and E. Wrench, *J. Chem. Soc.*, **1959**, 1324.
188. G. Baddeley, B. G. Heaton, and J. W. Rasburn, *J. Chem. Soc.*, **1960**, 4713.
189. M. S. Ahmad and G. Baddeley, *J. Chem. Soc.*, **1961**, 4303.
190. G. Baddeley and B. G. Heaton, *J. Chem. Soc.*, **1961**, 4306.
191. C. Arnal, J.-M. Bessiere, H. Christol, and R. Vanel, *Bull. Soc. Chim. France*, **1967**, 2479.
192. H. Christol and J.-M. Bessiere, *Bull. Soc. Chim. France*, **1968**, 2141.
193. J.-M. Bessiere and H. Christol, *Bull. Soc. Chim. France*, **1968**, 2147.
194. G. Baddeley and E. K. Baylis, *J. Chem. Soc.*, **1965**, 4933.
195. A. F. Boschung, M. Geisel, and C. A. Grob, *Tetrahedron Lett.*, **1968**, 5169.
196. R. C. Fort, Jr., and R. E. Hornish, *Chem. Commun.*, **1969**, 11.
197. R. E. Hornish, G. A. Liang, and R. C. Fort, Jr., paper presented at the 158th National American Chemical Society Meeting, New York, September 7–14, 1969.
198. F. Carey and H. Tremper, *J. Org. Chem.*, **34**, 4 (1969).
199. L. F. Fieser and M. Fieser, *Advanced Organic Chemistry*, Reinhold, New York, 1953, p. 326.
200. A. Belanger, J. Poupart, and P. Deslongchamps, *Tetrahedron Lett.*, **1968**, 2127.
201. A. Belanger, Y. Lambert, and P. Deslongchamps, *Can. J. Chem.*, **47**, 795 (1969).
202. H. Stetter and P. Goebel, *Ber.*, **96**, 550 (1963).
203. D. B. Denney and R. R. DiLeone, *J. Am. Chem. Soc.*, **84**, 4737 (1962).
204. G. R. Pettit and E. E. van Tamelen, *Org. React.*, **12**, 356 (1962).
205. E. E. v. Tamelen and E. A. Grant, *J. Am. Chem. Soc.*, **81**, 2160 (1959).
206. P. T. Lansbury, V. A. Pattison, and J. W. Diehl, *Chem. Ind. (London)*, **1962**, 653.
207. R. Scartazzini and K. Mislow, *Tetrahedron Lett.*, **1967**, 2719.
208. F. L. Lambert and K. Kobayashi, *J. Am. Chem. Soc.*, **82**, 5324 (1960).
209. F. L. Lambert, A. H. Albert, and J. P. Hardy, *J. Am. Chem. Soc.*, **86**, 3155 (1964).
210. J. W. Sease, P. Chang, and J. L. Groth, *J. Am. Chem. Soc.*, **86**, 3154 (1964).
211. J. Zavada, J. Krupicka, and J. Sicher, *Collect. Czech. Chem. Commun.*, **28**, 1664 (1963).

212. J. Zavada, J. Krupicka, and J. Sicher, *Collect. Czech. Chem. Commun.*, **30**, 3570 (1965).
213. R. C. Fort, Jr., and R. E. Franklin, *J. Am. Chem. Soc.*, **90**, 5267 (1968).
214. D. G. Pratt and E. Rothstein, *J. Chem. Soc. (C)*, **1968**, 2548.
215. G. A. Olah and M. B. Comisarow, *J. Am. Chem. Soc.*, **88**, 4442 (1966).
216. P. v. R. Schleyer, R. C. Fort, Jr., W. E. Watts, G. A. Olah, and M. B. Comisarow, *J. Am. Chem. Soc.*, **86**, 4195 (1964).
217. P. Kovacic and P. D. Roskos, *J. Am. Chem. Soc.*, **91**, 6457 (1969).
218. H. Stetter, M. Krause, and W.-D. Last, *Ber.*, **102**, 3357 (1969).
219. M. A. McKervey, D. Grant, and H. Hamill, *Tetrahedron Lett.*, **1970**, 1975.
220. R. C. Bingham and P. v. R. Schleyer, *J. Am. Chem. Soc.*, **93**, 3189 (1971).
221. R. C. Bingham, P. v. R. Schleyer, Y. Lambert, and P. Deslongchamps, *Can. J. Chem.*, **48**, 3739 (1970).
222. R. C. Bingham, W. F. Sliwinski, and P. v. R. Schleyer, *J. Am. Chem. Soc.*, **92**, 3471 (1970).
223. P. K. Freeman, R. B. Kinnel, and T. D. Ziebarth, *Tetrahedron Lett.*, **1970**, 1059.
224. R. C. Bingham and P. v. R. Schleyer, *Tetrahedron Lett.*, **1971**, 23.
225. J. E. Williams, V. Buss, L. C. Allen, P. v. R. Schleyer, W. A. Latham, W. J. Hehre, and J. A. Pople, *J. Am. Chem. Soc.*, **92**, 2141 (1970).
226. D. J. Raber, R. C. Bingham, J. M. Harris, J. L. Fry, and P. v. R. Schleyer, *J. Am. Chem. Soc.*, **92**, 5977 (1970); D. N. Kevill, K. C. Kolwyck, and F. L. Weitl, *J. Am. Chem. Soc.*, **92**, 7300 (1970).
227. D. J. Raber, R. C. Fort, Jr., E. Wiskott, C. W. Woodworth, P. v. R. Schleyer, J. Weber, and H. Stetter, *Tetrahedron*, **27**, 3 (1971).
228. D. M. Brouwer and H. Hogeveen, *Rec. Trav. Chim.*, **89**, 211 (1970).
229. P. v. R. Schleyer, L. K. M. Lam, D. J. Raber, J. L. Fry, M. A. McKervey, J. R. Alford, B. D. Cuddy, V. G. Keizer, H. W. Geluk, and J. L. M. A. Schlatmann, *J. Am. Chem. Soc.*, **92**, 5246 (1970); J. Burkhard, J. Vais, and S. Landa, *Z. Chem.*, **9**, 29 (1969).
230. P. Vogel, M. Saunders, W. Thielecke, and P. v. R. Schleyer, *Tetrahedron Lett.*, 1429 (1971).
231. L. A. Paquette, G. V. Meehan, and S. J. Marshall, *J. Am. Chem. Soc.*, **91**, 6779 (1969).
232. C. A. Grob, K. Kostka, and F. Kuhnen, *Helv. Chim. Acta*, **53**, 608 (1970).
233. R. C. Fort, Jr., R. E. Hornish, and G. A. Liang, *J. Am. Chem. Soc.*, **92**, 7558 (1970).
234. K. Becker, A. F. Boschung, and C. A. Grob, *Tetrahedron Lett.*, **1970**, 3831.
235. R. C. Fort, Jr., and K. L. Heinselman, unpublished observations, 1970.
236. H. C. Brown and W. C. Dickason, *J. Am. Chem. Soc.*, **91**, 1226 (1969).
237. D. N. Kevill and F. L. Weitl, *J. Am. Chem. Soc.*, **90**, 6416 (1968).
238. P. Beak, R. J. Trancik, and D. A. Simpson, *J. Am. Chem. Soc.*, **91**, 5073 (1969).

# CHAPTER 33

# Degenerate Carbonium Ions*

RONALD E. LEONE

*Research Laboratories, Eastman Kodak Company, Rochester, New York*

AND

J. C. BARBORAK AND PAUL V. R. SCHLEYER

*Department of Chemistry, Princeton University, Princeton, New Jersey*

| | |
|---|---|
| I. Introduction | 1838 |
| II. Reaction Processes Leading to Degeneracy | 1842 |
|    A. Wagner-Meerwein Rearrangement (1,2-Carbon Shift) | 1842 |
|    B. 1,2-Hydride and 1,3-Hydride Shifts | 1843 |
|    C. Double-Bond Participation: 1,2 Vinyl Shift and Homoallylic Rearrangement | 1844 |
|    D. Participation of a Cyclopropane Ring | 1845 |
|    E. Allyl Resonance | 1846 |
|    F. Other Processes Leading to Degeneracy | 1846 |
| III. $(CH)_n^+$ Series of Degenerate Carbonium Ions | 1846 |
|    A. $C_3H_3^+$ Ions | 1850 |
|    B. $C_5H_5^+$ Ions | 1851 |
|    C. $C_7H_7^+$ Ions | 1854 |
|       1. Tropylium Ion | 1854 |
|       2. 7-Norbornadienyl Cation and Bicyclo[3.2.0]heptadienyl Cation | 1855 |
|       3. Quadricyclyl or 3-Tetracyclo[3.2.0.0.$^{2,7}$0$^{4,6}$]heptyl Cation | 1860 |
|       4. Homoprismyl or 7-Tetracyclo[3.2.0.0$^{2,4}$.0$^{3,6}$]heptyl Cation | 1861 |
|    D. $C_9H_9^+$ Ions | 1863 |
|       1. 9-Homocubyl or 9-Pentacyclo[4.3.0.0$^{2,5}$.0$^{3,8}$.0$^{4,7}$]nonyl Cation | 1863 |
|       2. 9-Pentacyclo[4.3.0.0$^{2,4}$.0$^{3,8}$.0$^{5,7}$]nonyl Cation | 1869 |
|       3. Tetracyclo[4.3.0.0$^{2,4}$.0$^{3,7}$]non-8-en-5-yl Cation | 1872 |
|       4. Bicyclo[3.2.2]nona-3,6,8-trien-2-yl Cation | 1875 |
|       5. *cis*-8,9-Dihydro-1-indenyl Cation | 1878 |
|       6. Bicyclo[4.2.1]nona-2,4,7-trien-9-yl Cation | 1880 |
|       7. Barbaralyl or Tricyclo[3.3.1.0$^{2,8}$]nona-3,6-dien-9-yl Cation | 1882 |
|       8. Cyclononatetraenyl Cations | 1887 |
| IV. Other Degenerate Carbonium Ions | 1889 |
|    A. Isopropyl Cation and *sec*-Butyl Cation | 1889 |
|    B. Dimethyl-*tert*-butylcarbonium Ion and Related Cations | 1892 |
|    C. *Tert*-Amyl Cation | 1895 |
|    D. 2,4-Dimethylpentyl Cation | 1896 |

* Adapted in part from a review of the same title appearing in *Angew. Chem. Int. Ed. Engl.*, **9**, 860 (1970). We thank Verlag Chemie GMBH for permission to republish this article.

E. 1-*Tert*-Butyl-1,3,3-trimethylallyl Cation . . . . . 1897
F. Methylcyclobutyl Cation . . . . . . . . 1898
G. Cyclopentyl Cation . . . . . . . . . 1899
H. Methylcyclopentyl Cation . . . . . . . . 1899
I. Cyclobutenyl Cation . . . . . . . . . 1901
J. Pentamethyl- and Hexamethylcyclopentenyl Cations . . . 1902
K. Heptamethylbenzenonium Ion . . . . . . . 1903
L. 1,3,4,4-Tetramethylcyclohexenyl Cation . . . . . 1905
M. Dihydrophenanthrenium Cations. . . . . . . 1906
N. 3-Bicyclo[3.1.0]hexyl Cation . . . . . . 1906
O. 2-Norbornyl Cation . . . . . . . . . 1911
P. Bicyclo[3.1.0]hex-3-en-2-yl Cation . . . . . . 1915
Q. 2-Brexyl or 2-Tricyclo[4.3.0.0$^{3,7}$]nonyl Cation . . . 1921
R. Cyclopropylcarbinyl Cation Systems . . . . . 1923
   1. 8,9-Cyclo-2-adamantyl Cation . . . . . . 1923
   2. Tricyclo[3.2.1.0$^{2,7}$]octenyl Cation: A Degenerate
      Cyclopropylcarbinyl System . . . . . . 1925
   3. Cyclopropylcarbinyl Cation: Stereochemistry of Rearrangement . 1926
S. 4-Protoadamantyl Cation . . . . . . . . 1927
T. Adamantyl Cations . . . . . . . . . 1928
U. 4-Homoadamantyl or 4-Tricyclo[4.3.1.1$^{3,8}$]undecyl Cation . 1929
V. 8-Tetracyclo[4.3.0.0$^{2,4}$.0$^{3,7}$]nonyl Cation . . . . 1931
V. Conclusion . . . . . . . . . . . 1933
References . . . . . . . . . . . 1934

## I. INTRODUCTION

Chemists have always been especially interested in rearrangements because they violate the principle of minimum structural change (1). Originally rearrangements leading to changes in molecular or structural formula were examined. Clearly there must also be a class of degenerate rearrangements in which individual atoms and bonds become interchanged, but which only lead to products of the same gross structure (i.e., identical except for the special features that enable detection of the rearrangement) as that of the starting material. These transformations can only be followed by special techniques such as the loss of optical activity, the dispersal of an isotopic label, and the observation of equivalence by nmr. The most striking of these degenerate rearrangements are those capable of mixing all atoms of a given kind completely. Bullvalene is the best known organic molecule of this type; every Cope rearrangement returns to bullvalene and at only a little above room temperature all hydrogen and carbon atoms become equivalent on the nmr time scale (2).

Carbonium ions are particularly prone to rearrangement (3), and it would be expected that many examples of degeneracy could be found within this class of reaction intermediates. Lately considerable effort has been directed toward the study of carbonium ions, either as stable cations or as solvolysis intermediates, capable of undergoing degenerate rearrangements. This group of carbonium ions is reviewed here, with major emphasis placed on ions capable of multiple degeneracy. Balaban et al. have recently developed a mathematical treatment of multiple 1,2 shifts in carbonium ions on the basis of the theory of graphs (4).

A degenerate carbonium ion is defined as a carbonium ion that can rearrange through a finite energy barrier in such a fashion as to regenerate the same gross structure (as defined previously) as that of the starting ion. The degenerate rearrangement of such an ion is an automerization (5), i.e., reaction in which both the molecular formula and structural formula of the starting material are conserved (see Table I). The preservation of enantiomeric purity is not a requirement. Thus enantiomeric cations and their racemic mixtures, but not diastereomeric cations, are assumed to have the same gross structure.

We consider rearrangements whereby an optically active cation is racemized by an automerization or an isotopic label dispersed by an automerization to be examples of degeneracy. In fact, automerizations can often be detected by the techniques of using optically active or isotopically tagged precursors.

A number of carbonium ions, through resonance, possess equivalent ("homotopic") (6) atoms or groups. Examples are the termini of the allyl cation and all the carbon–hydrogen units in the tropylium ion. Such systems do not involve degenerate rearrangements in the sense of the previous definition; for the purposes of this chapter, however, it is never-

**TABLE I**

Definition of rearrangement, isomerization, and automerization (based on Ref. 5)[a]

| Molecular formula | Structural formula[b] | Name |
|---|---|---|
| Not conserved | Not conserved | Rearrangement |
| Conserved | Not conserved | Isomerization |
| Conserved | Conserved | Automerization |

[a] Because of historical precedent, we believe it is better to regard these categories as being *inclusive* rather than *exclusive*. Thus an isomerization is a special kind of rearrangement and an automerization is a special kind of isomerization (or rearrangement). For a recent, alternative nomenclature proposal *see* G. Binsch, E. L. Eliel, and H. Kessler, *Angew. Chem. Inter. Ed.*, **10**, 570 (1971).

[b] For convenience, enantiomers and isotopically labeled compounds, but not diastereomers, are considered here to have the same structural formula.

theless convenient to discuss some of these cases because resonance (e.g., allyl resonance) is also important in helping to achieve degeneracy in some cations. In these instances, resonance contributes to the overall rearrangement mechanism whereby the original ion structure is regenerated.

The problem of distinguishing between a degenerate cation rearrangement and equivalence through resonance can be illustrated by the allyl cation. The labeled allyl chloride 1 generates on solvolysis an allyl cation 2, which contains two positions made equivalent by resonance (7). The

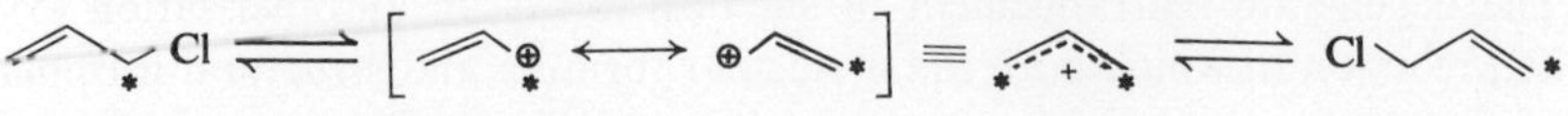

(1) ≈ 50%   (2) 1/2 1/2   (3) ≈ 50%

chloride product contains a label at the two positions shown in 1 and 3. The interconversion of the chlorides, 1 into 3, is degenerate, but cation 2 does not undergo a degenerate rearrangement (in the sense of our definition), since the equivalence of positions in 2, a *single* species, arises through resonance. However, automerizations are theoretically possible for the allyl cation. Three types of potentially degenerate processes serve as examples, although only the first has actually been observed: (*1*) *cis-trans* isomerization (8), (*2*) interchange of hydrogens, and (*3*) interchange of carbon atoms.

These last three processes differ from the one just described in that they

1. (4) ⇌ (4′)

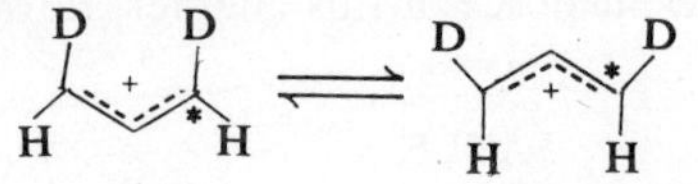

or (4) ⇌ (4″) or (4″)

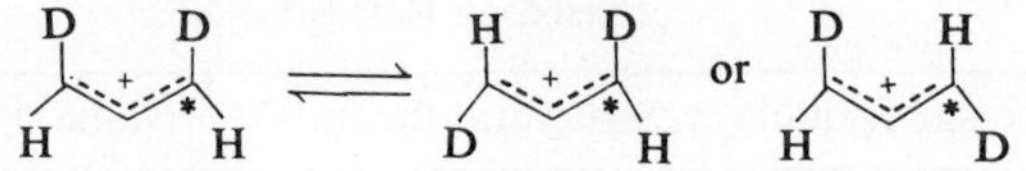

2. (4″) ⇌ (4‴) or

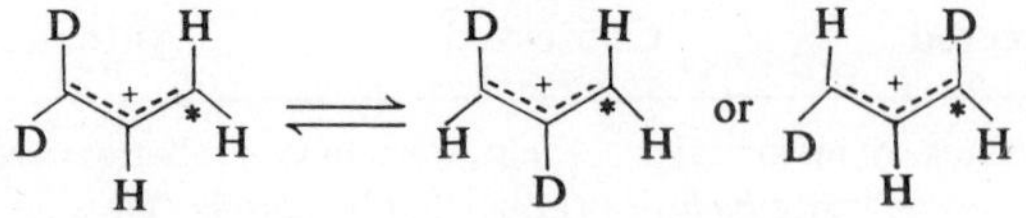

3. (4″) ⇌ (4″″)

require interchange of atoms and passage over an energy barrier to render positions equivalent (degenerate rearrangement), whereas the previous case involves no interchange of atoms and no energy barrier (equivalence through resonance).

Based on the definition of a degenerate carbonium ion presented here, it appears that the simplest degenerate rearrangements possible in cations are $5 \rightleftharpoons 5'$, $6 \rightleftharpoons 6'$, and $7 \rightleftharpoons 7'$. In these cases simple hydrogen inter-

$$H-C\equiv\overset{*}{C}^{\oplus} \rightleftharpoons {}^{\oplus}C\equiv\overset{*}{C}-H$$
$$(5) \qquad\qquad (5')$$

$$H_2C=\overset{\oplus}{\underset{*}{C}}H \rightleftharpoons H\overset{\oplus}{\underset{*}{C}}=CH_2$$
$$(6) \qquad\qquad (6')$$

$$\begin{array}{c}R\\ \diagdown\\ H\end{array}C=\overset{\oplus}{C}-R^* \rightleftharpoons R-\overset{\oplus}{C}=C\begin{array}{c}H\\ \diagup\\ R^*\end{array}$$
$$(7) \qquad\qquad (7')$$

changes produce degeneracy of the carbonium ions. Investigations have not yet been carried out to detect these possible degeneracies experimentally. However, the possibility of a degenerate 1,2 hydride shift in the ethyl cation **8** has been investigated (9). To achieve full degeneracy in **8** all five hydrogens must be exchanged. Olah and co-workers have recently demonstrated the existence of the ethyl cation and of hydrogen scrambling in the species by generating it from the complex of ethyl fluoride and antimony pentafluoride in strong acid solutions (10).

$$CD_3-CH_2^{\oplus} \rightleftharpoons {}^{\oplus}CD_2-CH_2D \rightleftharpoons \text{etc.}$$
$$(8) \qquad\qquad (8')$$

The potential degenerate rearrangements of the allyl cation represent a higher level of complexity. These allyl and ethyl cation examples illustrate the definition of a degenerate carbonium ion. More complicated degenerate cations are reviewed as outlined: we discuss two main categories of degenerate ions—those of the formula $(CH)_n{}^+$ where $n$ is odd (see Section III) and those of the type not having the $(CH)_n{}^+$ empirical formula (see Section IV).

A classification of degenerate carbonium ions may also be made on the basis of the atoms or groups undergoing interchange. In principle there are three kinds of ions: those exhibiting carbon atom degeneracy, those exhibiting hydrogen atom degeneracy, and those exhibiting combined carbon and hydrogen atom degeneracy. In practice all three types of degenerate carbonium ions have been observed.

Carbon atom degeneracy involves the interchange of carbon atoms with or without their respective attached hydrogen. Complete carbon degeneracy means that all carbon atoms are capable of interchanging and becoming equivalent. Partial carbon degeneracy implies that only some of the carbons in the ion become equivalent. In theory carbon degeneracy can be detected by $^{13}C$ scrambling within the ion. Experimentally, however, this form of degeneracy is usually determined by proton label scrambling. Carbon degeneracy should give rise to a single nmr signal for the protons attached to the interchangeable carbon atoms and scrambling of a hydrogen isotope label at the interchangeable positions. Recent developments in $^{13}C$-nmr spectroscopy have opened up the way to direct observation of carbon degeneracies.

Hydrogen atom degeneracy involves interchange of hydrogens while the carbon skeleton of the ion remains fixed. Complete hydrogen degeneracy means the interchange and equivalence of all hydrogen atoms on the carbon surface of the molecule. Such a process will result in a single peak for the $^1H$-nmr spectrum of the ion. Partial hydrogen degeneracy refers to the interchange of only some hydrogen atoms in the ion without altering the basic carbon framework.

Combination of both carbon and hydrogen atom degeneracies is possible, and this can also result in the equivalence of all carbons and hydrogens of the carbonium ion. Specific examples of the forms of degeneracy mentioned previously are discussed in detail later in this chapter. It has been observed experimentally that degenerate ions of the $(CH)_n{}^+$ series show degenerate properties only through carbon degeneracy and are often capable of complete mixing of all carbon atoms. Degenerate ions outside the $(CH)_n{}^+$ series can exhibit all types of degeneracy, but often these ions are not capable of complete degeneracy.

The interest in studying degenerate carbonium ions revolves around the following three questions:

1. Can degeneracy be achieved, and if so, to what extent?
2. What is the mechanism whereby the degeneracy is achieved?
3. Are there any special properties resulting from degeneracy? For example, are the carbonium ions especially stable or especially unstable?

The degenerate carbonium ions already studied are discussed in terms of these three considerations.

## II. REACTION PROCESSES LEADING TO DEGENERACY

### A. Wagner-Meerwein Rearrangement (1,2-Carbon Shift)

Probably the most important rearrangement process for degenerate carbonium ions is the Wagner-Meerwein rearrangement or 1,2-carbon shift (11,12). This process allows interchange of carbon atoms and consequently is the basis for many observed carbon automerizations. The ions listed are examples of degenerate ions arising from 1,2-carbon shifts:

2-norbornyl cation (9) (see Section IV-O)

9-homocubyl cation (10) (see Section III-D-1)

dimethyl-*tert*-butylcarbonium ion (11) (see Section IV-B)

## B. 1,2-Hydride and 1,3-Hydride Shifts

Migration of a hydrogen from one adjacent carbon atom to another (13) is responsible for the hydrogen degeneracy observed in some ions. A hydride shift can only contribute directly to degeneracy when the two carbon atoms involved bear a symmetrical relationship to each other, as in the following examples. Note that hydride shifts are not energetically favorable in the $(CH)_n^+$ ions, since such shifts would result in bridgehead or vinyl carbonium ions.

2-norbornyl cation (9) (see Section IV-O)

cyclopentyl cation (12) (see Section IV-G)

## C. Double-Bond Participation: 1,2 Vinyl Shift and Homoallylic Rearrangement

Double-bond participation refers to the electronic participation of a carbon–carbon double bond in stabilizing a positive charge at least two carbon atoms removed from the double bond (14,15). In some cases this participation may result in a 1,2-vinyl shift or a homoallylic rearrangement. Both of these processes can contribute to carbon degeneracy. A 1,2-vinyl shift was observed in the degenerate norbornadienyl cation **13** (see Section III-C-2) and may be potentially involved in the degeneracy of other ions e.g., **14** (see Section III-D-4). Homoallylic rearrangement is presumably also involved in other ion degeneracies such as **15** and **16** (see Sections III-D-3 and III-D-5, respectively).

### D. Participation of a Cyclopropane Ring

The bonds of a cyclopropane ring are capable of participating in charge delocalization (16,17). Such delocalization can give rise to either a degenerate homocyclopropylcarbinyl rearrangement (18) (cation **17** is the parent system) or a degenerate cyclopropylcarbinyl–cyclopropylcarbinyl rearrangement (19) (cation **18** is the parent system).

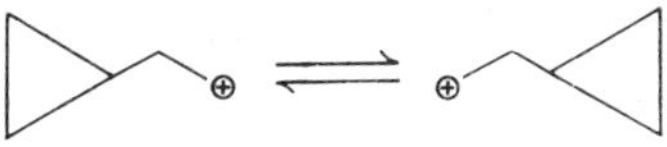

(17)

(18)

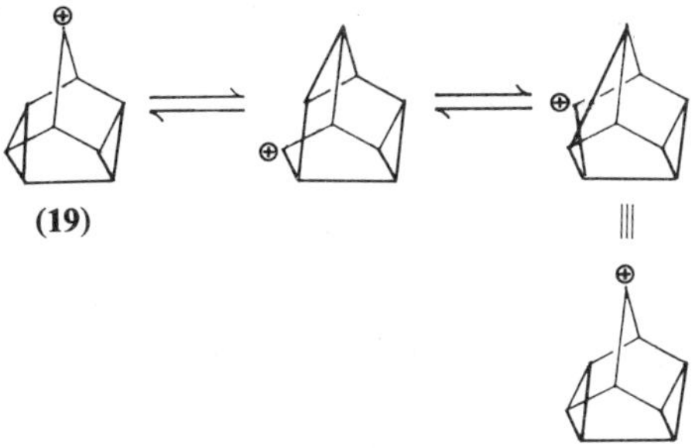

(19)

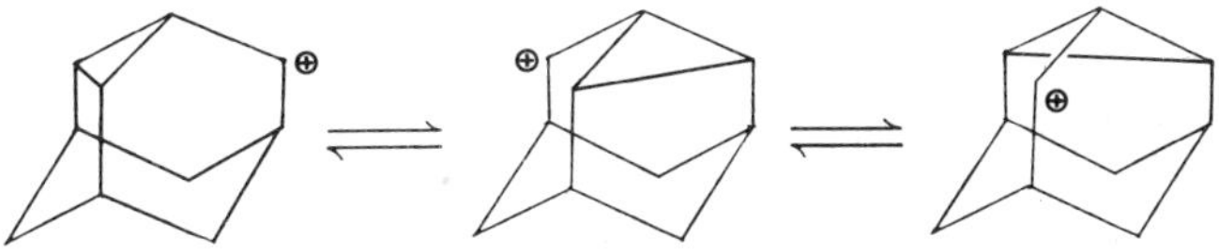

(20)

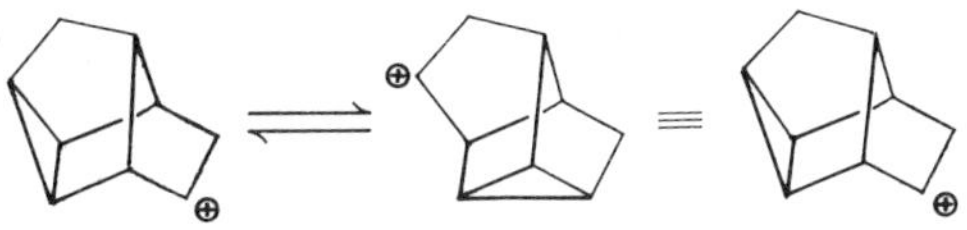

(21)

The 9-pentacyclo[4.3.0.0$^{2,4}$.0$^{3,8}$.0$^{5,7}$]nonyl cation (**19**) (see Section III-D-2) and the 8-tetracyclo[4.3.0.0$^{2,4}$.0$^{3,7}$]nonyl cation (**21**) (see Section IV-V) are examples of the first type, and the 8,9-cyclo-2-adamantyl cation (**20**) (see Section IV-R-1) is an example of the second type.

### E. Allyl Resonance

Allyl resonance (7), as applied to degenerate carbonium ions, refers to the simultaneous delocalization of positive charge over two allylic carbon atoms that are symmetrically disposed in the ion. The resonance can render two positions on an ion equivalent and can contribute to carbon degeneracy when coupled with one of the other rearrangement processes mentioned previously. Cations **14** and **16** shown in Section II-C serve as examples.

### F. Other Processes Leading to Degeneracy

Of the five processes we have discussed, four can be regarded as rearrangement processes: Wagner-Meerwein shifts, hydride shifts, double-bond migrations, and cyclopropane participations; the other, allylic resonance, contributes by symmetrization of the molecule. Another process not yet covered is the Cope rearrangement, which does not involve the cationic center but can act to scramble a label; one investigated case, that of the barbaralyl cation (Section III-D-7) serves as a possible example.

Newly described processes can also cause apparent degeneracy in a cation. These processes are not rearrangements in their nature, at least in their salient features; rather, they represent transition states or intermediates in those chemical processes that lead to degeneracy. Thus Hoffmann and others have suggested, on the basis of theoretical models, that $C_5H_5^+$ ions might exhibit properties that betray the existence of structures such as **65a**, whereas the $C_9H_9^+$ barbaralyl cation might owe its degeneracy to the threefold-symmetrical species **187a** (28,28a). These remarkable proposals are discussed in detail in Sections III-B and III-D-7, respectively.

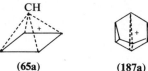

(65a)   (187a)

## III. $(CH)_n^+$ SERIES OF DEGENERATE CARBONIUM IONS

Degenerate carbonium ions having the empirical formula $(CH)_n^+$ where $n$ is odd are directly related to hydrocarbons of the form $(CH)_m$ where $m$ is even. Balaban has published a mathematical treatment for the derivation of all possible valence isomers of $(CH)_m$ hydrocarbons (20). Potentially degenerate carbonium ions can in theory be derived either by replacement of a carbon–carbon double bond in such a hydrocarbon by a $\rangle C^+ \!\!-\!\! H$

unit or by insertion of a $\diagup\!\!\!\diagdown\!\!\!\!\text{C}^+$—H unit into any carbon–carbon single bond.

The hydrocarbon isomers under consideration have four, six, eight, or ten carbon atoms and the corresponding carbonium ions contain three, five, seven or nine carbon atoms. Table II shows the relation between the carbonium ions and the hydrocarbons from which they can be derived. All of the ions appearing in the table have the potential for either complete or partial carbon degeneracy, although in fact this may not be observed. It should also be noted that certain ions, such as the cyclopropenyl cation **22** and the tropylium ion **26**, are not truly degenerate in the sense of our definition, since in these ions all the carbon atoms are equivalent through resonance.

All the $(CH)_n{}^+$ carbonium ions are similar in that they are potentially capable of mixing carbon atoms only by carbon–carbon bond breaking and formation. In these cases the hydrogens also remain attached to the carbon atoms, thus making the interchanging carbon–hydrogen units equivalent. This degeneracy is of the same form as that observed in bullvalene (**51**). Just as some of the $(CH)_m$ hydrocarbons represent interconvertible valence isomers, some of the $(CH)_n{}^+$ ions are also interconvertible.

As an alternative to the derivation of $(CH)_n{}^+$ ions from $(CH)_m$ hydrocarbons, potentially degenerate $(CH)_n{}^+$ ions can also be constructed by noting that the ion must contain a $\diagup\!\!\!\diagdown\!\!\!\!\text{C}^+$—H unit, plus an appropriate number of single bonds, double bonds, and rings to maintain the $(CH)_n{}^+$ empirical formula. All carbon atoms in the ion must bear a hydrogen atom; but vinyl and acetylenic cations will be excluded. Consequently, $C_3H_3{}^+$, $C_5H_5{}^+$, $C_7H_7{}^+$, and $C_9H_9{}^+$ ions can have only certain structures.

In the $C_3H_3{}^+$ group there is only one possibility, the cyclopropenyl cation **22** in which there is one ring and one double bond. For the $C_5H_5{}^+$ group there are three regular structures: the cyclopentadienyl cation **23**, monocyclic and containing two double bonds; the bicyclo[2.1.0]pentenyl cation **24**, bicyclic and containing one double bond; the homotetrahedryl cation **25**, tricyclic and containing no double bond (20). There are many possible structures for the $C_7H_7{}^+$ group, but only the more important ones are considered here. The tropylium ion **26** is monocyclic and can be thought of as having three double bonds. The norbornadienyl cation **13** contains two rings and two double bonds; similarly, the bicyclo[3.2.0]-heptadienyl cation **27** is bicyclic and contains two double bonds. The quadricyclyl cation **28** has four rings and no double bonds, and the homoprismyl cation **29** is tetracyclic with no double bonds.

Extension of this construction format to the $C_9H_9{}^+$ ions should be

## TABLE II
### (CH)$_m$ Hydrocarbons, $m = 4, 6, 8, 10$, and (CH)$_n^+$ Ions, $n = 3, 5, 7, 9$

| (CH)$_m$ Hydrocarbons | (CH)$_n^+$ Ions | (CH$_m$) Hydrocarbons |
|---|---|---|
| | C$_3$H$_3^\oplus$ (22) | C$_4$H$_4$ (33) |
| C$_4$H$_4$ (33) | C$_5$H$_5^\oplus$ (23) | C$_6$H$_6$ (35) |
| | (24) | (36) |
| (34) | (25) | (45) |
| C$_6$H$_6$ (35) | C$_7$H$_7^\oplus$ (26) | C$_8$H$_8$ (38) |
| (36) | (13) | (39) |
| (36) | (27) | (40) |
| (37) | (28) | (46) |

**TABLE II** (*Continued*)

| (CH)$_m$ Hydrocarbons | (CH)$_n^+$ Ions | (CH$_m$) Hydrocarbons |
|---|---|---|
|  (37) |  (29) | 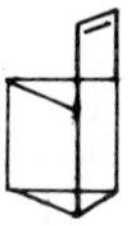 (47) |
| C$_8$H$_8$ 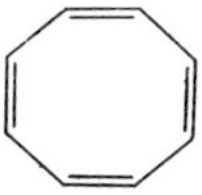 (38) | C$_9$H$_9^{\oplus}$  (30) | C$_{10}$H$_{10}$  (48) |
| 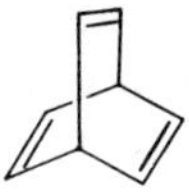 (39) | 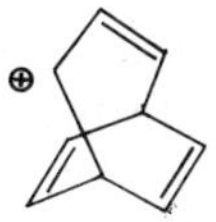 (14) | 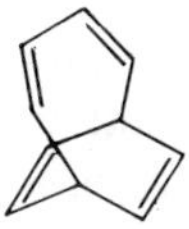 (49) |
| 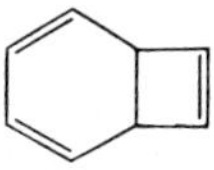 (40) | 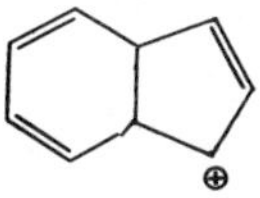 (16) | 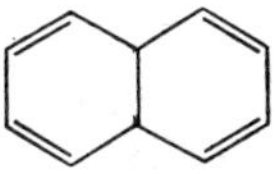 (50) |
| 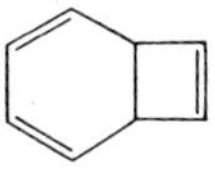 (40) | 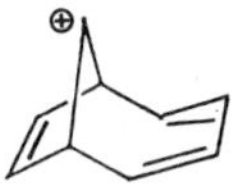 (31) | 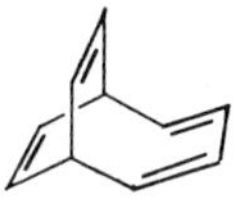 (49) |
|  (41) |  (32) |  (51) |
| 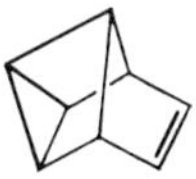 (42) | 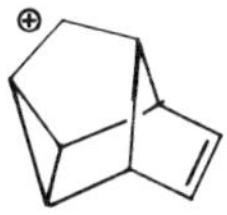 (15) | 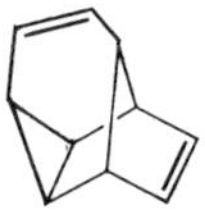 (52) |

(*continued*)

## TABLE II (continued)

| (CH)$_m$ Hydrocarbons | (CH)$_n{}^+$ Ions | (CH$^+$) Hydrocarbons |
|---|---|---|
| (43) | (19) | (53) |
| (44) | (10) | (54) |

apparent, and the multiplicity of possible structures increases with the larger number of carbon atoms. In this chapter only those C$_9$H$_9{}^+$ ions that have been studied to some degree are discussed (see Section III-D). Their structures are shown in Table II.

### A. C$_3$H$_3{}^+$ Ions

The simplest example of a (CH)$_n{}^+$ potentially degenerate carbonium ion is the C$_3$H$_3{}^+$ cyclopropenyl cation **22**; due to aromaticity, all three carbon atoms are equivalent. This system is not degenerate by the definition

(22)

advanced in Section I. In principle, however, cation **22** could exhibit true degenerate rearrangements if the carbon atoms or the hydrogen atoms were to interchange positions. Such a possibility could be detected by studying ion **55**, or merely a $^{13}$C and deuterium doubly-labeled alternative. Such an investigation has not yet been carried out. Recently cation **22** was obtained by two independent routes (21), and the carbonium ion salt

(55)   (55')   (55'')   etc.

**22** showed a single peak at $\tau - 0.87$ in fluorosulfonic acid (21a). These results indicate both the aromatic character of the ion and that all carbons and hydrogens are equivalent.

(56) $\xrightarrow[-20°]{SbCl_5 / CH_2Cl_2}$ (22') $SbCl_6^-$  Ref. 21a

(57) $\xrightarrow[-40°]{ClSO_3H}$ (22") $ClSO_3^- + CO + MeOH$  Ref. 21b

## B. $C_5H_5^+$ Ions

In the $C_5H_5^+$ family of cations there are several potentially degenerate carbonium ions. The first of these is the cyclopentadienyl cation **23**. Presumably all carbon atoms in this ion could become equivalent by resonance. However, as in the case of the cyclopropenyl cation, this is not an example of true carbon degeneracy. The cyclopentadienyl cation is thought not to be an aromatic system and is expected to be destabilized. The pentaphenyl (22a,23) and pentachloro (22b,23) derivatives have been prepared. The former has a singlet and the latter a triplet ground state (23).

A second degenerate ion possibility in the $C_5H_5^+$ group is the bicyclo-[2.1.0]pentenyl cation **24**. This ion is capable of rearranging to regenerate its own structure, but the carbon atoms do not interchange with each other. Cation **24** may also rearrange into the cyclopentadienyl cation **23** or the homotetrahedryl cation **25**.

These rearrangement possibilities indicate to some degree the interconvertibility of the $C_5H_5^+$ ions. However, all the cations (**23–25**) are destabilized by either ring strain or antiaromatic character.

(24) → (23)

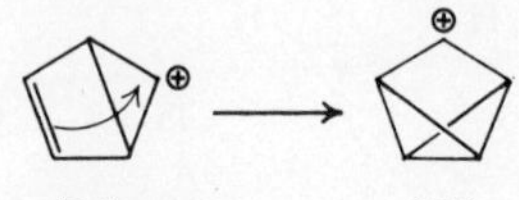

(24) → (25)

The first example of a truly degenerate carbonium ion in the $(CH)_n^+$ series is the homotetrahedryl or tricyclo[2.1.0.0$^{2,5}$]pentyl cation **25**. In this ion there is the possibility of complete carbon degeneracy resulting in the positional exchange of all carbon–hydrogen units through 1,2-carbon shifts. If each carbon–hydrogen unit were somehow uniquely identified, it has been calculated [(number of carbon atoms)! divided by the molecular symmetry number or $5!/2 = 60$], there would be 60 possible arrangements for **25** (24).

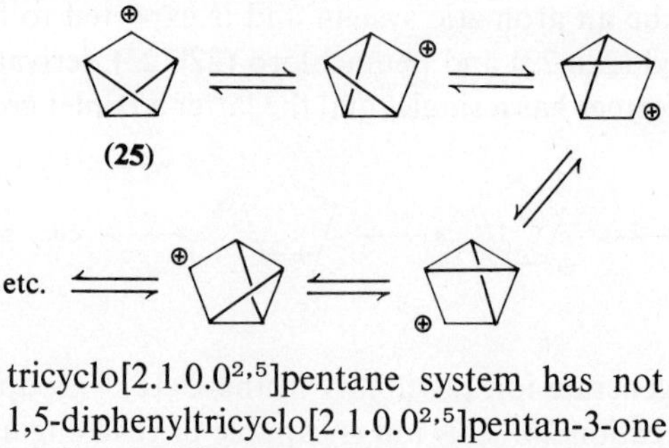

(25) ⇌ ⇌

etc. ⇌ ⇌

The parent tricyclo[2.1.0.0$^{2,5}$]pentane system has not yet been synthesized, but 1,5-diphenyltricyclo[2.1.0.0$^{2,5}$]pentan-3-one **58** (25), 1,5-diisopropyltricyclo[2.1.0.0$^{2,5}$]pentan-3-one **58a** (25), and 1,5-dimethyltricyclo[2.1.0.0$^{2,5}$]pentan-3-one **59** (26) have been prepared. Solvolysis of the *p*-nitrobenzoate **60** has been carried out (27), but no structure was

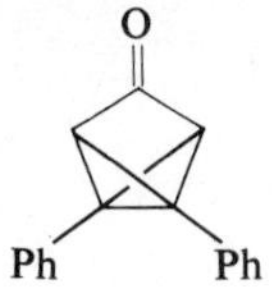

(58) Ph, Ph
(58a) *i*-Pr, *i*-Pr
(59) Me, Me
(60) R = PNB; Ph, Ph
(61) R = H

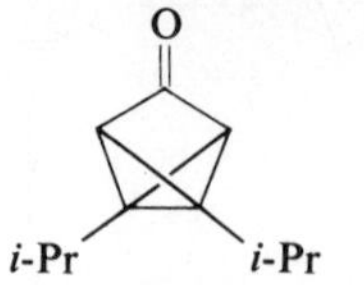

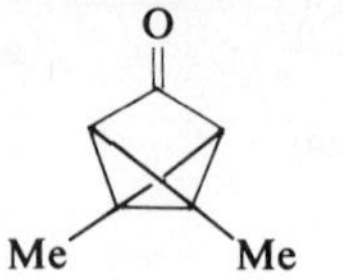

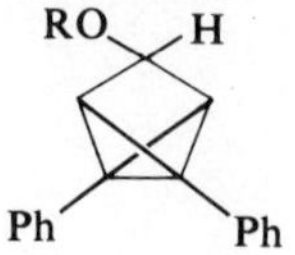

assigned with certainty to the solvolysis products. Solvolysis of **60** in 70% acetone indicated definitely only that ionization was accelerated and that the product was not 1,5-diphenyltricyclo[2.1.0.0$^{2,5}$]pentan-3-ol (**61**). The absence of **61** from the reaction products indicates that ion **62** cannot be the immediate precursor of the products.

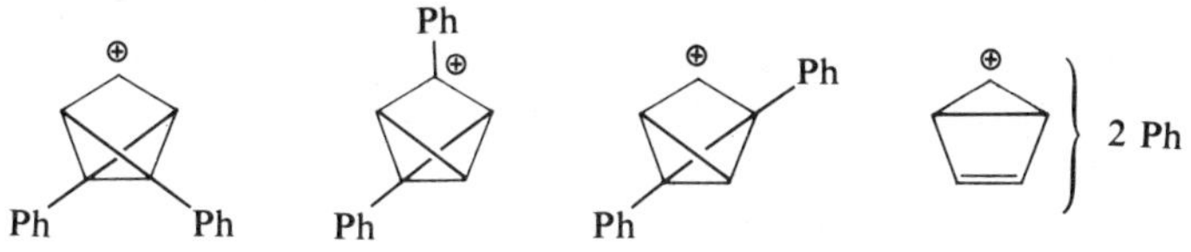

(**62**)  (**63**)  (**64**)  (**65**)

Possibly the products arise from ions like **63**, **64**, or **65**; however, further study is needed before the matter can be considered settled (27a).

In a fascinating theoretical treatment of the $C_5H_5^+$ ionic system, Stohrer and Hoffman have suggested the square pyramid **65a** as a possibly stable intermediate (28). Their calculations have shown that the cation **65b**

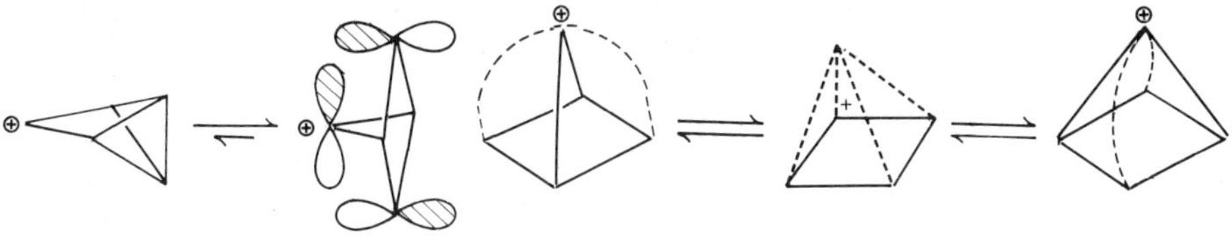

(**65b**)  (**65c**)  (**65a**)

may be unstable relative to the "bond stretch" isomer **65c** (only one of the three molecular orbital representations is shown here). Structure **65c** then can undergo a bond-switching process, as indicated diagrammatically, in a symmetry-allowed sequence via the square pyramid **65a**, which, Hoffmann suggests, represents an energy minimum. Formally, **65a** can be visualized as a $CH^+$ unit above a cyclobutadiene ring. Total degeneracy could be achieved by **65a** via the symmetry-allowed pseudorotation sequence $C_{4v} \rightarrow C_s \rightarrow C_{2v} \rightarrow C_s \rightarrow C_{4v}$ as indicated.

Substituted precursors of **65b** are known, but the pyramidal $C_5H_5^+$ ion could be approached from a number of directions. The cationic representation **65d** is expected to collapse to **65a**. Hoffmann believes that the

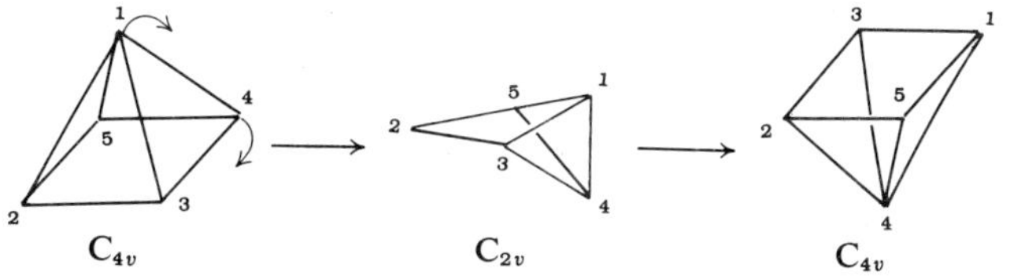

$C_{4v}$   $C_{2v}$   $C_{4v}$

cation **65e** might exist as the square pyramid **65a**. Although examples of **65e** have been investigated for R = Cl and R = φ (see above), albeit without considering the rearrangement formulated by Hoffmann, we suggest that the compound **65f** may have sufficient thermal stability to serve as a precursor for **65d**.

Another system, analogous to **65e** and synthetically readily accessible, is the fluorenyl cation **65g**. By choosing suitable substituents (X in **65g**), the possible rearrangement to **65g'** via **65h** could be observed. (However, see 98a.)

## C. $C_7H_7^+$ Ions

### 1. Tropylium Ion

Of the $C_7H_7^+$ group of potentially degenerate carbonium ions, the monocyclic tropylium ion **26** represents the simplest example. It has been shown to possess aromatic character and a planar $D_{7h}$ structure (29a). The carbon and hydrogen atoms are equivalent, and this is demonstrated in the ion's nmr spectrum, which consists of a single line. By our definition, this system is not truly degenerate, since equivalence arises through resonance; from the nmr spectrum alone, no information of such degeneracy can be found. The cation **26** could exhibit degeneracy if it could be shown that the carbon atoms interchange relative positions, as in the representations **26′, 26″,** and **26‴**.

Such a labeling study has in fact been carried out, and degeneracy has indeed been demonstrated. Siegel has studied the rearrangement of toluene, labeled with $^{13}$C in the 2 and 6 positions, as indicated in structure **26a**, in the mass spectrum (29b). Toluene has been shown to decompose in the mass spectrum via the tropylium ion to ethylene and a $C_5H_5{}^+$ species.

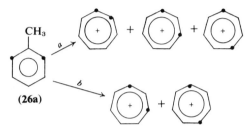

(26a)

Degeneracy would be demonstrated only if path *a* were followed, and the manifestation of this would be the production of $C_5H_5{}^+$, which bears no label. Siegel's results clearly revealed the presence of unlabeled $C_5H_5{}^+$, and hence the degeneracy of tropylium ion, at least under such high-energy conditions. It remains to be shown whether the same degeneracy exists in tropylium ion in solution.

## 2. 7-Norbornadienyl Cation and Bicyclo[3.2.0]heptadienyl Cation

The 7-norbornadienyl cation **13** undergoes a degenerate rearrangement that leads to the equivalence of five carbon atoms in the ion (30). In a search for "bridge flipping" (the interconversion of **13** to **13′** through **66**—Fig. 1) in this species, it was observed that the nmr spectrum at $-77°$ in $FSO_3H$ showed a broadening of the signals for protons bound to carbons C-7, C-1, C-4, C-5 and C-6, but not for those bound to C-3 or C-2. The nature of the rearrangement was determined by studying the nmr spectra of the ions derived from the labeled precursors **67–69**.

Fig. 1 NMR spectrum of **13**: H-1 and H-4: $\tau = 4.88$; H-5 and H-6: $\tau = 3.90$; H-7: $\tau = 6.73$: H-2 and H-3: $\tau = 2.54$.

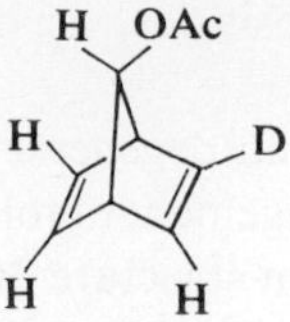

(67) 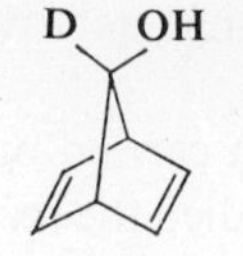 (68) 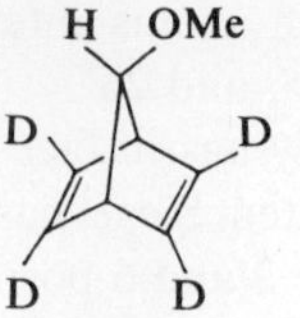 (69) 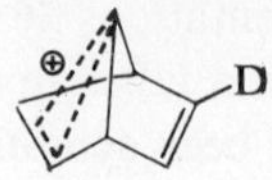 (70)

In FSO$_3$H at $-73°$, 67 exhibited the same four-signal spectrum as that of 13 with the intensity of the signal for H-5 and H-6 (unbound vinyl) one-half that of the signal for H-2 and H-3 (bound vinyl). When the cation 70 solution was warmed to $-47°$, the deuterium label scrambled over positions 1, 7, 4, 5 and 6 with a rate constant of $3 \times 10^{-4}$ sec$^{-1}$ but not over positions 2 and 3. This was evidenced in the nmr spectrum by the peak intensities approaching a ratio of 2:1, 6:1, 6:0.8 for the bound vinyl, unbound vinyl, bridgehead, and bridge protons, respectively. It was found that the deuterium label is incorporated sequentially at the different positions. The label appeared first at the bridgehead positions, 1 and 4, and then at the bridge position, 7. Similar rearrangements at similar rates were also observed for ions 71 and 72 derived from precursors 68 and 69 (see Table III). These rearrangements indicate the existence of partial carbon degeneracy in the 7-norbornadienyl cation.

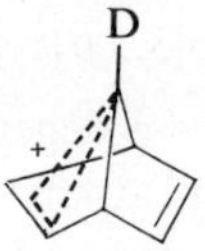

(71) 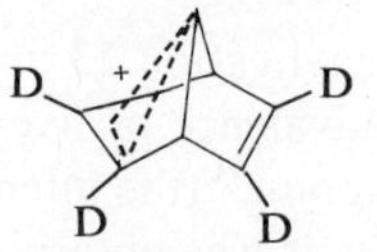 (72)

The mechanism proposed for the 5-carbon degenerate rearrangement is ring contraction to the bicyclo[3.2.0]heptadienyl cation 27 followed by

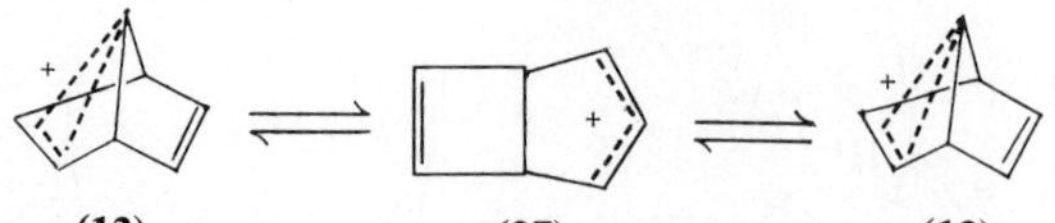

(13) ⇌ (27) ⇌ (13)

ring expansion back to the 7-norbornadienyl cation 13. For labeled ion 71 this would give rise to a sequence in which the bound vinyl group moves stepwise around the five-membered ring of 73 (see next page).

Evidence to support this mechanism comes from the nmr spectra of ions derived from the deuterated cis- and trans-bicyclo[3.2.0]heptadienols 74 and 75.

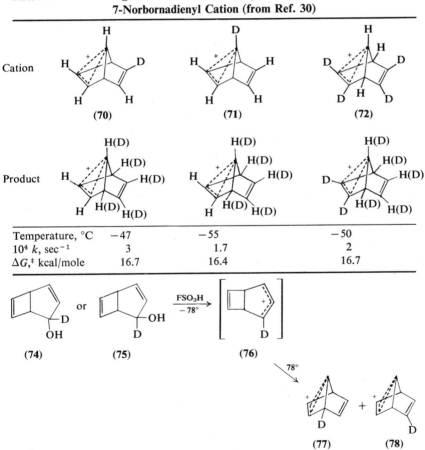

## TABLE III
Rates and Free Energies of Activation of Degenerate Rearrangements of the 7-Norbornadienyl Cation (from Ref. 30)

| Cation | (70) | (71) | (72) |
|---|---|---|---|
| Product | | | |
| Temperature, °C | −47 | −55 | −50 |
| $10^4$ $k$, sec$^{-1}$ | 3 | 1.7 | 2 |
| $\Delta G^{\ddagger}$, kcal/mole | 16.7 | 16.4 | 16.7 |

When either alcohol was extracted into $FSO_3H$ at $-78°$, and the carbonium ion solution was observed within 120 sec, the nmr spectrum was that of ion **77** plus ion **78** with no trace of signal for cation **76**.

This result agrees with a mechanism in which the first formed [3.2.0] cation ring expands to the [2.2.1] ion with either carbon of the cyclobutene vinyl function undergoing a 1,2 shift as the migrating vinyl group becomes the bound vinyl in **77** and **78**. The authors noted that neither ring contraction nor ring expansion of these systems occurs in solvolyzing solvents where carbonium ion lifetimes are very short compared to those in $FSO_3H$.

(13″) —HOAc→ (79)  Ref. 31

(80) OPNB —HOAc→ (81) OAc  Ref. 32

The 7-norbornadienyl cation **13** also undergoes a slower rearrangement, which results in C-2 and C-3 becoming equivalent with C-5 and C-6 (33). This equilibration of bound and unbound vinyl groups in **13** was studied by observing the nmr spectrum of cation **72** at higher temperatures. Cation **72** at $-50°$ has a deuterium atom at each bound vinyl position but

(72) $\xrightarrow[\Delta G^\ddagger = 16.7]{-50°\quad k = 2 \times 10^{-4} \text{sec}^{-1}}$ $\xrightarrow[\Delta G^\ddagger = 19.6]{-2.5°\quad k = 8 \times 10^{-4} \text{sec}^{-1}}$

only 40% of a deuterium atom at each unbound vinyl position as a result of degenerate five-carbon scrambling. At about 0° proton incorporation into the bound vinyl positions is observed, and the bound and unbound vinyl signals approach equal intensity.

The second rearrangement process requires 3 kcal/mole more activation energy than that for the five-carbon scrambling. Coupled with the more rapid degenerate five-carbon rearrangement, all seven carbon atoms of the 7-norbornadienyl cation **13** could become scrambled (complete carbon degeneracy). A possible mechanism for the second rearrangement process is bridge flipping and the value of $\Delta G^{\ddagger}$ for the rearrangement (19.6 kcal/mole) represents a lower limit on the energy barrier for this process.

Mechanisms other than bridge flipping are possible for the equilibration of the two vinyl groups in **13** (e.g., a 1,2 shift of an unbound vinyl carbon as shown); some 7-substituted norbornadienyl cations were studied to

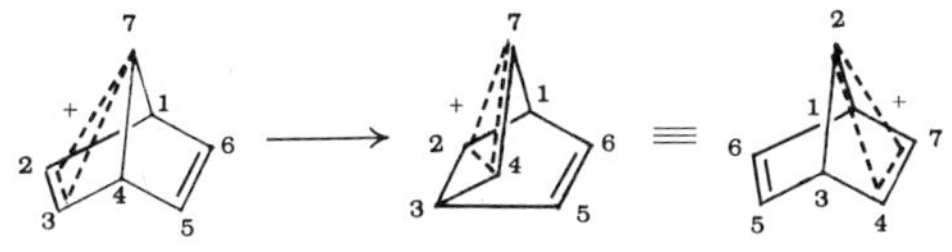

(13)

further test this hypothesis. The 7-methylnorbornadienyl cation **82** was investigated (33); the nmr spectrum of **82** in $FSO_3H$ gave evidence for an unsymmetrical species at $-45°$.

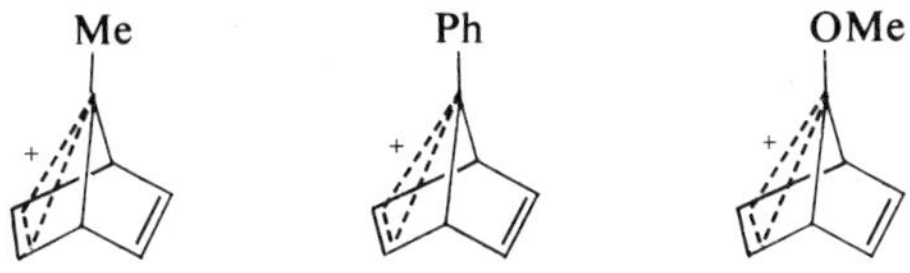

(82)  (83)  (84)

As the temperature was raised, the vinyl signals began to broaden and eventually coalesced at $-14°$. The methyl peak remained sharp at all times and the bridgehead signal sharpened to a pentuplet. These results indicated averaging of only the vinyl proton environments, which must have occurred by bridge flipping. The rate constant for this process at $-14°$ was

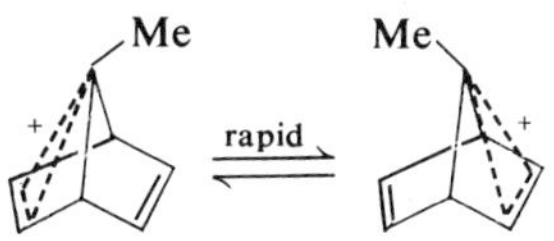

(82)  (82′)

189 sec$^{-1}$, and $\Delta G^{\ddagger}$ was 12.4 kcal/mole. The 7-phenylnorbornadienyl **83** and 7-methoxynorbornadienyl **84** cations were also studied under the assumption that substituents better able to accommodate a positive charge would stabilize a symmetrical cation more than an unsymmetrical cation (34). Both these cations at $-100°$ showed equivalence of all four vinyl protons. On examination of the chemical shifts of the phenyl hydrogens in **83**, it was concluded that the phenyl group was not sustaining considerable

positive charge. Hence it was concluded that **83** is best represented as unsymmetrical, yet undergoing rapid bridge flipping with a barrier of < 7.6 kcal/mole. Recent work on the 2-methyl-7-norbornadienyl cation **85** indicates that bridge flipping is not a favorable process when a substituent is located in one of the vinyl groups (35). It was observed that **85** decomposes rather than undergoing bridge flipping.

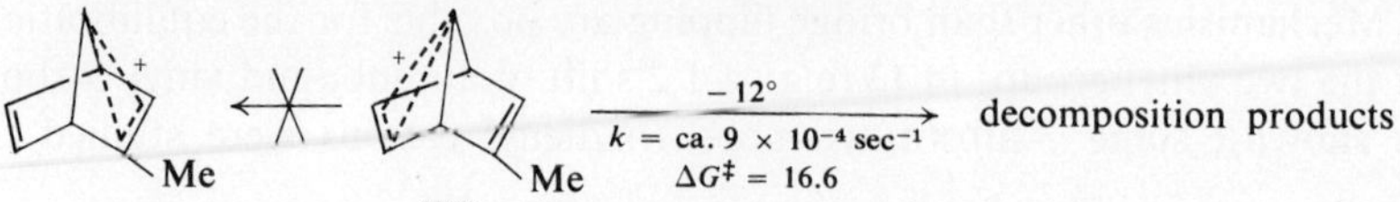

$$\text{(85)} \quad \xrightarrow[k = \text{ca. } 9 \times 10^{-4} \text{ sec}^{-1}]{-12°} \quad \text{decomposition products}$$
$$\Delta G^{\ddagger} = 16.6$$

## 3. Quadricyclyl or 3-Tetracyclo[3.2.0.0$^{2,7}$.0$^{4,6}$]heptyl Cation

The quadricyclyl cation **28** has the potential for complete carbon degeneracy if repetitive cyclopropylcarbinyl–cyclopropylcarbinyl rearrangements occur. The positive charge can be distributed to all carbons by such a mechanism involving both cyclopropyl rings. However, present

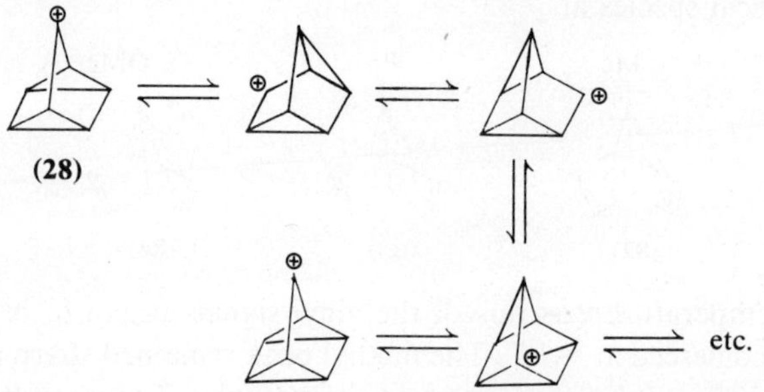

(28)

etc.

evidence supports structure **86** as best representing the carbonium ion intermediate due to the solvolysis of quadricyclane derivatives **87**. This charge delocalized cation would not necessarily exhibit the degenerate

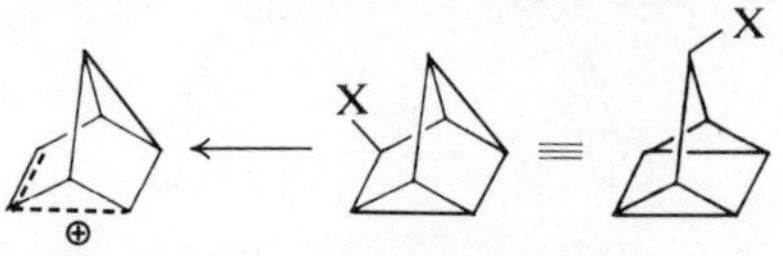

(86)   (87)

rearrangements as shown previously. Cation **86** was first proposed to account for the remarkable reactivity of quadricyclane derivatives (36,37). It was found that 7-quadricyclyl chloride **87**, X = Cl, was less reactive than *anti*-7-norbornenyl chloride **88** by a factor of only 100 and more reactive than 7-norbornyl chloride **89** by approximately $10^9$. Quadricyclane derivatives were found to be only about 30–40 times more reactive than the corresponding nortricyclane compounds **90**, which contain only one cyclopropane ring (36). The net effect of the second cyclopropane ring appears to be negligible. This is similar to the participation of only one carbon–carbon double bond in the 7-norbornadienyl cation (33).

The solvolysis of 7-deuterioquadricyclyl tosylate **91** at 25° for 2–3 days in acetic acid containing an excess of potassium acetate gave a 75% yield of a product mixture consisting of an approximately 50:50 ratio of **94** and **95** (38). Both the quadricyclyl acetate **94**, and the 7-norbornadienyl acetate **95** contained one deuterium, according to analysis of their nmr and mass spectra. This experiment indicates that the hydrogens of the quadricyclyl cation are not scrambled by rearrangement, thus ruling out a structure in which the positive charge is delocalized over two or more equivalent carbons. Hence the quadricyclyl cation represents an example of a nondegenerate ion. The results suggest the interconvertibility of the quadricyclyl and norbornadienyl cations **92** and **93**. The rearrangement of **92** to **93**, and subsequently to **95**, appears to be highly stereospecific.

### 4. Homoprismyl or 7-Tetracyclo[3.2.0.0$^{2,4}$.0$^{3,6}$]heptyl Cation

The homoprismyl cation **29** represents another potentially degenerate $C_7H_7^+$ carbonium ion. This ion has not yet been studied, but is unique in that it may undergo several possible rearrangements. It could exhibit degenerate properties through Wagner-Meerwein rearrangements as illustrated here. This kind of process would result in fourfold partial carbon degeneracy; the four carbons of the cyclobutane ring bearing the positive charge would become equivalent. A second kind of process, participation of the three-membered ring in a homocyclopropylcarbinyl rearrangement, could serve to delocalize the positive charge over three carbons.

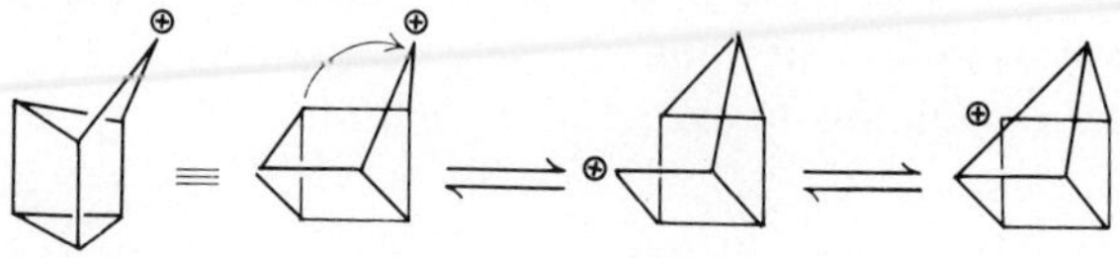

(29)

Combined with the Wagner-Meerwein rearrangement, this process would result in complete carbon degeneracy (charge dispersal over all seven carbon atoms). However, a third rearrangement process exists for **29**, that of cyclobutane ring participation. This process would serve to destroy the homoprismyl ring system and generate the quadricyclyl cation **28**.

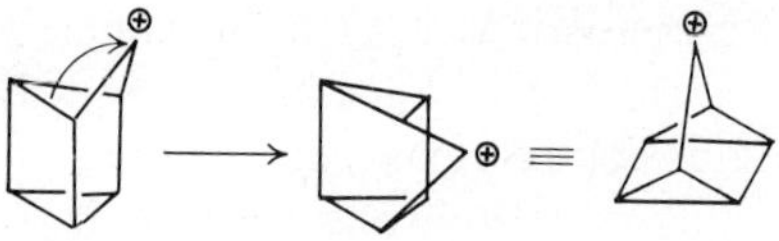

(29)                    (28)

The last rearrangement mentioned points out the relationship between the homoprismyl cation and other $C_7H_7^+$ ions. It is further interesting to note that **29** hypothetically could be formed by a photochemical cycloaddition within the bicyclo[3.2.0]heptadienyl cation.* The photochemical transformation of tropylium ion **26** to bicyclo[3.2.0]heptadienyl cation **27** has already been implicated by van Tamelen et al. in their isolation of the alcohol corresponding to **27** from aqueous, acidic solutions of **26** (41a). [The

---

* An analogous reaction is the photosensitized intramolecular cyclization of cyclopentadiene dimer to pentacyclo[5.3.0.0$^{2,5}$.0$^{3,9}$.0$^{4,8}$]decane; see Ref. 39.

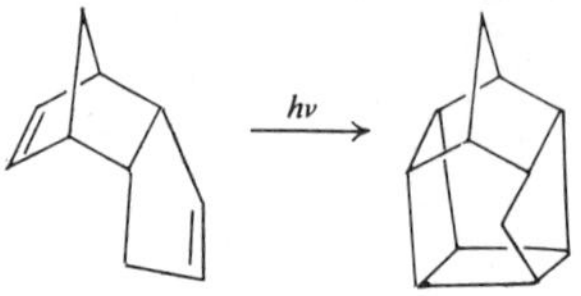

photochemical conversion of norbornadiene to quadricyclene has already been carried out (40)]. Irradiation of **26** in superacid gives **13** (41b).

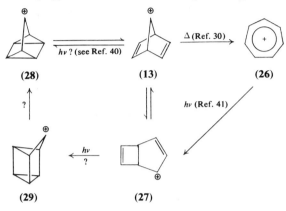

### D. $C_9H_9^+$ Ions

*1. 9-Homocubyl or 9-Pentacyclo[4.3.0.0$^{2,5}$.0$^{3,8}$.0$^{4,7}$]nonyl Cation*

The 9-homocubyl cation **10** represents one of the more interesting potentially degenerate carbonium ions within the $C_9H_9^+$ group. This cation, like the homotetrahedryl cation **25** (see Section III-B) and the homododecahedryl cation **96**, belongs to a series of homologous ions based on regular polyhedra (24). These three ions are similar in that they are all

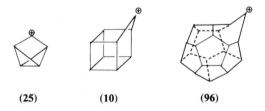

capable of complete carbon degeneracy by employing only 1,2 carbon shifts. All carbon–hydrogen units can exchange positions, thus making all carbon atoms and hydrogen atoms equivalent. The homododecahedryl cation **96** has not yet been investigated, and the parent hydrocarbon, dodecahedron, has not been synthesized.

Investigations by workers at Smith, Kline and French Laboratories and Princeton University were first to indicate that the predicted degeneracy of the 9-homocubyl cation **10** could be achieved under solvolytic conditions (24). The degeneracy of **10** was demonstrated by studying the deuterium labeling pattern obtained in the solvolysis products of the deuterated tosylate **102**. The synthesis of **102** involves photochemical cycloaddition of **98** to **99** and lithium tetradeuterioaluminate reduction of homocubanone **100** to form **101** as key steps. The nmr spectrum of **102** indicated essentially complete absence of the absorption peak for the

proton at C-9. The course of the rearrangements was followed by the appearance and integration of this signal in the spectra of the solvolysis products. Solvolysis of deuterated 9-homocubyl tosylate **102** exhibited good first-order kinetics in both acetic and formic acid, and in both cases the sole products isolated were 9-homocubyl acetate **103** and formate **104**, with the deuterium label scrambled. The acetolysis rate constant at 25° was approximately 400 times faster than that expected from the carbonyl frequency of **100**. This enhanced rate suggests that the 9-homocubyl cation **10** may have a bridged structure.

There are two mechanistic possibilities for deuterium scrambling in **10**. One possibility is a stereospecific process involving bridged or rapidly equilibrating ions, which would make five carbon atoms and their attached hydrogens equivalent, provided no stereochemical leakage occurred. This mechanism, which is illustrated, would scramble the deuterium label around only one five-membered ring.

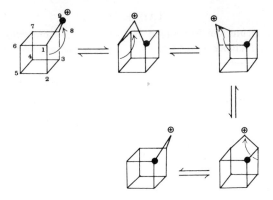

● = isotopic label

A second possibility, a nonstereospecific reaction path, would allow distribution of the positive charge and the deuterium label to all nine carbons. This would also permit the interchange of all carbon–hydrogen units in the manner exhibited by bullvalene (2).

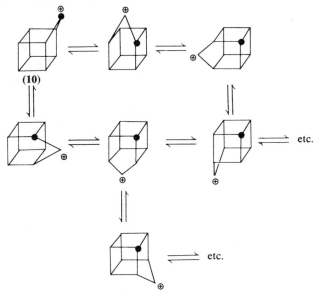

● = isotopic label

Table IV shows the labeling pattern calculated for each process at each rearrangement stage. The experimental results were the following: in buffered acetic acid at 125° for more than 10 half-lives, 36% deuterium remained at C-9 after solvolysis; in unbuffered acetic acid at the same temperature and time, 22.5% deuterium remained at C-9, indicating that

## TABLE IV
Percentage of Deuterium Remaining at C-9 in the 9-Homocubyl System as a Function of the Number of 1,2-Carbon–Carbon Rearrangements (from Ref. 24)

| Deuterium at C-9, % | Number of rearrangements | | | | | | | | |
|---|---|---|---|---|---|---|---|---|---|
| | 1 | 2 | 3 | 4 | 5 | 6 | 7 | 8 | ∞ |
| Stereospecific process | 50 | 37.5 | 31.3 | 27.3 | 24.8 | 23.1 | 22.1 | 21.4 | 20.0 |
| Nonstereospecific process | 50 | 37.5 | 31.3 | 27.0 | 23.7 | 21.2 | 19.3 | 17.7 | 11.1 |

more than five rearrangements had occurred; in refluxing unbuffered formic acid only 10% deuterium remained at C-9 with the other 90% deuterium distributed throughout the molecule. This preliminary experiment in formolysis indicated that the full degeneracy of **10** could be achieved but attempts to repeat this result have not succeeded (42).

Work by Pettit and Barborak on the homocubyl cation **10** further verified the findings described previously (43). These investigators em-

(111)

(112)

ployed both a different synthesis and a different deuterium labeling pattern to show that **10**, when generated in acetic acid, undergoes several stereospecific degenerate rearrangements before forming the acetate.

Two synthetic pathways *a* and *b* were used to prepare 9-homocubyl tosylate **110**. Substitution of perdeuteriocyclobutadienetricarbonyliron for **106** gave a tetradeuterio derivative of **110**. Path *a* afforded an equal mixture of the two tosylates **111** and **112**; path *b* gave only isomer **111**.

A sample of **111** and another consisting of an equal mixture of **111** and **112** was solvolyzed in acetic acid at 120° for 30 hr, and the acetates and unreacted tosylates were recovered. The percentage of deuterium at C-9 was measured to determine the extent of rearrangement. The amount of deuterium appearing at this position also depends on the stereospecificity of the rearrangement process. In a stereospecific process involving repeated migration of the carbon atom *trans* to the leaving group, **111** should show twice as much deuterium at C-9 as the mixture of **111** and **112**, since **112** cannot generate deuterium at C-9 in such a process. In a nonstereospecific rearrangement process both samples should give the same amount of deuterium at C-9. The acetate formed from **111** showed 26.1% deuterium at C-9, whereas that from **111** and **112** showed 12.3%. The rearrangement appears to be stereospecific under these conditions, suggesting, in combination with the previously cited results, that almost four Wagner-Meerwein shifts had occurred during solvolysis (see Table V).

The results just given compare favorably with those of the workers at Smith, Kline and French Laboratories and Princeton University. However, the recovered tosylates from **111** and **112** were also examined, and these contained deuterium at C-9 to a larger degree than was observed for the acetates. Approximately twice as much deuterium appeared at C-9 in the

## TABLE V
**Percentage of Deuterium at C-9 Accompanying 1,2 Shifts of 111 (from Ref. 43)**

| No. of 1,2 shifts | 0 | 1 | 2 | 3 | 4 | 5 | 6 | ∞ |
|---|---|---|---|---|---|---|---|---|
| Deuterium at C-9, % | 0 | 0 | 12.5 | 21.9 | 28.1 | 32.2 | 35.0 | 40.0 |

reaction of **111** as with the mixture of **111** and **112**. This result indicates stereospecific rearrangement through internal return of the tosylate ion pair and contrasts with the result of the first group of workers, which indicated that stereochemical leakage might at least occur in formolysis. Broussard and Pettit have most recently learned that, in addition to the stereospecific rearrangement process, the attacking nucleophile approaches on the side from which the leaving group departs, thus producing material in which stereochemistry is retained (44).

Studies have also been carried out on the bishomocubyl systems **113** and **114** (45). The solvolysis of these tosylates is of interest because the cations

**(113)** **(114)**

generated from **113** and **114** undergo rearrangements similar to the homocubyl cation **10**. In these cases the migration of the second methano carbon bridge serves as a means of detecting the rearrangements. The acetolysis (unbuffered) of tosylates **113** and **114** gave the acetates **116** and **119** as major products, which can be rationalized as arising via 1,2-alkyl shifts through cations **115**, **115′**, and **117**, followed by substitution with retention of configuration. Alternatively, the results can be explained by assuming that the bridged carbonium ions **120** and **121** are intermediates in the solvolyses of **113** and **114**, respectively. Recent work on the deuter-

**(113)** **(115)** **(115′)** **(116)**

ated tosylate **122** supports the intermediacy of the bridged ion **120** or its equivalent of two rapidly equilibrating classical ion pairs in the solvolysis of **113** (46). The results of the solvolyses of the bishomocubyl systems are in direct analogy with those for the homocubyl compounds.

### 2. 9-Pentacyclo[4.3.0.0$^{2,4}$.0$^{3,8}$.0$^{5,7}$]nonyl Cation

The 9-pentacyclo[4.3.0.0$^{2,4}$.0$^{3,8}$.0$^{5,7}$]nonyl cation **19** is another ion of the $C_9H_9^+$ group possessing the unique structural and symmetry properties that could result in multiple carbon degeneracy. Cation **19'**, derived from the solvolysis of **127**, is an analog of the *cis*-3-bicyclo[3.1.0]hexyl cation

(47) (see Section IV-N) and should be capable of threefold degeneracy by means of a homocyclopropylcarbinyl rearrangement **127 → 19′**. There is also the possibility of a bridge-flip rearrangement, which can again regenerate a cation identical with the original. The combination of both processes at the same time would make all positions interchangeable and all nine carbon atoms would become equivalent. An alternative way of

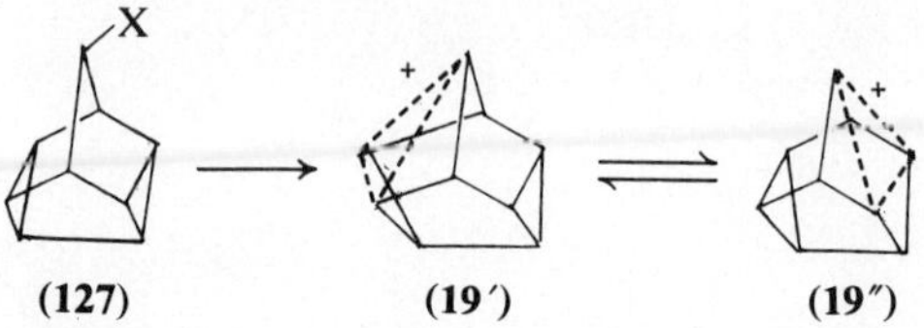

(127)   (19′)   (19″)

representing this possibility of complete degeneracy in **19** is by successive, discrete carbon–carbon bond shifts from the two participating cyclopropyl rings (all ions formed are identical).

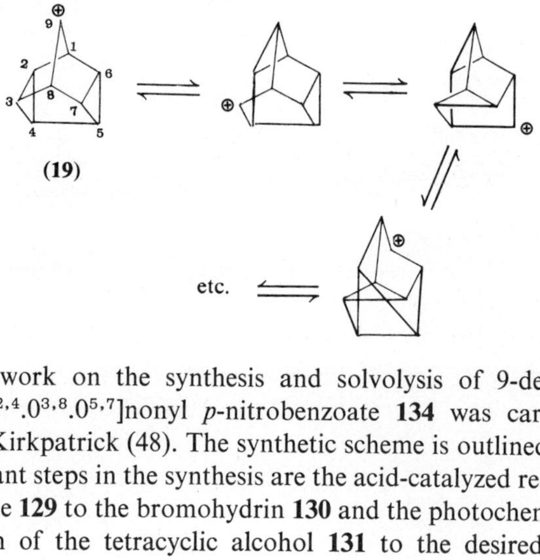

(19)

etc. ⇌

Recently, work on the synthesis and solvolysis of 9-deuteriopentacyclo[4.3.0.0$^{2,4}$.0$^{3,8}$.0$^{5,7}$]nonyl *p*-nitrobenzoate **134** was carried out by Coates and Kirkpatrick (48). The synthetic scheme is outlined below. The most important steps in the synthesis are the acid-catalyzed rearrangement of the epoxide **129** to the bromohydrin **130** and the photochemical valence isomerization of the tetracyclic alcohol **131** to the desired pentacyclic alcohol **132**.

The solvolysis of the unlabeled *p*-nitrobenzoate **135** in 65% aqueous acetone at 125.0° exhibited good first-order kinetics with a rate constant of $7.0 \times 10^{-5}$ sec$^{-1}$. Under the same conditions, 7-norbornadienyl *p*-nitrobenzoate (**136**) solvolyzed with a first-order rate constant of $5.8 \times 10^{-4}$ sec$^{-1}$ (49). Therefore, $k_1(\mathbf{135})/k_1(\mathbf{136})$ equals 0.12 (17).

The large rate enhancement observed for **135** cannot be attributed to an overall relief of strain upon cyclopropane ring opening, since **132** is the major reaction product. The acceleration most probably is a result of the geometry in this system, which is particularly favorable for homocyclopropylcarbinyl participation in the solvolysis transition state. This hypothesis was previously advanced for the similar *endo-anti*-8-tricyclo-[3.2.1.0$^{2,4}$]octyl *p*-nitrobenzoate (**137**) (16), which solvolyzes approximately 80 times faster than **135**.

Solvolysis of **134** for 4–5 half-lives regenerated **132**, in which $^1$H–nmr absorption at position C-9 had reappeared. The extent of the deuterium scrambling was determined by the $^2$H– and $^1$H–nmr spectra of the product. The $^2$H–nmr spectrum showed two peaks at τ 5.8 and 8.2 in a ratio of

1:2.05. This indicated that one-third of the deuterium label remained at C-9 and the other two-thirds was distributed between two other equivalent positions. Quantitative integration of the $^1$H–nmr spectrum of scrambled **132** showed the presence of 0.67 proton at C-9. Also, the signal at $\tau$ 8.2 was reduced from an area of 2 to 1.3 protons, and the remainder of the spectrum was essentially unchanged. These results provide evidence for the triply degenerate carbonium ion intermediate **19**. Further rearrangement through participation of the second cyclopropane ring did not occur under conditions employed.

### 3. Tetracyclo[4.3.0.0$^{2,4}$.0$^{3,7}$]non-8-en-5-yl Cation

The tetracyclo[4.3.0.0$^{2,4}$.0$^{3,7}$]non-8-en-5-yl cation **15** is capable of partial carbon degeneracy through a combination of cyclopropylcarbinyl and homoallylic rearrangements. Transformation of **15** into ion **138** is readily conceivable; **138**, with a $C_2$ symmetry axis through C-2 and the

midpoint of the C-6–C-7 bond, should be convertible into **15'** or back into **15** by homoallylic rearrangement. If ion **15** contains a deuterium label at C-5, repetitive operation of the foregoing mechanism will eventually result in deuterium scrambling to the four positions C-4, C-5, C-8, and C-9 because of the equivalence of these four carbon atoms. The five carbon atoms of the central five-membered ring, C-1, C-2, C-3, C-6, and C-7, would also become equivalent by such a process, but they would bear no deuterium label.

Schleyer and Leone were the first to study cation **15** (50). The synthesis is primarily based on the homo Diels-Alder reaction of dimethyl acetylenedicarboxylate with **139** to form **140**, which contains the desired tetracyclic ring system. Acid cleavage of the *tert*-butyl ether **142** to form acetate **143** is also an important step. Alcohol **131** is the same as that synthesized independently by Coates and Kirkpatrick (see Section III-D-2) (48). The saturated derivative **146** was also prepared for comparison purposes.

The solvolysis of **145** in acetic acid buffered with an excess of sodium acetate at 110° for 3 hr (over 10 half-lives) led to the formation of 5-deuterioacetate (**147**) as the sole product, with the deuterium completely unscrambled. Attempts to study the solvolysis under more vigorous

# DEGENERATE CARBONIUM IONS

(139) → (140) MeO₂C, CO₂Me — 2 steps → (141) HO₂C, CO₂H

(131) OH ← (143) OAc ← (142) O-t-Bu

(144) O — 2 steps → (145) D, OTs  (146) OTs

conditions with unbuffered acetic or formic acid failed because of solvent addition to the double bond. Table VI provides a summary of the acetolysis rate constants for the undeuterated tosylate **148**, the saturated derivative **146**, and nortricyclyl tosylate **149**.

The results obtained are somewhat surprising, since the solvolysis of nortricyclyl tosylate **149** produces some rearranged product, norborn-2-

**TABLE VI**
**Acetolysis of 148, 146, and 149 (from Ref. 50)**

|  | (148) | (146) | (149) |
|---|---|---|---|
| $k$, sec$^{-1}$ at 75° | $1.85 \times 10^{-4}$ | $7.38 \times 10^{-3}$ | $1.62 \times 10^{-3}$ |
| $k_{rel}$ | 0.11 | 4.6 | 1.0 |
| $v_{CO}$, cm$^{-1}$ | 1755 | 1765 | 1762 |

en-5-yl acetate **150** (51). Also, work by Freeman and Balls on the related 8-tetracyclo[4.3.0.0$^{2,4}$.0$^{3,7}$]nonyl cation (**21**) indicated facile rearrangement of the following type (see Section IV-V) (52). Since no deuterium scrambling occurred in the solvolysis of **145**, we can conclude that "leakage" to **138** and subsequent homoallylic rearrangement was unfavorable. Apparently cyclopropyl ring opening in **15** does not occur because the added two-carbon bridge (C-8–C-9) causes more strain in the norbornene than in the nortricyclene system. Both the inductive effect and the steric effect of the carbon–carbon double bond may cause the solvolysis rate of **148** to be significantly slower than that of **146**. The carbonyl frequency of ketone **144** indicates that there is a significant rate acceleration in the ionization of **148**, probably because of cyclopropylcarbinyl delocalization.

Work by Klumpp serves independently to verify the results described above on cation **15** (53). By employing a synthesis related to that of Schleyer and Leone, the tetracyclic compounds **151**, **145**, and **152** were prepared. The solvolytic studies indicated were carried out in an attempt to generate a degenerate cation corresponding to **15**. No deuterium scrambling was observed in any of the cases studied. Klumpp feels that the cause for non-rearrangement is a high activation energy for the transfer of positive charge between two equivalent cyclopropylcarbinyl-homoallyl configurations within the same molecule (**15** and **15'**). The Grignard reagent from **152** was also prepared and hydrolyzed to give **154**. Interestingly enough, it was found that some deuterium ($\approx 35\%$) scrambled to C-8 and C-9 under these conditions. An exact interpretation of this result still requires further investigation.

The results obtained by both groups clearly show that the tetracyclo-[4.3.0.0$^{2,4}$.0$^{3,7}$]non-8-en-5-yl cation (**15**) is an unusual example of a non-degenerate ion, despite the possibilities for degenerate rearrangement. It may be significant that the Grignard reagent of this system rearranged (presumably through a carbanion type intermediate), but further studies to determine the mechanism of deuterium scrambling are necessary.

### 4. Bicyclo[3.2.2]nona-3,6,8-trien-2-yl Cation

The bicyclo[3.2.2]nona-3,6,8-trien-2-yl cation **14**, a vinylog of the 7-norbornadienyl cation, can potentially exhibit degenerate properties through 1,2-vinyl shift mechanisms or through the equivalent homoallylic barbaralyl cations **32**. Combination of either of these rearrangement processes with allyl resonance (which delocalizes the positive charge between C-2 and C-4) would result in the eventual distribution of a deuterium at C-2 to all nine carbon atoms of the ion.

Goldstein and Odell sought to determine whether cation **14** might exhibit unusual stability or instability as a six $\pi$-electron "bicyclotropylium" ion (**155**) (54). To this end, bicyclo[3.2.2]nona-3,6,8-trien-2-one (**159**) was first synthesized. Reduction of **159** afforded bicyclo[3.2.2]nona-3,6,8-trien-2-ol (**160**), which when treated with ethereal AlCl₃ gave a 58% yield of tricyclo[3.3.1.0$^{2,8}$]nona-3,6-dien-9-ol or "barbaralol" (**161**).

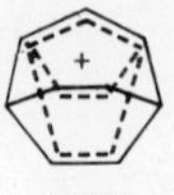

(155)

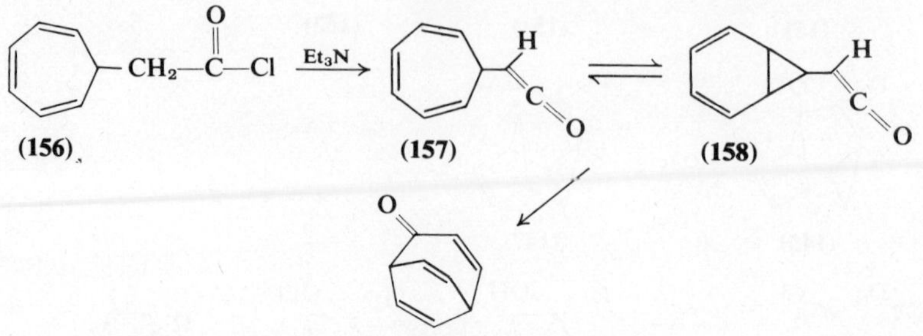

(156) → Et₃N → (157) ⇌ (158)

(159)

Similarly, reduction of **159** with LiAlH₄–AlCl₃ gave 30% barbaralol (**161**), 30% barbaralane (**162**), and 13% indan (**163**). No bicyclo[3.2.2]nona-2,6,8-triene (**164**) was found in the product mixture. These observations suggested not only that the barbaralane skeleton is more stable than the

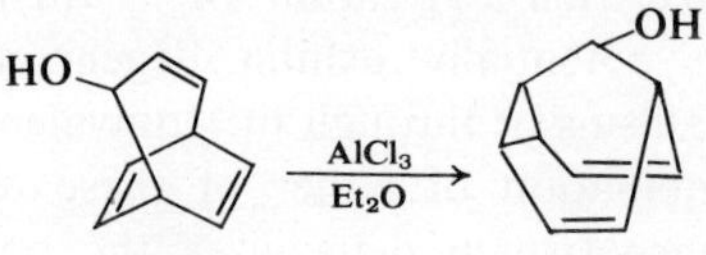

(160) —AlCl₃/Et₂O→ (161)

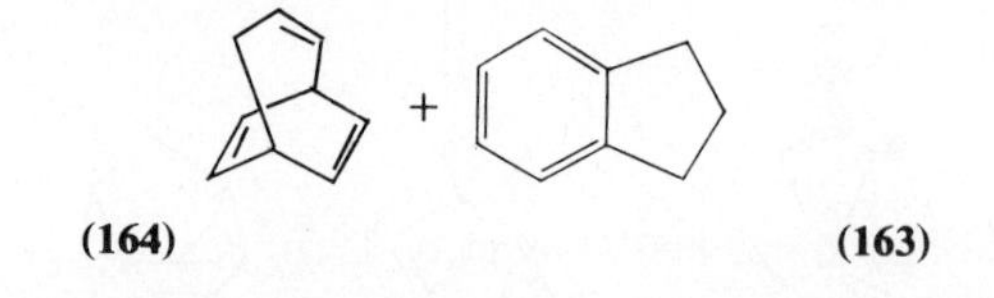

(159) —LiAlH₄/AlCl₃→ (161) + (162)

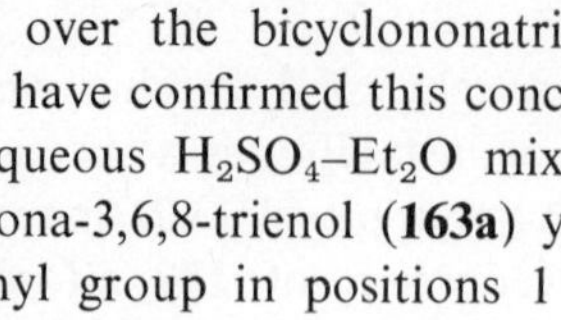

(164) + (163)

bicyclo[3.2.2]nona-2,6,8-trienyl skeleton, but also that the barbaralyl cation **32** is preferred over the bicyclononatrienyl cation **14**. Recent extensive investigations have confirmed this conclusion. Under very mild conditions, e.g., 5% aqueous $H_2SO_4$–$Et_2O$ mixtures for 5 min at 25°, 2-methylbicyclo[3.2.2]nona-3,6,8-trienol (**163a**) yields a mixture of barbaralols with the methyl group in positions 1 and 2 (**163h** and **163i**, respectively) (58b). On the other hand, the alcohol **163a** gives the 9-methyl-9-barbaralyl cation **163b** at −135° in $FSO_3H$–$SO_2ClF$–$CD_2Cl_2$;

the same ion can be obtained directly from **163c** (55,56). Similarly, under the same conditions, **160** gives the 9-barbaralyl cation **32** (57). (The behavior of ions **32** and **163b** is discussed in Section III-D-7.)

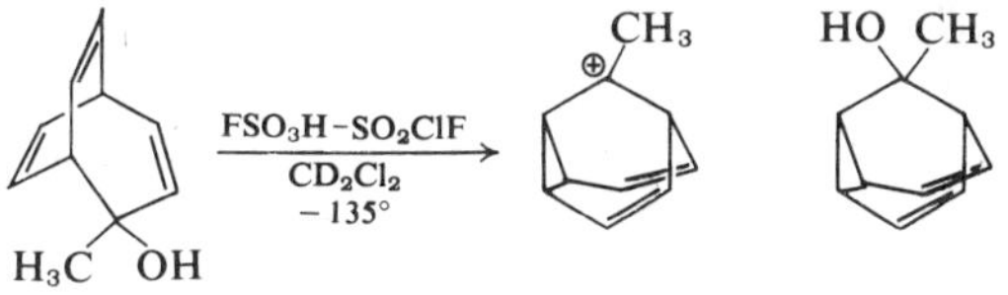

(163a)   (163b)   (163c)

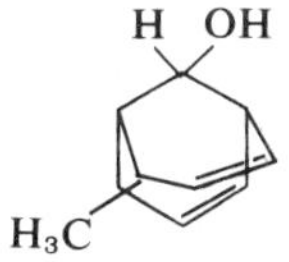

(163h) 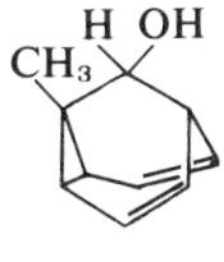   (163i)

Additional information is revealed by solvolysis of deuterium labeled bicyclo[3.2.2]nona-3,6,8-trien-2-yl 3,5-dinitrobenzoate (**163d**) (58a) and the corresponding *p*-nitrobenzoate (59). From **163d** only 9-barbaralol

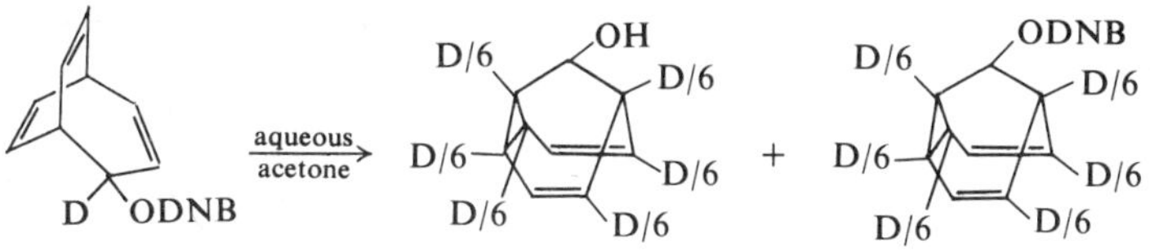

(163d)   (163e)   (163f)

**163e** and the return product **163f** are obtained: in both these products the deuterium label has become virtually statistically scrambled to six *but to only six* of the *nine* available positions, as shown.* This demonstrates that the conversion of cation **14** to **32** is not a simple, direct process. (The possible mechanism whereby the observed scrambling takes place is discussed in Section III-D-7.)

Experimental results seem to indicate that cation **14** possesses no unusual stability as might be depicted by formulation **155**. The solvolysis rates of bicyclo[3.2.2]nona-3,6,8-trien-2-yl derivatives, e.g., **163d**, are not accelerated, and, perhaps, are even slower than one might anticipate for allyl-stabilized systems (58,59). Goldstein has given a theoretical justification for cation **14** not possessing exceptional stability in terms of the concept of "bicycloaromaticity" (60). By using MO symmetry arguments.

---

* It should be kept in mind that barbaralyl derivatives undergo rapid degenerate Cope rearrangement (67) which contribute to the extent of deuterium scrambling observed in the solvolysis products **163e** and **163f**.

it was shown that the 2-bicyclo[3.2.2]nonatrienyl cation **14** should be *destabilized* and *antibicycloaromatic*; however, under these conditions the corresponding anion **165** should be *stabilized* and *bicycloaromatic*. As an experimental verification of this prediction, anion (**165**) was recently prepared by Winstein and Grutzner, who interpreted its nmr spectrum to indicate a delocalized structure (**166**) (61).

(**165**)    (**166**)

Another revealing observation was made by Winstein and Grutzner concerning anion **165**. Their preliminary experiments indicated the occurrence of a degenerate nine-carbon rearrangement in **165** (61). The sequence outlined below was carried out. The hydrocarbon **164** exhibited essentially complete deuterium scrambling to all carbon atoms. The nmr spectrum of **169** immediately after preparation showed no scrambling. This carbon degeneracy could be formulated as in the case of cation **14** as proceeding through a series of 1,2-vinyl shifts or through a barbaralyl intermediate.

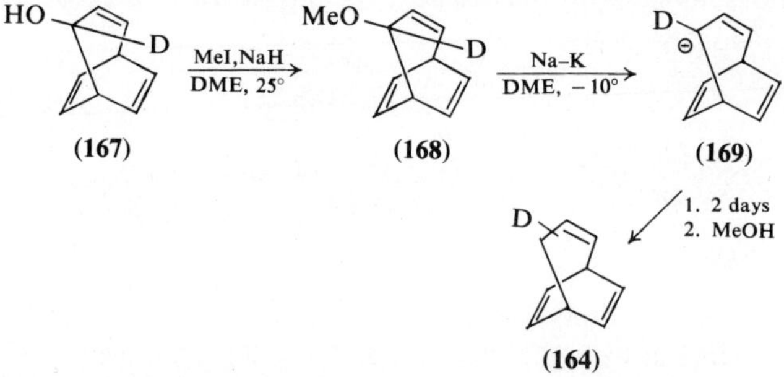

(**167**) $\xrightarrow[\text{DME, 25°}]{\text{MeI, NaH}}$ (**168**) $\xrightarrow[\text{DME, }-10°]{\text{Na-K}}$ (**169**)

$\xrightarrow[\text{2. MeOH}]{\text{1. 2 days}}$ (**164**)

This degeneracy is still under investigation, but, together with the work of Klumpp on the rearrangements of **170** (53), it may indicate that the *anions* of certain potentially degenerate systems can exhibit degenerate properties, although the *cations* do not.

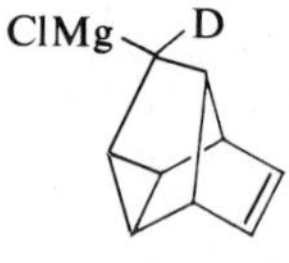

(**170**)

## 5. cis-8,9-Dihydro-1-indenyl Cation

The *cis*-8,9-dihydro-1-indenyl cation **16** could show degenerate properties through a homoallylic rearrangement followed by a cyclopropylcarbinyl–cyclopropylcarbinyl rearrangement. The combination of these processes with allyl resonance in **16** and **16'** would result in the equivalence of carbon atoms 1–7. This is represented in summary by ion intermediate **171**, in which the seven equivalent carbons form a tropylium ion type ring with a two-carbon bridge rotating above the ring.

Cation **16** has been shown to be unremarkable under conditions of solvolysis. The methanolysis of 1-chloro-8,9-dihydroindene (**172**) has been carried out, and the product was 1-methoxy-8,9-dihydroindene (**173**) (62).

More recently, solvolysis of both *exo* **173a** (63) and *endo* **173b** (63) derivatives of 1-hydroxy-*cis*-8,9-dihydroindene have been studied. In aqueous acetone the product is 1-*exo*-hydroxy-*cis*-8,9-dihydroindene (**173c**). Studies of labeled **173d** under the same conditions give a scrambling pattern in the product **173e**, which indicates no degeneracy in the intermediate ion **16** other than that expected on the basis of allylic resonance (63).

In particular, no evidence for the formation of the bishomotropylium ion (**173f**) *on solvolysis* is found, and the same is true for the solvolysis of **173g** (63). However, ion **173f** is formed in $FSO_3H-SO_2ClF-CD_2Cl_2$ solution at $-125$ to $-135°$ from both **173c** and the alcohol related to **173g** (63). Under these conditions, **173f** appears to be one of the most stable $C_9H_9^+$ cations, and it is the product of rearrangement of the 9-

barbaralyl cation **32** (57). Even the tertiary 9-methyl-9-barbaralyl cation **163b** rearranges to a methyl-substituted bishomotropylium ion (55,56).

**(173a)** → (aqueous acetone) **(173c)** ← **(173b)**

**(173d)** → **(173e)**   **(173f)**

**(173g)**

### 6. Bicyclo[4.2.1]nona-2,4,7-trien-9-yl Cation

The bicyclo[4.2.1]nona-2,4,7-trien-9-yl cation **31**, another vinylog of the 7-norbornadienyl cation, may undergo two types of rearrangement processes that could lead to degeneracy. One of these would be the 1,2-vinyl shift of C-7 or C-8 to form intermediate ion **174**. After resonance interconversion to **174'**, cation **174** could then ring expand back to cation **31'**. Repeated operation of this process, in direct analogy with the

**(31)**   **(174)**   **(174')**   **(31')**

mechanism of degeneracy observed for the 7-norbornadienyl cation (see Section III-C-2), would result in the equivalence of atoms C-1–C-6 and C-9. A second type of rearrangement in **31** would be a 1,2-vinyl shift of C-2 or C-5. This would result in the formation of the dihydroindenyl cation **16**, which can exhibit degenerate properties as described in Section III-D-5. The dihydroindenyl rearrangements mix atoms C-2–C-8.

There is an intriguing alternative degenerate rearrangement for cation **174**. Stability in this ion could derive from the homotropyliumlike structure **174″**, which might result in total label scrambling via the fluxional process shown below.

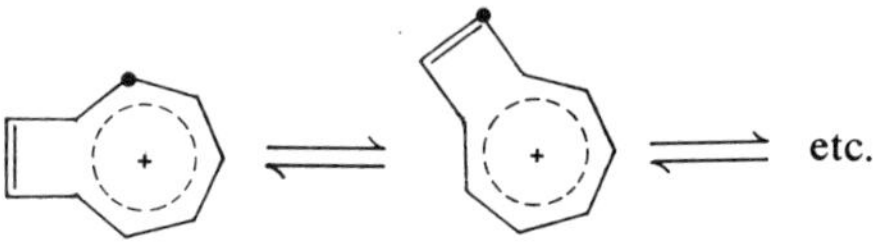

(**174″**) ⇌ (**174‴**) ⇌ etc.

Work on the parent bicyclo[4.2.1]nona-2,4,7-trien-9-yl cation **31** has not yet been carried out. However, some experiments have been performed on the 9-phenyl substituted system, synthetically accessible through the work of Cantrell and Shechter. These workers prepared compound **176** from the reaction of dilithium cyclooctatetraenediide **175** with benzoyl chloride (64). Kende and Bogard obtained alcohol **177** by alkaline

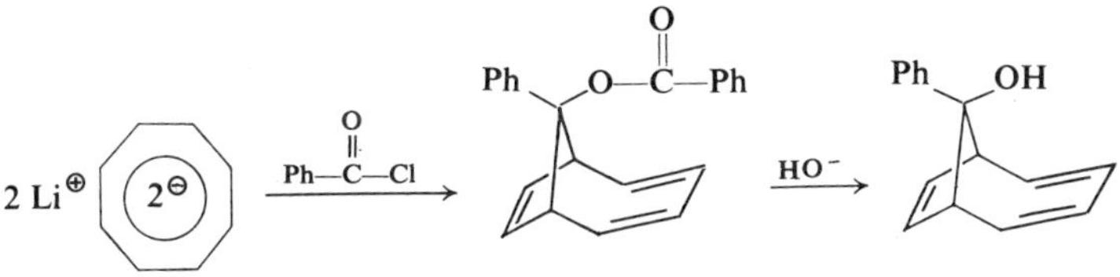

(**175**)  (**176**)  (**177**)

hydrolysis of **176** (65). Alcohol **177**, when treated with 2 eq of thionyl chloride and 1 eq of pyridine gave a high yield of 1-chloro-9-phenyl-*cis*-

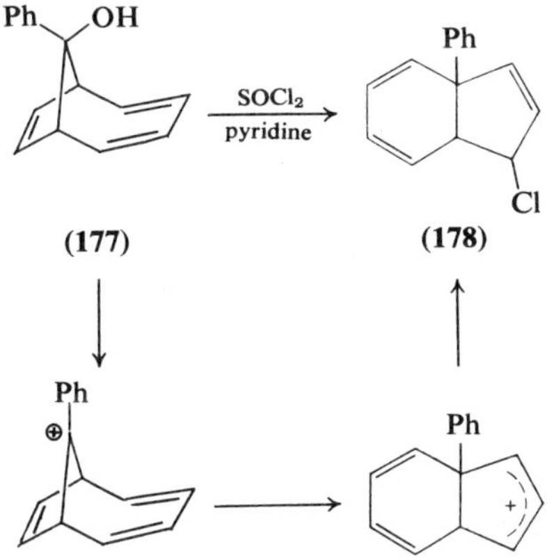

(**177**) →(SOCl$_2$, pyridine)→ (**178**)

↓ ↑

(**179**) → (**180**)

8,9-dihydroindene (**178**) (65). Presumably, the mechanistic path through ions **179** and **180** is involved. This hypothesis was verified by a labeling experiment (66). The results seem to indicate that the bicyclo[4.2.1]nonatrienyl cation **31** would rearrange to the dihydroindenyl cation **16**, but that the reverse process is not favorable. No deuterium scrambling was observed in product **182**, which indicates that the potential degenerate rearrangements in the dihydroindenyl cation **16** may not occur. The interconversion of two $C_9H_9^+$ type ions is also demonstrated by the foregoing experiments. Work on **31** has just been reported (65a)

### 7. *Barbaralyl or Tricyclo[3.3.1.0$^{2,8}$]nona-3,6-dien-9-yl Cation*

The barbaralyl cation **32** is unique among the potentially degenerate carbonium ions within the $C_9H_9^+$ group. This cation has the capacity to exhibit degeneracy through a combination of degenerate carbonium ion rearrangements and degenerate Cope rearrangements. The latter possibility exists because the barbaralyl cation **32**, like bullvalene (**2**), contains a homotropylidene structure. This rearrangement permits the interconversion and equivalence of carbon atoms C-1 and C-5, C-2 and C-4, and C-8 and C-6. Some of the potential carbonium ion rearrangements of **32** that could lead to degeneracy are shown in Scheme I.

Mechanism 1 involves the interconversion of the 9-barbaralyl **32** and the bicyclo[3.2.2]nonatrienyl **14** cations. Since **14** has the potential of undergoing complete carbon degeneracy as described in Section III-D-4, this mechanism would predict complete carbon scrambling in **32** also. Such degeneracy may be even easier to achieve. Because of symmetry in the allylic ion **14**, many of the carbons are equivalent. If the degenerate Cope rearrangement in **32** is rapid, the mere conversion of **32** into **14** and back again will lead eventually to complete degeneracy.

Mechanism 2 employs a 1,2-vinyl shift to form ion **183**, which could give rise to either the dihydroindenyl **16** or the bishomotropylium **184** cations. If this process were reversible, **32** would eventually become completely degenerate. This mechanism seems unlikely, since precursors to **16** and to **183** give no barbaralyl products either under solvolysis (62,63) or stable

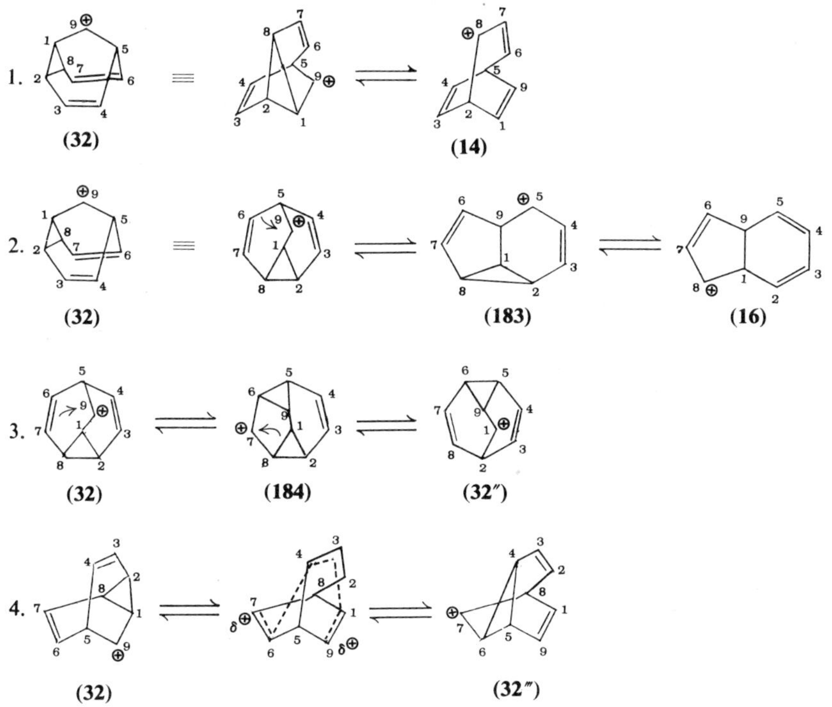

Scheme I.

ion conditions (57,63). Mechanism 3 involves direct formation of the bishomotropylium ion **184** by homoallylic rearrangement from **32**. As shown (and as implied by **171**), ion **184** has the potential of extensive automerization, but this process is not rapid under stable ion conditions at $-125°$ (57). Furthermore, experimental observations show that **184** is more stable than **32** and does not reconvert, at least in superacid media (55,57). Thus, even though mechanism 3 can in principle give rise to complete scrambling in the 9-barbaralyl cation **32**, this does not in fact seem to take place.

Mechanism 4 is a synchronous process involving the concerted migration of three bonds. This results in the direct reformation of the 9-barbaralyl cation **32'''**, but with some of the carbon atoms mixed. After several such rearrangements, carbons 3, 7, and 9 would become equivalent, as would the six remaining carbon atoms (keeping the degenerate Cope rearrangement 32 ⇌ 32' in mind) (67). A fifth mechanistic possibility has recently come to light as the result of theoretical work of two groups (28a); discussion of this mechanism for degeneracy is deferred for a moment, until actual solvolysis results have been given.

Only recently has the experimental picture of the barbaralyl cation 32 become clear. The first evidence was obtained by solvolysis (68a). Contrary to an earlier suggestion (67), 9-barbaralyl tosylate 185 was found to be quite reactive; in acetic acid at 16.5° it solvolyzed with a first-order rate constant of $2.28 \times 10^{-3}$ sec$^{-1}$ (68a). Solvolysis of deuterated tosylate 186

(185)    (186)

in acetic acid (sodium acetate buffer, 30 min at 25°) and in aqueous acetone (80% acetone, 1 hr at 18°) gave barbaralyl acetate 187 and barbaralol 161, respectively, as the main products with the deuterium label scrambled to every position (68a). Multiple integrations of the nmr spectra of the products from several solvolytic runs gave the average percentage deuterium distributions as shown in Table VII.

The results obtained reveal that the potential for carbon degeneracy in the barbaralyl cation 32 can be realized. However, the label distributions are not uniquely consistent with any one mechanism postulated previously. The first three mechanisms would predict a statistical deuterium distribution after many rearrangements. Mechanism 4 would predict 33.3% deuterium at C-9 and 66.6% deuterium at C-3, C-7 after many rearrange-

## TABLE VII
**Solvolysis Products of Barbaralyl Tosylate 186 (Mean Values from Several Experiments)**

| | Position | Solvolysis in | | Statistical deuterium distribution, % deuterium |
|---|---|---|---|---|
| | | HOAc D, % | Aqueous acetone ± 3% | |
| (187) R = Ac | 9 | 28.4 | 53.2 | 11.1 |
| (161) R = H | 1, 5 | 7.4 | 5.2 | 22.2 |
| | 2, 8, 4, 6 | 18.2 | 1.7 | 44.4 |
| | 3, 7 | 46.1 | 39.9 | 22.2 |

ments. The observed experimental values do indicate a high proportion of deuterium at C-3, C-7, and C-9, particularly in the product derived from solvolysis in aqueous acetone.

In the first publication treating the barbaralyl cation, it was suggested that the scrambling pattern was accounted for most economically by the structure **187a**, which, at least formally, is the composite of the equilibrating structures **187b-187d** (68a). The structure **187a** has the necessary threefold symmetry which scrambles deuterium among $C_3$, $C_7$, and $C_9$

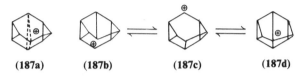

**(187a)**     **(187b)**     **(187c)**     **(187d)**

(or among $C_1$, $C_2$, $C_4$, $C_5$, $C_6$, and $C_8$ when the barbaralyl ion is derived from the bicyclo[3.2.2]trienyl system).

Hoffmann et al. have recently considered the barbaralyl cation theoretically, reaching the remarkable conclusion that **187a** represents not just a possible structure, but the most stable alternative (28a)! The salient point of Hoffmann's analysis is this: *p*-orbitals on $C_3$, $C_7$, and $C_9$ overlap with the Walsh orbitals of the cyclopropane rings

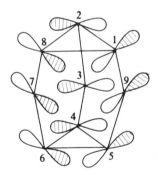

Note that conjugation is possible only along the long axis of the molecule. The orbitals on $C_3$, $C_7$, and $C_9$ are mutually antibonding, and hence no bonding is implied by the dotted lines of **187a**. Furthermore, the calculations suggest that structures **187b–187d**, fanciful in any case, are inappropriate; collapse of **187a** to one of **187b** or **187c** is symmetry forbidden. Finally, the interconversion of **187a** to **187e** is symmetry allowed; Hoffmann believes that, because of similarities of structure and electronic configuration between **187a** and **187e**, they are indistinguishable and in fact represent the same species. A similar conclusion has been reached on the basis of CNDO/2 calculations by Yoneda, Winstein, and Yoshida (28a).

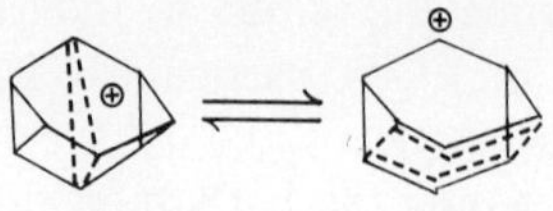

(187a) (187e)

The existence of the threefold degenerate rearrangement process in the 9-barbaralyl cation **32** has been confirmed in two additional ways. As described in Section III-D-4, solvolysis of deuterium-labeled bicyclo-[3.2.2]nonatrien-2-yl derivatives, e.g., **163d**, gave rise to 9-barbaralyl products with the deuterium scrambled to six, but only to six positions, e.g., **163e** and **163f**, in accord with the threefold-symmetrical structure proposed by Hoffmann et al. (28).

The "methyl-labeled" bicyclononatriene **163a** behaves somewhat differently in $FSO_3H$–$SO_2ClF$ solution at low temperature (55,56). The expected ion **188** is not observed; rather the more stable tertiary species **163b** is first produced. This ion undergoes a degenerate rearrangement ($\Delta G^{\ddagger} = 7.3$ kcal/mole at $-121°$), which renders the "left-hand" and "right-hand" sides of the molecule equivalent (56). Although this equiva-

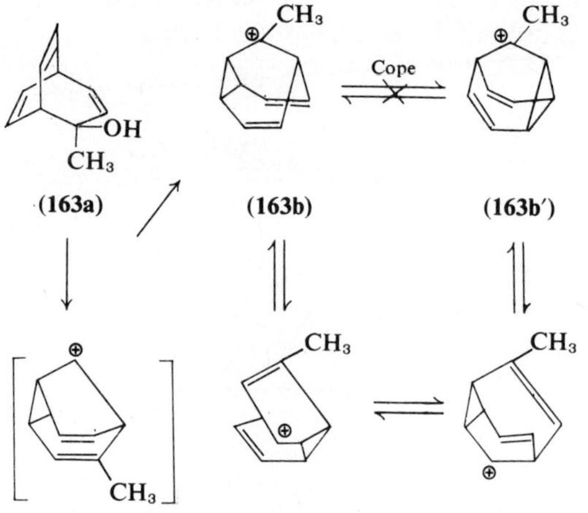

(163a)   (163b)   (163b′)

(188)   (189)   (189′)

lence might easily be pictured as a degenerate Cope rearrangement, **163b** ⇌ **163b′**, analysis of the nmr spectrum shows that this is not actually taking place.* Rather, the interconversion probably occurs via **189** and **189′**, i.e., mechanism 4 (56). At slightly higher temperatures ($\Delta G^{\ddagger} = 11$

---

* Hoffmann and co-workers have recently given reasons to question this result. See Ref. 28.

kcal/mole at $-116°$) ion **163b** rearranges irreversibly to a methyl-substituted bishomotropylium ion (55,56).

The crowning achievement is the direct observation of a single-line nmr spectrum for the 9-barbaralyl cation **32** (57). This demonstrates that even at $-135°$ the degenerate rearrangements leading to complete scrambling of all nine (CH) units are fast on the nmr time scale! For bullvalene, the same result is not achieved until the temperature is 200° higher (2)! Mechanism 1 (via the less stable ion **14**) is the process implicated for complete degeneracy; the process suggested by mechanism 4 must be even faster. At $-125°$, **32** rearranges irreversibly to **184–173f**.

In connection with the barbaralyl cation **32**, the carbonium ion **190** derived from homobullvalene is of interest. Cation **190**, a $C_{11}H_{11}^+$ vinylog of the barbaralyl cation **32**, also has the potential to combine

(**190**)

degenerate carbonium ion rearrangements with degenerate Cope rearrangements to make **190** completely degenerate. Through cyclopropylcarbinyl–cyclopropylcarbinyl rearrangements, cation **190** could be viewed as structure **190'**, or its equivalent, which has $C_{3v}$ symmetry. This formulation makes cation **190** a carbonium ion equivalent of bullvalene, which also has $C_{3v}$ symmetry (2). Combination of this charge delocalization with allyl resonance and the threefold degenerate Cope rearrangement would distribute the positive charge to every carbon atom of cation **190**. The monohalocarbene adducts of bullvalene could presumably serve as solvolytic precursors to cation **190**. However, recent attempts using this route have not been successful (57a).

(**190**)                                           (**190'**)

### 8. Cyclononatetraenyl Cations

In connection with the study of the thermal transformation of 9-chlorobicyclo[6.1.0]nonatriene (**190a**) to 8,9-dihydroinden-7-yl chloride (**190b**) by Barborak et al., the solvolysis of **190a** in various mixtures of

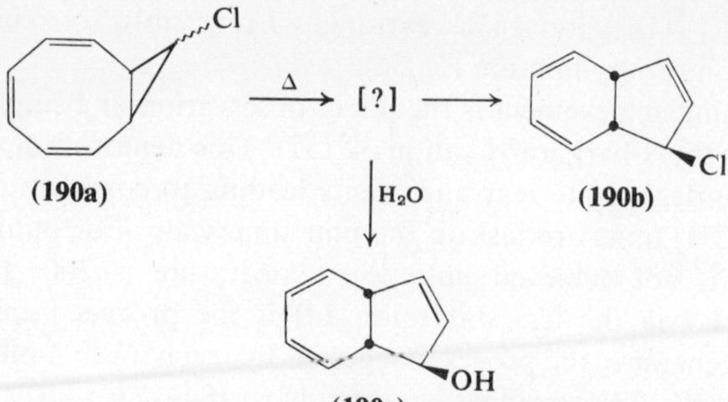

(190a) →Δ [?] → (190b)

↓ H₂O

(190c)

aqueous acetone was undertaken (68b). It was found that the solvolysis rate was nearly equal to that of thermolysis and that it did not vary significantly when concentrations of the nucleophile, water, were changed; thus it was suggested that the thermolysis proceeded in a rate-determining step to an unstable intermediate which, in the absence of nucleophile, collapsed to **190b** but in its presence was captured in a fast step to produce the alcohol **190c**.

Deuterium labeling, as indicated in structure **190d**, was used to investi-

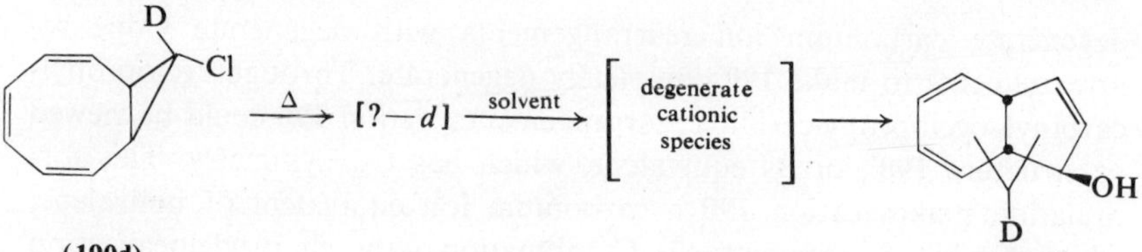

(190d) →Δ [?−d] →solvent [degenerate cationic species] → (190e)

gate the mechanism of solvolysis: the dihydroindenyl product **190e** showed complete deuterium-label scrambling, thus implicating either a symmetrical cationic species or one undergoing rapid degenerate rearrangement. The cyclononatraenyl cation **190f** was proposed as a possibility.

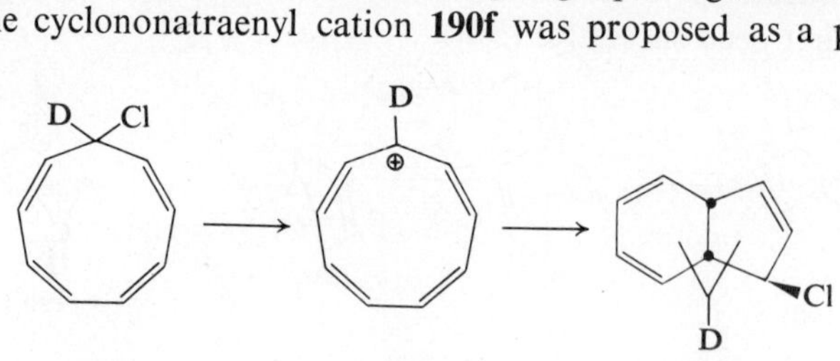

(190g) → (190f) → (190h)

That this suggestion is a reasonable one was shown more recently by Anastassiou (68c). Dissolution of **190g** in a polar solvent such as $SO_2$ led to chloride product **190h** in which the label had been completely scrambled at a rate much greater than the thermal isomerization of **190g** to **190b**, which proceeds without label scrambling (68b). Furthermore, the rate of isomerization was much greater than that of chlorobicyclo[6.1.0]-nonatrienes **190a**.

The ionization mechanism and the exact nature of the intermediate **190f** are still subjects of speculation. Since **190f** itself might not exhibit great stability and hence may not be expected to form rapidly from **190g**, we look for another possibility, and an attractive one is represented by the structures **174″**, mentioned earlier (Section III-D-6) in another context. Structure **174″** is essentially a homotropylium cation; fluxionality

in this species rationalizes the label scrambling.

## IV. OTHER DEGENERATE CARBONIUM IONS

A number of degenerate carbonium ions not having a $(CH)_n^+$ empirical formula have been studied in detail. These ions can exhibit forms of carbon degeneracy, hydrogen degeneracy, and combinations of both. Most ions of this type do not have the potential for complete degeneracy, except under conditions of long life and high acidity, and in this sense they contrast with ions of the $(CH)_n^+$ series. Also, there is no apparent systematization within this group of ions. Many of these cations have been generated in highly acidic solvents and studied directly by nmr.

### A. Isopropyl Cation and *sec*-Butyl Cation

Saunders and Hagen have studied the isopropyl cation **191** and found that this ion exhibits degenerate properties in nmr solvents (69a). The cation was prepared from isopropyl chloride **192** as shown. At temperatures below 0°, ion **191** has two peaks, a doublet for the methyl groups at high

$$\underset{(192)}{H_3C-\underset{Cl}{\overset{H}{\underset{|}{\overset{|}{C}}}}-CH_3} \xrightarrow[\substack{-90° \\ in\,vacuo}]{SbF_5,\,SO_2ClF} \underset{(191)}{H_3C-\underset{\oplus}{\overset{H}{\underset{}{\overset{|}{C}}}}-CH_3}$$

field and a septet for the methine hydrogen at lower field strength. Over the range from 0 to 40°, the spectrum changed in a manner indicating interchange between the two types of protons present. The simplest mechanism for this interchange is reversible rearrangement to $n$-propyl cation **193**.

$$\underset{(191)}{H_3C-\underset{\oplus}{CH}-CH_3} \xrightarrow{1,2-H\sim} \underset{(193)}{H_3C-CH_2-\underset{\oplus}{CH_2}} \xrightarrow{1,2-H\sim} \underset{(191')}{H_3C-\underset{\oplus}{CH}-CH_3}$$

The activation energy determined for this mechanism was $16.4 \pm 0.4$ kcal/mole. Another process that might be involved is rearrangement of the intermediate $n$-propyl cation **193** to protonated cyclopropane **194** before return to **191**. Olah and White furnished strong supporting evidence for such an intermediate in their study of the isopropyl cation obtained from 2-chloropropane **192** with 50%-$^{13}C$ enrichment of the C-2 atom (69b). Scrambling of the isotopic label to the 1-position was observed by nmr.

Proton scrambling in protonated cyclopropane **194** is well documented, so complete mixing of all protons could occur before return to **193** (70).

$$H_3C-CH_2-\underset{\oplus}{CH_2} \rightleftharpoons \underset{(194)}{\triangle} \rightleftharpoons \underset{(194')}{\triangle} \rightleftharpoons \underset{(194'')}{\triangle}$$

(193)      (194)      (194')      (194'')

Degenerate carbon scrambling in the 1-chloropropane to 2-chloropropane isomerization has also been accounted for by invoking a protonated cyclopropane species. Lee and Woodcock have observed isotopic label scrambling during partial isomerization of 1-$^{14}C$-1-chloropropane (**195**) to 2-chloropropane catalyzed by $AlCl_3$ (71). After 90% conversion, recovered **195** showed 7 and 22% rearrangement of the label from C-1 to C-2 and C-3, respectively.

$$\underset{(195)}{Cl^{14}CH_2-CH_2-CH_3} \longrightarrow \underset{71\%\quad 7\%\quad 22\%}{Cl^{(14)}CH_2-^{(14)}CH_2-^{(14)}CH_3}$$

Equilibrating protonated cyclopropane intermediates were proposed, but because of complications, a decision could not be made between corner- and edge-protonated alternatives.

In a similar type of investigation, Karabatsos et al. have reflected on the nature of the proposed protonated cyclopropane intermediate in these rearrangements (72). In the reaction of labeled *n*-propanol with $ZnCl_2$ and HCl, a small portion of the process ($\sim 2\%$) can be interpreted in terms of intervention of a protonated cyclopropane path. For propanol (**196**), labeled as shown in the 2-position,

$$CH_3-\underset{\underset{D}{|}}{\overset{\overset{D}{|}}{C}}-CH_2OH \xrightarrow[ZnCl_2]{HCl} CH_3-\underset{\underset{D}{|}}{\overset{\overset{D}{|}}{C}}-CH_2Cl$$

(**196**)

If a corner-protonated species such as **197**

(**197**)    (**198**)

had intervened over the edge-protonated **198**, extensive label scrambling should have been observed. Superficially, these results appear to challenge the observations of many investigators, who have observed extensive label scrambling in similar systems (70a). The differences, presumably, are traceable to variations in the systems studied, as well as the conditions used. The question of the relative stabilities of edge- and corner-protonated cyclopropanes is still unsettled. Theoretical calculations at the *ab initio* (70b) and semiempirical (70c) levels do not give consistent results.

Saunders and his co-workers also studied the *sec*-butyl cation **199**, prepared by reaction of 2-chlorobutane **200** with strongly acidic nmr solvents at $-110°$ in a vacuum line (73a). The nmr spectrum at low temperature consisted of two peaks (plus a sharp singlet due to *tert*-butyl cation **202** impurity, formed in the preparation). These two peaks

$$CH_3\overset{\oplus}{\underset{1}{C}}H\underset{2}{C}H_2\underset{3}{C}H_3\underset{4}{}$$

(**199**)

$$H_3C-\underset{\underset{Cl}{|}}{\overset{\overset{H}{|}}{C}}-CH_2CH_3 \xrightarrow[\substack{-110° \\ in\ vacuo}]{SbF_5-SO_2ClF} \qquad \updownarrow \qquad + \text{ some } CH_3-\underset{\underset{CH_3}{|}}{\overset{\overset{CH_3}{|}}{C}}{}^{\oplus}$$

(**200**)    (**202**)

$$CH_3CH_2\overset{\oplus}{\underset{3}{C}}HCH_3\underset{4}{}$$
$$\underset{1}{}\quad\underset{2}{}$$

(**199'**)

were assigned to the 1 and 4 and 2 and 3 protons of **199** averaged by rapid degenerate 3,2-hydride shifts (see above). The activation energy for this process has an upper limit of 6 kcal/mole. As the cation **199** sample was warmed from $-110$ to $-40°$, the two peaks broadened and then coalesced to a singlet. The *tert*-butyl peak remained unchanged in this temperature range. Above $-40°$ conversion of **199** into **202** occurred fairly rapidly.

The single absorption peak for all positions in **199** at higher temperatures indicates complete equivalence of all hydrogens. One process for this could be reversible rearrangement to the primary *n*-butyl cation **203**; this is one of the possible mechanisms advanced for the isopropyl cation **191**. However, the energy barrier for the complete proton equivalence in **199** is much lower than in **191**; this tends to disfavor such a mechanism for **199**.

$$CH_3\overset{\oplus}{C}HCH_2CH_3 \underset{}{\overset{1,2-H\sim}{\rightleftarrows}} \overset{\oplus}{C}H_2CH_2CH_2CH_3 \underset{}{\overset{1,2-H\sim}{\rightleftarrows}} CH_3\overset{\oplus}{C}HCH_2CH_3$$
$$(199) \qquad\qquad (203) \qquad\qquad (199'')$$

A better mechanism appears to be cyclization to a protonated methylcyclopropane **204**, followed by proton scrambling and reopening to **199**. This mechanism would account for the activation energy of only 7.5

$$CH_3\overset{\oplus}{C}HCH_2CH_3 \rightleftarrows (204) \rightleftarrows (204') \rightleftarrows (204'')$$

(199)   (204)   (204')   (204'')

$\pm$ 0.1 kcal/mole for complete mixing of all hydrogens (cf. value of 16.4 $\pm$ 0.4 kcal/mole obtained for the process of complete hydrogen mixing in the isopropyl cation).

## B. Dimethyl-*tert*-butylcarbonium Ion and Related Cations

Olah and Lukas observed that when 2,2,3-trimethylbutane **205** was dissolved in $FSO_3H-SbF_5-SO_2ClF$ at $-60°$, the dimethyl-*tert*-butylcarbonium ion **11** was generated (74). The nmr spectrum of **11** showed a single sharp singlet at $\delta$ 2.86. The equivalence of all five methyl groups in **11** was attributed to fast 1,2-methyl shifts. This carbonium ion serves as a

(205)   $\xrightarrow[-60°]{-H^\ominus}$   (11)   $\rightleftarrows$   (11')

good example of twofold partial carbon degeneracy. Since only one peak was observed in the spectrum of **11**, a static nonclassical ion structure such as **206**, which would exhibit two distinct methyl resonances, is ruled out.

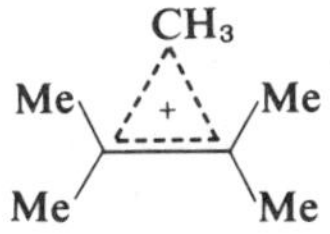

(206)

Saunders and Vogel, however, have invoked **206** as a transcient species to account for extensive deuterium scrambling in the ion **11**-$d_3$ (75). Below $-40°$, **11**-$d_3$ in $SbF_5$–$SO_2ClF$ exhibited only the equilibrium with

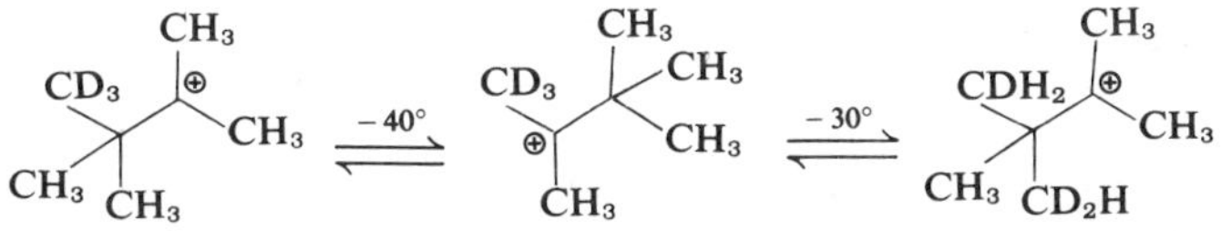

(11-$d_3$)  $\xrightarrow{-40°}$  (11-$d_3'$)  $\xrightarrow{-30°}$  (11-$d_3''$)

**11**-$d_3'$. At $-30°$, however, other deuterated species, e.g., **11**-$d_3''$, were observed in the nmr spectrum. The investigators proposed that corner-protonated species such as **206**-$d_3$ were involved, as outlined below.

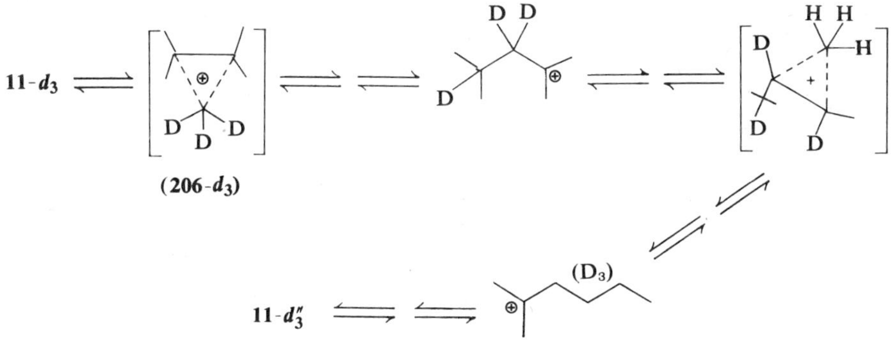

(206-$d_3$)

11-$d_3''$

($D_3$)

Both the dimethylisopropylcarbonium ion **207** and the dimethyl-fluoroisopropylcarbonium ion **208** are similar to cation **11**. Ion **207**, generated from 2,3-dimethylbutane (**209**) in $FSO_3H$–$SbF_5$, exhibited only a doublet at $\delta$ 3.32 ($J = 4.2$ Hz) for the four methyl groups (74). The equivalence of all methyl groups was best interpreted in terms of the rapidly equilibrating ions **207** and **207'** in which the single tertiary hydrogen undergoes a degenerate 1,2 shift.

[Structures 209 → 207 ⇌ 207′ with Me groups, reaction conditions −H⁻/−20°]

Ion **207** has been used by Saunders and co-workers as a probe for equilibrium deuterium isotope effects (76). When one methyl group was replaced by a -CH₂D, it was found that deuterium preferred the structure in which -CH₂D was attached remotely rather than to the carbon bearing the positive charge. Replacing two methyl groups with -CD₃, as in structure **207″**, resulted in the hitherto undetected degenerate rearrangement to the ion **207‴**, again presumably via protonated cyclopropane intermediates.

[Equilibrium scheme showing 207″ ⇌ [protonated cyclopropane intermediates] ⇌ 207‴]

Ion **208**, obtained from 2,3-difluoro-2,3-dimethylbutane (**209**), showed a single doublet at δ 3.10 ($J = 11$ Hz) for the proton resonance spectrum

[Structures 209 → 208 ⇌ 208′ with SbF₅/SO₂, −90°]

(77). This observation is compatible with ion **208** existing as the doubly degenerate equilibrating cation pair **208** and **208′**. Rapid 1,2-fluorine shifts are responsible for the equivalence of the four methyl groups. Structures **210** and **211** were ruled out on the basis of chemical shift comparisons of the hydrogen and fluorine in **207** and **208** with known species of similar structure.

[Structures (210) and (211) — bridged cyclopropyl cations with H and F respectively]

Brouwer has observed the formation of 2-methylpentyl cation **212** from 2,3-dimethyl butyl cation **207**, as well as the interconversion of **212** with its degenerate isomer **212'**, via the 3-methylpentyl cation **213**. Additionally, Brouwer found that the reaction kinetics for these rearrangements are not greatly dependent on the nature of the acidic solvents, which were composed of various mixtures of HF, $FSO_3H$, $SO_2$, or $SO_2ClF$ (78).

The cation **213** was itself found to undergo degenerate rearrangements by Olah and White (69b), who used $^{13}C$ nmr spectroscopy on ions obtained from $^{13}C$-enriched precursors. A protonated-cyclopropane species was postulated as the vehicle for degenerate rearrangement.

### C. *Tert*-Amyl Cation

Saunders and Hagen studied the temperature-dependent nmr spectrum of the *tert*-amyl cation **214** and observed that this ion exhibited partial degeneracy (79). Cation **214** was prepared by allowing 2-chloro-2-methylbutane **215** to react with an excess of antimony pentafluoride in sulfuryl chlorofluoride in a vacuum line at $-90°$. Below $-40°$ the nmr spectrum consisted of peaks at $\tau$ 7.70 (triplet), 5.58 (triplet), and 5.25 (multiplet) assigned to protons *a*, *c*, and *b*, respectively. As the cation solution was warmed above $-40°$, the triplets for the methyl protons broadened and then coalesced, but the multiplet for the methylene protons remained unchanged. Upon recooling the sample, the original spectrum was reobtained. The spectral changes demonstrated interchange of the two types of methyl groups without mixing in the $CH_2$ protons. This is best explained by a mechanistic scheme in which first a 1,2-hydride shift occurs, then a degenerate 1,2-methyl shift, followed by another hydride shift. Mechanisms involving primary carbonium ions are considered less likely

$$CH_3\underset{\oplus}{C}(CH_3)-CH_2CH_3 \underset{}{\overset{1,2-H\sim}{\rightleftharpoons}} CH_3\underset{\oplus}{C}(CH_3)-CHCH_3 \overset{1,2-H\sim}{\rightleftharpoons}$$

(214)  (216)

$$CH_3\underset{\oplus}{CH}CH(CH_3)CH_3 \underset{}{\overset{1,2-H\sim}{\rightleftharpoons}} CH_3CH_2\underset{\oplus}{C}(CH_3)-CH_3$$

(216′)  (214′)

and could also involve complete mixing of all types of hydrogens. Brouwer and Mackor also studied the *tert*-amyl cation **214** in HF–SbF$_5$ solution and observed the same reversible line broadening at temperatures above $-20°$ for the different methyl groups (80). Brouwer (81) examined the system in detail and independently came to the same conclusions as Saunders and Hagen concerning the equilibrium process. Recent higher-temperature studies have shown that above 130° the methyl and methylene signals coalesce (82).

## D. 2,4-Dimethylpentyl Cation

The nmr spectrum of 2,4-dimethylpentyl cation **217**, obtained from the corresponding chloride and the solvent mixture FSO$_3$H–SbF$_5$–SO$_2$ClF, has been studied by Brouwer and Van Doorn (83). The absorptions due to the methyls are broad at $-117°$, and as the temperature increases, the signals broaden further until they coalesce at $-80°$. At the same time, the methylene signal remains unchanged, indicating that the degeneracy, which accounts for methyl-signal broadening, takes place via a single 1,3-hydride shift rather than consecutive 1,2-shifts. A barrier of 6.5 kcal/mole was determined. The authors favored as a transition state a

(217)  (217′)

species in which a partial bond exists between the migrating hydride and the carbon atoms involved (**218**), rather than one in which there is partial bonding between the carbon atoms as well **(219)** as in a protonated cyclopropane. Saunders has reinvestigated this problem and has obtained

**(218)**  **(219)**

more accurate barriers to 1,3-hydride shifts in **217** (73b), namely, 9–10 kcal/mole. Barriers of 12–14 kcal and <6 kcal for 1,4- and 1,5-hydride shifts, respectively, were determined in the 2,5-dimethylhexyl and 2,6-dimethylheptyl cations (73b).

### E. 1-*t*-Butyl-1,3,3-trimethylallyl Cation

Sorensen and Ranganayakulu (84) have reinvestigated the mechanism of the cyclopropylcarbinyl–allylcarbinyl–allyl rearrangements leading to the 1-*t*-butyl-1,3,3-trimethylallyl cation, **222**, originally studied by Poulter and Winstein (85).

In the work of Sorensen and Ranganayakulu, specifically labeled diene

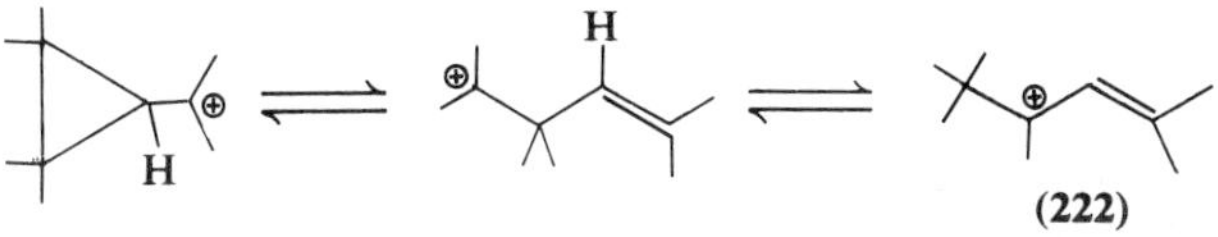

**(222)**

**223** was treated with the acidic solvent mixture $FSO_3D$–$SbF_5$ at $-80°$ and the allylic cation **222′**, which immediately formed exhibited a "*completely, and evenly* scrambled deuterium label between the 1-methyl and *t*-butyl groups," thus implicating the cyclopropylcarbinyl ion **224** as intermediate.

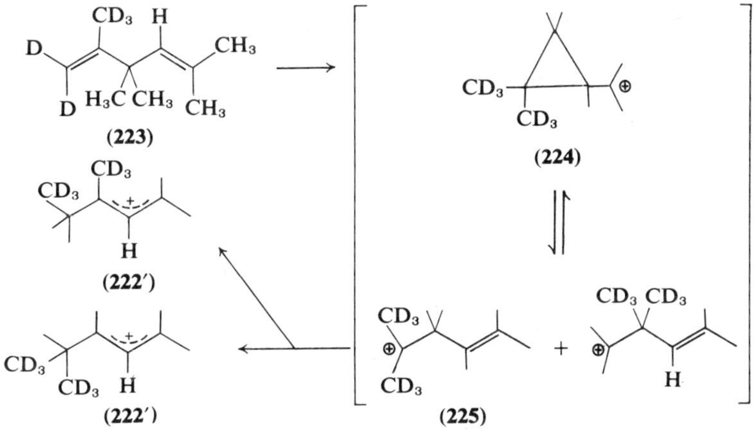

**(223)**

**(224)**

**(222′)**

**(222′)**

**(225)**

Much slower than this four-methyl scrambling process is one in which *all six* methyls in **222** equilibrate (84,85). This can be accounted for by a cyclopropylcarbinyl-cyclopropylcarbinyl rearrangement in intermediate **224**.

### F. Methylcyclobutyl Cation

Saunders and Rosenfeld have generated the methylcyclobutyl cation **226** at low temperature in $SbF_5$–$SO_2ClF$ solvent (86). At 80°, two singlets in a 2:1 ratio were observed, whereas at $-25°$ splitting was observed appropriate for coupling between groups of three and six equivalent protons. Various possible rationales were considered: 1,2-hydride shifts to equilibrate ion **226** with secondary cyclobutyl ions

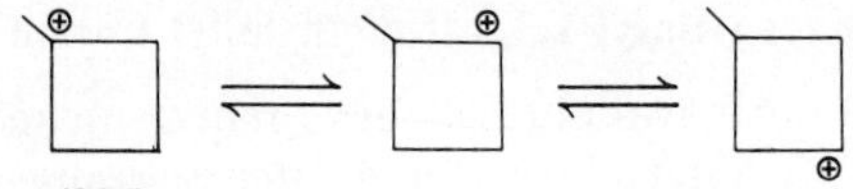

(226)

equilibrating cyclopropylcarbinyl ions

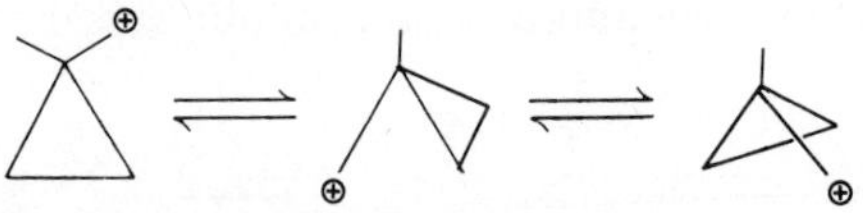

or rapid equilibrium between ion **226** and the corresponding cyclopropylcarbinyl cation **227**

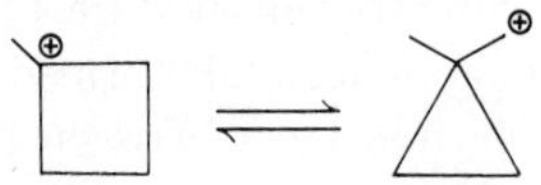

(226)    (227)

However, only the last possibility, the equilibrium between ions **226** and **227**, satisfied all the experimental observations. Additionally, an identical cation mixture was obtained in a similar manner from 1-chloromethyl-1-methylcyclopropane **228**

(228)    (227)    (226)

At temperatures above $-25°$, the cation mixture (**226**) and (**227**) isomerized irreversibly to the methyl cyclopropyl carbonium ion **229**.

Similar results have been obtained by Olah et al. (19d), who also studied the 1-phenylcyclobutyl cation, a classical static species.

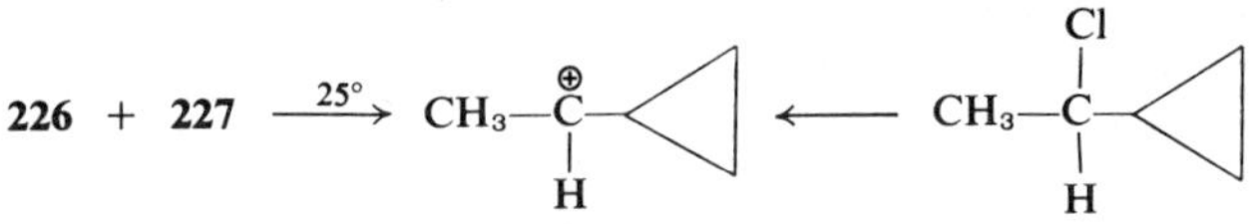

$$226 + 227 \xrightarrow{25°} CH_3-\overset{\oplus}{\underset{H}{C}}-\triangleleft \longleftarrow CH_3-\underset{H}{\overset{Cl}{C}}-\triangleleft$$

(229)             (230)

### G. Cyclopentyl Cation

Olah and Lukas investigated the cyclopentyl cation **12** by examining the nmr spectrum of the ion generated in highly acidic solvents (87). They found that when cyclopentane (**231**) was dissolved in $FSO_3H$–$SbF_5$ or HF–$SbF_5$ diluted with $SO_2ClF$ below $-10°$, the cyclopentyl cation **12** was formed by hydride abstraction. The ion exhibited a single peak at $\delta$ 4.75 and the nmr spectrum remained unchanged down to $-130°$. This indicated that **12** was undergoing a series of rapid 1,2-hydride shifts that resulted in complete hydrogen degeneracy. No study of possible carbon degeneracy in ion **12** has been reported. The ESCA spectrum of **12** shows it to be a classical species (87a).

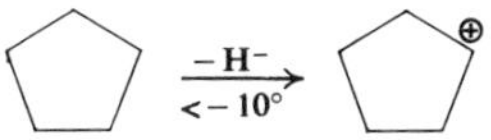

$$\bigcirc \xrightarrow[<-10°]{-H^-} \bigcirc^{\oplus}$$

(231)       (12)

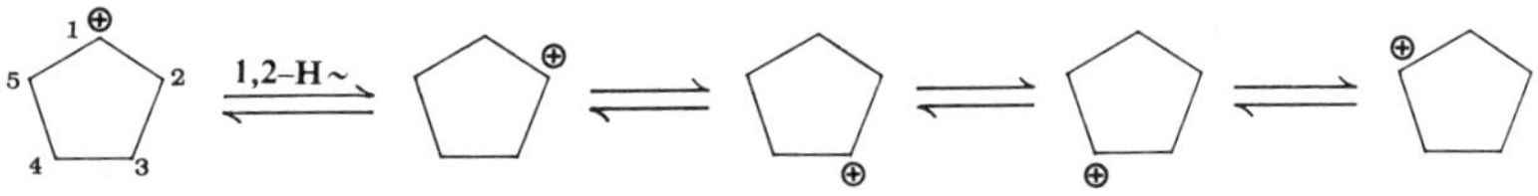

(12) $\xrightarrow{1,2-H\sim}$ ⇌ ⇌ ⇌ ⇌

Cation **12** was also generated from cyclopentyl bromide **232** and cyclopentene **233** as shown (87). Quenching of these solutions of **12** with methanol gave cyclopentyl methyl ether **234** in high yield.

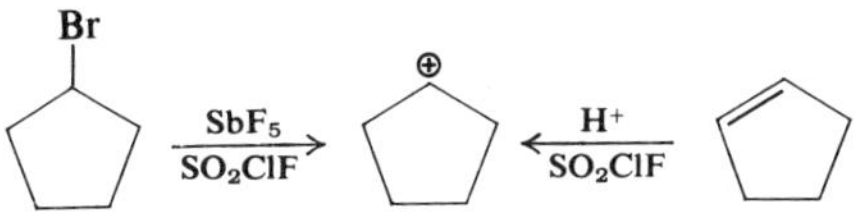

(232) $\xrightarrow{SbF_5 / SO_2ClF}$ (12) $\xleftarrow{H^+ / SO_2ClF}$ (233)

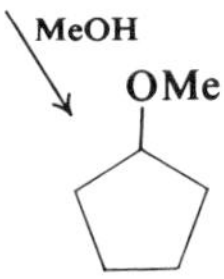

↓ MeOH

(234) OMe

### H. Methylcyclopentyl Cation

Saunders and Rosenfeld have studied the methyl-substituted cyclopentyl cation **235** (82). Cation **235** was prepared from cyclohexyl chloride and excess SbF$_5$–SO$_2$ClF, and the nmr spectra were observed at elevated temperatures. At +110°, the methyl and ring protons coalesced to a

singlet, representing rapid exchange between the methyl and ring hydrogens. In addition, cation **235**, generated from 1-methylcyclopentyl chloride enriched with $^{13}$C in the methyl group, moved $^{13}$C into the ring. This observation could only be explained by invoking equilibrium with the cyclohexyl cation **236**, which shared the transcient species **237** in common with methylcyclopentyl cation **235**.

Degenerate rearrangements were also observed in the related 1,3-dimethyl-1-cyclopentyl cation **238** by Hogeveen and Gaasbeek, who generated the ion, as a mixture with 1-methylcyclohexyl **239** and 1,2-dimethyl-1-cyclopentyl cations, by treatment of norbornane (**241**) with HF–SbF$_5$ solution (88).

The degenerate rearrangement involved interconversion of ions **238** and **238'**, presumably via a succession of 1,2-hydride shifts

The methylcyclohexyl cation (**239**) also exhibits degenerate rearrangement properties by way of a succession of 1,2-hydride shifts, in a manner analogous to that observed in the methylcyclopentyl cation.

## I. Cyclobutenyl Cation

A redistribution of –$CD_3$ groups in a pentamethylcyclobutenium ion has revealed degenerate rearrangements. Koptyug et al. observed that the protonation of 4-methylene-1,2,3,3-tetramethylcyclobut-1-ene **(241)** in 85% $D_2SO_4$ resulted in total hydrogen–deuterium exchange in the 1- and 3-methyl groups of the resulting cyclobutenyl cation **(242)**; converting the

medium to 100% $D_2SO_4$ by introduction of $SO_3$ brought about degenerate scrambling of –$CD_3$ and –$CH_3$ groups in positions 1, 2, and 3, but the geminal methyls in position 4 did not enter into rearrangement (89). Thus, 1,2-methyl shifts of the type prevalent in the related methylated cyclopentenyl cations **248** and **248a** and in the heptamethylbenzenonium cation **249** failed to take place in the present case. The authors proposed two possible mechanistic pathways

A precedent for path *b* is found in work by Breslow and co-workers (90), who solvolyzed the cyclopropenylcarbinyl system **245** and observed rearrangement to cyclobutenyl products **(246)**

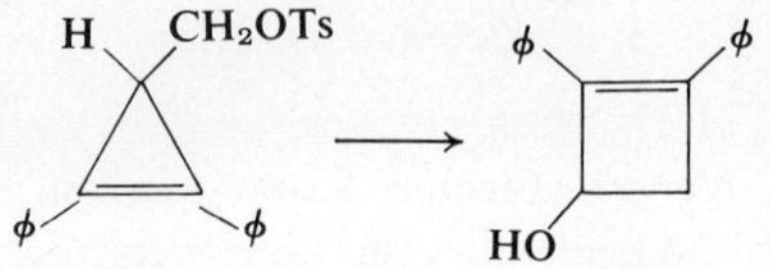

(245) (246)

Pathways *a* and *b* may well be essentially identical, since both classical species **243** and **244** may be visualized as the single delocalized structure **247**.

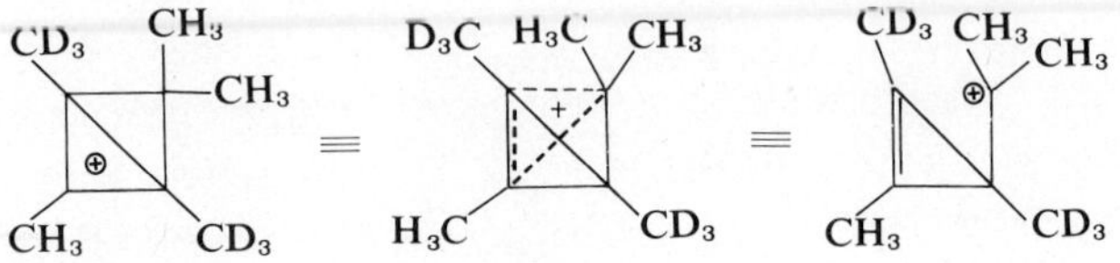

(247)

Finally, it can be speculated that the reason for the failure to observe 1,2-methyl shifts is the necessity of a cyclobutadienoid transition state **247a**, which raises the barrier to migration above that for other rearrangement possibilities.

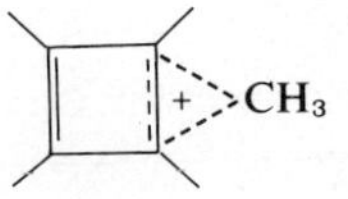

(247a)

## J. Pentamethyl- and Hexamethylcyclopentenyl Cations

The pentamethyl- and hexamethylcyclopentenyl cations were studied in detail by Brouwer and Van Doorn (91), and the rearrangements 1–3 were observed.

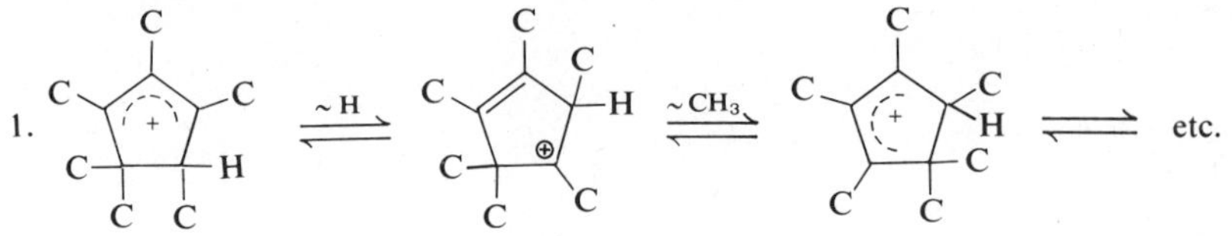

(248) (248′)

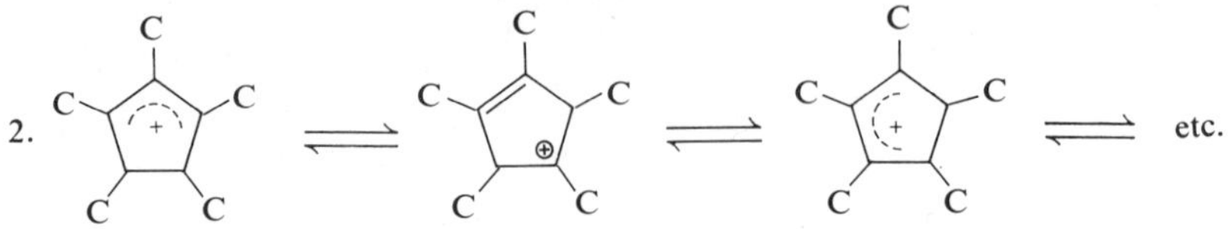

(248a)

3. [structure equilibrium]

It was found that reaction 1, involving both hydride and methyl shifts, was faster than both reactions 2 and 3, even though hydride shifts are ~$10^5$–$10^6$ times faster than methyl shifts in benzenonium ions (which would have led us to expect that reactions 3 and 1 should have had similar rates, whereas reaction 2 should have been dramatically faster). The authors suggested that steric crowding in cation **248** resulted in the observed rate acceleration, since the relief of strain in going to **248'** would partially compensate for loss of conjugation. The same steric argument should be applicable to the heptamethylbenzenonium ion **249**, where methyl shifts are more rapid, as well as to the pentamethylcyclobutenyl cation **242**, in which methyl shifts do not occur. An entirely satisfactory explanation for all of these cases is not yet at hand.

## K. Heptamethylbenzenonium Ion

The heptamethylbenzenonium ion **249** represents one of the earliest studied examples of a degenerate carbonium ion. Cation **249** was originally synthesized more than ten years ago by the exhaustive methylation of trimethylbenzenes **250** with methyl chloride in the presence of aluminum chloride or by the reaction of hexamethylbenzene **251** with methanol in the presence of concentrated sulfuric acid (92). The nmr spectrum of ion **249** at 25° showed four resonance signals in the ratio of 1:2:2:2 with the three

lowest field peaks assigned to the methyl groups labeled (4), (3), and (2), respectively, and the highest field peak due to the *gem* dimethyl groups labeled (1). Saunders studied the nmr spectrum of **249** at higher temperatures and observed that the three low field peaks coalesced into a broad singlet and the high field peak remained a singlet, although broadened, at ≈70° (93). This was interpreted in terms of **249** undergoing a series of rapid, degenerate 1,2-methyl shifts.

Alternatively, a "random" shift mechanism could be imagined in which a methyl group leaving the saturated carbon might jump from one ring carbon to another in a series of steps before becoming reattached to one of the ring carbons. The intermediate for such a process is represented by **252**. Recently, Derendyaev et al. studied the nmr spectrum of **249** with the aid of multiple-resonance techniques and established that the "random" migration of a methyl group did not play a significant role in the rearrangement of **249** (94). It was concluded that the intramolecular 1,2-methyl shift mechanism was responsible for the observed spectral changes. Analogous 1,2-hydride shifts have recently been demonstrated in the parent benzenonium ion (95).

Hogeveen and Volger have investigated the protonation of hexamethyl-dewarbenzene (**253**), which yields **253'** at 3° (96). At lower temperatures (−90°), however, the nmr spectrum suggests the presence of ions **254** and **255** in a ratio of 1:3, respectively. Warming the solution to 0° results in rapid equilibration of the species **254** and **255**, via the presumed mechanism illustrated, which involves a series of cyclopropylcarbinyl interconversions

## L. 1,3,4,4-Tetramethylcyclohexenyl Cation

Sorensen and Ranganayakulu have reexamined the 1,3,4,4-tetramethylcyclohexenyl (**256**)—1-*t*-butyl-3-methyl-cyclopentenyl (**257**) cation interconversion, and in so doing, they discovered a degenerate rearrangement taking place in cation **256** (97).

By means of isotopic labeling, the investigators showed that an equilibrium existed; by careful analysis of products and reaction kinetics, they were able to demonstrate that the overall reaction scheme could be represented by the diagram above, with cation **258** invoked as a transitory species.

### M. Dihydrophenanthrenium Cations

Shubin et al. have investigated the degenerate rearrangement in 9-ethyl-9,10-dimethylphenanthrenium ion (**259**) and have proposed (98) the following possible mechanism:

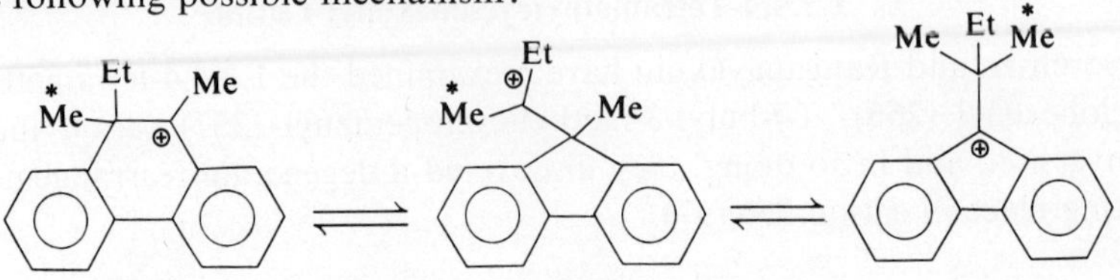

(**259**)    (**260**)

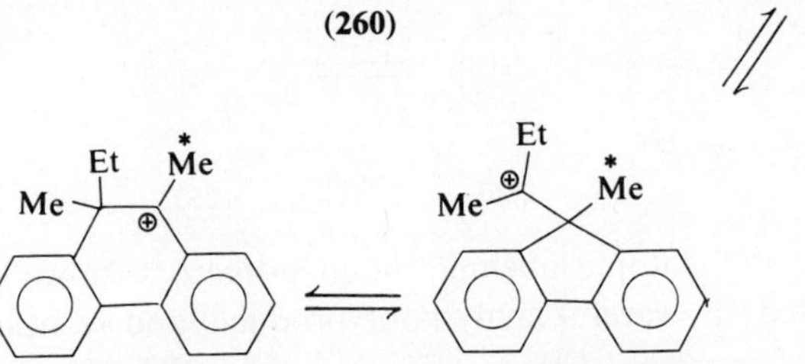

The authors also recognized that a simple 1,2-ethyl shift, or sets of 1,2-methyl shifts, would account for the observed result just as well. The proposed ring-contraction mechanism requires passing from an ion that is both tertiary and benzylic (**259**) to one that is only tertiary; lacking concrete evidence for it, and given the alternative tertiary, benzylic ⇌ tertiary, benzylic rearrangement possibility, the ring-contraction mechanism should be disregarded in favor of the simpler mechanism (98a). The latter mechanism was chosen in a similar system (**261**) (99). The migrating abilities of various aryl groups versus methyl were studied in the series (**261**, X = H, Me, Cl, F, CF$_3$).

The migrational tendency of p-X-phenyl groups followed the order Me > F ~ H > Cl > CF$_3$. Degeneracy in this system could best be accounted for by mechanism *1* rather than mechanism *2* since no NMR evidence for the latter intermediate was found.

### N. 3-Bicyclo[3.1.0]hexyl Cation

The 3-bicyclo[3.1.0]hexyl cation **262** could exhibit threefold carbon degeneracy by means of the homocyclopropylcarbinyl rearrangements shown. Alternatively, this ion could exist in the trishomocyclopropenyl cation form **262'**, in which positions 1, 3, and 5 as well as 2, 4, and 6 are equivalent to each other, owing to the charge-delocalized structure. According to the definition advanced in Section I, the 3-bicyclo[3.1.0]hexyl cation can only exhibit true degenerate rearrangements if it exists as the rapidly equili-

brating set of ions. Structure **262'** is a homoaromatic analog of the cyclopropenyl cation and would simply exhibit threefold equivalence of positions.

Winstein and his co-workers have studied thoroughly the solvolysis of *cis*-3-bicyclo[3.1.0]hexyl tosylate (**263**) and *trans*-3-bicyclo[3.1.0]hexyl tosylate (**264**) (47,100). These investigations provide strong evidence that the acetolysis of the *cis* isomer proceeds primarily through the trishomocyclopropenyl cation **262'**, whereas that of the *trans* isomer proceeds primarily by an ordinary solvent-assisted process **262a**.

Evidence for this mechanistic scheme is the following:

1. The rate of solvolysis of **263** is faster than that of **264**.
2. The value of $k_\Delta$ is approximately 50 times that of $k_s$ for the solvolysis of **263**.
3. The solvolysis of **263** exhibits a special salt effect, whereas that of **264** does not.
4. The exclusive product of solvolysis of **263** is the *cis*-acetate **265** with essentially complete retention of configuration. The solvolysis of **264** gives several products, some of them derived from ion **266**.
5. Solvolysis of *cis*-3-deuteriobicyclo[3.1.0]hexyl tosylate (**267**) gives product *cis*-acetate **268**, with one-third of a deuterium distributed at positions 1, 3, and 5.

Similarly, solvolysis of the 6,6-dideuterated isomer **269** gives a product *cis*-acetate with 1.35 cyclopropane methylene protons. This result is within experimental error of the value 1.33 predicted for the trishomocyclopropenyl intermediate. The *trans* isomer product **270** showed no cyclopropane methylene protons.

Corey and Uda have studied the solvolysis of *cis*-1,5-diphenylbicyclo-[3.1.0]hex-3-yl tosylate (**271**) to form cation **273** (via **272**?) (101). Acetolysis of **271** gave a rate slower than that of **263** with the formation of

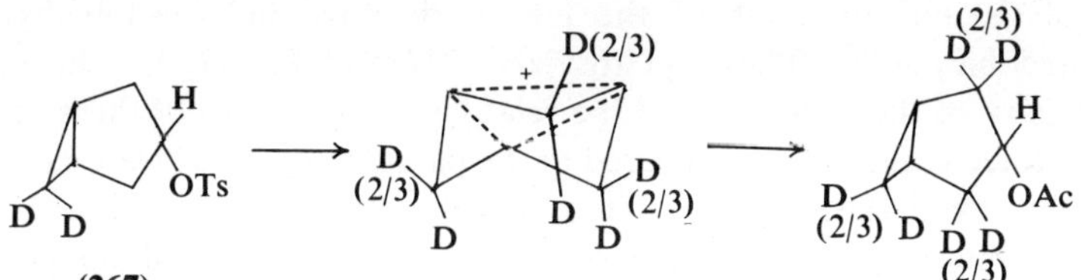

**(267)** → **(2/3)** → **(OAc structure)**

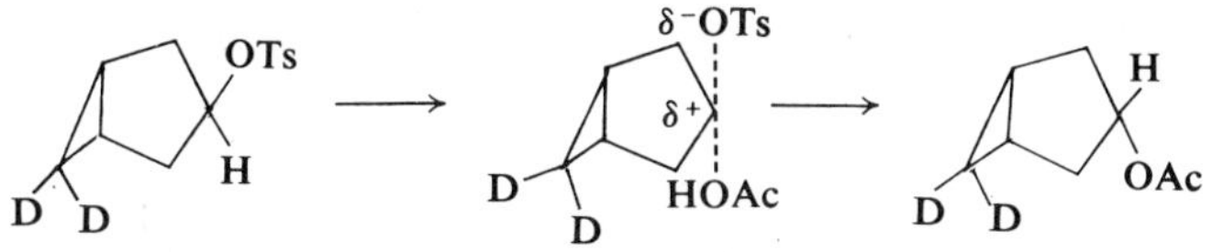

**(269)** → **(δ⁻OTs / HOAc)** → **(270)**

274, 275, and 276 as products. These authors feel that these results are

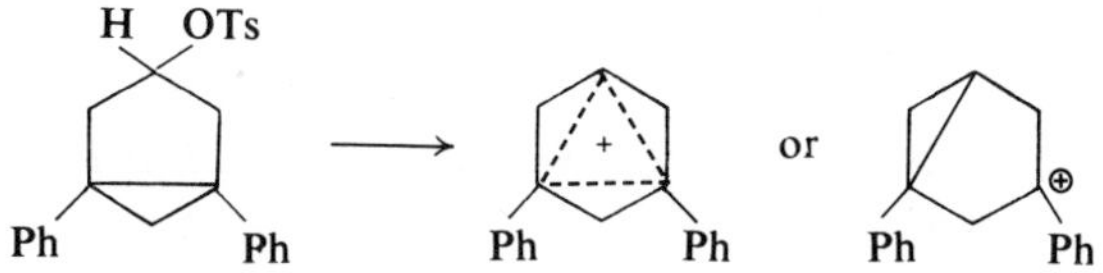

**(271)** → **(272)** or **(273)**

inconsistent with the trishomocyclopropenyl cation formulation 272.

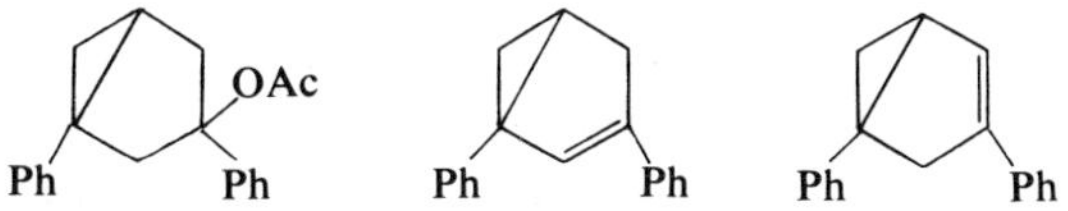

**(274)** **(275)** **(276)**

However, phenyl substituent rate effects are notably complicated by opposing inductive and resonance effects (102). Winstein has reported the

**(277)** $\xrightarrow{\text{SOH}}$ 90% + 10%

$k_{277}/k_{263} = 5.1$

**(278)** $\xrightarrow{\text{SOH}}$ 97% + 3%

$k_{278}/k_{263} = 7.0$

solvolysis and products of the *cis*-monomethyl and *cis*-dimethyl-substituted bicyclo[3.1.0]hex-3-yl tosylates **277** and **278** (100). The stereochemistry of the products and the relative rate increases of these methyl-substituted compounds compared to the parent system **263** were said to be consistent with the formation of the trishomocyclopropenyl cations **279** and **280** as intermediates in these solvolyses. However, based on previous

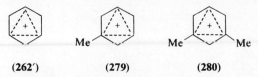

(262')    (279)    (280)

examples for rate enhancements due to methyl substituents on charged delocalized structures (16,103), the value of 7.0 for $k_{278}/k_{263}$ is considerably lower than expected. The value should be closer to $(5.1)^2 = 26.0$ for the intermediate **280**.

In connection with the 3-bicyclo[3.1.0]hexyl cation, the 3-tricyclo-[7.1.0.0$^{5,7}$]decyl cation (**281**) is also of interest (100). Ion **281** contains two

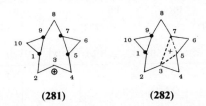

(281)    (282)

*cis*-3-bicyclo[3.1.0]hexyl moieties within the same cation. If delocalization similar to that observed in the trishomocyclopropenyl cation **262'** were to take place, ion **282** would be formed with positions 6 and 4 equivalent. On the other hand, if **281** were to exist as a rapidly equilibrating set of ions as shown, then the positions 2, 4, 6, 8, and 10 would be equivalent and this would constitute an example of fivefold degeneracy. Winstein et al. (104) studied the solvolysis of the tetradeuterated 3-tricyclo[7.1.0.0$^{5,7}$]decyl tosylate (**283**) and isolated the products.

(281) ⇌ ⇌ ⇌ etc.

The fully protonated starting tosylate showed 2.0 cyclopropane methylene protons at τ 10.3 in the nmr spectrum, whereas the tetradeuterated tosylate

(283) gave acetolysis products **285** and **286** with 0.52 protons at τ 10.3. This is in accord with ion intermediate **284** shown below, but it is not consistent with a rapidly equilibrating set of ions arising from **281**. The latter mechanism would predict 1.20 cyclopropane methylene protons at τ 10.3. These results indicate participation of only one cyclopropane ring in the 3-tricyclo[7.1.0.0$^{5,7}$]decyl cation, even though participation by both rings would seem to be possible. This is a situation similar to that observed by Coates and Kirkpatrick for the 9-pentacyclo[4.3.0.0$^{2,4}$.0$^{3,8}$.0$^{5,7}$]-nonyl cation **19** (48).

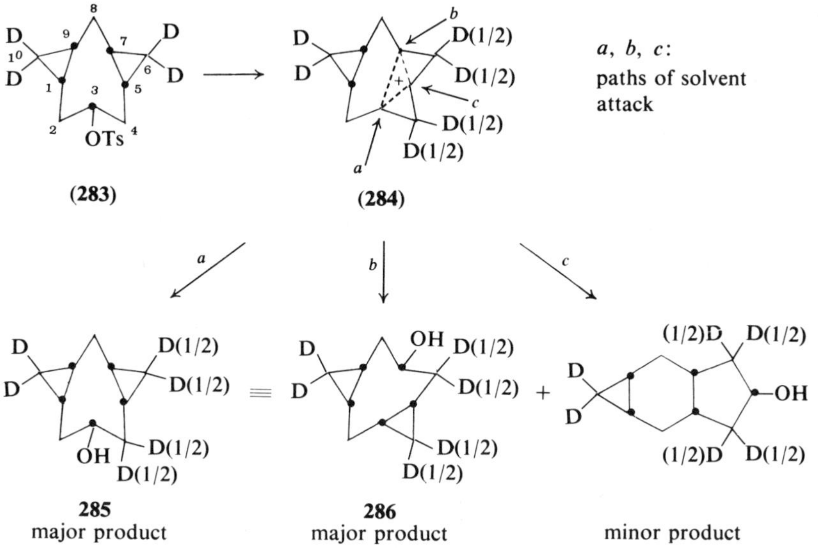

## O. 2-Norbornyl Cation

Schleyer et al. were the first workers to report the direct observation of the nmr spectrum (60 MHz) of the 2-norbornyl cation **9** (105). The ion was prepared by dissolving either 2-*exo*-norbornyl chloride **287** or 2-*exo*-norbornyl fluoride **288** in SbF$_5$ or in SbF$_5$–liquid SO$_2$. The nmr spectrum in the temperature range $-5$ to $+37°$ consists of a single broad band at δ 3.75, indicating that all the protons (and probably all the carbons) are equivalent. At $-60°$ the spectrum separates into three bands of relative areas 4 (δ 5.35), 1 (δ 3.15), and 6 (δ 2.20). Further cooling to $-120°$ produced no other changes in the spectrum of **9**. Evidence for **9** in solution was obtained by quenching the cation solution with water; this produced 2-*exo*-norbornanol **289**. (The ion **9** is shown in classical form for convenience.)

The foregoing results were interpreted in terms of the 2-norbornyl cation (9) undergoing three reaction processes: (*a*) a Wagner-Meerwein rearrangement **9** → **9′**, occurring rapidly over the temperature range studied; (*b*) a 6,2-hydride shift (**9**) → **9″**, also occurring rapidly over the temperature range studied; and (*c*) a 3,2-hydride shift **9** → **9‴**, which is fast at room temperature but slow below −23° (temperature of coalescence of the three observed peaks) (106).

At −60° the 3,2-hydride shift is not occurring rapidly enough to be significant. The three-peak nmr spectrum can be interpreted in terms of positions made equivalent by the Wagner-Meerwein rearrangement and the 6,2-hydride shift. The peak of area 4 is due to protons at C-1, C-2, and C-6; the peak of area 1 is due to the one proton at C-4; the peak of area 6 is due to protons at C-3, C-5, and C-7. Addition of the 3,2-hydride shift at higher temperatures makes all positions equivalent by interchanging C-1 with C-4, C-2 with C-3, and C-5 with C-6. Thus, by this interpretation the 2-norbornyl cation **9** represents an example of combined carbon and hydrogen degeneracies which results in the equivalence of all carbon and hydrogen atoms.

Jensen and Beck observed that the best-defined nmr spectrum (60 MHz) of the 2-norbornyl cation **9** was obtained by treating 2-norbornyl bromide **290** with gallium tribromide in $SO_2$ solvent (107). The nmr spectrum of **9** obtained by these workers exhibited a fine structure not recorded by Schleyer, Olah, and co-workers. The spectrum recorded by Jensen and Beck at $-80°$ consisted of resonance signals at δ 5.2, 3.1, and 2.1 with relative intensities of 4:1:6 and observed multiplicities of 7, 1, and 6, respectively. These workers feel that the observed slow 3,2-hydride shifts are not consistent with a classical ion (**9**) being the stable species observed in the nmr studies, since secondary, secondary-hydride shifts should occur in classical **9** with an activation energy lower than that observed. They believe the observed spectrum and rates of proton exchange can be adequately accounted for on the basis that alkyl and hydrogen bridged ions **291** and **292** occur in the system as stable forms, or one of these forms is a transition state and the other a stable species.

Olah et al. generated the norbornyl cation in strong acid solutions from the precursors shown in Scheme II (108).

These workers obtained a 100-MHz nmr resonance spectrum of this cation which showed the fine structure reported by Jensen and Beck. They also obtained a Raman spectrum of the cation in strong acid solution and concluded that the ion was best represented as a protonated nortricyclene (108). These authors felt that either a corner-protonated nortricyclonium ion **291** or an edge-protonated nortricyclonium ion **292** was most compatible with the nmr and Raman spectra. On this basis the observed high-energy barrier for a 3,2-hydride shift in the 2-norbornyl cation **9** arises from the necessary transition of either ion **291** or **292** to the classical

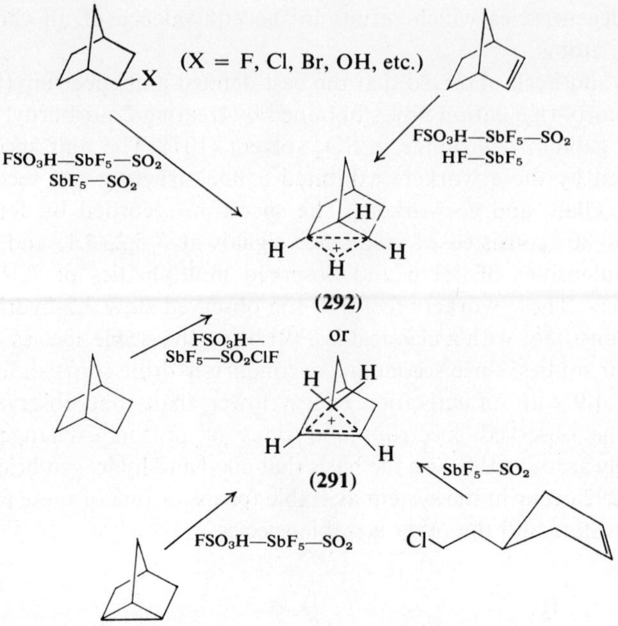

Scheme II.

2-norbornyl cation **9** before the 3,2-hydride shift occurs. Ion **291** or **292** at low temperatures would then represent what was originally assigned for the 2-norbornyl cation **9** as the result of a combination of Wagner-Meerwein rearrangements and 6,2-hydride shifts.

Recently Olah and White examined the $^{13}$C-nmr spectrum of the norbornyl cation generated in SbF$_5$-SO$_2$ from 2-*exo*-norbornyl chloride (**287**) at $-70°$ (109,110). This spectrum consisted of three peaks: (*1*) a pentuplet ($J = 53.3$ Hz) at 101.8 ppm (from $^{13}$CS$_2$) for three equivalent carbons associated with four equivalent protons; (*2*) a triplet ($J = 140.2$ Hz) at 162.5 ppm for three equivalent carbons each associated with two equivalent protons; and (*3*) a doublet ($J = 153$ Hz) at 156.1 ppm for one carbon associated with one proton.

This spectrum is most consistent with rapidly interconverting edge-protonated nortriclonium ions (**292**, **292'**, and **292"**) or with rapidly interconverting corner-protonated nortriclonium ions (**291**, **291'**, and **291"**). The equilibrating classical 2-norbornyl cation **9** is not in accord with the low-temperature Raman, proton, and carbon nmr spectra.

To determine whether the edge-protonated ion **292** or the corner-protonated species **291** is the more stable cation, Olah and White recorded the 100-MHz proton nmr spectrum of the norbornyl cation at temperatures down to −156° (109,110). At −120°, the spectrum was identical with that described previously. In the temperature range −128 to −150° the low-field peak due to the four equilibrating protonated cyclopropane ring protons broadened and then separated into two resonances each of relative area 2 at δ 3.05 and 6.59. The high-field resonance, due to the six methylene protons, also broadened, developing a shoulder at δ 1.70. The peak at δ 2.82 due to the single bridgehead proton remained unchanged. This spectrum is best explained (in conjunction with the foregoing data) in terms of the nonclassical corner-protonated ion (**291**) structure. Cation **291** represents the most stable ion in strong acid solution with ions like **292** as transition states between forms of **291**. A similar conclusion has been reached even more recently on the basis of an ESCA spectrum of the 2-norbornyl cation (87a).

An example of a carbonium ion system which is neither an equilibrating classical ion nor a fully delocalized nonclassical one was studied by Olah et al. (111). The cation obtained in $SbF_5$–$SO_2$ or $FSO_3H$–$SbF_5$–$SO_2$ from 1,2-dimethyl-2-*endo*-norbornanol (**293**) or *endo*- and *exo*-2-chloride (**294**) was probed with combined techniques of $^{13}C$ and $^1H$ nmr and Raman spectroscopy. The ion was assigned the degenerately rearranging unsymmetrically bridged structures **295** and **295′**.

(**295**)   (**295′**)

(**293**)   (**294**)

The result attractively demonstrates the existence of a large variety of possible intermediates, from classical to nonclassical, that can convincingly be invoked for rearranging cationic species.

### P. Bicyclo[3.1.0]hex-3-en-2-yl Cation

Swatton and Hart were the first to recognize the possibility that an appropriately substituted ion with the bicyclo[3.1.0]hex-3-en-2-yl cation (**296**) structure could exhibit fivefold carbon degeneracy as illustrated

(112). These investigators reported experiments that were best interpreted

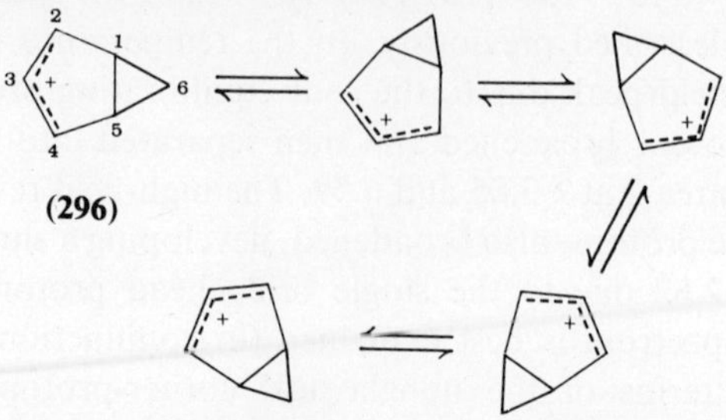

**(296)**

in terms of carbonium ions like **296** in which four or five such structures rapidly equilibrate (112). When a solution of hexamethylbicyclo[3.1.0]-hex-3-en-2-one **297** in 97% sulfuric acid was allowed to stand at 22.6° for 30 min and then hydrolyzed, a nearly quantitative yield of hexamethyl-2,4-cyclohexadienone (**298**) was obtained. To study the mechanism of this

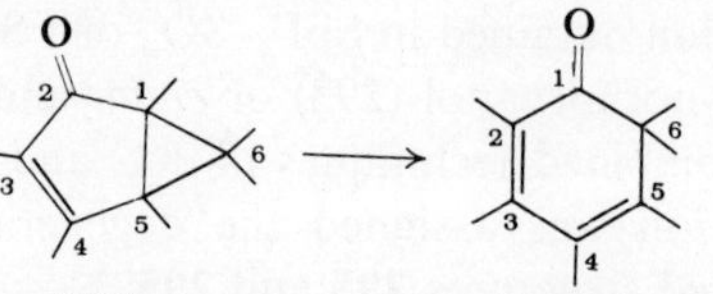

**(297)** → **(298)**

reaction, the compound with two $CD_3$ groups (**299**) was rearranged to give **300**, with the proton integrations for each methyl group as shown. This result is consistent with Scheme III in which **302** and **302'** are rapidly equilibrated and each can collapse by proton loss to **303** and **303'** ($-CD_3$ groups marked by an asterisk). Rapid equilibration of **302** and **302'** prior

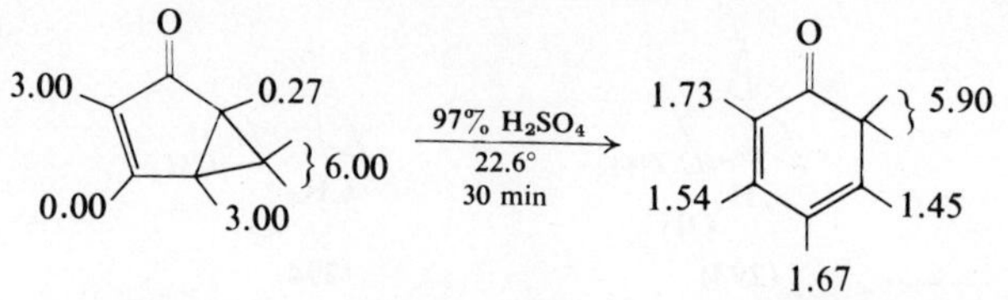

**(299)** $\xrightarrow{\substack{97\% \ H_2SO_4 \\ 22.6° \\ 30 \ min}}$ **(300)**

The figures indicate the number of protons at the respective positions.

to collapse should lead to a 1:1 mixture of **303** and **303'** with the label distribution as indicated. This is in good agreement with the values obtained for product **300**. The rapid migration of the cyclopropane ring around four of the five sides of the cyclopentenyl ring is thus established.

It is pertinent here to mention the work of Shubin et al., who investigated

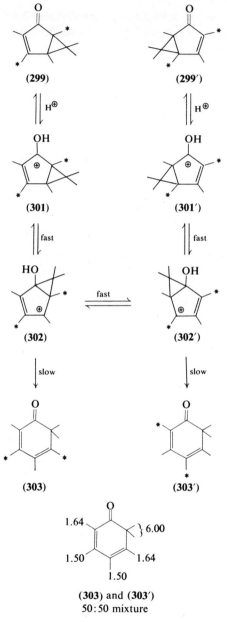

Scheme III. (Perdeuterated methyl groups are asterisked.)

the interconversion of 2,3,4,5,6,6-hexamethylhexa-2,4-dien-1-one (**304**) and 2,3,4,4,5,6-hexamethylhexa-2,5-dien-1-one (**305**), the methyls initially labeled with deuterium as shown (113).

These authors suggested a series of 1,2-methyl shifts as a rationale for the observations, and this mechanism, rather than the intervention of **301** and **302** would appear to account better for the preferential appearance of $CD_3$ groups in the 3, 4, and 5-positions in the main product, **305**. Similarly, it has been found that the treatment of 2,3,4,4,5,6-hexamethylhexa-2,5-dien-1-one labeled with $^{13}C$ in the carbonyl carbon **306** with $BF_3$–etherate at 130° results in a randomization of the isotopic label (114) among the ring carbons.

The investigators have proposed random label scattering via the hexamethylbenzenonium ion **307**, but such a mechanism requires the unlikely occurrence of 1,2-OR shifts in order to convincingly account for the observed scrambling. However, other alternatives are not obvious.

R = H, $CH_3$, or $BF_3$

Childs and Winstein studied the heptamethylbicyclo[3.1.0]hexenyl cation (**308**) and observed fivefold degenerate scrambling as predicted by Swatton and Hart (115). Cation **308** was prepared by the irradiation of a solution of the heptamethylbenzenonium ion **249** in $FSO_3H$ at $-78°$. The reverse process, **308** to **249**, occurs smoothly on warming the solution of ion **308** up to −9°. Cation **308** displays a temperature-dependent nmr spectrum, and to permit lower temperature observations (**308**), was formed photochemically in a 2:1 SO$_2$ClF–FSO$_3$H mixture. In this solvent the nmr spectrum of (**308**) could be observed over the temperature range −110 to −9°.

Below −100° the spectrum is unexceptional. At higher temperatures the signals for the five methyl groups on the cyclopentenyl ring broaden, then coalesce at about −87° and eventually become a sharp statistically averaged singlet ($\tau$ 7.90) at −48°. Although these five methyl groups become averaged at elevated temperatures, the two methyl groups on the cyclopropane methylene carbon atom C-6 remain distinct and show no sign of averaging up to −9°.

The observed fivefold degenerate scrambling in **308** involves migration of the cyclopropane methylene carbon C-6 around the cyclopentene ring as indicated. The key step is a concerted sigmatropic 1,4 shift of C-6 which, in the ring system **308**, must be suprafacial (116).

The symmetry of the highest occupied molecular orbital of this system, shown in **309**, predicts inversion of the bridging C-6 carbon in each shift. Thus the starred methyl group should always remain "outside" and the unstarred methyl group should remain "inside" during the fivefold degenerate rearrangement. This agrees with the experimental observation that the two C-6 methyl groups maintain their identities when the other five methyl groups average rapidly. Recent work by Hart et al. on the stereochemistry of the rapid equilibration of protonated bicyclo[3.1.0]-hexenones also indicate that these interconversions occur without interchange in the positions of substituents at C-6 (117).

Similar results were obtained by Koptyug et al. in their study of hexamethylbicyclo[3.1.0]hexenyl cation (118). Thus it is reported that treatment of 5-α-chloroethylpentamethylcyclopenta-1,3-diene (**310**) with $AlCl_3$ in $CH_2Cl_2$ at $-60°$ leads to cation **311**, whereas protonation of the diene **312** gives cation **313**.

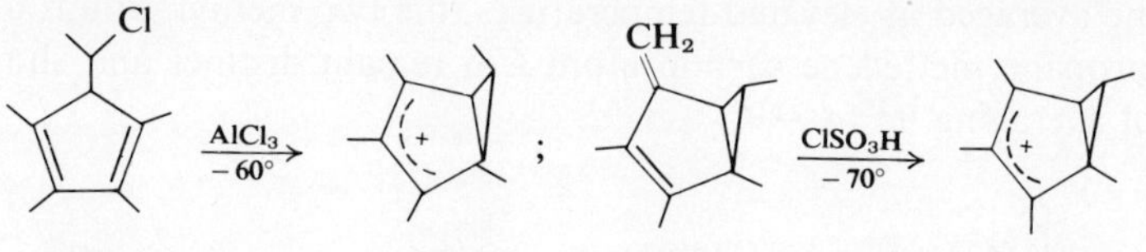

(310)    (311)    (312)    (313)

The stereochemical assignments of the $C_6$-methyl groups in cations **311** and **313** seem reasonable on the basis of this work. The two cations were not interconvertible. A similar fivefold degenerate rearrangement has been found in the $AlCl_3$-complexes of 5-acyl-1,2,3,4,5-pentamethylcyclopentadienes (118a).

Investigations by Zimmerman and Crumrine on the interconversion of **314** into **315** (119) and by Brennan and Hill on the interconversion of **316**

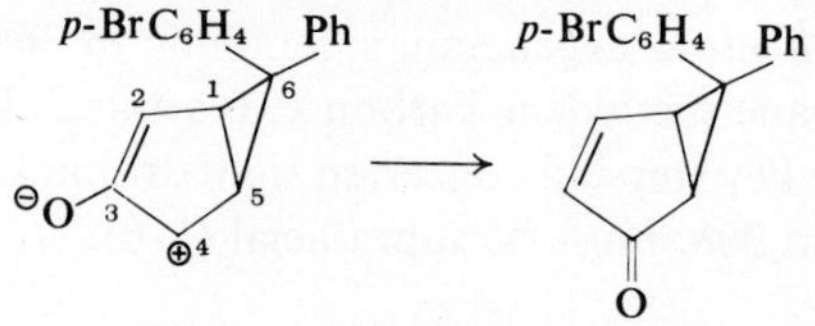

(314)    (315)

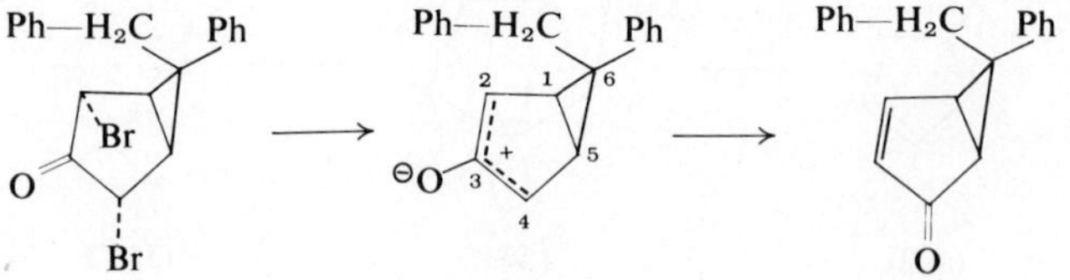

(316)    (317)

into **317** (120) indicate that these 1,4-sigmatropic rearrangements proceed with inversion of configuration at C-6.

Vogel et al. have recently studied the isotopically labeled parent system in strong acid medium (121). Thus it was found that slow label scrambling could be observed for the cation **296-d**, generated from the alcohol **318** at temperatures below that which would result in thermal isomerization to the

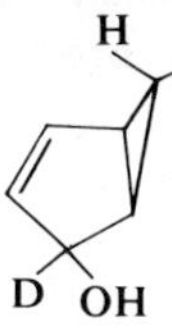

(318) 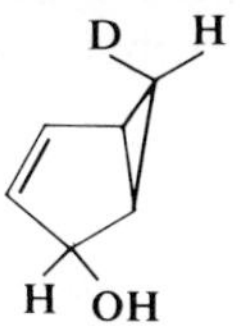    (319) 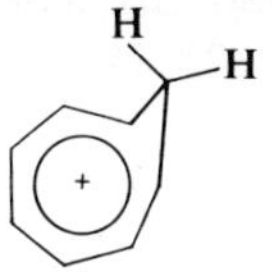    (320)

benzenonium ion; at the same time, the alcohol **319**, in which the label appeared in the *endo*-6 position, exhibited no increase in the nmr absorption for proton incorporation at this position. Again in this case, therefore, degeneracy is achieved by a circumambulation of the cyclopropane ring about the five-membered ring. The instability of the parent cation **296** relative to the vinylogous homotropylium cation **320** reflects the effects of antiaromaticity in the system.

### Q. 2-Brexyl or 2-Tricyclo[4.3.0.0$^{3,7}$]nonyl Cation

Nickon and co-workers pointed out the possibility that the 2-brexyl cation **321** could exhibit multiple degenerate properties (122). Ion **321**, by consecutive Wagner-Meerwein rearrangements, could transfer the positive charge originally at C-2 to all five carbons of the central five-membered ring, which is identified in **321** by heavy dots. Repetition of this process many

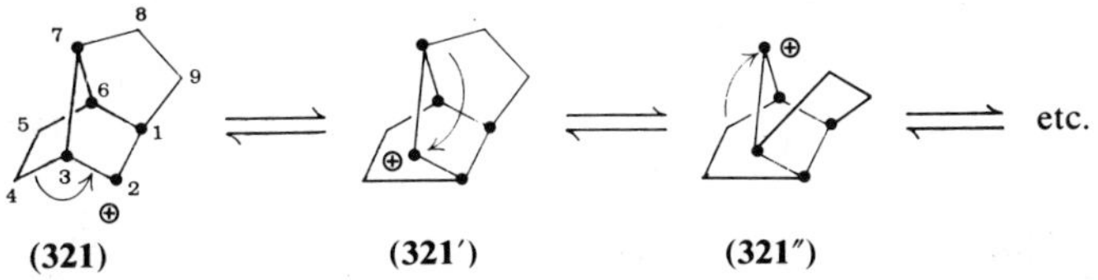

(321)    (321′)    (321″)    etc.

times would result in carbon atoms 1, 2, 3, 6, and 7 becoming equivalent to each other and in carbon atoms 4, 5, 8, and 9 becoming equivalent to each other also. Thus the theoretical nmr spectrum of cation **321** would contain two peaks of areas 5 and 8 if the mechanism just described were operative. A second rearrangement possibility exists for **321**, that of a 4,2-hydride shift, which would lead to the 4-brexyl cation **322**, an ion not capable of

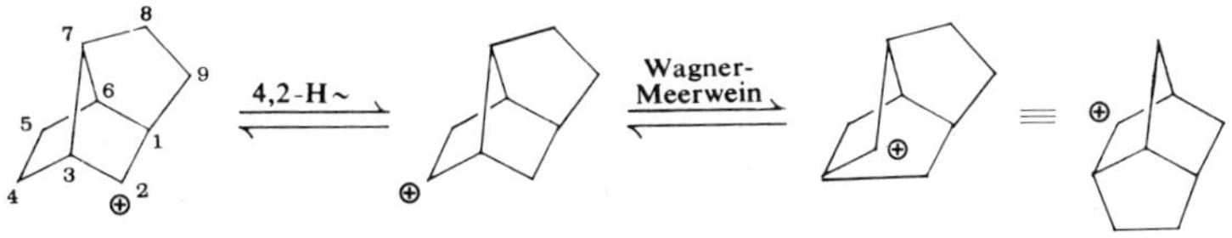

(321)    (322)    (323)

multiple degeneracy. Wagner-Meerwein rearrangement in **322** produces the 2-brendyl cation **323**.

Nickon and his associates have synthesized brexan-2-one **324** and brexan-2-ol **325** (122). The solvolysis of 2-brexyl brosylate **326** was

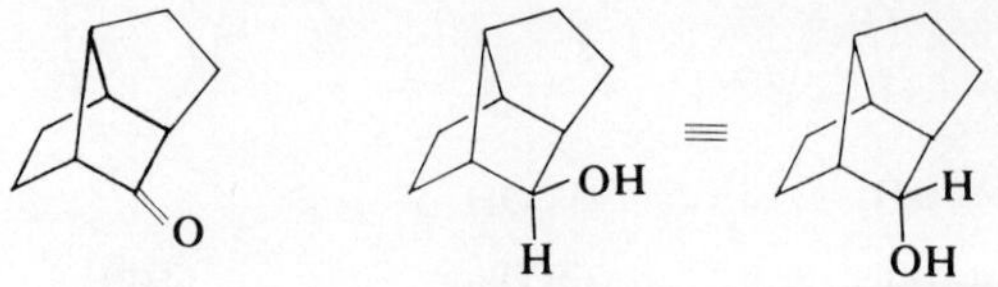

**(324)** **(325)**

carried out to give the product mixture shown (123). Bly and co-workers have reported the solvolysis of β-(*syn*-7-norbornenyl)ethyl brosylate **(330)**

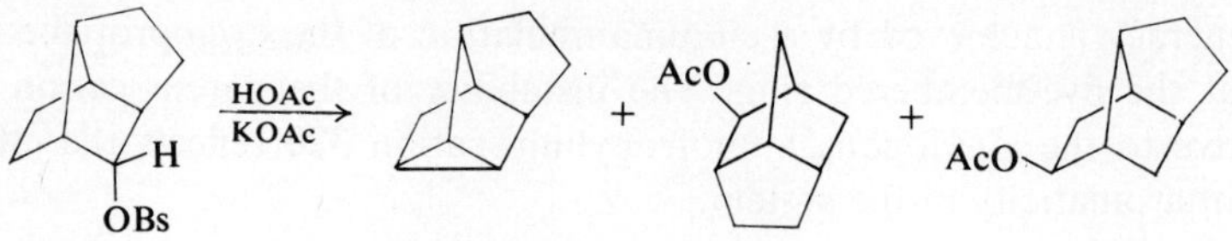

**(326)** **(327)** **(328)** **(329)**
  6.4%  51.3%  42.3%

to give the same set of products (124). Both sets of results can be explained in terms of initial formation of the 2-brexyl cation **321**, which then rapidly

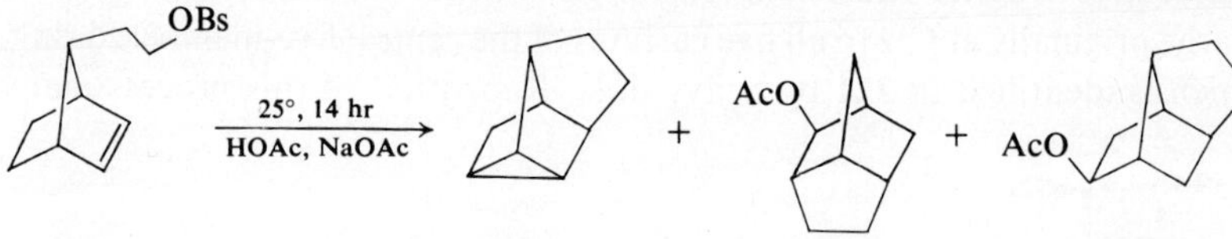

**(330)** **(327)** **(328)** **(329)**
  22%  42%  36%

**(330)** **(321)** **(322)** **(329)**

**(326)** **(323)** **(328)**

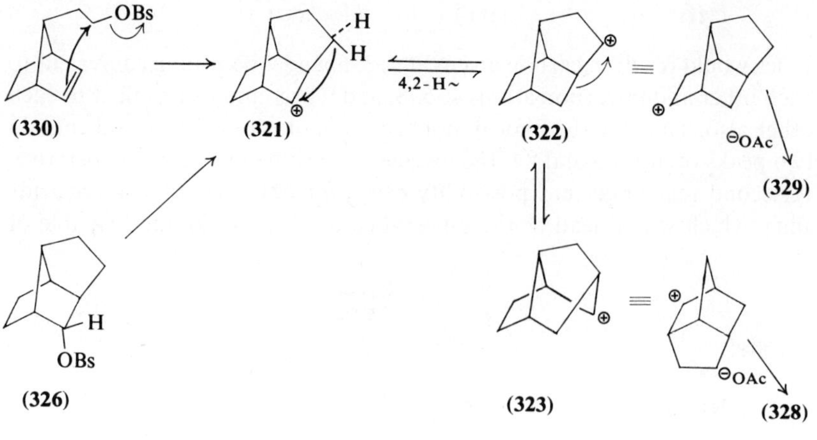

undergoes the 4,2-hydride shift to form cation **322**; **322** then yields **323** by Wagner-Meerwein rearrangement. The two acetate products arise from these ions. Lack of formation of acetate **331**, which would arise from the 2-brexyl cation **321**, indicates that this cation is not as stable as ions **322** and **323**. Bridged and protonated cyclopropane ion pair intermediates afford a more sophisticated rationalization of these results.

(331)

Nickon et al. studied the acid-catalyzed acetolysis of deltacyclane (**327**) and also found only products **328** and **329** (125). The brendyl acetate (**328**) is the more stable of the two products and predominates almost exclusively if the acetolysis mixture is allowed to equilibrate. Once again no product

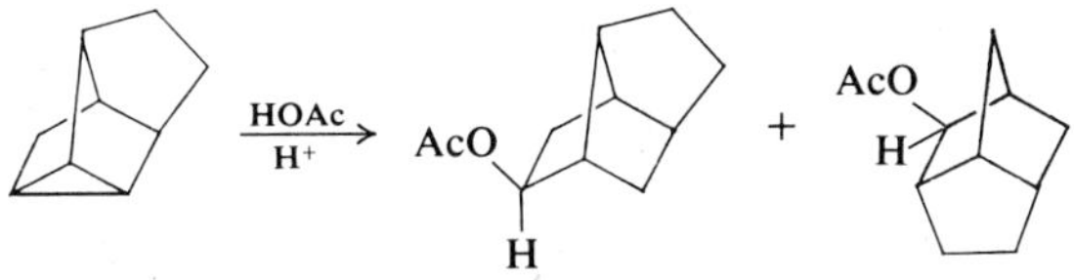

(327) (329) (328)

**331** due to the 2-brexyl cation was observed. This seems to indicate that the proposed degeneracy for the 2-brexyl cation (**321**) may never be observed because of the instability of the ion relative to other tricyclic cation isomers.

### R. Cyclopropylcarbinyl Cation Systems

#### 1. 8,9-Cyclo-2-adamantyl Cation

The degenerate cyclopropylcarbinyl–cyclopropylcarbinyl rearrangement, **18** ⇌ **18′** (19), also takes place in more elaborate systems.

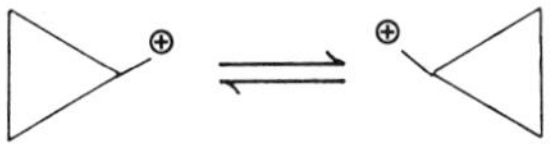

(18) (18′)

The 8,9-cyclo-2-adamantyl cation **332** contains the proper symmetry such that continuous cyclopropylcarbinyl–cyclopropylcarbinyl rearrangements could lead to threefold carbon degeneracy as indicated (classical ions shown for convenience). Baldwin and Foglesong studied the solvolysis of derivatives leading to the formation of cation **332** and observed the nearly complete equivalence of positions 2, 8, and 9 (126,127). This

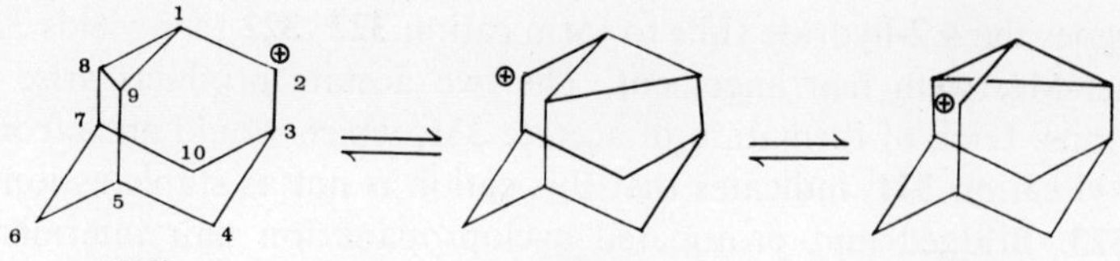

(332)   (332′)   (332″)

provides evidence for the occurrence of degenerate cyclopropylcarbinyl–cyclopropylcarbinyl rearrangements within this system.

Reduction of 8,9-cyclo-2-adamantanone (127) 333 with sodium borohydride gave alcohol 334, which was converted to its 3,5-dinitrobenzoate 335. Solvolysis of 335 in 60% aqueous acetone afforded a 95% yield of

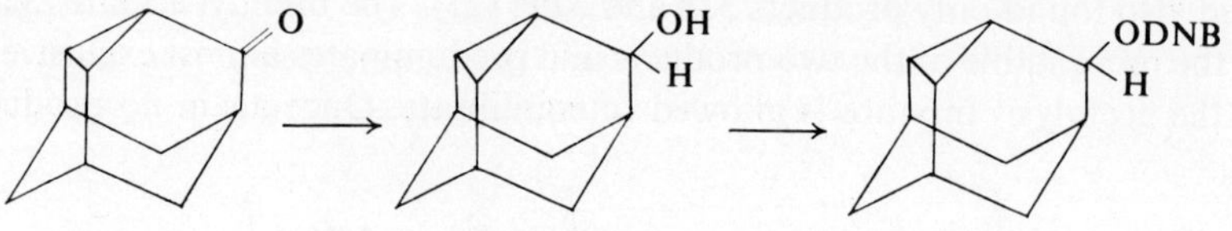

(333)   (334)   (335)

334. The 2-deuterio and 2-tritio analogs of 334 were prepared with NaBD$_4$ and NaBT$_4$. Solvolysis of the labeled 3,5-dinitrobenzoates 336 and

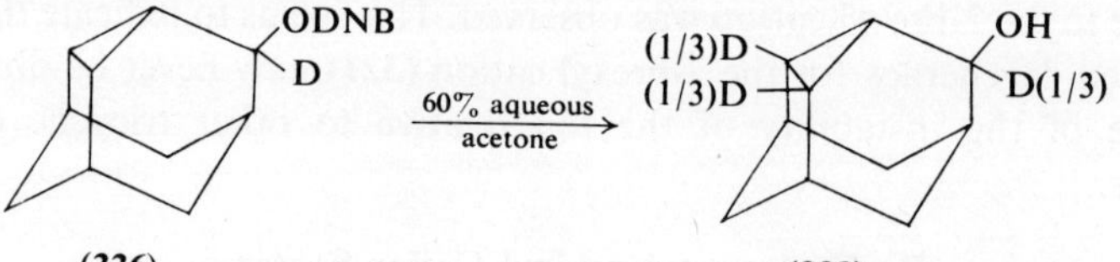

(336)   (338)

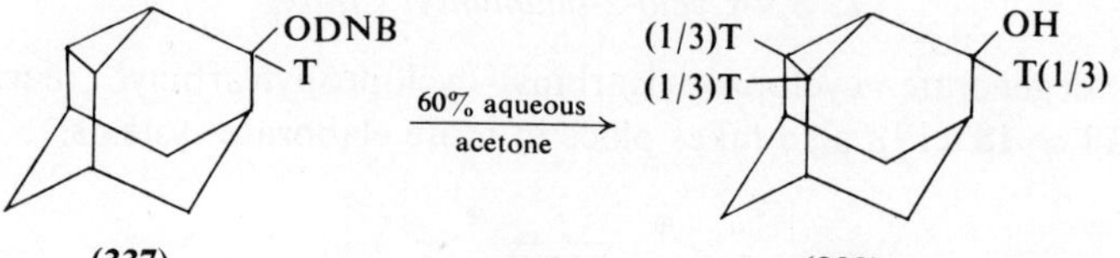

(337)   (339)

337 gave product alcohols 338 and 339 with the label distributions as shown. The deuterium contents were determined by mass spectrometry and the tritium activities were determined by liquid scintillation counting.

These results indicate that the 2, 8, and 9 positions of the 8,9-cycloadamantyl system attain a high degree of equivalence during the solvolysis reaction. The solvolysis rate constant for 335 indicates a substantial rate acceleration above that expected for classical ion 332. Based on these two observations, the authors feel that the 8,9-cyclo-2-adamantyl cation is best represented as a set of rapidly equilibrating, charge-delocalized, "bisected" (16) cyclopropylcarbinyl cations **340**, **340′**, and **340″**.

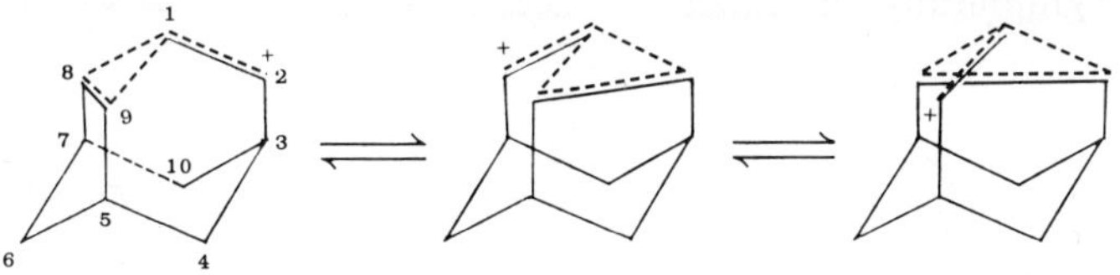

(340)   (340′)   (340″)

### 2. Tricyclo[3.2.1.0$^{2,7}$]octenyl Cation: A Degenerate Cyclopropylcarbinyl System

An interesting variation of the degenerate cyclopropylcarbinyl rearrangement has recently been observed by Hart and Love (128); the essence of the rearrangement appears in the hypothetical sequence

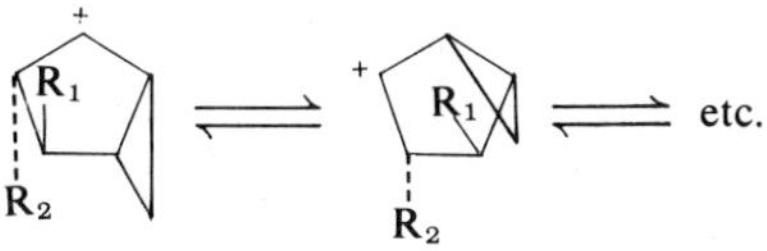

in which cyclopropylcarbinyl–cyclopropylcarbinyl interconversion is accompanied by concomitant movement of groups $R_1$ and $R_2$ to adjacent positions. This rearrangement mode was demonstrated for the benzobicyclo[3.2.1]octadiene system **341** during treatment with a large excess of $CF_3COOD$ at 60°, in which all but the *anti*-methyl group on the bridge exchanged; the rate of exchange for the five methyl groups was uniform. In contrast, the ketone **342** exhibited exchange for only the asterisked methyl group.

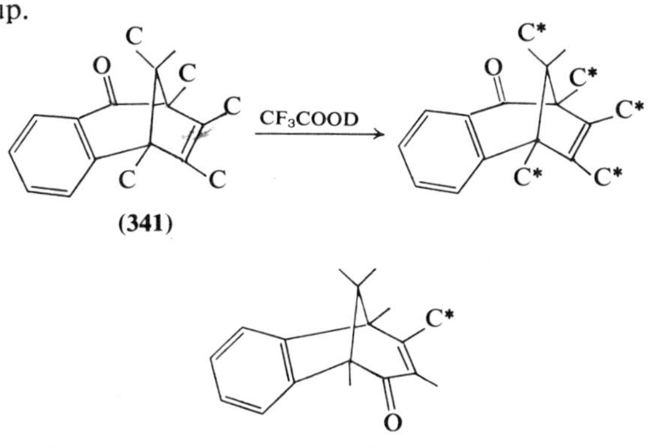

(341)

(342)

* = deuterium content

These observations suggest a rearrangement process in which the carbon atoms making up the benzo- and carbonyl positions in effect revolve relative to the five-membered ring containing the free olefin moiety, which may be graphically illustrated

Exchange takes place via the alkene corresponding to each cation.

### 3. Cyclopropylcarbinyl Cation: Stereochemistry of Rearrangement

Recent work has shown that the degenerate cyclopropylcarbinyl–cyclopropylcarbinyl cation rearrangement proceeds with retention of stereochemistry exhibited in the starting species. Wiberg and Szeimies have deuterated bicyclohexane **343** in DOAc and analyzed the deuterium distribution in the acetate products (19a).

Almost all the deuterium was found in the *cis*-2-, *cis*-3-, and carbinyl positions, indicating high stereoselectivity in rearrangement. Because of unavoidably large error limits in locating deuterium in the product, Schleyer and Majerski investigated a similar rearrangement using the hexadeuterated system **288**. The location of a single proton among six

deuterons by pmr spectroscopy is obviously much more accurate than the inverse—the location of one deuteron among six protons by difference (19c).

After products had been separated (allylcarbinol and cyclobutanol are also produced), it was found that equal quantities of hydrogen appeared in the *cis*-2-, *cis*-3-, and carbinyl positions; hence, within experimental error, the degenerate rearrangement was shown to be stereospecific. The proposed delocalized species **345** most economically explains the results of the two investigations, as well as the observation that cyclobutyl products [with stereospecifically *cis* arrangement of substituent and label (19c)] are also produced.

Olah and co-workers have now refined their pmr and cmr study of the stable cyclopropylcarbinyl, methylcyclopropylcarbinyl, and dimethylcyclopropylcarbinyl cations in super acid solvents (19d). While the last two cations are indicated to have a static bisected structure at low temperatures, the parent cyclopropylcarbinyl cation shows degeneracy: at −80° there are only three sets of pmr signals due to the methine hydrogen, the three *cis* and the three *trans* methylene protons. By an analysis of the carbon chemical shifts, it was concluded that the preferred structure was not bisected and that rapid equilibration occurred through a puckered cyclobutyl cation.

## S. 4-Protoadamantyl Cation

The cations derived from suitable derivatives of 4-*exo*-protoadamantanol **346** and the *endo*-epimer **347** have been studied by Schleyer and

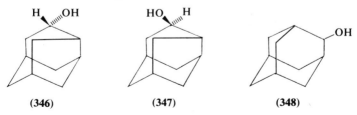

Lenoir (129). Although the dinitrobenzoate of **346** in aqueous acetone gave only 2-adamantanol (**348**), the tosylate derived from **347** reacted $10^4$ times more slowly (estimated considering usual differences between rates of solvolysis of tosylates and dinitrobenzoates), to yield 20% of **347** and 80% of **348**. In order to probe the mechanism of the latter reaction, the

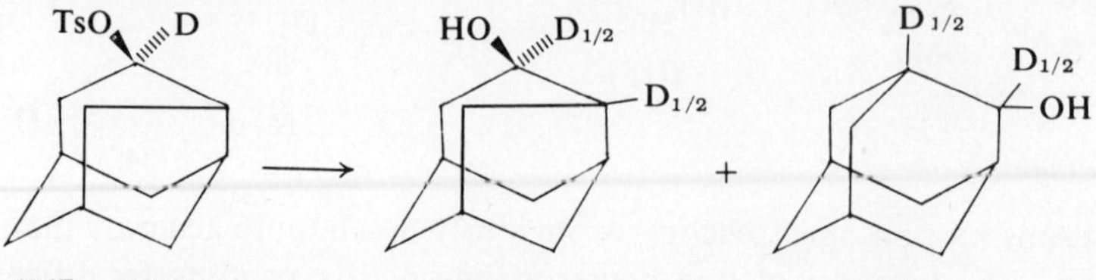

deuterated compound **347-OTs** was synthesized and solvolyzed, and the products were analyzed. It was found that deuterium had been scrambled evenly between two positions in the two products **347-d** and **348-d**. The mechanism suggested by Lenoir and Schleyer consistent with these observations involves the bridged ion **349**

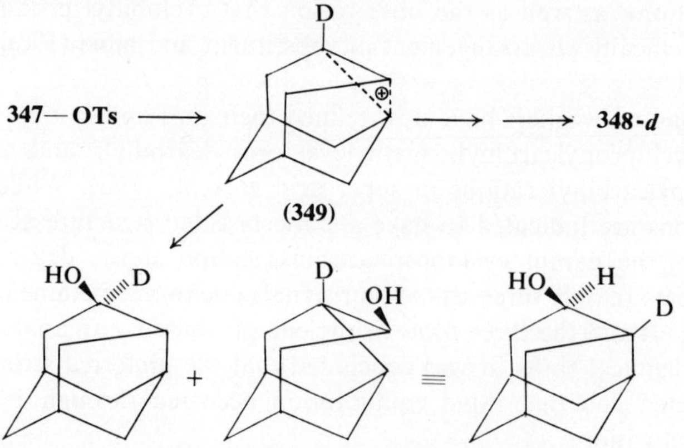

### T. Adamantyl Cations

Degenerate rearrangements involving stable adamantyl cations have been unsuccessfully sought by a number of workers (91b,130). The reason for this failure, presumably, is cation instability under sufficiently vigorous conditions. By using specifically labeled (2-$^{14}$C)-adamantane (**350**, X = H), Majerski et al. were able to demonstrate that degenerate carbon scrambling did in fact occur when the hydrocarbon was treated with AlBr$_3$ at elevated temperatures (131). Such a skeletal reorganization was rationalized according to the following scheme, which includes as a key step a degenerate 4-protoadamantyl-4-protoadamantyl rearrangement.

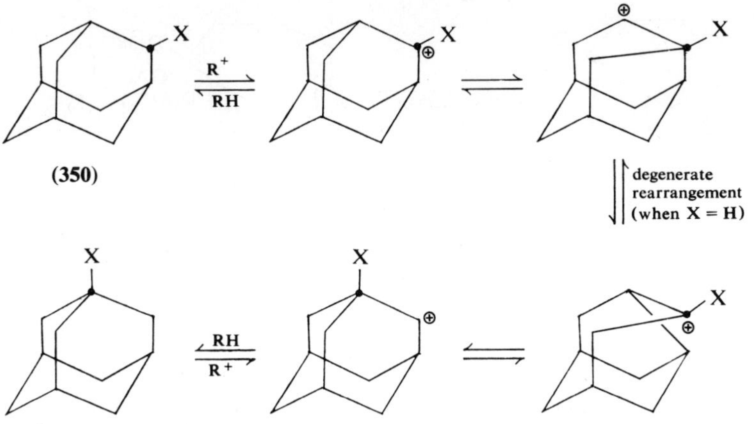

(350)

degenerate rearrangement (when X = H)

(351)

An exactly analogous mechanistic process was discovered by Majerski et al., who showed, by means of isotopic labeling, that in the 2-methyl to 1-methyladamantane rearrangement (**350**, X = $CH_3$ to **351**, X = $CH_3$), the methyl group did not migrate, but rather remained attached to the carbon to which it was initially bonded (132). A simple 1,2-methyl shift is prevented because of orthogonality between the carbon–carbon bond of the "migrating" methyl and the $p$ orbital of the initially formed carbonium ion (91b,130b,c).

## U. 4-Homoadamantyl or 4-Tricyclo[4.3.1.1$^{3,8}$]undecyl Cation

The 4-homoadamantyl cation **352**, as noted by Nordlander, is capable of exhibiting combined carbon and hydrogen degeneracy (133). Cation **352** can be degenerate with respect to both Wagner-Meerwein rearrangement and vicinal hydride shifts. These two processes can be distinguished

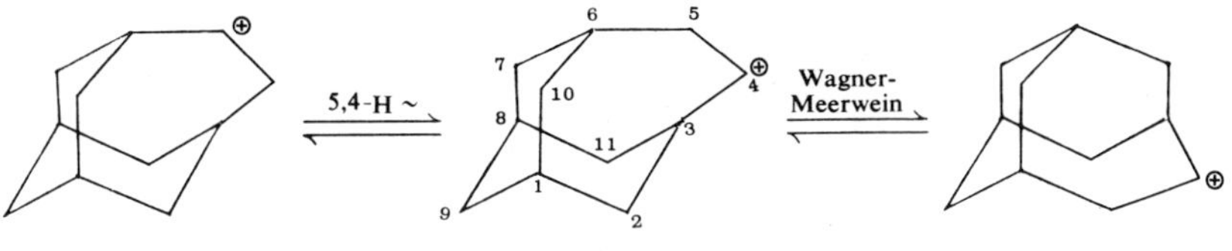

(352′)   (352)   (352″)

on the basis of deuterium label redistribution; a deuterium label originally at C-4 in **352** would appear at C-5 in **352′** but at C-3 in **352″**. Combination of both 1,2-carbon shifts and 1,2-hydride shifts could lead to elevenfold (complete) degeneracy in **352** if classical cations are involved. The degeneracy may be limited by ion pairing or by bridged-cation formation.

Nordlander and his associates have synthesized the deuterated 4-homoadamantyl tosylate **353** and have solvolyzed it with the results

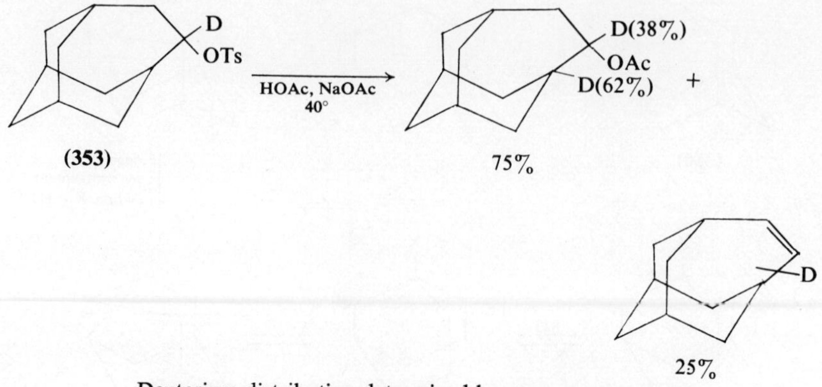

(353) → HOAc, NaOAc, 40° → [product with D(38%), OAc, D(62%)] 75% + [alkene with D] 25%

Deuterium distribution determined by mass spectrometry.

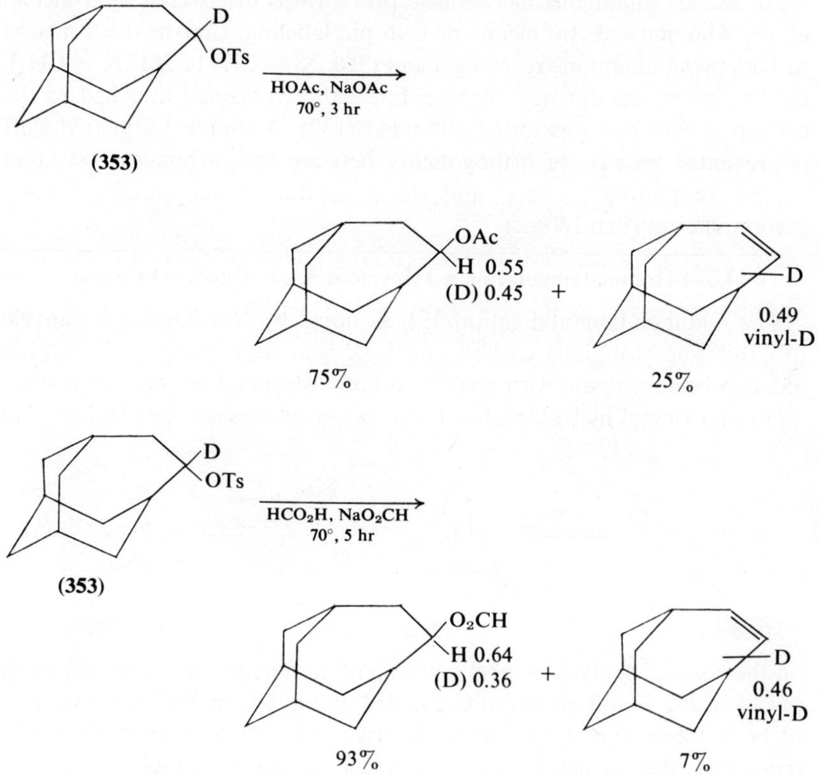

(353) → HOAc, NaOAc, 70°, 3 hr → [OAc, H 0.55, (D) 0.45] 75% + [alkene D, 0.49 vinyl-D] 25%

(353) → HCO$_2$H, NaO$_2$CH, 70°, 5 hr → [O$_2$CH, H 0.64, (D) 0.36] 93% + [alkene D, 0.46 vinyl-D] 7%

Scheme IV. Deuterium distribution as determined by NMR.

indicated (133). These results indicate a 1,2-carbon shift as the primary process responsible for label redistribution (this predicts 50% deuterium at C-3 and C-4), but intervention of hydride shifts could be responsible for part of the label scrambling. A similar study at Princeton involved an independent synthesis and solvolysis of **353** with the results shown in Scheme IV (134). These data indicate a deuterium distribution quite close to the predicted values for the occurrence of a degenerate 1,2-carbon shift in **352**. The formolysis of **353** has also been carried out and the results are in accord with those obtained on acetolysis (135). The acetolyses of the dideuterated and trideuterated tosylates **354** and **355** were also studied at Princeton, and the acetate and olefin products showed label scrambling

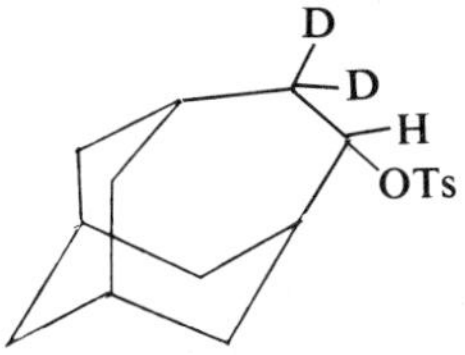

(354) 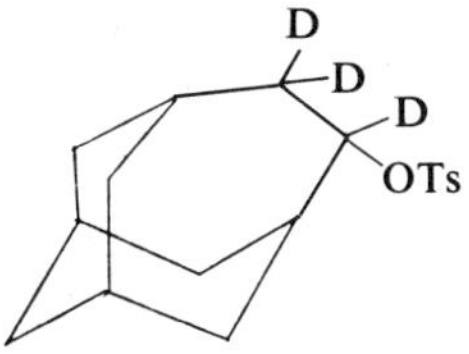  (355)

results inconsistent with a mechanism involving only a 1,2-carbon shift (135). It has been tentatively concluded that hydride shifts are involved in the rearrangements of cation **352**.

## V. 8-Tetracyclo[4.3.0.0$^{2,4}$.0$^{3,7}$]nonyl Cation

The 8-tetracyclo[4.3.0.0$^{2,4}$.0$^{3,7}$]nonyl cation **21** is capable of twofold carbon degeneracy if homocyclopropylcarbinyl rearrangements of the type illustrated occur. Alternatively, it is possible that this cation could exist in

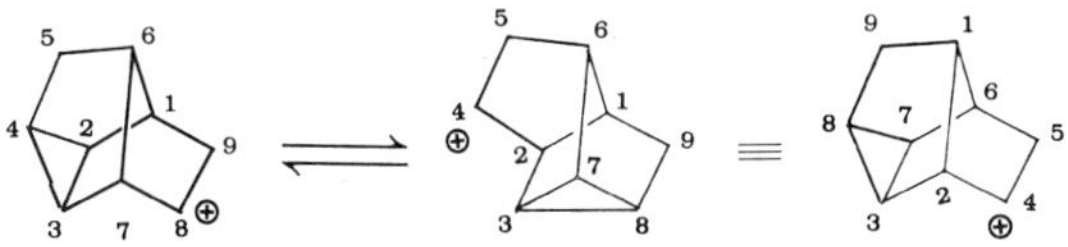

(21)  (21')

the nonclassical form **356**, in which carbon atoms 8 and 4 are simply equivalent due to charge delocalization.

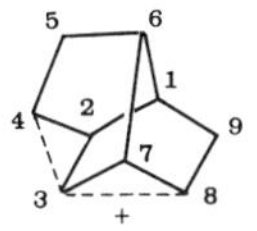

(356)

Freeman and Balls synthesized the tetracyclic ketone **357** and were able to prepare the derivatives **358**, **359**, and **360** for solvolysis (52). The acetolysis of **358** and **359** gave only one acetate product, **361**. Acetolysis of the

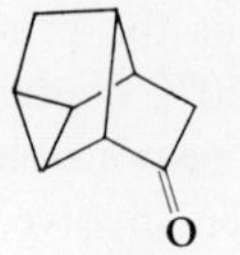

(357) 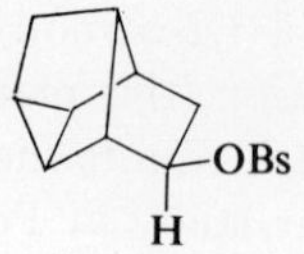 (358) 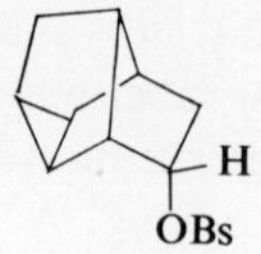 (359) 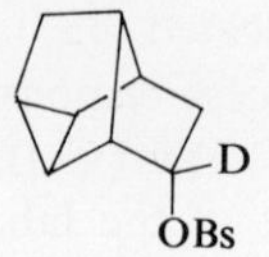 (360)

labeled compound **360** produced *exo*-acetate which, according to infrared and nmr spectroscopy, contained ca. one-half a deuterium at C-8 and ca. one-half a deuterium in the cyclopropane ring. These results are consist-

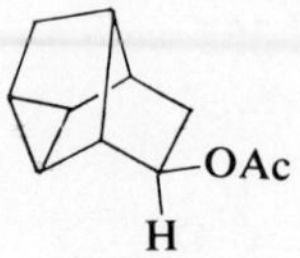

(361) 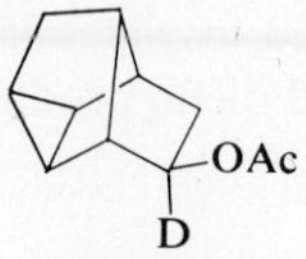 (362) 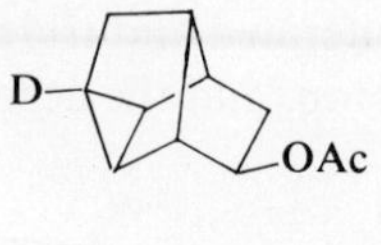 (363)

ent with the formation of a nearly equimolar mixture of acetates **362** and **363**, in which the deuterium is distributed between C-8 and C-4.

The rate constants for the acetolysis of **358** and **359** at 25° are 2.79 × $10^{-4}$ sec$^{-1}$ and 4.74 × $10^{-6}$ respectively. From these values anchimeric assistance factors of $10^{5.9}$ for **358** and $10^{3.7}$ for **359** were calculated. Since the rate factor for anchimeric assistance in the 2-*exo*-norbornyl cation is only $10^{3.3}$, it appears that both the *exo* and *endo* isomers **358** and **359** of the tetracyclo[4.3.0.0$^{2,4}$0$^{3,7}$] nonyl system solvolyze with anchimeric assistance. It can be argued that this disfavors a rapid equilibration of classical cations **21** and **21′** as the mechanism for the deuterium label scrambling, since a classical ion derived from the *endo* compound **359** should exhibit little anchimeric assistance. However, it is not clear why the *endo* isomer shows an accelerated rate; the nature of participation has not been identified.

When either alcohol **364** or **365** is dissolved in FSO$_3$H–SO$_2$ at −78°, the same carbonium ion solution is observed. The nmr spectrum of this

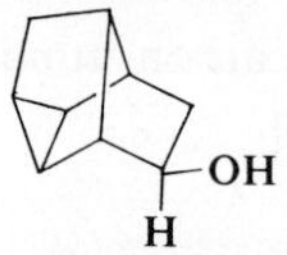

(364) 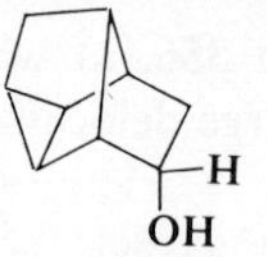 (365)

solution does not change significantly from −55 to −10° and is characteristic of a cation showing absorptions at τ 4.83, 7.08, 7.50, 7.83, and 8.23 in the ratio 2:2:2:1:4, respectively. Quenching the cation solution with aqueous potassium hydroxide produces **364**. This spectrum is best interpreted in terms of ion structure **356** with the proton resonance signals indicated (133). These results indicate a 1,2-carbon shift as the primary process responsible for label redistribution (this predicts 50% deuterium at C-3 and C-4), but intervention of hydride shifts could be responsible for part of the label scrambling. A similar study at Princeton involved an independent synthesis and solvolysis of **353** with the results shown in Scheme IV (134). These data indicate a deuterium distribution quite close to the predicted values for the occurrence of a degenerate 1,2-carbon shift in **352**. The formolysis of **353** has also been carried out and the results are in accord with those obtained on acetolysis (135). The acetolyses of the dideuterated and trideuterated tosylates **354** and **355** were also studied at Princeton, and the acetate and olefin products showed label scrambling

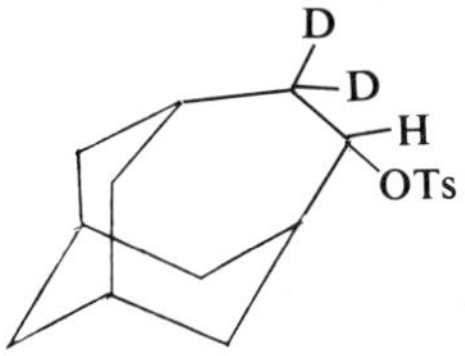

(354) 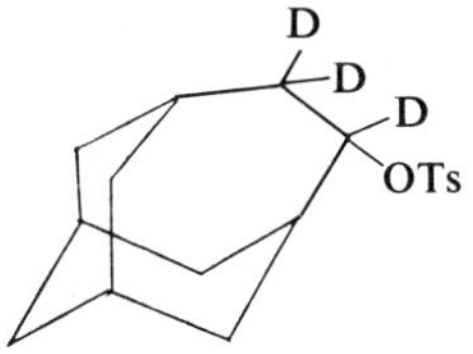         (355)

results inconsistent with a mechanism involving only a 1,2-carbon shift (135). It has been tentatively concluded that hydride shifts are involved in the rearrangements of cation **352**.

## V. 8-Tetracyclo[4.3.0.0$^{2,4}$.0$^{3,7}$]nonyl Cation

The 8-tetracyclo[4.3.0.0$^{2,4}$.0$^{3,7}$]nonyl cation **21** is capable of twofold carbon degeneracy if homocyclopropylcarbinyl rearrangements of the type illustrated occur. Alternatively, it is possible that this cation could exist in

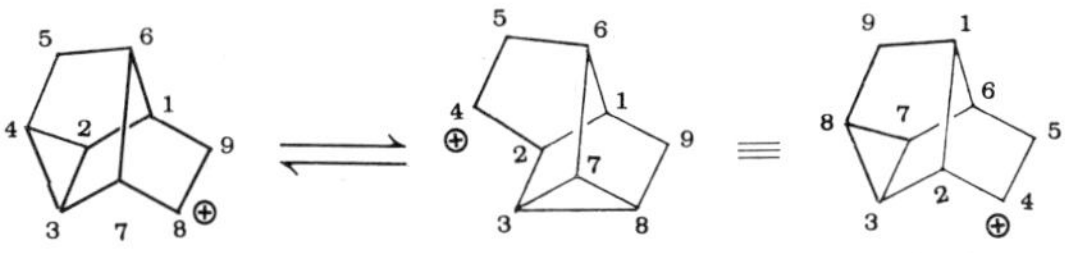

(21)         (21')

the nonclassical form **356**, in which carbon atoms 8 and 4 are simply equivalent due to charge delocalization.

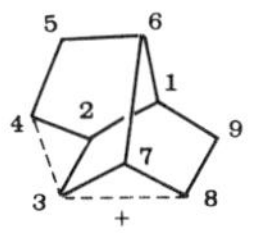

(356)

Freeman and Balls synthesized the tetracyclic ketone **357** and were able to prepare the derivatives **358**, **359**, and **360** for solvolysis (52). The acetolysis of **358** and **359** gave only one acetate product, **361**. Acetolysis of the

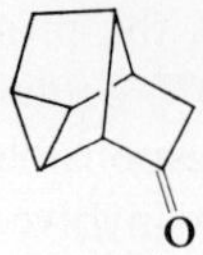

(357) 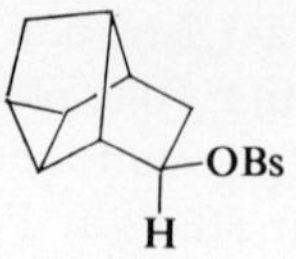 (358) 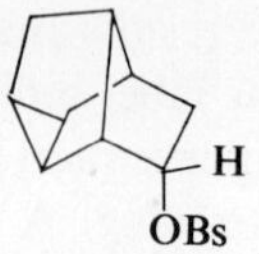 (359) 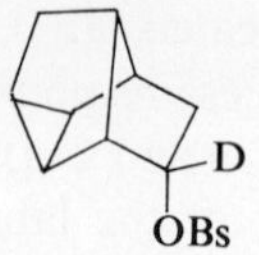 (360)

labeled compound **360** produced *exo*-acetate which, according to infrared and nmr spectroscopy, contained ca. one-half a deuterium at C-8 and ca. one-half a deuterium in the cyclopropane ring. These results are consist-

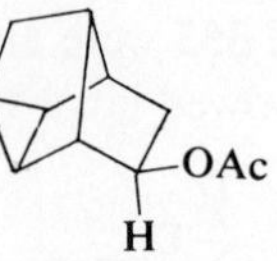

(361) 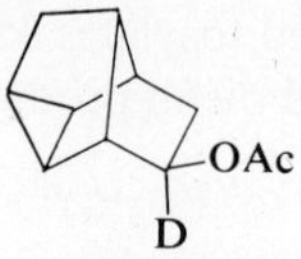 (362) 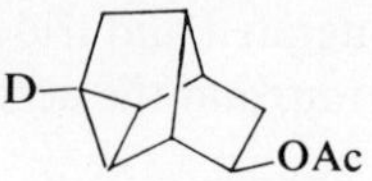 (363)

ent with the formation of a nearly equimolar mixture of acetates **362** and **363**, in which the deuterium is distributed between C-8 and C-4.

The rate constants for the acetolysis of **358** and **359** at 25° are 2.79 × $10^{-4}$ sec$^{-1}$ and 4.74 × $10^{-6}$ respectively. From these values anchimeric assistance factors of $10^{5.9}$ for **358** and $10^{3.7}$ for **359** were calculated. Since the rate factor for anchimeric assistance in the 2-*exo*-norbornyl cation is only $10^{3.3}$, it appears that both the *exo* and *endo* isomers **358** and **359** of the tetracyclo[4.3.0.0$^{2,4}$0$^{3,7}$] nonyl system solvolyze with anchimeric assistance. It can be argued that this disfavors a rapid equilibration of classical cations **21** and **21'** as the mechanism for the deuterium label scrambling, since a classical ion derived from the *endo* compound **359** should exhibit little anchimeric assistance. However, it is not clear why the *endo* isomer shows an accelerated rate; the nature of participation has not been identified.

When either alcohol **364** or **365** is dissolved in FSO$_3$H–SO$_2$ at −78°, the same carbonium ion solution is observed. The nmr spectrum of this

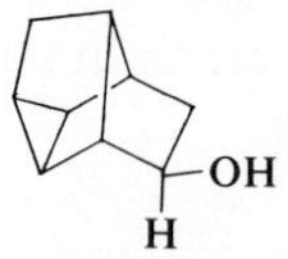

(364) 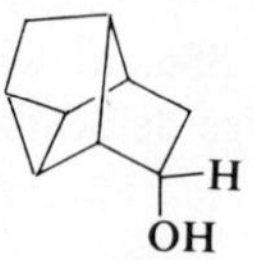 (365)

solution does not change significantly from −55 to −10° and is characteristic of a cation showing absorptions at τ 4.83, 7.08, 7.50, 7.83, and 8.23 in the ratio 2:2:2:1:4, respectively. Quenching the cation solution with aqueous potassium hydroxide produces **364**. This spectrum is best interpreted in terms of ion structure **356** with the proton resonance signals assigned as shown. Based on the rate data, the deuterium label scrambling, and the nmr spectrum of the foregoing carbonium ion, it appears that ion

$\tau = 8.23$
$\tau = 4.83$
$\tau = 8.23$
$\tau = 7.83$
$\tau = 4.83$

(356)

*a* protons $\tau = 7.08$ or $7.50$
*b* protons $\tau = 7.50$ or $7.08$

**356** best represents the structure of the carbonium ion derived from the solvolysis of 8-tetracyclo[4.3.0.0$^{2,4}$.0$^{3,7}$]nonyl precursors.

## V. CONCLUSION

It was mentioned in the Introduction that interest in degeneracy in carbonium ions centered around three questions:
1. Can degeneracy be achieved, and, if so, to what extent?
2. By what mechanism does a cation become degenerate?
3. Are there special properties arising from degeneracy?

These questions, to some degree at least, can now be answered.

Degeneracy in carbonium ion rearrangements has been observed often. Degenerate cations have been generated both under solvolytic conditions as well as in highly acidic solvents. Under the latter conditions, increased ion stabilities and lifetimes often permit observation of degeneracy that does not occur in the presence of stronger nucleophiles. The extent to which degeneracy occurs varies greatly, from the numerous examples of twofold degeneracy to instances in which degeneracy is extended over several atoms, as in the solvolytically generated homocubyl cation **10**, or over all carbon atoms, as has been recently observed for the barbaralyl cation **32** in fluorosulfonic acid solvents at $-135°$ on the NMR time scale.

The basic mechanisms for most of the examples of degeneracy generally seem to be unexceptional. Thus in the great majority of cases Wagner-Meerwein 1,2 shifts or hydride transfers can conveniently account for the observed phenomena. Double-bond and cyclopropyl ring participation are invoked in other cases. Most recently, however, especially through theoretical analyses, other possibilities have been suggested to explain apparent anomalies in these rearrangements. The best example is the barbaralyl cation. At normal temperatures under conditions of solvolysis, the barbaralyl cation exhibits threefold degeneracy. While this behavior can be explained by concerted homoallylic–cyclopropylcarbinyl interconversions, theoretical work now suggests the much more attractive possibility that the cation structure has threefold symmetry. Although no presently available result supports it, Hoffmann's suggestion of the pyramidal $C_5H_5^+$ structure certainly will stimulate attempts at verification.

It cannot be said unequivocally that degeneracy does or does not confer stabilization or the special properties on a cation. The stability or instability observed in degenerate cations can more convincingly be contributed to other influences. For example, the potentially degenerate bicyclo[3.2.2]-nona-2,5,8-trien-4-yl cation apparently is destabilized, judging from the fact that efforts to detect its intermediacy have been negative. This instability, we believe, is better attributed to longicyclic antiaromaticity discussed by Goldstein and Hoffmann (136). The homocubyl cation appears to be stabilized, but surely this can be adequately explained by nonclassical resonance than by some special effect ascribable to degeneracy; on the other hand, the tetracyclo[4.3.0.0$^{2,4}$.0$^{3,7}$]-non-8-en-5-yl cation (15) is stabilized but exhibits no degenerate properties. Degeneracy in each species appears to be the result of a unique combination of factors, both structural and dynamic, which bring it about. Based on the data at hand, we cannot say that the energetics and dynamics of a cation are affected by the existence of degeneracy. Nevertheless, we hope it is clear from this review that molecules exhibiting degeneracy afford a fertile field for study.

## REFERENCES

1. G. W. Wheland, *Advanced Organic Chemistry*, 3rd ed., Wiley, New York, 1960, p. 48.
2. W. v. E. Doering and W. Roth, *Tetrahedron*, **19**, 715 (1963); J. F. M. Oth, R. Merényi, G. Engel, and G. Schröder, *Tetrahedron Lett.*, **1966**, 3377; G. Schröder, J. F. M. Oth, and R. Merényi, *Angew. Chem.*, **77**, 774 (1965); *Angew. Chem. Inter. Ed.*, **4**, 752 (1965).
3. For examples, see Y. Pocker, in Vol. I, *Molecular Rearrangements*, P. D. Mayo, Ed., McGraw-Hill, New York, 1963, p. 1; J. A. Berson, *ibid.*, p. 111; R. Breslow, *ibid.*, p. 133.
4. A. T. Balaban, D. Farcăsiu, and R. Bănică, *Rev. Roum. Chim.*, **11**, 1205 (1966); cf. A. T. Balaban, *Rev. Roum. Chim.*, **15**, 1960 (1970).
5. A. T. Balaban and D. Farcăsiu, *J. Am. Chem. Soc.*, **89** 1958 (1967).
6. K. Mislow, personal communication.
7. P. B. D. de la Mare in Reference 3, p. 27.
8. P. v. R. Schleyer, T. M. Su, M. Saunders, and J. C. Rosenfeld, *J. Am. Chem. Soc.*, **91**, 5174 (1969); J. M. Bollinger, J. M. Brinich, and G. A. Olah, *ibid.*, **92**, 4025 (1970); N. C. Deno, R. C. Haddon, and E. N. Kowak, *ibid.*, **92**, 6691 (1970).
9. C. C. Lee and M. K. Frost, *Can. J. Chem.*, **43**, 526 (1956); J. D. Roberts and J. A. Yancey, *J. Am. Chem. Soc.*, **74**, 5943 (1952); P. C. Myhre and E. Evans, *ibid.*, **91**, 5641 (1969); H. H. Jaffé and S. Billets, *ibid.*, **94**, 674 (1972).

10. G. A. Olah, J. R. DeMember, R. H. Schlosberg, and Y. Halpern, *J. Am. Chem. Soc.*, **94**, 156 (1972).
11. G. Wagner and W. Brickner, *Ber. Dtsch. Chem. Ges.*, **32**, 2302 (1899).
12. H. Meerwein, *Justus Liebigs Ann. Chem.*, **396**, 200 (1913); *ibid.*, **405**, 129 (1914); *ibid.*, **417**, 255 (1918).
13. For a review, see J. L. Fry and G. J. Karabatsos, in *Carbonium Ions*, Vol. II, G. A. Olah and P. v. R. Schleyer, Eds., Interscience, New York, 1970, Chapter 14.
14. C. W. Shoppee, *J. Chem. Soc.*, **1946**, 1147; M. Simonetta and S. Winstein, *J. Am. Chem. Soc.*, **76**, 18 (1954).
15. S. Winstein, H. M. Walborsky, and K. Schrieber, *J. Am. Chem. Soc.*, **72**, 5795 (1950); S. Winstein, M. Shatavsky, C. Norton, and R. B. Woodward, *ibid.*, **77**, 4183 (1955).
16. P. v. R. Schleyer and G. W. Van Dine, *J. Am. Chem. Soc.*, **88**, 2321 (1966); P. v. R. Schleyer and V. Buss, *ibid.*, **91**, 5880 (1969), and further literature cited there; see also J. C. Martin and B. R. Ree, *ibid.*, **91**, 5882 (1969); B. R. Ree and J. C. Martin, *ibid.*, **92**, 1660 (1970); V. Buss, R. Gleiter, and P. v. R. Schleyer, *ibid.*, **93**, 3927 (1971).
17. (a) H. Tanida, T. Tsuji, and T. Irie, *J. Am. Chem. Soc.*, **89**, 1953 (1967); (b) M. A. Battiste, C. L. Deyrup, R. E. Pincock, and J. Haywood-Farmer *ibid.*, **89**, 1954 (1967).
18. R. R. Sauers and R. W. Ubersax, *J. Org. Chem.*, **31**, 495 (1966).
19. (a) K. B. Wiberg and G. Szeimies, *J. Am. Chem. Soc.*, **90**, 4195 (1968); *ibid.*, **92**, 571 (1970); (b) G. A. Olah, D. P. Kelly, C. L. Jeuell, and R. D. Porter, *ibid.*, **92**, 2544 (1970); (c) Z. Majerski and P. v. R. Schleyer, *ibid.*, **93**, 665 (1971); (d) G. A. Olah, C. L. Jevell, D. P. Kelly, and R. D. Porter, *ibid.*, **94**, 146 (1972).
20. A. T. Balaban, *Rev. Roum. Chim.*, **11**, 1097 (1966). Cf. Reference 4.
21. (a) R. Breslow, J. T. Groves, and G. Ryan, *J. Am. Chem. Soc.*, **89**, 5048 (1967); (b) D. G. Farnum, G. Mehta, and R. G. Silberman, *ibid.*, **89**, 5048 (1967); (c) R. Breslow and J. T. Groves, *ibid.*, **92**, 984 (1970).
22. (a) R. Breslow, J. W. Chang, and W. A. Yager, *J. Am. Chem. Soc.*, **85**, 2033 (1963); (b) R. Breslow, R. Hill, and E. Wasserman, *ibid.*, **86**, 5349 (1964).
23. R. Breslow, H. W. Chang, R. Hill, and E. Wasserman, *J. Am. Chem. Soc.*, **89**, 112 (1967).
24. P. v. R. Schleyer, J. J. Harper, G. L. Dunn, V. J. DiPasquo, and J. R. E. Hoover, *J. Am. Chem. Soc.*, **89**, 698 (1967).
25. S. Masamune, *J. Am. Chem. Soc.*, **86**, 735 (1964).
26. W. v. E. Doering and M. Pomerantz, *Tetrahedron Lett.*, **1964**, 961.
27. S. Masamune, K. Fukomoto, Y. Yasunari, and D. Darwish, *Tetrahedron Lett.*, **1966**, 193.
27a. For new experimental work see S. Masamune, M. Sakai, H. Ona, and A. J. Jones, *J. Am. Chem. Soc.*, in press.
28. W. -D. Stohrer and R. Hoffmann, *J. Am. Chem. Soc.*, **94**, 1661 (1972).
28a. R. Hoffmann and W.-D. Stohrer, *Bull. Chem. Soc. Japan*, **45**, 2513 (1972).
29. S. Yoneda, S. Winstein, and Z. Yoshida, *ibid.*, **45**, 2510 (1972); (a) K. M. Harmon, this volume, Chapter 29, p. 1579; (b) cf. A. S. Siegel, *J. Am. Chem. Soc.*, **92**, 5277 (1970).
30. R. K. Lustgarten, M. Brookhart, and S. Winstein, *J. Am. Chem. Soc.*, **89**, 6350 (1967).

31. P. R. Story and M. Saunders, *J. Am. Chem. Soc.*, **84**, 4876 (1962).
32. P. R. Story and B. J. A. Cooke, *Chem. Commun.*, **1968**, 1080.
33. M. Brookhart, R. K. Lustgarten, and S. Winstein, *J. Am. Chem. Soc.*, **89**, 6352 (1967).
34. M. Brookhart, R. K. Lustgarten, and S. Winstein, *J. Am. Chem. Soc.*, **89**, 6354 (1967).
35. R. K. Lustgarten, M. Brookhart, and S. Winstein, *J. Am. Chem. Soc.*, **90**, 7364 (1968).
36. H. G. Richey, Jr. and N. C. Buckley, *J. Am. Chem. Soc.*, **85**, 3057 (1963).
37. P. R. Story and S. R. Fahrenholtz, *J. Am. Chem. Soc.*, **86**, 527 (1964).
38. P. R. Story and S. R. Fahrenholtz, *J. Am. Chem. Soc.*, **88**, 374 (1966).
39. G. O. Schenck and R. Steinmetz, *Chem. Ber.*, **96**, 520 (1963).
40. W. G. Dauben and R. L. Cargill, *Tetrahedron*, **15**, 197 (1961); G. S. Hammond, N. J. Turro, and A. Fischer, *J. Am. Chem. Soc.*, **83**, 4674 (1961).
41. (a) E. E. v. Tamelen, T. M. Cole, R. Greeley, and H. Schumacher, *J. Am. Chem. Soc.*, **90**, 1372 (1968); (b) H. Hogeveen and C. J. Gasbeek, *Rec. Trav. Chim.*, **89**, 1079 (1970).
42. W. G. Dauben, personal communication.
43. J. C. Barborak and R. Pettit, *J. Am. Chem. Soc.*, **89**, 3080 (1967).
44. R. Pettit and J. A. Broussard, unpublished results; see J. A. Broussard, dissertation, University of Texas at Austin, 1970.
45. W. L. Dilling and C. E. Reineke, *Tetrahedron Lett.*, **1967**, 2547; W. J. Dauben and D. L. Whalen, *J. Am. Chem. Soc.*, **93**, 7244 (1971).
46. W. L. Dilling, R. A. Plepys, and R. D. Kroening, *J. Am. Chem. Soc.*, **91**, 3404 (1969).
47. S. Winstein, E. C. Friedrich, R. Baker, and Y. Lin, *Tetrahedron Suppl.*, **8** [II], 621 (1966), and literature cited there.
48. R. M. Coates and J. L. Kirkpatrick, *J. Am. Chem. Soc.*, **90**, 4162 (1968); *ibid.*, **92**, 4883 (1970).
49. S. Winstein and C. Ordronneau, *J. Am. Chem. Soc.*, **82**, 2084 (1960).
50. P. v. R. Schleyer and R. E. Leone, *J. Am. Chem. Soc.*, **90**, 4164 (1968).
51. S. J. Cristol, M. K. Seifert, D. W. Johnson, and J. B. Juvale, *J. Am. Chem. Soc.*, **84**, 3918 (1962).
52. P. K. Freeman and D. M. Balls, *Tetrahedron Lett.*, **1967**, 437; P. K. Freeman, D. M. Balls, and J. N. Blazevich, *J. Am. Chem. Soc.*, **92**, 2051 (1970).
53. G. W. Klumpp, *Rec. Trav. Chim. Pays-Bas*, **87**, 1053 (1968).
54. M. J. Goldstein and B. G. Odell, *J. Am. Chem. Soc.*, **89**, 6356 (1967).
55. P. Ahlberg, D. L. Harris, and S. Winstein, *J. Am. Chem. Soc.*, **92**, 2146 (1970); P. Ahlberg, D. L. Harris, M. Roberts, P. Warner, P. Seidl, M. Sakai, D. Cook, A. Diaz, J. P. Dirlam, H. Hamberger, and S. Winstein, *ibid.*, **94**, 7063 (1972).
56. P. Ahlberg, J. B. Grutzner, D. L. Harris, and S. Winstein, *J. Am. Chem. Soc.*, **92**, 3478 (1970).
57. P. Ahlberg, D. L. Harris, and S. Winstein, *J. Am. Chem. Soc.*, **92**, 4454 (1970).
57a. M. Goldstein, R. C. Krauss, and S.-H. Dai, *J. Am. Chem. Soc.*, **94**, 680 (1972).
58. (a) J. C. Barborak and P. v. R. Schleyer, *J. Am. Chem. Soc.*, **92**, 3184 (1970); (b) J. C. Barborak, unpublished.
59. J. B. Grutzner and S. Winstein, *J. Am. Chem. Soc.*, **92**, 3186 (1970).
60. M. J. Goldstein, *J. Am. Chem. Soc.*, **89**, 6357 (1967). See ref. 136.
61. J. B. Grutzner and S. Winstein, *J. Am. Chem. Soc.*, **90**, 6562 (1968).
62. T. J. Katz and P. J. Garratt, *J. Am. Chem. Soc.*, **86**, 5194 (1964).

63. D. Cook, A. Diaz, J. P. Dirlam, D. L. Harris, M. Sakai, S. Winstein, J. C. Barborak, and P. v. R. Schleyer, *Tetrahedron Lett.*, **1971**, 1404.
64. T. S. Cantrell and H. Shechter, *J. Am. Chem. Soc.*, **87**, 3300 (1965).
65. A. S. Kende and T. L. Bogard, *Tetrahedron Lett.*, **1967**, 3383.
65a. T. A. Antkowiak, D. C. Sanders, G. B. Trimitsis, J. B. Press, and H. Shechter, *J. Am. Chem. Soc.*, **94**, 5366 (1972).
66. A. S. Kende, personal communication.
67. W. v. E. Doering, B. M. Ferrier, E. T. Fossel, J. H. Hartenstein, M. Jones, Jr., G. Klumpp, R. M. Rubin, and M. Saunders, *Tetrahedron*, **13**, 3943 (1967).
68. (a) J. C. Barborak, J. Daub, D. M. Follweiler, and P. v. R. Schleyer, *J. Am. Chem. Soc.*, **91**, 7760 (1969); (b) J. C. Barborak, T.-M. Su, P. v. R. Schleyer, G. Boche, and W. Schneider, *J. Am. Chem. Soc.*, **93**, 271 (1971); (c) A. G. Anastassiou and E. Yakali, *Chem. Commun.*, **1972**, 92.
69. (a) M. Saunders and E. L. Hagen, *J. Am. Chem. Soc.*, **90**, 6881 (1968); (b) G. A. Olah and A. M. White, *ibid.*, **91**, 5801 (1969).
70. (a) For reviews on protonated cyclopropanes see C. J. Collins, *Chem. Rev.*, **69**, 543 (1969), C. C. Lee, *Progr. Phys. Org. Chem.*, **7**, 129 (1970), and Reference 13; (b) L. Radom, J. A. Pople, V. Buss, and P. v. R. Schleyer, *J. Am. Chem. Soc.*, **93**, 1813 (1971); **94**, 311 (1972); (c) N. Bodor and M. J. S. Dewar, *ibid.*, **93**, 6685 (1971); N. Bodor, M. J. S. Dewar, and D. H. Lo, *ibid.*, **94**, 5303 (1972).
71. C. C. Lee and D. J. Woodcock, *J. Am. Chem. Soc.*, **92**, 5992 (1970).
72. G. K. Karabatsos, C. Zioudrou, and S. Meyerson, *J. Am. Chem. Soc.*, **92**, 5996 (1970).
73. (a) M. Saunders, E. L. Hagen, and J. Rosenfeld, *J. Am. Chem. Soc.*, **90**, 6882 (1968); (b) M. Saunders, private communication.
74. G. A. Olah and J. Lukas, *J. Am. Chem. Soc.*, **89**, 4739 (1967).
75. M. Saunders and P. Vogel, *J. Am. Chem. Soc.*, **93**, 2559 (1971).
76. M. Saunders, M. H. Jaffe, and P. Vogel, *J. Am. Chem. Soc.*, **93**, 2558 (1971); M. Saunders and P. Vogel, *ibid.*, **93**, 2561 (1971).
77. G. A. Olah and J. M. Bollinger, *J. Am. Chem. Soc.*, **89**, 4744 (1967).
78. D. M. Brouwer, *Rec. Trav. Chim. Pays-Bas*, **88**, 9 (1969).
79. M. Saunders and E. L. Hagen, *J. Am. Chem. Soc.*, **90**, 2436 (1968).
80. D. M. Brouwer and E. L. Mackor, *Proc. Chem. Soc.*, **1964**, 147.
81. D. M. Brouwer, *Rec. Trav. Chim. Pays-Bas.*, **87**, 210 (1968).
82. M. Saunders and J. Rosenfeld, *J. Am. Chem. Soc.*, **91**, 7756 (1969).
83. D. M. Brouwer and J. A. Van Doorn, *Rec. Trav. Chim. Pays-Bas*, **88**, 573 (1969).
84. T. S. Sorensen and K. Ranganayakulu, *Tetrahedron Lett.*, **1970**, 659.
85. C. D. Poulter and S. Winstein, *J. Am. Chem. Soc.*, **91**, 3649, 3650 (1969).
86. M. Saunders and J. Rosenfeld, *J. Am. Chem. Soc.*, **92**, 2548 (1970); cf. ref. 19d.
87. G. A. Olah and J. Lukas, *J. Am. Chem. Soc.*, **90**, 933 (1968).
87a. G. A. Olah, G. D. Mateescu, and J. L Riemenschneider, *J. Am. Chem. Soc.*, **94**, 2529 (1972).
88. H. Hogeveen and C. J. Gaasbeek, *Rec. Trav. Chim. Pays-Bas*, **88**, 1305 (1969).
89. V. A. Koptyug, I. A. Shleider, and I. S. Isaev, *Zh. Org. Khim.*, **7**, 852 (1971); V. A. Koptyug, I. A. Shleider, I. S. Vasilyena, and A. I. Rezvukhin, *ibid.*, **7**, 1089 (1971); I. A. Shleider, I. S. Isaev, and V. A. Koptyug, *ibid.*, **8**, 1337 (1972).
90. R. Breslow, J. Lockhart, and A. Small, *J. Am. Chem. Soc.*, **84**, 2793 (1962).
91. (a) D. M. Brouwer and J. A. Van Doorn, *Rec. Trav. Chim. Pays-Bas*, **89**, 333 (1970); (b) cf. D. M. Brouwer and H. Hogeveen, *ibid.*, **89**, 211 (1970).

92. W. v. E. Doering, M. Saunders, H. G. Boyton, H. W. Earhart, E. F. Wadley, W. R. Edwards, and G. Laber, *Tetrahedron*, **4**, 178 (1958).
93. M. Saunders, in *Magnetic Resonance in Biological Systems*, A. Ehrenberg, B. G. Mahmstrom, and T. Vanngard, Eds., Pergamon Press, New York, 1967, p. 90.
94. B. B. Derendyaev, V. I. Mamatyuk, and V. A. Koptyug, *Tetrahedron Lett.*, **1969**, 5.
95. G. A. Olah, R. H. Schlosberg, D. P. Kelly, and G. D. Mateescu, *J. Am. Chem. Soc.*, **92**, 2546 (1970).
96. H. Hogeveen and H. C. Volger, *Rec. Trav. Chim. Pays-Bas*, **88**, 353 (1969).
97. T. S. Sorensen and K. Ranganayakulu, *J. Am. Chem. Soc.*, **92**, 6539 (1970).
98. V. G. Shubin, D. V. Korchagina, B. G. Derendjaev, V. I. Mamatyuk, and V. A. Koptyug, *Zh. Org. Khim.*, **6**, 2066 (1970).
98a. D. V. Korchagina, B. G. Derendyaev, V. G. Shubin, and V. A. Koptyug, *ibid.*, **7**, 2582 (1971).
99. V. G. Shubin, D. V. Korchagina, G. I. Borodkin, B. G. Derendjaev, and V. A. Koptyug, *Chem. Commun.*, **1970**, 696.
100. S. Winstein, *Aromaticity*, Int. Symp., Sheffield, England, The Chemical Society, London, *Spec. Publ.* No. 21, p. 16.
101. E. J. Corey and H. Uda, *J. Am. Chem. Soc.*, **85**, 1788 (1963).
102. G. D. Sargent, *Quart. Rev. Chem. Soc.*, **20**, 367 (1966); *Carbonium Ions*, Vol. III, G. A. Olah and P. v. R. Schleyer, Eds., Wiley-Interscience, New York, 1972, Chapter 24.
103. P. D. Bartlett and G. D. Sargent, *J. Am. Chem. Soc.*, **87**, 1297 (1965); P. G. Gassman and D. S. Patton, *ibid.*, **91**, 2160 (1969).
104. S. Winstein, P. Bruck, P. Radlick, and R. Baker, *J. Am. Chem. Soc.*, **86**, 1867 (1964).
105. P. v. R. Schleyer, W. E. Watts, R. C. Fort, Jr., M. D. Comisarow, and G. A. Olah, *J. Am. Chem. Soc.*, **86**, 5679 (1964).
106. M. Saunders, P. v. R. Schleyer, and G. A. Olah, *J. Am. Chem. Soc.*, **86**, 5680 (1964).
107. F. R. Jensen and B. H. Beck, *Tetrahedron Lett.*, **1966**, 4287.
108. G. A. Olah, A. Commeyras, and C. Y. Lui, *J. Am. Chem. Soc.*, **90**, 3882 (1968).
109. G. A. Olah and A. M. White, *J. Am. Chem. Soc.*, **91**, 3954 (1969).
110. G. A. Olah, A. M. White, J. R. DeMember, A. Commeyras, and C. Y. Liu, *J. Am. Chem. Soc.*, **92**, 4627 (1970).
111. G. A. Olah, J. R. DeMember, C. Y. Lui, and R. D. Porter, *J. Am. Chem. Soc.*, **93**, 1442 (1971).
112. D. W. Swatton and H. Hart, *J. Am. Chem. Soc.*, **89**, 5075 (1967).
113. V. G. Shubin, V. P. Chzhu, A. I. Rezvukhin, A. A. Tabatskaya, and V. A. Koptyug, *Izv. Akad. Nauk SSSR, Ser Khim.*, **1967**, 2365.
114. I. S. Isaev, T. G. Egorova, I. A. Shleider, E. T. Lippmaa, T. I. Pehk, and V. A Koptyug, *Dokl. Akad. Nauk SSSR*, **189**, 1258 (1969).
115. R. F. Childs and S. Winstein, *J. Am. Chem. Soc.*, **90**, 7146 (1968).
116. R. B. Woodward and R. Hoffmann, *J. Am. Chem. Soc.*, **87**, 2511 (1965).
117. H. Hart, T. R. Rodgers, and J. Griffiths, *J. Am. Chem. Soc.*, **91**, 754 (1969).
118. V. A. Koptyug, L. I. Kuzubova, I. S. Isaev, and V. I. Mamatyuk, *Chem. Commun.*, **1969**, 389.
118a. M. Zeya and R. F. Childs, *J. Am. Chem. Soc.*, **94**, 289 (1972).
119. H. E. Zimmerman and D. S. Crumrine, *J. Am. Chem. Soc.*, **90**, 5612 (1968).
120. T. M. Brennan and R. K. Hill, *J. Am. Chem. Soc.*, **90**, 5614 (1968).

121. P. Vogel, M. Saunders, N. M. Hasty, Jr., and J. A. Berson, *J. Am. Chem. Soc.*, **93**, 1551 (1971).
122. A. Nickon, H. Kwasnik, T. Swartz, R. O. Williams, and J. B. DiGiorgio, *J. Am. Chem. Soc.*, **87**, 1613, 1615 (1965).
123. A. Nickon, private communication.
124. R. S. Bly, R. K. Bly, A. O. Bedenbaugh, and O. R. Vail, *J. Am. Chem. Soc.*, **89**, 880 (1967).
125. A. Nickon, G. D. Pandit, and R. O. Williams, *Tetrahedron Lett.*, **1967**, 2851.
126. J. E. Baldwin and W. D. Foglesong, *J. Am. Chem. Soc.*, **89**, 6372 (1967).
127. J. E. Baldwin and W. G. Foglesong, *J. Am. Chem. Soc.*, **90**, 4303 (1968).
128. H. Hart and G. M. Love, *J. Am. Chem. Soc.*, **93**, 6264 (1971).
129. D. Lenoir and P. v. R. Schleyer, *Chem. Commun.*, **1970**, 941; cf. L. A. Spurlock and K. P. Clark, *J. Am. Chem. Soc.*, **92**, 3829 (1970).
130. (a) See, e.g., H. W. Whitlock, Jr., and M. W. Siefken, *J. Am. Chem. Soc.*, **90**, 4929 (1968), and references therein; (b) P. Vogel, M. Saunders, W. Thielecke, and P. v. R. Schleyer, *Tetrahedron. Lett.*, **1971**, 1429; (c) P. v. R. Schleyer, L. K. M. Lam, D. J. Raber, J. L. Fry, M. A. McKervey, J. R. Alford, B. D. Cuddy, V. G. Keizer, H. W. Geluk, and J. L. M. A. Schlatmann, *J. Am. Chem. Soc.*, **92**, 5246 (1970).
131. Z. Majerski, S. H. Liggero, P. v. R. Schleyer, and A. P. Wolf, *Chem. Commun.*, **1970**, 1596.
132. Z. Majerski, P. v. R. Schleyer, and A. P. Wolf, *J. Am. Chem. Soc.*, **92**, 5732 (1970).
133. J. E. Nordlander, F. Ying-Hsiush Wu, S. P. Jindal, and J. B. Hamilton, *J. Am. Chem. Soc.*, **91**, 3962 (1969).
134. P. v. R. Schleyer, E. Funke, and S. H. Liggero, *J. Am. Chem. Soc.*, **91**, 3965 (1969).
135. S. H. Liggero, unpublished results. J. E. Nordlander, private communication.
136. M. J. Goldstein and R. Hoffmann, *J. Am. Chem. Soc.*, **93**, 6193 (1971).

# Index

Absorption spectra, 1508
Acetic acid, 1512
Acetonitrile, 1520, 1522, 1587
Acid-solvent systems, 1699
   acid compound, increase of characteristic cation, 1701
   acidity function $H_X$, 1700
   $AsF_5$, $BF_3$ and $PF_5$, 1702
   autoprotolysis equilibrium, 1701
   base increase of characteristic anion, 1701
   convenient freezing point, self-ionization, 1703
   cryoscopic and conductometric methods, 1703
   fluoride ion acceptors, 1702
   1:1M $FSO_3H$-$SbF_5$--"magic acid," 1703
   Hammett acidity function $H_0$ weak bases, 1700
   $H_0$ values for pure protonic acids, 1701
   Lewis acid $SbF_5$, 1702
   oleum system, 1702
   v factor, number of kinetically separate dissolved particles, 1703
Activation parameters, 1567
Adamantyl Cations, ($2$-$^{14}$C)-adamantane, 1928
   with $AlBr_3$, 1928
   $2$-$^{14}$C-Adamantane, degenerate carbon scrambling, 1928
   2-methyl to 1-methyladamantane rearrangement, 1929
   simple 1,2-methyl shift because of orthogonality, 1929
Additive polysubstitution, 1554
Alcohols, ionization of, 1533
Alkoxy substituent, 1587
Alkoxytropenylium ions, 1588
Alkyltropenylium ions, α carbon of, 1627
   salts, 1627
Allyl resonance, 1846
Allyl substituent, 1587
Amidocarbonium ions, ∂-acylaminohalides, 1672
   amidocarbonium ions, 1671
   amido cations, 1670
   amido function, 1670
   amidomethylamines, 1672
   amidomethylol amides, 1670, 1671
   amidomethylol compounds, 1672
   amino- and amidocarbonium ions, 1672
   Einhorn reaction, 1671
   Friedel-Crafts reactions, 1672
   nitrilium ion species, 1672
   $S_N^2$–$S_N^1$ reactions, 1672
   Tscherniac reaction, 1671
Amine oxidations, N-acyloxyammonium salts, 1653
   amine nitrosation, 1653
   aminium cation radical, 1655
   azacarbonium ion intermediates, 1653
   azacarbonium ion species, 1653
   azacationic species, 1655
   cleavage of amines with chlorine dioxide, 1655
   dibenzylammonium fluoroborate, 1654
   hydride transfer from tertiary amines, 1654
   nitrosative cleavage of tertiary amines, 1653
   oxidation of tertiary aliphatic amines, 1653
   substituted tetrazenes oxidation of, 1655
   ternary iminium salt, 1654
   three-electron π bond, 1655
   triphenyl-carbonium ion, 1654
Aminoacylium ions, 1652
   acylium ions, 1652
   amine oxidations of α-dialkylamino compounds, 1653
   α-aminoacyl derivatives, 1652
   aminocarbonium ion, 1652
   azide ion, 1652
   benzoyl chloride, 1652
   carbamylium ion, 1652
   dimethylcarbamyl cation ($Me_2\overset{+}{N}CO$), 1652
   dimethylcarbamyl chloride, 1652
   protonation of amides, 1652
   tetramethyl diamido phosphorochloridate, 1652
Aminocarbonium ions, aminocarbonium, 1649

aminomethylcarbonium ion, 1649
dielectric constant, 1648
ethoxymethylpiperidine, 1649
N-hydroxymethylamine, 1649
N-hydroxy-methylpiperidine, 1649
Mannich reactions, 1649
methylenebispiperidine, 1649
new polarographic reduction, 1649
reactivity increased, 1648
second alkylation, 1649
α-Aminohalides, aminonitrile with silver ion, 1663
azacarbonium ions $R_2 NCH_2{}^+$, 1663
carbon attack by nucleophile, 1663
chloroformamidines are covalent, 1663
crystalline azacarbonium salts, 1662
immonium salts, 1663
in the infrared, 1662
reactions of formamidinium with organolithium, 1663
Aminonitrilium ions, aminonitrilium cations, 1682
delocalized species, 1682
tautomeric shift, 1682
Aminonitrilium ions, from haloazines, aldazines, 1684
fragmentation, halozines, 1685
hydrazidic halides, 1684
mechanisms of solvolysis of monochloroaldazines, 1684
oxadiazoles, 1684, 1685
Aminonitrilium ions from hydrazidic halides, anchimerically assisted expulsion, 1687
N-arylhydrazidic halides, 1685
N-arylhydrazidic halides hydrolysis, 1686
bromination kinetics to syn-anti isomerizations, 1688
competition for the cationic center, 1687
delocalization, 1687
heterocyclic hydrazidic halides, 1688
imidazolidenes, 1688
ortho-substituents, 1686
1,3,4-oxadiazoles, 1688
Δ S+values, 1685
tetrazole ring, (with a p value of −2.2), 1687
tetrazole ring fragmentation, 1687
N-tetrazolyhydrazidic halides, formation of triazolyl azides, 1686, 1687
2-tetrazolyl ring positions, 1688
1,2,4-triazoles, 1688

triazolotetrazoles, 1687
Aminonitrilium ions from ketazines, 1683
amino carbonium ions, 1683
chlorine, 1683
dichloroazo compounds, 1683
dissociation to carbonium ions, 1683
kinetics and mechanism of, 1683
nucleophilic replacements, 1683
rate of decomposition, 1683
rate of determining step, 1683
stereospecific aspects, of ketazine halogenation, 1683
of substitution, 1683
Aminonitrilium ions nitrilimines, conjugate bases of aminonitrilium ions, 1688
1,3-dipolar ion, 1689
by electron-donating substituents, 1689
electron-withdrawing substituents, 1689
Huisgen 1, 3-dipolar reactions, 1688
hydrazidic halide to form azocarbonium ion, 1689
p-Aminotrityl, $H_R$ methods, 1535
tert-Amyl cation, 1895
tert-amyl cation in HF-SbF$_5$ solution, 1896
2-chloro-2-methylbutane, 1895
degenerate 1,2-methyl shift, 1896
1,2-hydride shift occurs, 1896
Anion interchange, 1591
Antimony pentachloride, 1518
Antimony trichloride, 1521
Anucleophilic anions, 1587
Arylcarbinol basicity, for stability measurements, 1530
Aryl substituent, 1587
Aryl-substituted polycarbonium ions, 1587
Aryltropenylium ions, 1583
chargetransfer interactions, 1628
9-Arylxanthyls, 1544
Asymmetric 9-arylxanthyl, 1571
Atropisomerism, 1570, 1571
Azide ion ($N_3-$), 1517
Azobenzene accepts hydride from cycloheptatriene, 1590

Barbaralyl or tricyclo[3.3.1.0$^{2,8}$]nona-3,6-dien-9-yl cation, 1882
antibonding, 1885
9-barbaralyl cation, 1883, 1886, 1887
9-barbaralyl tosylate, 1884
bicyclo[3.2.2]nonatrienyl cations, 1882
bishomotropylium, 1883

bullvalene, 1887
complete scrambling, 1887
cyclopropyl-carbinyl-cyclopropylcarbinyl rearrangements, 1887
degenerate carbonium ion rearrangements, 1882
degenerate Cope rearrangement, 1882, 1886, 1887
degenerate rearrangements, 1886, 1887
deuterium-labeled bicyclo[3.2.2]nonatrien-2-yl, 1886
deuterium label scrambled to every position, 1884
homobullvalene, 1887
homotropylidene structure, 1882
methyl-labeled bicyclononatriene, 1886
rearranges to methyl-substituted bishomotropylium ion, 1887
symmetry allowed, 1885
symmetry forbidden, 1885
synchronous process involving concerted migration of three bonds, 1883
threefold degenerate rearrangement, 1886
vinylog of barbaralyl cation, 1887
1,2-vinyl shift to dihydroindenyl, 1883
Walsh orbitals of cyclopropane rings, 1885
Basicity of arylcarbinols, 1530
Basicity (pK) of the solvent, 1510
Basicity, solvents of intermediate, 1512
Basic solvent, 1511
Benhydryl chloride, 1521
Benzoic acid, 1585
p-benzoquinone dibenzenesulfonimide, 1590
Benzyl cation, 1503
Benzyl chloride, 1521
Bicyclo[3.1.0]hex-3-en-2-yl cation, 1915
5-acyl-1,2,3,4,5-pentamethylcyclopentadienes, 1920
benzenonium ion, 1921
cations not interconvertible, 1920
of 5-α-chloroethylpentamethylcyclopentadienes, 1920
circumambulation of cyclopropane ring, 1921
concerted sigmatropic 1,4 shift of C-6 suprafacial, 1919
cyclopentene ring, 1919
endo-6 position, no increase in NMR absorption, 1921
fivefold degenerate rearrangement, 1920

fivefold degenerate scrambling, 1919
hexamethylbicyclo[3.1.0]-hex-3-en-2-one, 1916
hexamethylbicyclo[3.1.0]hexenyl cation, 1920
2,3,4,5,6,6-hexamethylhexa-2,4-dien-1-one, 1918
hexamethyl-2,4-cyclohexadienone, 1916
parent cation instable to vinylogous homotropylium cation, 1921
randomization of the isotopic label, 1918
random label scattering via hexamethylbenzenonium, 1918
rapid migration of cyclopropane ring around four of five sides of cyclopentenyl ring, 1916
1,4-sigmatropic rearrangements proceed with inversion of configuration label scrambling, 1920
stereochemical assignments, 1920
symmetry of highest occupied molecular orbital, 1919
temperature-dependent NMR spectrum, 1919
3-Bicyclo[3.1.0]hexyl cation, 1906, 1908, 1910
cis-acetate, 1908
3-bicyclo[3.1.0]hexyl cation, 1907
cis-3-bicyclo[3.1.0]hexyl moieties in trishomocyclopropenyl cation, 1910
cis-3-bicyclo[3.1.0]hexyl tosylate, 1907
cis-3-deuteriobicyclo[3.1.0]hexyl tosylate, 1908
cis-dimethyl-substituted bicyclo[3.1.0]hex-3-yl tosylates, 1910
cis-1,5-diphenylbicyclo[3.1.0]hex-3-yl tosylate, 1908
cis isomer, 1907
complete retention of configuration, 1908
homocyclopropylcarbinyl rearrangements, 1906
inductive and resonance effects, 1909
mechanistic scheme, 1908
participation of only one cyclopropane ring in 3-tricyclo-[7.1.0.0$^{5,7}$]decyl cation, 1911
9-pentacyclo[4.3.0.0$^{2,4}$.0$^{3,8}$.0$^{5,7}$]nonyl cation, 1911
phenyl substituent rate effects, 1909
rapidly equilibrating set of ions, 1907
rate enhancements due to methyl substituents, 1910

rate of solvolysis, 1908
solvolysis of the 6,6-dideuterated isomer, 1908
special salt effect, 1908
threefold carbon degeneracy, 1906
trans-3-bicyclo[3.1.0]hexyl tosylate, 1907
trans isomer, 1908
3-tricyclo-[7.1.0.0$^{5,7}$]decyl cation, 1910
3-tricyclo-[7.1.0.0$^{5,7}$]decyl tosylate, 1910
trishomocyclopropenyl cation, 1906, 1909, 1910
Bicyclo[3.2.2]nona-3,6,8-trien-2-yl cation, 1875
  allyl-stabilized systems, 1877
  anions, 1878
  barbaralane bicyclo[3.2.2]nona-2,6,8-triene, 1876
  barbaralane skeleton more stable than bicyclo[3.2.2]nona-2,6,8-trienyl, 1876
  barbaralol, 1876
    9-barbaralyl cation, 1877
    9-methyl-9-barbaralyl cation, 1876
  barbaralyl intermediate, 1878
  bicycloaromaticity, 1877
  bicyclo[3.2.2]nona-3,6,8-trien-2-ol with AlCl$_3$=barbaralol, 1875
  bicyclo[3.2.2]nona-3,6,8-trien-2-one, 1875
  to bicyclononatrienyl cation, 1876
  2-bicyclo[3.2.2]nonatrienyl cation, 1878
  bicyclotropylium ion, 1875
  cations, 1878
  destabilized and antibicycloaromatic, 1878
  cis-8,9-dihydro-1-indenyl cation, 1878
  Cope rearrangement, 1877
  degenerate nine-carbon rearrangement, 1878
  degenerate properties, 1878
  equivalent homoallylic barbaralyl cations, 1875
  homoallylic rearrangement by cyclopropylcabinyl-cyclopropylcarbinyl rearrangement, 1878
  1-exo-hydroxy-cis-8,9-dihydroindene, 1879
  methanolysis of 1-chloro-8,9-dihydroindene, 1879
  methyl-substituted bishomotropylium ion, 1880
  no degeneracy, of allylic resonance, 1879
    bishonotropylium ion, 1879
  solvolysis of deuterium labeled bicyclo-[3.2.2]nona-3,6,8-trien-2-yl 3,5-dinitrobenzoate, 1877
  solvolysis rates of bicyclo[3.2.2]nona-3,6,8-trien-2-yl derivatives not accelerated, 1877
  tertiary 9-methyl-9-barbaralyl cation, 1880
  1,2-vinyl shifts, 1878
Bicyclo[4.2.1]nona-2,4,7-triene-9-yl cation, 1880
  bicyclo[4.2.1]nonatrienyl cation rearrange to dihydroindenyl cation, 1882
  1-chloro-9-phenyl-cis-8,9-dihydroindene, 1882
  cyclooctatetraenediide with benzoyl chloride, 1881
  dihydroindenyl, 1880
  labeling experiment, 1882
  7-norbornadienyl cation, 1880
  9-phenyl substituted system, 1881
Biphenyl derivatives, 1571
2-Brexyl cation, 1921
  acid-catalyzed acetolysis of deltacyclane, 1923
  2-brendyl cation, 1921
  brexan-2-one, 1922
  to 4-brexyl cation, 1921
  multiple degenerate properties, 1921
  β-(syn-7-norbornenyl)ethyl brosylate, 1922
  possible second rearrangement of, 4,2-hydride shift, 1921
  solvolysis of 2-brexyl brosylate, 1921
  by Wagner-Meerwein rearrangements, 1921
Bridgehead carbonium ions, adamantane, strain free, 1796
  alkyl groups, 1798
  bicyclo[2.2.2]octyl bridgehead, 1796
  bicyclo[1.1.1]pentyl and bicyclo[2.1.1]hexyl, 1797
  cyclobutyl carbonium ion, 1797
  deuterium isotope effect on dipole moments of saturated hydrocarbons, 1798
  electron withdrawing methyl, 1799
  homoadamantane, 1796
  hybridization, 1799

# SUBJECT INDEX

inductive versus field, 1799
methyl groups retarded solvolysis, 1797
nonbonded repulsions, 1797
removal of hydrogen, 1799
torsional strain, 1796
transmission of electrical effects through bridged molecules, 1797
Bromotropenylium ion, 1588
9-Bromotrypticene, 1547
Brønsted acidity measurements, 1530
*t*-Butyl halides, 1587
*t*-Butyl-norcaradienepercarboxylate, 1590
1-*t*-Butyl-1,3,3-trimethylallyl cation, 1897
 completely and evenly scrambled deuterium label, 1897
 cyclopropylcarbinyl-allylcarbinyl-allyl rearrangements, 1897
 cyclopropylcarbinyl ion, 1897
4,4′,4″-*t*-Butyltrityl cation by p$K_R$+, 1540
4,4′,4″-*t*-Butyltrityl chloride, 1527

Carbonium ion, internal stabilization, 1542
 Lewis acidity of, 1509
 oxidation, 1535
 ring closure, 1535
 a suitable covalent precursor, 1539
 sulfonation, 1535
 trityl, diphenylmethyl, 1518
Carbonium ions, hydride exchange, 1542
 hydride transfer reactions, 1542
Carbonium stabilization, 1535
Carboranyl, 1587
Carboxy substituent, 1587
Carboxytropenylium bromide, 1583
Cation, covalent and cationic species, 1539
 methyl, methoxyl, fluoro, dimethylamino, 1539
Cation propeller interconversion, 1567
Charge delocalization, 1503, 1507, 1511
Charge distribution, 1503
Charge localization, 1511
Chemical and physical properties, 1581
Chirality in arylcarbonium ions, 1561
9-Chloro-9-phenyl-10,10-dimethyldihydroanthracene, 1547
3-Chloro-1,2,3,4-tetraphenyl-cyclobutenium pentachlorostannate, 1506
Complex formation, 1511
Condensed ring systems, bridgehead with minimal strain, 1816
 decalin and hydrindan, 1814
 differences in product composition, function of substrate stereochemistry, tight ion pair, 1816
 location of gegenion, 1816
 trigonal atom, 1816
Conductiometric techniques, 1514
Conductivity, 1508
Conformational requirements, trityl cation, 1554
Conformation, trityl cation, 1554
Cryoscopic data, 1530
Cryoscopy, 1508
Crystalline, covalent, 1507
Cumyl halide solvolysis, 1548
Cyano substituent, 1587
Cyclobutenyl cation, 1901
 -$CD_3$ groups in pentamethylcyclobutenium ion degenerate rearrangements, 1901
 cyclobutenyl products, 1901
 cyclopropenylcarbinyl system, 1901
 heptamethylbenzenonium cation, 1901
 mechanistic pathways, 1901
 methylated cyclopentenyl cation, 1901
 4-methylene-1,2,3,3-tetramethylcyclobut-1-ene, 1901
 1,2-methyl shifts of cyclobutadienoid transition state, 1902
Cyclobutenyl cation, tetraphenylfuran, 1542
Cycloheptatriene, 1581, 1585
Cycloheptatriene, in presence of *t*-butyl chloride, 1589
 reaction, of boron, aluminum, ferric, stannic and titanium chlorides, 1589
 with platinum (IV) halides hexahaloplatinic acids, 1590
Cycloheptatriene-ethyl azodicarboxylate, 1590
Cyclononatetraenyl cations, 1887
 chlorobicyclo[6.1.0] nonatriene to 8,9-dihydroinden-7-yl, 1887
 cyclononatraenyl cation, 1888
 deuterium labeling, 1888
 dihydroindenyl, 1888
 fluxionality, 1889
 homotropylium cation, 1889
 mechanism of solvolysis, 1888
 thermolysis proceeded in a rate-determining step, 1888
Cyclopentadienylium cation, 1587
Cyclopentyl cation, complete hydrogen degeneracy, 1899
 cyclopentane, 1899
 cyclopentene, 1899

cyclopentyl bromide, 1899
cyclopentyl cation, 1899
cyclopentyl methyl ether, 1899
rapid 1,2-hydride shifts, 1899
Cyclopropylcarbinyl cation systems, 1923
  anti-methyl group on bridge exchanged, 1925
  benzo-bicyclo[3.2.1]octadiene system, 1925
  8,9-cyclo-2-adamantanone, 1924
  8,9-cyclo-2-adamantyl cation, 1923
  cyclopropylcarbinyl-cyclopropylcarbinal rearrangement, 1923
  2-deuterio and 2-tritio analogs, 1924
  dimethylcyclopropylcarbinyl cations, 1927
  3,5-dinitrobenzoate, 1924
  3,5-dinitrobenzoates, 1924
  exhibited exchange for methyl group, 1925
  five-membered ring containing the free olefin moiety, 1926
  high stereoselectivity in rearrangement, 1926
  methylcyclopropylcarbinyl cations, 1927
  2, 8, and 9 positions of 8,9-cycloadamantyl system attain high degree of equivalence during solvolysis, 1924
  puckered cyclobutyl cation, 1927
  tricyclo[3.2.1.0$^{2,7}$]octenyl cation, 1925

d and l enantiomers, 1561
Dauben reaction, 1586
Deamination, apocamphyl amine, 1800
  of bridgehead amines, 1799
  decomposition of $N$-nitrosoamides, 1800
  diazonium ion, 1800
  Favorskiilike route, 1799
  $S_N1$ process, 1800
Degenerate carbonium ions, allyl cation, 1840
  allyl chloride, 1840
  automerization, 1839
  Bullevalene copy rearrangement, 1838
  carbon atom degeneracy, 1841
  $(CH)_n^+$, 1841
  combined carbon and hydrogen atom degeneracy, 1841
  degenerate cation rearrangement equivalence through resonance, 1840
  degenerate 1,2 hydride shift in ethyl cation, 1841
  degenerate rearrangements, 1839
  dispersal of an isotopic label, 1838
  hydrogen atom degeneracy, 1841
  interchange, of carbon atoms, 1840
    of hydrogens, 1840
  isomerization, 1839
  loss of optical activity, 1838
  multiple 1,2, shifts in carbonium ions, theory graphs, 1838
  observation of equivalence by NMR, 1838
  potential degenerate rearrangements, 1841
  principle of minimum structural change, 1838
  rearrangements, 1838, 1839
  resonance equivalent ("homotopic") groups, 1839
  simplest degenerate rearrangements, 1841
  simple hydrogen interchanges, 1841
Diarylcarbonium ions, 1587
Diastereomers, 1565
Diastereomeric fluorines, 1566
Diastereotopic probe, 1565
Diazonium ion, 1801
Dibromide, 1583
1,6-Dibromo-2,4-cycloheptadiene, 1583
Dihydrophenanthrenium cations, degeneracy in this system, 1906
  9-ethyl-9, 10-dimethylphenanthrenium ion, 1906
  1,2-methyl shifts, 1906
  migrational tendency, 1906
  ring-contraction mechanism, 1906
  simple 1,2-ethyl shift, 1906
1,2-Dihydroxytropenylium chloride (tropolone hydrochloride), 1592
4-Dimethylaminotrityl carbinol, $pK_R^+$, 1536
4-Dimethylaminotrityl carbinol, relative ability, 1536
Dimethyl-*tert*-butylcarbonium ion and related cations, corner protonated species, 1893
  2,3-difluoro-2,3-dimethylbutane, 1894
  dimethyl-*tert*-butyl-carbonium ion, 1892
  2,3-dimethylbutane, 1893
  dimethylfluoroisopropylcarbonium, 1893
  dimethylisopropylcarbonium ion, 1893
  doubly degenerate equilibrating cation pair, 1894
  equilibrium deuterium isotope effects, 1894
  equivalence of all methyl groups, 1893

# SUBJECT INDEX

2-methylpentyl cation from 2,3-dimethyl butyl cation, 1895
rapidly equilibrating ions, 1893
2,2,3-trimethylbutane, 1892
twofold partial carbon degeneracy, 1893
2,4-Dimethylpentyl cation, 1896
consecutive 1,2-shifts, 1896
1,3-hydride shift, 1896
1,4- and 1,5-hydride shifts, 1897
4,4'-Dimethyltrityl cations, 1539
Diphenyl carbinols, 1530
Diphenyl carbonium ion, 1558
Diphenyl carbonium ions, stability of, 1533
Diphenylmethyl cation, 1503, 1508
Diprotonated alkoxy alcohols, aliphatic, 1733
by NMR diprotonated, 1733
Dissociation energy, 1515
Dissociation, $K_2$, 1527
Dissymmetrically solvated species, 1561
Dissymmetric solvation, 1562
Donor-acceptor complex, 1510

Electrical properties, 1508
Electron affinity, 1515
Electron-diffraction, 1505
Electronic factors, 1548
Electronic spectrophotometry, 1528
Electronic spectroscopy, 1519
Electrophilic character, 1518
Electrophilic substituent constants, 1548
Emf measurements, electrolytic reduction, 1537
radicals, 1537
reversible electrode potentials, 1537
triarylmethyl cations, 1537
Emf method, trityl radicals, 1538
Enamine salts, enamine salt formation, 1665
immonium compound, 1665
infrared spectra, 1665
polyhalogenated enamines, 1666
ultraviolet, 1665
Enamine studies, ambident nucleophiles reacting with electrophilic reagents, 1657
immonium-azacarbonium ion, 1659
immonium salts, 1658
kinetics of hydrolysis, 1659
N-C-protonated species, 1658
protonation of enamine, 1658
slow proton transfer, 1659

spectroscopic nmr evidence, 1658
vinyl quaternary ammonium salts, 1658
Enantiomeric pair, 1567
Equilibration, acid-base type, 1518
Exact structural information, 1505
Exchange processes, acylium ion, 1804
1-adamantanol adamantane and adamantanone, sulfuric acid as oxidizing agent, 1807
bis-(methylene) compound, 1805
bridgehead carboxylic acid 1-adamantanol, 1804
bridgehead carboxylic acid -adamantyl bromide, 1804
bridgehead carboxylic acid kinetic control, 1804
bridgehead cations, 1802
carbon monoxide, 1804
carboxylation of isopropyl adamantane or related alcohol to give adamantane acid, 1804
congressane, 1803
direct introduction of amine function, aluminum chloride and nitrogen trichloride, 1805
formamido-adamantane, 1805
homoadamantane ring opening and reclosure, 1805
homoadamantane, route to bridgehead carbonium ion, 1805
hydrated to carboxylic acid, 1804
hydride-halide exchange, 1803
hydride transfers, 1803
intermolecular hydride transfers, 1803
intramolecular hydride transfers, 1803
involving change disporportionation, 1806
ionic bromination, 1803
Koch carboxylation, 1804
Koch reaction, 1805
"magic acid," 1802
methylene-bicyclo[3.3.1] nonanone, 1805
noradamantane, 1803
Ritter amidation, 1805
2-substituted adamantane, 1807
yields of bridgehead adamantane, 1803
Extent of ionization, 1508
Extreme hydrolytic sensitivity, 1517

Formic acid, high ionization power of, 1512, 1513
Fragmentation, 1-bromo-3-aminoadamantanes, 1813

fragmentation products, 1814
"frangomeric effect," 1813
quinuclidine, 1813
Fragmentation reactions, γ-aminochlorides, 1650
  γ-aminohalides, 1649
  α-amino ketoxime, 1652
  antiparallel electron pair, 1650
  azacarbonium ions, 1649
  Bechmann rearrangements, 1652
  4-bromoquinuclidine, 1650
  α,α-dimethyl-γ-aminopropyl chlorides, 1650
  fragmentation, 1650, 1652
  homomorph, 1649
  synchronous or one-step fragmentation, 1650
Free energy values, trityl chlorides, 1521
Friedel-Crafts, role of Lewis acids, 1523

Greater coplanarity of xanthyl cation, 1547

Halocarbons, 1587
Halogens, 1587
Halotropenylium ions, chlorotropenylium ion, 1629
  hydrolysis increases going from iodine to fluorine, 1628
Heptamethylbenzenonium ion, 1,2 hydride shifts, 1904
  intramolecular 1,2-methyl shift, 1904
  with methyl chloride, 1903
  protonation of hexamethyldewarbenzene, 1905
  random migration of methyl group, 1904
  rapid degenerate 1,2 methyl group, 1903
  trimethylbenzenes, 1903
Hexachlorostannate, 1524
Hexacoordinate-antimonate mixed anion, 1525
Hexafluoroantimonate, 1586, 1587
Hexafluoroarsenate, 1586
High ionization power of $m$-cresol, 1513
4-Homoadamantyl or Tricyclo[4.3.1.1$^{3,8}$]-undecyl cation, 1,2 carbon shift, 1931
  deuterated 4-homoadamantyl tosylate, 1929
  deuterium label, 1929
  deuterium label redistribution, 1929
  exhibits combined carbon and hydrogen degeneracy, 1929

formolysis, 1931
4-homoadamantyl cation, 1929
Wagner-Meerwein rearrangement and vicinal hydride shifts, 1929
9-Homocubyl cation, bishomocubyl compounds, 1868
  bridged carbonium ions, 1868
  1,2 carbon shifts, 1863
  complete carbon degeneracy, 1863
  course of rearrangements of solvolysis, 1864
  dodecahedron, 1863
  extent of rearrangement, 1867
  homocubanone, 1864
  homocubyl cation, 1868
  homocubyl compounds, 1868
  homodecahydryl cation, 1863
  homotetrahedryl cation, 1863
  labeling pattern, 1865
  nonstereospecific reaction path, 1865
  perdeuteriocyclobutadienetricarbonyl iron substitution, 1865
  recovered tosylates, 1868
  retention of configuration, 1868
  stereochemical leakage, 1864, 1865
  stereospecific internal return, 1868
  synthetic pathways to prepare 9-homocubyl tosylate, 1865
Homoprismyl or 7-tetracyclo[3.2.0.0$^{2,4}$.0$^{3,6}$] heptyl cation, complete carbon degeneracy, 1862
  conversion of norbornadiene to quadricyclene, 1862
  cyclobutane ring participation, 1862
  homoprismyl cation, 1862
  photochemical cycloaddition, 1862
  photochemical transformation to bicyclo[3.2.0.]heptadienyl cation, 1862
  rearrangement possibilities, 1861
  Wagner-Meerwein rearrangements, 1862
Hydride transfer, 1586
Hydrogen bond, 1512
Hydrogen dichloride anion, 1524
Hydroxytropenylium ions, cation-to-anion hydrogen bonding, 1629
  chromogenic transition, 1629
  hydrogen bonding, 1629

Imine protonation, aminocarbonium ion, 1661
  Diels-Alder reactions, 1660
  immonium form, 1661

NMR spectrum, 1660
protonated aldimines and ketimines, 1660
p-substituted dimethylaniline, 1660
Immonium salts, alkylation of aldimines or ketimines, 1663
crystalline perchlorates of azacarbonium ions, 1663
dialkylideneammonium cation, 1665
dichloromethylenedimethylammonium, 1665
N-isopropylidenedimethylaminium perchlorate, 1664
N-isopropylidene pyrrolidinium perchlorate, 1664
NMR spectra, 1663
ternary iminium salts, 1663, 1664
Infrared spectroscopy, 1506
Internal energy effects of trityl cations, 1538
Iodide, 1586
Ionic compounds, 1507
Ionic hexachloroantimonate, 1518
Ionic nature, 1506
Ionic perchlorate, 1518
Ionizability, 1509, 1517
Ionization, 1520
Ionization constants, 1523
Ionization equilibria, trityl chlorides, 1523
Ionization equilibrium constants, 1514, 1520, 1522
Ionization, $K_1$, 1527
Ionizing power of the solvent, 1509
Ionogen, 1508
Ionogenic alcohol, 1519
Ionogenic reactions, 1581
Ionophore, 1508
Ionophoric salt, 1519
Ion-pairing, 1504, 1509
Ion pairs, 1519
Ion pair, solvent separated, 1509
Ions, free, 1519, 1520
Irradiation of cycloheptatriene-t-butyl bromide mixtures with gamma rays, 1590
Isopropyl cation and sec-butyl cation, averaged by rapid degenerate 3,2 hydride shifts, 1892
2-chlorobutane, 1891
complete equivalence of all hydrogens, 1892
corner-and edge-protonated alternatives, 1890
cyclization to protonated methyl-cyclopropane, 1892
labeled n-propanol, 1891
primary n-butyl cation, 1892
proton scrambling, 1892

Kinetic and thermodynamic properties, 1551

Labile groups, 1587
Leaving group, size, 1544
Leaving groups of large steric bulk, 1517
Lewis acid, acidities of and carbonium ion, 1518
halide affinity of, 1519
influence on ionization, equilibrium constant $K_{eq}$ of p-methyltrityl chloride, 1522
Lewis acids, 1502, 1509, 1522
influence of, 1517
Lewis bases, 1510
Liquid sulfur dioxide, 1514
Low-frequency dielectric constants, 1508

Malachite green, interrelated equilibrium constants, 1537
Mannich reactions, amino-methylol, 1647
kinetics of Mannich reaction, 1647
mechanism of reaction, 1645
methylene bisamines, 1647
Mass spectra azacations, 1661
4,4',4''-Methoxytrityl chloride, 1528
Methyl cation, 1503
Methylcyclobutyl cation, 1-chloromethyl-1-methylcyclopropane, 1898
1,2 hydride shifts with secondary cyclobutyl, 1898
methyl cyclobutyl cation, 1898
methylcyclopropyl carbonium ion, 1898
Methylcyclopentyl cation, cyclohexyl cation, 1900
cyclohexyl chloride, 1899
degenerate rearrangement via succession of 1,2 hydride shifts, 1900
1,3-dimethyl-1-cyclopentyl cation, 1900
1,2,-dimethyl-1-cyclopentyl cations, 1900
1,2 hydride shifts, degenerate rearrangement, 1900
1-methylcyclohexyl, 1900
methylcyclohexyl, 1900
methylcyclohexyl cation, 1900
1-methylcyclopentyl chloride with $^{13}C$, 1900

methyl and ring hydrogens, 1900
methyl-substituted cyclopentyl cation, 1900
norbornane, 1900
rapid exchange, 1900
transient species, 1900
3-Methyl-4-dimethylaminotrityl cation, 1550
Methyllithium, 1590
*para*-Methyltrityl chloride, 1522
Methyltropenylium ions, 1588
1-Methyl-2-tropenyliumyl-1,2-dicarba-closo-dod-ecaborene, 1591
Molecular orbital calculations of resonance energy and spectral properties of ions, 1582
Monocarbonium ions, 1587
Monohalocarbenes on benzene by chlorination of cycloheptatriene, 1590
Mono-*p*-methyltrityl chloride, 1523
Multiple substitution, 1552

Nitrilium ions from imidic halides, metallic halides, 1679
  nitrilium salts, 1679
Nitrilium ions from nitriles, by arylation, aryl fluorides, 1676
  quinazolines, 1676
  using diazonium fluoroborates, 1676
  by direct alkylation with oxonium salts, N-acyl compounds, 1674
  dialkyl cyanamides, 1674
  3,5-diazapyrylium salts, 1674
  N-ethyl acetonitrilium fluoroborate, 1673
  intramolecular cyclization, 1673
  nitrilium ions, synthesis of, 1673
  oxonium salts, 1673
  by protonation, basicity sequence, 1678
  Gattermann Houben-Hoesch reactions, 1677
  hydrogen halides, 1676
  infrared spectrum, 1677
  nitrilium salt, 1677
  NMR spectra, 1677
  NMR spectroscopy, 1677
  protonated series, 1678
  protonation of nitriles in $FSO_3H\text{-}SbF_5\text{-}SO_2$, 1678
  reverse order, 1678
  synthesis of substituted oxazoles, 1678
  by reaction with metallic cations, absorption by CN group, 1679
  chelation of methylamine to hexakis nitrilium ions, 1679
  complexes between nitriles, 1678
  complexes of acetonitrile with boron trihalides, 1679
  dimethylcyanamide and acetonitrile, 1679
  enhanced electrophilic character, 1679
  Houben-Hoesch reaction, 1678
  increase in stretching frequency, 1679
  mineral acid, 1678
  stretching frequency, 1678
  triple-bonded, 1679
  by the Ritter reaction, amidomethylate, 1675
  aziridinium to imidazolinium salts, 1676
  Gilkmans, 1675
  mechanism studies, 1674
  methylenebisamines, 1675
  nitrilium salt, 1674
  nitrilium species, 1676
  polyamides, 1675
  stereo-specific addition to olefins, 1675
  synthetic exploitations, 1674
  ultraviolet and infrared, 1675
Nitrilium ions from oximes, Beckmann fission or fragmentation reaction, 1681
  Beckmann rearrangement, 1680
  Beckmann transformations, 1681
  N-chloroketimines, 1682
  crystalline nitrilium salts, 1681
  elimination process, 1681
  imido esters, 1681
  infrared absorption, 1682
  kinetics of rearrangement, 1681
  oxime-*o*-sulphonic, 1681
  ring size, influence of substituent effects, 1681
  solvent variation, 1681
  tetrazoles, 1681
  transition state for formation of bridged nonclassical cation, 1681
NMR, 1508
NMR investigations, 1560
NMR spectroscopy, carbon magnetic resonance, 1709
  charge density, 1706
  chemical shift and charge, 1708
  cis-trans isomerization, 1708
  close correlation of chemical shift with

# SUBJECT INDEX

charge density, 1709
determination of rates of inter-and intramolecular processes of rotational barriers, 1706
diluents $SO_2$ and $SO_2ClF$, 1706
double-bond character of oxygen, 1708
exchange rate, 1706
external standard, 1707
heat of protonation, 1707
hindered rotation, 1708
$H_3O^+$ peak, 1707
internal standard, 1707
internuclear double resonance, 1710
lifetime of proton, 1706
low temperature medium high acidity, 1706
NH protons, shift and bond order of nitrogen protonated nitriles, 1708
$^{17}O$ and $^{14}N$ or $^{15}N$ resonance, 1710
proton magnetic resonance, 1706
protons attached to sulfur at higher field, 1708
rapid mixing, 1708
rate of exchange, 1706
sample tubes with HF, 1707
suitable reference standard, 1707
tetramethyl ammonium ion, 1707
vicinal coupling constants, 1708
Nomenclature, 1508
Noncentrosymmetric space, centrosymmetric space, 1563
Nondimethylamino trityls, 1551
7-Norbornadienyl cation, "bridge flipping," 1855
  carbonium ion lifetimes, very short, 1858
  degenerate rearrangement, 1855, 1856
  equilibration of bound and unbound vinyl, 1858
  five-carbon scrambling, 1859
  mechanisms, 1859
  mechanism, 1,2 shift, 1858
  7-methylnorbornadienyl cation, 1859
  7-phenylnorbornadienyl, 1859
  rapid bridge flipping, 1860
  rates and free energies of activation of degenerate rearrangements, 1857, 1858
  second rearrangement process, 1858
  1,2 shift of unbound vinyl cation, 1859
  unsymmetrical species, 1859
  vinyl signals, 1859
2-Norbornyl cation, alkyl and hydrogen bridge ions, 1913
  $^{13}C$ NMR spectra of, 1914
  corner-protonated nortricyclonium ion, 1913
  edge-protonated nortricyclonium ion, 1913
  endo- and exo-2-chloride, 1915
  ESCA of the 2-norbornyl cation, 1915
  fine structure from Raman spectra, 1913
  four equilibrating protonated cyclopropane ring protons, 1915
  3,2 hydride shift, 1912
  6,2 hydride shift, 1912
  nonclassical corner-protonated ion, 1915
  2-exo-norbornanol, 1911
  2-norbornyl bromide with gallium tribromide, 1913
  norbornyl cation from precursors, 1913
  by 2-exo-norbornyl chloride, 1911
  by 2-exo-norbornyl fluoride, 1911
  protonated nortricyclene, 1913
  proton NMR spectrum of norbornyl cation, 1915
  rapidly interconverting edge-protonated nortricyclonium ions, 1914
  secondary-secondary hydride shifts, 1913
  Wagner-Meerwein rearrangement, 1912
Nortricyclane compounds, carbon-carbon double bond participation of, 1861
  7-deuterioquadricyclyl tosylate, solvolysis of, 1861
  hydrogens of the quadricyclyl cations scrambled by rearrangement, 1861
  net effect, 1861
  nondegenerate ion, 1861
  stereospecific, 1861
Nortricyclyl tosylate, cation example of a nondegenerate ion, 1874
  cyclopropylcarbinyl delocalization, 1874
  deuterium scrambled to C-8 and C-9, 1874
  Grignard reagent, 1874
  homoallylic rearrangement, 1874
  inductive and steric effects, 1874
Nucleophilicity, 1509, 1517
Nucleophilicity, of anion, 1517

Optically active biphenyls, 1570
Ortho substituents, 1545
Other degeneracy, aromatic character, 1851
  barbaralyl cation, 1846
  bond-switching process, 1853

bullvalene, 1847
  (CH)$_m$ possible valence isomers, 1846
  (CH)$_n$+ possible valence isomers, 1846
  Cope rearrangement, 1846
  cyclopentadienyl cation, 1851
  cyclopropane participations, 1846
  cyclopropenyl cation, 1847
  double-bond migrations, 1846
  fluorenyl cation, 1854
  hydride shifts, 1846
  ionization was accelerated, 1853
  60 possible rearrangements, 1852
  ring strain or antiaromatic, 1852
  truly degenerate carbonium ion in (CH)$_n$+, 1852
  Wagner-Meerwein shifts, 1846
Other degenerate carbonium ions, sec-butyl cation, 1889
  carbon degeneracy, 1889
  complete degeneracy, 1889
  hydrogen degeneracy, 1889
  isopropyl cation, 1889
  isopropyl chloride, 1889
  n-propyl cation to protonated cyclopropane, 1890
  proton interchange, 1890
  proton scrambling in protonated cyclopropane, 1890
Ousene-type compounds, closo-borane, 1630
  [7.12$^1$]-1,2- and [7.12$^1$]-1,7-dicarbahemiousenium, 1630
  [7.7.12$^{1,7}$]-1,7-dicarbaousenium ion, 1630
  electronic, infrared, and 1H and $^{11}$B NMR spectra Z values correlations and polarographic reduction potentials, 1630
  [7.10$^2$] and [7.12] hemiousenide ions, 1630
  [7.7.10$^{xy}$] ousene, 1630
Oxonium ion, 1504
Oxonium, protonated species, 1513
Oxonium-type complex, 1510, 1512

Participation of a cyclopropane ring, charge delocalization, 1845
  degenerate cyclopropylcarbinyl-cyclopropyl-carbinyl rearrangement, 1845
  homocyclopropylcarbinyl rearrangement, 1845
Pentabromostannate, 1586
Pentachlorobromostannate, 1586

Pentachlorostannate, 1524
9-Pentacyclo[4.3.0.0$^{2,4}$.0$^{3,8}$.0$^{5,7}$]nonyl cation, acceleration, 1870
  analog of cis-3-bicyclo 3.1.0 hexyl cation, 1869
  bridge flip rearrangement, 1870
  carbon-carbon bond shifts from two participating cyclopropyl rings, 1870
  favorable for homocyclopropylcarbinyl participation, 1871
  large-rate enhancement, 1870
  solvolysis of p-nitrobenzoate, first order kinetics, 1870
  three fold degeneracy by homocyclopropyl-carbinyl rearrangement, 1870
9-Pentacyclo[4.3.0.0$^{2,4}$.0$^{3,8}$.0$^{5,7}$]nonyl cation endo-anti-8-tricyclo[3.2.1.-.0$^{2,4}$]octyl p-nitrobenzoate, acetolysis rate constants, indicated facile rearrangement, 1874
  acetolysis rate constants, norborn-2-en-5-yl acetate, 1873
  acetolysis rate constants, nortricyclyl tosylate, 1873
  deuterium completely unscrambled, 1872
  deuterium scrambling, 1872
  homoallylic rearrangements of tetracyclo-[4.3.0.0$^{2,4}$.0$^{3,7}$]non-8-en-5-yl cation, 1872
  solvolysis deuterium scrambling, 1871
  triply degenerate carbonium ion intermediate, 1871
Pentamethyl and hexamethylcyclopentenyl cations, benzenonium ions, 1903
  hydride and methyl shifts, 1903
  loss of conjugation, 1903
  relief of strain, 1903
  steric crowding, 1903
Perchlorate, n-alkyl, 1524
Perchlorate, ester R$_3$COClO$_3$, 1524
Perchloric acid in glacial acetic acid, 1585
Phenyl-p-biphenyl-α-naphthyl carbinol, 1569
1-Phenyl-p-biphenyl-α-naphthyl methyl benzoate, 1569
Phenyl-p-biphenyl-α-naphthylthioglycolic acid, 1561
Phenylpyridinium bromide, 1511
Phenyltropenylium ion, 1583
9-Phenylxanthenyl carbonium ions, 1525
Phosgene, 1590

Phosphoric acid in glacial acetic acid, 1585
Phosphorus pentachloride, 1588
Phosphorus pentachloride mechanism, 1588
Photoconductivity, 1508
Picrysulfonate, 1586
$pK_R+$, values conversion to $\Delta F°$, 1531
Planar model, 1555
Plane-propeller structure, 1557
Polar solvent, 1511
Polyamino compounds, azacarbonium species, 1670
  formamides, alkylated, 1670
  formamidine salts, 1670
  guanidinium cations, 1670
  guanidinium type, 1670
  isothiouronium salts, 1670
  triazidomethylium salt, 1670
  triply substituted guanidine, 1670
Polyethylenyl, 1587
Polymerization of cycloheptatriene, 1588
Polymethoxy-substituted triaryl carbinols, 1530
Potassium permanganate, 1585
Propeller interconversion by two ring flip mechanism, 1573
Propeller like, 1505
4-Protoadamantyl cation, 4-*exo*-protoadamantanol, 1927
  *endo*-spimer, 1927
Protolysis equilibria, carbinols, 1537
Protolysis equilibria, monoamino trityl, 1537
Protonated alcohols, chemical shift data, 1711
  exchange rate, 1711
  hydrogen sulfates, 1710
  proton on oxygen, shielding of, 1711
  spectrum, 1711
Protonated aldehydes, allylic-like coupling between the proton on oxygen and the hydrogens on the α carbon, 1735
  calculated $\pi$ electron density, 1738
  carbon-bond character, 1735
  charge distribution, 1838, 1739
  chemical shifts, 1735, 1738
  $^{13}C$ spectroscopy, 1738
  coupling constants, 1743
  deshielding on oxygen, 1735
  double-bond character, 1735
  $^{19}F$ chemical shifts, 1740
  hybridization and bond order considerations, 1738
  inductive and resonance interaction, 1741
  IR data, 1734
  methyl ethyl ketone, 1742
  methyl groups cause an upfield shift, 1738
  more than one isomer exists, 1742
  NMR spectra of protonated benzaldehyde, 1739
  nonempirical molecular orbital calculations, 1743
  protonated acetaldehyde, 1742
  protonated acetone, $^{13}C$ chemical shift, 1738
  protonated acetophenone, 1739
  protonated formaldehyde, 1741
  protonated formaldehyde, $^{13}C$ chemical shift, 1738
  protonated formic acid, 1.5 kcal/mole, 1743
  size of the coupling between existence of, isomers for protonated carbonyls, 1735
  structure of protonated carbonyls, 1735
  UV data, 1734
  vicinal coupling constants, 1741
Protonated aliphatic glycols, *n*-butyl alcohol leaves, 1714
  chemical shift, 1714
  cleavage of protonated *n*-propyl alcohols, 1714
  cyclization of diprotonated glycols, reactions, 1714
  diprotonated glycols, 1714
  diprotonated species at lower levels, 1714
  formation of allylic carbonium ions, 1714
  hydride shift through bridged intermediate, 1715
  pinacolone rearrangements, 1714
  protonated dimethyltetrahydrofurans, 1716
  protonated diols, 1714
  rate of cleavage, 1715
  reactivity of protonated alcohols, 1714
  stable allylic carbonium ion, 1715
Protonated amides and thioamides, amides in acid systems, 1748
  barrier to free rotation, 1748
  chemical shifts, 1751
  cryoscopic measurements of amides, 1749
  double bond character, 1748
  infrared evidence, 1749
  IR spectra, 1750

IR spectra of N-acyltrialkylammonium
  halides, 1750
IR spectroscopy, 1749
long-range allylic coupling found, 1753
monoprotonated thioamides, 1749
NMR evidence, 1750
NMR of N, 1751
NMR spectroscopy, 1749
N-protonation, with rapid exchange, 1751
  with slow proton exchange, 1751
O-protonation, with fast exchange, 1751
  with slow exchange, 1751
oxygen protonation, 1753
  of amides, 1749
peptide link, 1748
positive charge, delocalization of, 1753
protonated amides, 1753
protonated thioacetamide, 1753
protonated thioacetanilide, 1753
resonance stabilization, 1754
restricted rotation in protonated amides, 1748
salts of acetamide, 1749
small couplings, 1753
stronger acid systems, 1753
studies, recent IR, 1750
sulfur protonation, 1753
thioamides, 1749
Protonated amino acids, by $^{13}$C spectroscopy, 1774
cryoscopic studies, 1774
extent of diprotonation, 1774
first protonation site, 1774
L-leucine, 1774
partial diprotonation of the base, 1774
polyprotonated, 1774
protonation of both amino group and carboxyl function, 1774
proton NMR spectroscopy, 1774
Protonated carbamic acid and derivatives,
  alkyl carbamates protonated form by NMR, 1767
alkyl carbamates via alkyl-oxygen fission, 1767
bear a W relation, 1767
cleave alkyl-oxygen scission, 1767
N,N-diisopropylcarbamate esters, 1768
four-bond coupling, 1767
free rotation, 1767
hindered coupling, 1768
hindered rotation, 1767, 1768
protonated carbamic acids, 1767

unstable/lose carbon dioxide, 1767
Protonated carboxylic acids, aliphatic, activation parameters of configurational equilibrium, 1757
alkyl-oxygen or acyl oxygen scission, 1758
four-bond coupling, 1757
increasing size of R group, 1758
interaction between lone pairs and two heteroatoms, 1758
methyl formate, 1758
NMR spectra of two isomers, 1758
ortho methyl groups, 1760
protonated thioformic acid, 1757
solvation of ions, 1758
spectroscopy, UV, IR, NMR and Raman, 1754
steric requirements, 1758
Protonated carboxylic acids, esters, alkyl cleavage, 1761
t-butyl cation, 1760
cleaves via alkyl-oxygen fission, 1760
ethyl acetate, 1761
methyl formate, 1761
by NMR acyl-oxygen fission, 1760
possible modes of cleavage, 1760
protonated anhydrides, 1762
protonated carbon monoxide, 1761
protonated carboxylic acids, 1762
rapid intermolecular exchange, 1762
stability of oxocarbonium ions, 1761
tertiary alcohols cleave, 1760
Protonated dialkyl carbonates, dialkoxy carbonium ion, 1764
di-t-butyl cleaves, 1765
electrophile with carbamic acids, 1767
fast exchange by NMR, 1766
ortho esters of carbonic acids, 1766
preparation of the ions, 1766
protonated alkyl hydrogen carbonates, 1765
protonated carbonic acid, 1764
protonated on carbonyl group, 1764
protonated trithiocarbonic acid, 1766
thio analogs of carbonic acid, 1766
Protonated diaminocarboxylic acids, α-amino acids, second asymmetric center, 1775
guanidine group, 1775
tetraprotonated, 1775
Protonated N, N-diiodopropylcarbamate esters, acidity function behavior amides, 1768

SUBJECT INDEX 1955

hydration requirements, 1768
N protonated, 1768
N protonation, 1768
O protonation, 1768
rearrangement from O-protonated to N-protonated ions, 1768
thermodynamically least stable ion, 1768
thermodynamic stabilities, 1768
Protonated enamines, C-protonated species, 1771
hydrolysis to isobutraldehyde, 1771
N-protonated observed, 1771
N-protonated species, 1771
site of protonation, 1771
Protonated ethers, bridged species, "free" complexes between two ethers, 1722
$n$-butyl methyl ether, 1724
carbon protonated, 1730
cleavage of protonated, 1724
cryoscopic methods, 1724
effects, inductive and resonance, 1724
greater bascity of cyclic aliphatic, 1721
kinetic investigation of, 1724
oxygen protonated, 1730
rapid tertiary cleavage, 1725
relief of internal strain, oxonium salts, 1721
secondary group cleavage, 1725
Protonated formylurea, irreversible, 1770
mechanism of triprotonation, 1770
selenourea, diprotonated, 1770
sites of protonation, 1770
triply protonated, 1770
triprotonated, irreversible, 1770
Protonated glycine, 1774
Protonated imines, aldehydes, 1771
ketimines, hindered rotation, 1771
ketones, 1771
$N$-propylidinemethylimine, barriers to rotation, 1771
Protonated ketones, barrier to rotation, 1745
changes in conformation, 1746
cleavage, 1747
conformation about the α-carbon-carbonyl carbon bond, 1746
energy of activation for syn-anti conversion, 1745
equilibrium, 1744
groups, different alkyl, 1744
hydroxy proton within plane, 1745
intermolecular proton exchange, 1745
isomer assignment, 1744

isomer distribution, 1746
isomer in greater abundance, 1745
long-range, allylic like coupling, 1745
preferred conformation, 1746
protonated aldehydes rotation, 1746
proton on oxygen resonances, 1744
rapid rearrangement, 1747
reactivity of protonated carbonyl, 1746
Protonated ketoximes, Beckmann rearrangement, 1772
coupled to NH proton, 1772
dissolution of nitrosocyclohexane, 1772
$N$-methylacetonitrilium ion, 1772
nitrogen protonation of oximes, 1772
protonated cyclohexanone oximes, 1772
restricted rotation about C=N bond, 1772
Protonated L-alanine, 1775
Protonated L-cystine, 1775
Protonated leucine, 1774
Protonated L-isoleucine, 1775
Protonated L-leucine, 1775
Protonated L-methionine, 1775
Protonated nitriles, chemical shift, 1773
NMR studies of nitriles, 1773
protonated nitriles, slowly hydrated, 1773
Protonated nitro compounds, barrier to rotation, 1773
charge delocalization, 1773
cleave, of carbonium, 1774
of hydronium, 1774
of nitrosonium, 1774
cryoscopic measurement in sulfuric acid, 1773
fluorodimethyl, 1774
2-fluoro-2-nitropropane, 1774
$FSO_3H$-$SbF_5$ protonation on oxygen, 1773
nitromethane ionized nitrobenzene 40% ionized, 1773
1-nitro-2-methylpropane, 1774
ortho protons, 1773
restricted rotation, 1773
Protonated organic compounds, conformation of protonated compound, 1698
dihydroxymethyl carbonium ion, 1698
dimethylhydroxy carbonium ion, 1698
electronic distribution in positively charged species, 1698
as intermediates, 1698
nuclear magnetic resonance, of nitrogen, 1698
of oxygen, 1698

of sulfur protoantion, 1698
site of protonation, 1698
super acid systems, low nucleophilicity, 1698
super acid systems, high acidity, 1698
trihydroxycarbonium ion, 1698
Protonated organic compounds, experimental methods, cryoscopic and conductometric, 1699
electronic-ultraviolet and visible spectroscopy, 1699
Raman spectroscopy, 1699
spectroscopic methods, IR, NMR, 1699
Protonated peptides, simple, anhydrous HF good solvent for proteins, 1775
carbohydrates, 1776
chlorophyll, 1776
glycylglycine, 1775
glycylglycylglycine, 1775
glycylglycylglycine, tri-, tetra-, and pentaprotonated, 1775
many retentions of biological activity, 1776
no cleavage of peptide linkages, 1775
Vitamin $B_{12}$, 1776
Protonated sulfides, more stable than protonated thiols, 1733
NMR spectroscopy of protonated thiane-3,3,5,5-$d_4$, 1733
protonated di-*t*-butyl sulfides, 1733
protonated secondary sulfides, extraordinary stability, 1733
stable to cleavage, 1733
to trimethylcarbonium ions, 1733
Protonated thiols, cleaves, protonated primary thiols, 1717
decomposition, 1717
protonated hydrogen sulfide, 1717
resonance at higher field, 1717
stability of, 1717
trimethylcarbonium, 1717
Protonated urea and guanidine, charge delocalization, barrier to rotation, 1769
cryoscopic measurements, 1769
diprotonated, 1769
diprotonated, tetramethylurea, 1769
limit-20, kcal, 1769
lower carbon-nitrogen bond order, barrier to rotation, 1770
resonance stabilization of charge, 1769
restricted rotation, 1769
spectroscopic data, IR, UV, NMR, 1769

Protonated valine, 1774
Protonation of oxygen, 1516
in reactions with ether, alkyl phenyl ethers, exclusive oxygen protonation, 1729
carbon protonation, 1729
C-protonated species, 1729
NMR spectrum, data, 1729
O-protonated species, 1729
oxygen protonation, 1729
of phenols and alkyl phenyl ethers, 1729
strength of acid, 1729
temperature-dependent rate of exchange, 1729
UV data, 1729
Protonation of saturated dicarboxylic acids, by cryoscopy, 1763
cyclic anhydrides, 1763
1,2-dihydroxycyclobutenedione, protonated, 1763
diprotonated cyclic, 1764
by NMR, 1763
NMR of cyclobutenium di-cation, 1763
oxocarbonium ion, 1764
structure, 1764

Quadricyclyl cation, cyclopropylcarbinyl-cyclopropylcarbinyl rearrangements, 1860
degenerate rearrangements, 1860
nortricyclane compounds, 1860
potential complete carbon degeneracy, 1860
quadricyclane derivatives, 1860
solvolysis of quadricyclane derivatives, 1860
Quinonoid structure, 1504

Rapid exchange process, 1504
Rates of solvolysis, rearrangements reactions, Curtis of Schmidt reaction, 1811
Hoffmann reaction, 1811
migration to electron-deficient atom, 1811
neighboring carbon participation, 1811
of neopentyl-like structures, 1811
Reactions of α-dialkylamino compounds, α-aminoethers compared to acetals, 1655
α-aminonitriles and Grignard reagents, 1656
azacarbonium ions, 1656
azacations as transient species, 1656
bisulfite treatment, 1657

# SUBJECT INDEX

cryoscopic studies, 1657
α-dialkylaminoesters generation of azacations, 1655
α-dialkylaminomethylsulfonic acids, 1656
α-dialkylaminosulphonic acids with iodine, 1656
 equilibrium conditions, 1656
 immonium ion, 1656
 iodine oxidations, 1657
Rearrangement reactions, neopentyl type rearrangements, 1809
 strain inherent, 1809
 strain relief, 1809
Relative solvolysis rates for $p$-nitrobenzoates, 1818
Relief of strain, 1517
Resonance, 1503
 energy, 1560
 theory, 1504
Retention of optical activity, 1561
Reversible hydration, 1530
Ring flip, 1563
 one, 1565, 1567
 two or three, 1565, 1567
Ring twist, 1563
Rotating platinum electrode, 1585
Rotational barrier, 1566

Saturation effect, 1550, 1551
Saturation ratio, 1552
Selenium dioxide, 1585
Sesquixanthyl cation, 1547
Silver assisted solvolysis, 1519
Silver salt, of Lewis acid, 1519
Single-crystal optical, 1508
Sites, basic and acidic of solvent, 1513
$S_N1$ exchange on the NMR time scale, 1540
$S_N1$ fast and slow exchange methods, 1540
$S_N1$ ionization, 1518
$S_N1$ mechanism, 1502
$S_N1$, $S_N2$ exchange, 1540
Solvation equilibria, 1504
Solvent influence, 1522
Solvents, low basicity, 1513
 low dielectric constant, 1513
 methylene chloride, acetonitrile, nitromethane, 1542
Solvolysis rates, 1-adamantyl carbonium prepared by decarbonylation, 1823
 adamantyl cation, 1824
 amination reaction of adamantane, 1825
 barrier to 1,2 hydride shift greater observation of rearrangement of tricyclo-[5.2.1.0$^{4,10}$] decyl alcohol, 1825
 bridgehead free radical stability parallels the carbonium ion stability, 1823
 "cage effect," 1824
 chemical shift of the $\gamma$ proton, 1824
 chromic acid cleavage of secondary benzylic alcohols, 1822
 conformationally isomeric carbonium ions, 1827
 decomposition of bridgehead chloroformates, 1829
 deoxidation, 1827
 empty orbital with rear lobes of carbon-hydrogen bonds, 1824
 fragmentation-recombination mechanism, 1825
 half-wave potentials, polarographic potentials, 1822
 high tosylate/bromide ratios, 1827
 intermolecular hydride shifts, 1825
 ionic chlorination, 1825
 lack of solvent assistance, 1825
 one-electron displacement on bromine, 1823
 phenyl apocamphyl sulfide, 1822
 PMR spectrum, 1824
 Raney nickel desulfurization, 1822
 rapid racemization of sulfonium salt, 1822
 rationalized by strain analysis, 1829
 rear-side solvent participation, 1825
 $S_N2$-like and $S_N1$-like mechanism, 1823
 solvolytic data, 1825
 steric effects, 1827
 unfavorable hyperconjugation, 1825
Solvolysis rates, of decalyl chlorides, 1819
 of decalyl-$p$-nitrobenzoates, 1819
 of hydrindanyl-$p$-nitrobenzoates, 1819
 of perhydropentalenyl-$p$-nitrobenzoates, 1820
Specific solvation, 1509
Spectrophotometric methods, 1530
Spectrophotometric, trityl alcohols, 1531
Spectroscopic methods, IR, 1704
 NMR, 1704
 UV, 1704
 visible, 1704
Spectroscopic techniques, 1514
Stabilization energies, by standard cell potential, 1537
Stabilization energies, standard free energy, 1537

of trityl cations, 1538
Stabilization energy, 1548
Stable arylcarbonium salts, 1502
Steric bulk, 1587
Steric factors, 1509
Steric interactions, 1546
cis-Stilbene, asymmetrical conformation, 1558
Stream of oxygen, 1585
Structure change, 1526
Structure stability, 1525
Substituent effects, 1522
p-Substituted diphenylmethyl cations, 1550
Substituted ethynyl, 1587
Sulfur dioxide, 1524
Sulfuric acid, concentrated, 1519
Symmetrical conformations, cis-stilbene, 1558
Symmetrical phenyl twisting, 1559
Symmetrical propeller structure, 1560
Syn-triphenylcyclopropenium perchlorate, 1505

Techniques of preparing salts, 1581
8-Tetracyclo[4.3.0.0$^{2,4}$.0$^{3,7}$] nonyl cation, acetolysis, anchimeric assistance factors, 1932
  acetolysis, 8-tetracyclo[4.3.0.0$^{2,4}$.0$^{3,7}$]-nonyl precursors, 1932
    tetracyclo[4.3.0.0$^{2,4}$.0$^{3,7}$] system, 1933
  homocyclopropylcarbinyl, 1931
    nonclassical form, 1931
1,3,4,4-Tetramethylcyclohexenyl cation, 1-t-butyl-3-methyl-cyclopentenyl cation interconversion, 1905
  isotopic labeling, 1905
Tetraphenylcyclobutenyl cation, 1508
Thioalkoxy substituent, 1587
Three-bladed propeller conformation, 1556
Tin tetrachloride, 1523
Total electronic charge delocalization, 1551
Trapping of 8-hydrindanyl carbonium ion, 1820
  ion-pair mechanism, 1821
  1-norbornyl hypochlorite, 1821
  test for intermediacy of carbonium ions, 1821
  thionyl halides with alcohols, 1821
Tri-p-aminotrityl perchlorate, 1563
Triaryl carbonium ions, 1502
Triaryl chloride, 1524

Triarylmethyl halides, 1502
Tribromide, 1591
Triiodide, 1591
Tri-p-methoxycarbonium ions, 1524
Tri-p-methoxyphenylcarbonium ions, 1524
Tri-p-methoxytrityl hydrogen dichloride, 1505
Tri-p-methoxytrityl perchlorates, 1506
Tri-p-methoxytrityl tetrafluoroborate, 1506
4,4',4''-Trimethyltrityl cations, 1539
Triphenylcarbonium fluoroborate, 1586
Triphenylcarbonium hexafluoroantimonate, 1586
Triphenylcarbonium hexafluoroarsenate, 1586
Triphenylcarbonium hexafluorophosphate, 1586
Triphenylcarbonium ion salts, 1587
Triphenylcyclopropenium, 1544
Triphenylmethane, 1505, 1554
Triphenylmethyl bromide, 1554
Triphenylmethyl (trityl) cation, 1501
Triphenyl perchlorate, 1586
Tri-p-tolyl carbonium ions, 1524
Tri-p-tolylmethyl chloride, 1512
Trityl acetate, 1515
Trityl alcohols, ionization of, 1513
  measurements of, 1532
  stability from $pK_R^+$, 1532
Trityl azide, 1508
Trityl bromide, 1511
Trityl carbinols, 1535
Trityl cation, 1503, 1508, 1512, 1516
Trityl chloride, 1514, 1518, 1520, 1521, 1522, 1523
Trityl chlorides, stability by spectroscopy, 1532
Trityl chloroborates, 1521
Trityl ethers, 1512
Trityl fluoride, 1515
Trityl fluoroborate, 1505, 1506, 1512
Trityl formate, 1517
Trityl halides, 1512, 1517
Trityl halides, ionization-dissociaiton of, 1514
Trityl nitrate, 1517
Trityl perchlorate, 1505, 1506, 1508, 1524, 1563
Trityl, SnCl$_5$, 1523
Trityl, steric requirement, 1544
Trityl tetrahaloborate, 1525
Trityl tosylate, 1517

# SUBJECT INDEX

Trityl trifluoroacetate, 1506
Tropenium ion, 1580, 1581
Tropenyl ethers, 1591
Tropenyl group, 1581
Tropenylium bromide, 1583
Tropenylium fluoroborate, 1587
Tropenylium halides, 1591
  benzaldehyde, 1601
  charge transfer interactions, 1601, 1602
  disproportionate, 1601
  heat of hydrogenation data, 1601
  stability of, 1601
Tropenylium hydrogen halides, anion-to-cation electron transfer, 1594
  charge transfer, 1594
  weak transfer, 1594
Tropenylium ion, 1580, 1590
Tropenylium ions, acidities of, 1618, 1619
  allyltropenylium, 1621
  carbon hydrogen in plane deformation bonds, 1618
  carbon hydrogen stretching frequency, 1618
  chlorotropenylium, 1621
  cyano or carboranyl substituents, 1617
  decarboxylation of acids, 1609
  dimerization salts, 1597
  electrodonating substituents, 1617
  electronic spectra, 1619
  fluorotropenylium, 1621
  hydride transfer, 1616
  increase in acidity, 1617
  iodotropenylium, 1621
  IR, 1593
  Kosower-Z value plots, 1621
  2-methylphenyltropenylium ion, 1623
  4-methylphenyltropenylium ion, 1623
  NMR coupling constants, 1617
  NMR spectra, 1622, 1623
  nucleophilic interchange, 1616
  physical properties of substituted, 1617
  polarographic reduction, 1624
  polymorphism, 1603
  Raman spectra, 1593
  reaction of tropenyl methyl ether, 1606
  reduction, 1597, 1603
  tropenylation, 1608
  tropenyl bromide with Grignard or lithium, 1607
  UV spectra, 1593
  x-ray crystallographic, 1617
Tropenylium ions, alkylation of carbonyl
  and activated methylene compounds, coupling reactions, 1609
  decarboxylation of acids, 1609
  electrochemical reduction, 1609
  reduction to tropenyl radical, 1609
  tropenylation, 1608
  α-tropenyl carbonyl compounds, 1607
Tropenylium ion salts, aromatic ring current, 1596
  conductivity and polarographic reduction, 1596
  diamagnetic susceptibility, 1596
  polarographic reduction, 1596
  ring current, 1596
  strong electrolyte, 1596
Tropenylium ions, oxidation, by chromic acid to benzaldehyde, 1585
  hydrogen peroxide, 1585
  silver oxide to benzaldehyde, 1585
Tropenylium ions, reactions and properties of substituted tropenylium ions, nucleophilic attack of a $p$ atom, 1627
Tropenylium ion stability by $pK_{R^+}$, x-ray crystallographic investigations, 1592
Tropenylium ion structure, 1591
Tropenyl methyl, 1583
Tropenyl methyl ether, 1591
Tropilidene, 1581
Tropolone, 1585
Tropyl group, 1581

Unsymmetric screw helix, 1556

van't Hoff $i$-factor, 1530
Vibrational spectroscopy, carbonyl protonation, antisymmetric carbon-oxygen stretching band, 1705
  NMR, application of, 1705
  protonation, of carbonyl compounds, 1705
  of carbonyl oxygen, 1705
  site of protonation, 1705
  X-H bending mode, frequencies, 1705
  X-H stretching, X-H and X-D deformation frequencies, 1705
Vilsmeier-Haack complexes, chloroformylation, 1666
  complex NMR spectra, 1668
  formylation of indole, 1669
  formylation species, 1668
  influence of solvent, 1667
  ion pair, 1668

NMR spectra, 1667
UV spectra, 1669
Vilsmeier-Haack reaction, 1666
Vilsmeier reaction, 1668
Vinyl substituent, 1587
Visible-ultraviolet spectroscopy, 1508

Wagner-Meerwein rearrangement, automerizations, 1842
 double bond participation, 1844
 1,2 hydride shifts, 1843
 1,3 hydride shifts, 1843
 interchange of carbon atoms, 1842
 rearrangement, homoallylic, 1844
 1,2-vinyl shift, 1844

Walden inversion, abnormal front-side bimolecular substitution, 1783
 geometrically derived instability, 1783
 infrared and proton magnetic resonance, 1784
 isoelectronic trimethyl- and trihalo derivatives of boron, 1783
 planar cation, 1784
 theoretical calculations, 1784

Xanthene thioglycolic acid, 1570
Xanthenyl cation, 1508
Xanthyl, 1546
X-ray diffraction study of trityl perchlorate, 1560